(TOWARD) DISCOVERY OF LIFE BEYOND EARTH AND ITS IMPACT

KAVLI-IAU SYMPOSIUM 387

COVER ILLUSTRATION:

'Yellow Sun Rocket' 2014, by Helen Schell (Artist & Space Specialist) – The Human Spaceship

IAU SYMPOSIUM PROCEEDINGS SERIES

INTERNATIONAL ASTRONOMICAL UNION

UNION ASTRONOMIQUE INTERNATIONALE

International Astronomical Union

IAU

(TOWARD) DISCOVERY OF LIFE BEYOND EARTH AND ITS IMPACT

PROCEEDINGS OF THE 387th SYMPOSIUM OF THE INTERNATIONAL ASTRONOMICAL UNION DURHAM, UNITED KINGDOM 15–19 APRIL 2024

Edited by

HERMINE LANDT
Durham University, UK

MARTIN DOMINIK
University of St Andrews, UK

and

CAROL OLIVER
University of New South Wales, Australia

THE KAVLI FOUNDATION

CAMBRIDGE
UNIVERSITY PRESS

CAMBRIDGE UNIVERSITY PRESS
University Printing House, Cambridge CB2 8BS, United Kingdom
1 Liberty Plaza, Floor 20, New York, NY 10006, USA
10 Stamford Road, Oakleigh, Melbourne 3166, Australia

First published 2026

Printed in Great Britain by Henry Ling Limited, The Dorset Press, Dorchester, DT1 1HD

Typeset in System LaTeX 2ε

A catalogue record for this book is available from the British Library Library of Congress Cataloguing in Publication data

This journal issue has been printed on FSC™-certified paper and cover board. FSC is an independent, non-governmental, not-for-profit organization established to promote the responsible management of the world's forests. Please see www.fsc.org for information.

ISBN 9781009545860 hardback
ISSN 1743-9213

For EU product safety concerns, contact us at Calle de José Abascal, 56, 1°, 28003 Madrid, Spain, or email eugpsr@cambridge.org

Table of Contents

Preface

In early 2022, I had received funding from Durham University to make the first steps towards establishing a multi-disciplinary astrobiology-focused space research centre, the Centre for Cosmic Impact (CCI). In this capacity, I visited in July 2022 the SETI Institute, Mountain View, California, USA, who have been the leaders in the search for life beyond Earth since decades. I had just been impressed by the professional and vibrant outreach and communication programme at SETI, when the IAU call for applications for 2024 symposia reached me. During my background research for the CCI, I had also learned that, broadly speaking, there are two very different and rather disparate communities striving to find life beyond Earth; the scientists occupied with bio-signatures and those with techno-signatures, with both engulfed by an even larger multi-disciplinary community made up of anthropologists, theologians, lawyers, philosophers, policy makers, science journalists, artists, and others, who continuously analyse and debate what such a profound discovery, should it be made, will mean for humanity as a whole. A Kavli-IAU symposium seemed the perfect opportunity not only to give this high-aim endeavour an international voice but to provide a platform for unification. If we could just get to know each other better and talk, we could 'team work to make the dream work'.

When I then received the good news in May 2023 that our symposium proposal had been accepted by the IAU, it was clear that the organising committee needed not only to be multi-disciplinary itself but to contain members from the whole spectrum of astrobiology. We decided early on to structure the five symposium days thematically and we unequivocally wanted to incorporate sufficient room for panel discussions each day. Two workshops should provide the necessary hands-on experience for the (in-person and online) participants to deal with the realpolitik of a discovery scenario. But incorporating the aspect of communication, which would ultimately serve as the glue essential to bind the disparate communities, was the tricky part. A special day set apart for it would not do since communication needed to be 'weaved into the fabric of the symposium' to work efficiently. I believe that, since we got this component right, the experiment - on a small scale - was successful. It is vital that we try this experiment on a big scale in the future. The Post-Detection Hub at the University of St Andrews and the planned CCI at Durham University could provide this multi-disciplinary forum, which is such a crucial part in humanity's search for life beyond Earth but has not been sufficiently incorporated yet in other current institutes and centres devoted to this goal.

Finally, I would like to acknowledge additional financial support for the Kavli-IAU symposium 2024 from the Breakthrough Initiatives and from NASA through grant no. 80NSSC24K0656. The innovative virtual poster gallery that permitted the exhibition of a large number of interactive posters was designed and run by the UK-based charity 4wardFutures. I am also grateful to Diana Worrall, IAU Assistant General Secretary, for her unwaivering support and flexibility that made decision-making for me easy and helped me keep the symposium preparations on track. The timely publishing of these proceedings benefitted greatly from the smooth operations by Lina Canas at the IAU and Molly Haigh at Cambridge University Press and, of course, the high-quality submissions by the symposium participants. Even though the reader might not have been able to take in the inspiring and congenial atmosphere present at the symposium venue, The Hub at Durham University, I hope that this spirit will resonate through the articles in this volume.

Durham, UK, May 2025

Hermine Landt
(Chair of the Organising Committee for the Kavli-IAU Symposium 2024)

Editors

Hermine Landt
Durham University, UK

Martin Dominik
University of St Andrews, UK

Carol Oliver
University of New South Wales, Australia

Organizing Committee

Hermine Landt (Chair), Durham University, UK.
Heda Agic, Durham University, UK.
Daniel Angerhausen, ETH Zuerich, Switzerland.
Federica Bianco, University of Delaware, USA.
Chris Cowie, Durham University, UK.
Martin Dominik, University of St Andrews, UK.
Jacob Haqq-Misra, Blue Marble Space Institute of Science, USA.
Sohan Jheeta, NoRCEL, UK.
Andjelka Kovačević, University of Belgrade, Serbia.
Nigel Mason, University of Kent, UK.
Chris McKay, NASA Ames Research Center, USA.
Lucas Mix, Durham University, UK.
Carol Oliver, University of New South Wales, Australia.
Mazlan Othman, Academy of Sciences Malaysia, Malaysia.
Istvan Praet, Durham University, UK.
George Profitiliotis, TU Delft, The Netherlands.
Margaret Race, SETI Institute, USA.
Richard Wilman, Durham University, UK.

List of Participants

Abdul Qadir, Y.
Abel, B.
Agata, H.
Agic, H.
Albergaria Pereira, D.
Ansbro, E.
Arnould, J.
Awiphan, S.
Azua-Bustos, A.
Balsalobre Ruza, O.
Baxter, S.
Bowler, S.
Bursa, M.
Cao, D.
Caprotti, A.
Caprotti, M.
Chan, M.
Chatzichristou, E.
Cheung, S.
Clements, D.
Creaner, O.
Croft, S.
Davenport, J.
Davis, G.
de Paulis, D.
Decin, L.
Denning, K.
Doebler, N.
Dominik, M.
Dravins, D.
Dubin, B.
Edmunds, M.
Elliott, J.
Evans, J.
Fairweather, J.
Ferus, M.
Finer, E.
Fröhlich, V.
Fu, X.
Garcia Martinez, P.
Genevieve, K.
Goumas, M.
Haqq-Misra, J.
Harrington, A.
Iliev, L.
Impey, C.
Jafari, S.
Jheeta, S.
Josselin, E.
Kajrekar, P.
Kee, T.
Kennedy, C.
Kershenbaum, A.
Kestler, I.
Knuth, K. H.
Kovačević, A.
Laine, P.
Landt, H.
Lee, R.
Leeuw, L.
Linde, P.
Lystbæk Vestergård, G.
Ma, S.
Malik, S.
Martire, G.
Mason, N.
Mason, L.
Mateo-Martí, E.
McGee, S.
McGruder, C.
Milligan, A.
Mix, L.
Narasimha Ponnaganti, R.
Nováková, J. N.
Nicholson, A.
Oliver, C.
Pavlidou, V.
Pérez-Mercader, J.
Perkins, D.
Persson, E.
Piper, C.
Pitrou, P.
Praet, I.
Profitiliotis, G.
Raha, N.
Ren, T.
Rummel, J.
Rytsarskii, A.
Saleh, G.
Salisbury, J.
Schulze-Makuch, D.
Schwarz, A.
Sellers, L.
Shaterzadeh Yazdi, S.

Smith, H.
Solano, E.
Spillmann, P.
Steves, B.
Tang, W.
Teays, T.
Tennen, L.
Tinetti, G.
Utila, M.
Vainio, O.-P.
Vickers, P.
Vidal, C.
Wan, J.
Wandel, A.
Wandia, K.
Ward, M.
Webb, S.
Westall, F.
Whitehouse, D.
Wilman, R.
Wisniowski, T.
Zucker, S.

(Toward) Discovery of Life beyond Earth and its Impact
Proceedings IAU Symposium No. 387, 2026
H. Landt, M. Dominik & C. Oliver, eds.
doi:10.1017/S1743921324002035

Astrobiology in Interdisciplinary Context

Lucas Mix
Durham University
email: lucas.mix@durham.ac.uk

Abstract. Astrobiology and SETI search for life-not-as-we-know-it (yet). Researchers aspire to an understanding not yet found. With that in mind, knowledge about three types of context can inform and improve the search. Observational context: knowing our location as humans (anthropism) can mitigate biases introduced by our desire to over-value humanity (anthropocentrism). Historical context: knowing the history of the search, not just the last 10–30 years, can clarify concepts and significance. Disciplinary context: understanding the rules by which different disciplines operate can reveal the limitations of a single perspective (astronomy, biology, ecology, sociology, law, medicine, philosophy, theology...). Astrobiology and SETI tell epic stories about the place of life in the cosmos. They locate humans morally as well as physically. We should ask why we tell them and how our context shapes the telling, because the stories shape human interactions with the world.

Keywords. Anthropism, Anthropocentrism, Context, Interdisciplinarity, Value

---

1. Value in Astrobiology

The 2024 Kavli-IAU symposium (387) provided a rare opportunity for cross-disciplinary conversation "Toward Discovery of Life Beyond Earth and Its impact." Set in the cathedral city of Durham, UK, the event brought together natural scientists, social scientists, humanities scholars, and communicators to think through the search for extraterrestrial life and intelligence, to explore the idea of life-not-as-we-know-it, and the impact of discovering such life.

Astrobiology is the study of life at planetary scale. How do we understand life in cosmological context? How do we understand Earth life as a single, coherent system? And how do we look for other such systems beyond Earth? SETI, meanwhile, is an acronym for the search for extra-terrestrial intelligence. SETI focuses on life humans can communicate with. I will refer to SETI as a subset of astrobiology, the search for intelligence as one way of searching for life. Both search for life-not-as-we-know-it.

It may be more accurate to say that astrobiologists search for life-not-as-we-know-it-yet. That 'yet' highlights the act of recognition. We may not know it as life now. We may not even know it as life when we find it. But at some point, we will see a commonality between 'us' and 'them'. Humans value life on Earth. We hope to find that value mirrored in space, another form of life out there. Astrobiology is aspirational. Researchers aspire to an understanding not yet found. They look for data in hopes of a new and deeper understanding.

Astrobiology is also radically interdisciplinary. Concepts of life and intelligence span multiple fields and carry implications for value and morality as well as fact. These concepts are not only present in multiple, radically diverse discourses, they are foundational for many (e.g., evolutionary theory, process philosophy, and ecotheology). The search evokes interest (and funding) because it produces – or hopes to produce – data and theories that matter to a wide variety of people. And, because they matter to those people, communication around methods, results, and interpretation attract a high level of scrutiny.

In short, people care. The physical context of humanity cannot be isolated from the metaphysical context of humanity: ethics, morality, and spirituality. Scientists can try to minimize bias in observation and theorizing, but value commitments inevitably find their way in. This happens transparently at the stages of science called 'motivation', 'outreach', and 'impact,' but it also occurs more subtly at other points (Hilgartner 1990; Midgley 1992).

I say this not to undermine science, but to encourage the vigilance necessary to do science well. Scientists aspire to impartiality but recognize the limits of human perspective. An acknowledged gap can be accounted for and narrowed. An unacknowledged gap remains uncorrected. Thus, attending to metaphysical context is a necessary first step toward pursuing a deeper impartiality.

With that tension in mind – between aspiration and practice – I'd like to share a few reflections based on my time working with the Science Organizing Committee for this symposium, and more broadly based on 28 years working on interdisciplinary communication in astrobiology. They were discussed throughout the event and appear throughout this volume as practical, philosophical, and theological angles to consider as we examine the search, possible detection scenarios, and how we communicate about both.

2. Observational Context

Astrobiology is grounded literally as well as figuratively. Experimental design and interpretation always occur on Earth. Most experiments are on the surface of Earth or in Earth orbit. Even interstellar spacecraft and Martian rovers were launched from Earth, use Earth materials, and remain in constant contact with earthly supervisors. Perhaps someday, research will be carried out by AI or human-machine hybrids (Chiang 2000), but for now it is irrevocably human. This may seem trivially true, but it has important consequences. As such, it is shaped by human biology and human perspectives, moulded by our evolutionary history and constrained by our astronomical location.

Observational context balances universal intent. A desire to be unbiased coupled with adamantine discipline in the scientific method and rigorous peer review (by human peers) can never fully erase the impact of our humanity on our knowledge. I have referred to this elsewhere as the difference between anthropocentrism and anthropism (Mix 2021). The former is an unvirtuous desire to centralize and elevate humanity within our cosmic context – both physical and metaphysical. The latter is a virtuous recognition of our location. And a necessary step in correcting for such biases. Anthopic reasoning will be required to mitigate anthropocentrism (though it can never fully eliminate it).

Often, we (humans) assume that life will be like us. Questioning that assumption can lead to new discoveries. Multicellular organisms (like humans) are rare relative to unicellular organisms (like bacteria). Knowing that, astrobiologists can look for a broader range of chemistries for life (Thweatt et al. 2024). Yellow stars (like the Sun) are rare relative to other star types (namely red dwarfs). Knowing that, astrobiologists can broaden the search for habitable planets (Gillon et al. 2017). Only by noting bias (anthropic reasoning) can the bias (anthropocentric) be removed.

Attempts to transcend our perspective will always be limited in this way – be they universal theories of life, agnostic biosignatures, or attempts to define exotic life. Thus, scepticism about defining life is justified (Cleland 2012, 2019). And yet, the aspirational endeavour should not be abandoned. We can think clearly about unclear things (Mix 2015). We can argue from known toward unknown, if we do so with humility, attending to data and to prior assumptions.

Jerry Soffen, project scientist for the Viking Missions, and Barry Blumberg, the first director of the NASA Astrobiology Institute, both compared astrobiology to a cathedral. It is a massive, interdisciplinary endeavour, requiring thousands of people and likely to last centuries. It will not be completed in one lifetime, but we can begin to see something amazing taking shape, something that rises above the conceptual skyline and points to something beyond it.

3. Historical Context

Astrobiology represents the latest branding of an age-old search. Twentieth century approaches went by bioastronomy, exobiology, and xenobiology (Sullivan and Baross 2007, pp. 7–66). Even before planets were understood to be rocks in the sky, scholars debated a plurality of worlds (Dick 1984; Crowe 2008). Other habitable zones with other inhabitants were regularly discussed elsewhere on the sphere of earth (e.g., the perioeci of Crates), as well as outside the sphere (e.g., the planets in Lucian's *Iacromennipus*), inside the sphere (i.e., the underworld), and simply elsewhere (e.g., Faerie in Spencer's *Faerie Queen*). Speculation on these other worlds continues to inform the rhetoric of space exploration (Mix 2022). Then, as now, observational data provided a jumping off point for discussion of deep philosophical questions about human identity, location, and significance.

The questions are not new, but the methodologies are. Astrobiology arose in the 1990s in response to early detection of exoplanets and claims of fossils inside a Martian meteorite. It is characterized by particular attention to exoplanets and extremophiles, molecular phylogenetics in deep time, and robotic exploration of the solar system. It draws on an ever-expanding mountain of data – from telescopes and rovers to genomic sequencing and synthetic biology – processed with the latest in informatics, machine learning, and other big-data techniques.

Astrobiologists frequently claim they are addressing life-not-as-we-know-it questions for the first time. They are either considering something entirely novel or bringing in data for the first time. In Andy Weir's bestseller *Hail Mary*, the protagonist is thrown out of academia for considering life in media other than water (p38). Meanwhile the Astrobiology Primer (version 3) says that "there is a growing recognition in the astrobiology community that using Terran-based assumptions for the chemistry of life may not account for potential lifeforms" suggesting that this was not pursued vigorously in the past (Grefenstette et al. 2024, S-186). And yet, astrobiologists and their predecessors *have* asked just about every iteration of the alternative chemistry question, from Christian Huygens (*Cosmotheoros* 1698 – Book I, p28) to Primer versions 1 and 2 (Mix et al. 2006; Domagal-Goldman et al. 2016). This has been an active area of research for over a century.

This failure of memory blinds researchers, reviewers, and readers from careful comparison with earlier work. The particular genius of modern science comes from its ability to work with cumulative data. The rhetoric of novelty inhibits both memory and creativity because it constantly invites researchers to reinvent the wheel.

The critical questions for any astrobiologist considering life-not-as-we-know-it will be the detail questions. Not "is it new?" but "how is it new?" Is there a new apparatus, a new methodology, or a larger data set? Alternatively, does this experiment represent a new philosophy, theory, or language? In the latter cases, conceptual analysis will be as important as data analysis. Consider, for example, the work of Lynn Margulis in promoting the endosymbiotic theory (Sagan 1967). Margulis credited Merechowsky and Minchin with having both the idea and the data fifty years earlier; she synthesized it in a new way based on new ways of thinking about sub-cellular structure and function.

The history of a question shapes how it gets asked and which answers are considered. Thus, historical location shapes observation just as much as physical location. By knowing the history, researchers can correct for observer bias while maximizing the data with which they operate.

4. Interdisciplinary Context

Astrobiology is communal and interdisciplinary. It is not done alone, or even side by side, but necessarily in diverse groups. It is integrative, and thus requires work across traditional boundaries. Conceptual context plays a critical role, particularly in papers reporting discovery of life-not-as-we-know-it. These papers often cite work across fields and, frequently, feature co-authors from around the university. And yet, they can still suffer from insularity. Papers that would not have survived peer review in their primary field, end up published as interdisciplinary discoveries.

Consider claims for arsenic based life. Wolfe-Simon and colleagues (2011) isolated an arsenic-tolerant bacterium which they believed was using arsenic in place of phosphorus in key metabolic reactions – notably as a component in proteins and nucleic acids. This would represent a remarkable deviation from the biochemistry of life-as-we-know-it. Subsequent work by teams with experience culturing extremophiles demonstrated that this could not be the case (Erb et al. 2012; Reaves et al. 2012). Other commentators, focusing on chemical energetics, pointed out the implausibility of proposed reactions (Benner 2011). Each felt that the initial paper's claims reached conclusions without tests considered standard in their own field. Similar stories might be told around Bialy and Loeb's (2018) claim for 'Oumuamua as a technosignature (Guzik et al. 2020). Or Greaves and colleague's (2021) claim for phosphine as a potential biosignature in the clouds of Venus (Jordan et al. 2022). (See Clements' chapter in this volume for current thinking, new data, and lessons learned about phosphine on Venus.)

It has been argued that extraordinary claims require extraordinary proofs. Indeed, exactly this argument was raised in response to the McKay et al. (1996) claim for finding fossils in ALH84001 – a foundational argument for 'astrobiology' under that label (McMahon 2020). Astrobiology presses at the margins. Goalpost shifting is both common and, perhaps inevitable, but that is not the issue here.

The goalposts are firmly established for known games, that is for established disciplines. Players who wander between fields, need to know the rules of play for the field they are on. And they need to recognize the field when they play in it. (NB: I do not think McKay et al. missed a trick in proposing they had found 'nano-bacteria'. If anything, they may have been moved in the opposite direction by a current vogue in microbiology, later dismissed. I raise the issue only to distinguish goalpost shifting from moving to a different, yet well established, field of play).

These are not cautionary tales about bad science. Some authors were more careful than others, but they all attempted to make claims that cut across numerous disciplines. "What gets us into trouble is not what we don't know; it's what we know for sure that just ain't so." This quote is commonly attributed to Mark Twain and, significantly, historians have been unable to trace it to him. We must learn to ask for advice, not only about the things we don't know, but about things we think we know that may be viewed differently in different areas of the academy. Where experts exist, they should be asked. That requires knowing who they are and how to find them. It also requires a clear and well-bounded understanding of your own expertise.

5. Building to Last and to Impress

Like a cathedral, astrobiology dominates the intellectual skyline; it attracts attention by creating a focal point, rising amidst diverse structures, and pointing beyond the town to something else. It tells stories about life, the universe, and the place of one within the other. These epic narratives locate humans morally as well as physically (Midgley 1992; Zakariya 2017; Mix 2022). Because they are epic, they are interesting. But, because they are epic, they challenge us to ask why we tell them and how our context shapes the presentation.

Astrobiology addresses the most fundamental questions of how we understand the universe and our place within it. It incorporates the growing body of scientific evidence on questions of human embodiment: intelligence enacted in biology, biology grounded in geology, and geology situated in astronomy, all mediated by chemistry and motivated by thermodynamics. Astrobiology can be revolutionary when it connects all these pieces, using the latest data and data processing techniques.

Nor is astrobiology limited to the natural sciences. The concrete goal of searching for life elsewhere brings in engineering and gives us common ground. The constant challenge of asking the questions in new ways brings in philosophy and the social sciences. Meanwhile, engagement with the big questions forces us to deal with the history of those questions and the history of answers to them. Astrobiology is inescapably contextual.

Astrobiologists and SETI researchers do better research and communicate better when they understand this context and the broad interdisciplinary history of the questions. Conferences like this one are a good start. They bring together diverse expertise and allow participants across fields to understand one another's work. The 2024 symposium brought together researchers in anthropology, astronomy, biology, chemistry, futurism, geology, linguistics, philosophy, sociology, and theology with practitioners of space law, space policy, science communication, and science journalism to discuss how all the pieces fit together and how we talk about them. The organizers thought carefully about what is known and unknown and how we think about the difference. We tried to work outward from one to the other. Workshops and practical discussion of policy and communication were interwoven with research presentations throughout the week, because the message can never be fully disentangled from the medium. I can also recommend the Astrobiology Primer version 3.0 released last year (Schaible et al. 2024). Written by early career scientists, the Primers come out about once a decade and attempt a quick but broad introduction, free from jargon and open to new ideas. We have an opportunity to distribute and decolonize space research – to bring in broader expertise and wider perspectives. Astrobiology will thrive if it can acknowledge its own diversity and remember its own history.

We don't know what we'll find – which is, after all, the point. We look because we don't know what's out there – yet. Looking will change the way we look at the universe. That is why the public cares. It is why astrobiologists care. It is also why we must be ever so careful in how we communicate about astrobiology. People can do amazing things when willing to devote lifetimes to a project. A cathedral is just such an endeavour. So is astrobiology. It shapes the way we see the world and our place in it. It shapes the intellectual landscape. And, if it points to something beyond the world, so much the better.

References

Benner, S. A. 2011, *Science*, 332, 1149
Bialy, S., Loeb, A. 2018, *Ap. Lett.*, 868, L1
Chiang, T. 2000, *Nature*, 405, 517
Cleland, C. E. 2012, *Synthese*, 185, 125
Cleland, C. E. 2019, *The Quest for a Universal Theory of Life*
Crowe, M. 2009, *Extraterrestrial Life Debate, Antiquity to 1915*
Dick, S. J. 1984, *Plurality of Worlds*

Domagal-Goldman, S. D., Wright, K. E., Adamala, K., *et al.* (eds.) 2016, *Astrobiology,* 16, 561
Erb, T. J., Kiefer, P., Hattendorf, B., *et al.* 2012, *Science,* 337, 467
Gillon, M., Triaud, A., Demory, BO. *et al.* 2017, *Nature*, 542, 456
Greaves, J. S., Richards, A. M. S., Bains, W., *et al.* 2021, *Nat. Astron.*, 5, 655
Grefenstette, N., Chou, L., Colón-Santos, S., *et al.* 2024, *Astrobiology* 24, S-186
Guzik, P., Drahus, M., Rusek, K., *et al.* 2020, *Nat. Astron*, 4, 53
Hilgartner, S. 1990, *Soc. Stud. Sci.*, 20, 519
Jordan, S., Shorttle, O., Rimmer, P. B. 2022, *Nat. Commun.*, 13, 3274
McKay, D. S., Gibson Jr., E. K., Thomas-Keprta, K. L., *et al.* 1996, *Science*, 273, 924
McMahon, S. 2020, In Smith, K. C. and Mariscal, C. (eds.), *Social and Conceptual Issues in Astrobiology*, 117
Midgley, M. 1992, *Science as Salvation*
Mix, L. J. 2015, *Astrobiology,* 15, 15
Mix, L. J. 2017, *Int. J. Astrobiol.* 17, 189
Mix, L. J. 2021, *Dialog*, 60, 86
Mix, L. J. 2022, In: Cockell, C. (ed.), *The Institutions of Extraterrestrial Liberty*, 233
Mix, L. J., Armstrong, J. C., Mandell, A. M., *et al.* (eds.) 2006, *Astrobiology*, 6, 735
Reaves, M. L., Sinha, S., Rabinowitz, J. D., *et al.* 2012, *Science*, 337, 470
Sagan, L. 1967, *J. Theor. Biol.,* 14, 225
Schaible, M. J., Szeinbaum, N., Bozdag, G. O., *et al.* 2024, *Astrobiology,* 24, S-4
Sullivan III, W. T., Baross, J. (eds.) 2007, *Planets and Life*
Thweatt, J. L., Harman, C. E., Araújo, M. N., *et al.* 2024, *Astrobiology,* 24, S-124
Weir, A. 2021, *Project Hail Mary*
Wolfe-Simon, F., Blum, J. S., Kulp, T. R., *et al.* 2011, *Science*, 332, 1163
Zakariya, N. B. 2017, *A Final Story*

(Toward) Discovery of Life beyond Earth and its Impact
Proceedings IAU Symposium No. 387, 2026
H. Landt, M. Dominik & C. Oliver, eds.
doi:10.1017/S1743921324002059

Life out there: expectations and reality

Frances Westall
CNRS-Centre de Biophysique Moléculaire, Orléans, France
email: frances.westall@cnrs.fr

Abstract. We are in the age of the direct and indirect search for extraterrestrial life. The major question is: what are we looking for and to what extent can life on Earth provide an analogy for extraterrestrial life. We will address these questions by examining the emergence of life and its evolution through geological time, based on consideration of the Earth as an exoplanet. We conclude that, with the exception of a couple of hundred million years during the Great Oxydation Event 2.3-2.1 Ga, it would not have been possible for an extraterrestrial observer to detect traces of terrestrial life until the advent of planktonic eukaryotic algae after the Neoproterozoic Oxydation Event about 800 Ma. Thus, for most of the history of life on Earth, there would be no externally observable evidence of its existence. We address the implications of this for the search for extraterrestrial life.

Keywords. Exoplanets, extraterrestrial life, biosignatures, evolution of life on Earth, Earth as an exoplanet

---

Introduction

The expanding possibilities brought by new space telescopes to search for habitable planets and, perhaps even, traces of life are exciting in prospect, fuelling the imagination. They are accompanied by incredible advances in technology on Earth in artificial intelligence. Mankind is endowed with intelligence and an active imagination and writers of science fiction from the 20th century are finding some of their stories coming true.

The hypothesis that life exists on other planets is predicated on a number of fundamental, Earth-centric considerations: (1) What is life? (2) How did life appear on Earth? (3) How did life evolve on Earth and how did the geological evolution of the Earth influence biological evolution? We will address these questions, using the Earth as an analogue of an exoplanet and considering at what stage signatures of life would have been detectable to extraterrestrial observers (cf. Westall et al. 2025). I will conclude with a short discussion on the consequences of these reflections for the search for extraterrestrial life.

In this contribution, my approach will be based on the geological record of the Earth and the evolution of terrestrial life, and the implications for biosignatures that could be recognised from space. In this respect, it could be considered as a minimalist approach to the consideration of life in the Universe.

What is life?

With respect to the question, what is life, the NASA Astrobiology working definition states that life is a self-sustaining chemical system capable of Darwinian evolution. This definition has been revisited by Vitas and Dobovišek (2019) to address a thermodynamical aspect of life as a far from equilibrium system. It also includes the flow of information from the environment to the living system. Thus, the definition of Vitas and Dobovišek (2019) is the following: "Life is a far from equilibrium self-maintaining chemical system capable of processing, transforming and accumulating information acquired from the environment." This is the definition that I will be using.

How did life appear on Earth?

There are many scenarios concerning the emergence of life on Earth that have been reviewed previously (e.g. Westall et al. 2018, 2023a; Colón-Santos et al. 2024). Here, I will address the pertinent constraints for the emergence of life, *i.e.* the critical ingredients of life ("as we know it"), and briefly summarise the different scenarios suggested.

Life as we know it is composed of a variety of classes of organic molecules, namely amino acids, fatty acids, carbohydrates and nucleobases, that are assembled together to form the fundamental composites of a cell: proteins, phospholipids, polysaccharides, and an information transferring molecule (RNA/DNA). The principal elements of these critical molecules are the C H N O P S elements, as well as a number of transition elements used as cofactors and as important structural ingredients of the larger molecules. These cofactors include Fe, Zn, Mn, Co, Cu, Mo, Ni, W, and V (Zerkle et al. 2005). Zerkle et al. (*op.cit.*). However, note that Cu and Mo utilisation appeared with the oxygenation of the Earth through time and would not have been prevalent at the origin of life.

A wide variety of organic molecules was brought to the early Earth through accretion of extraterrestrial materials, such as meteorites (carbonaceous chondrites that comprise up to 3-4% organic matter, Sephton and Botta 2008; Schmitt-Kopplin et al. 2010; Martins and Pasek 2024) and micrometeorites (Maurette and Brack 2006; Dartois et al. 2013). While about 70% of extraterrestrial organic matter is insoluble (IOM), up to 30% is soluble (DOM). Meteorites and micrometeorites falling into the early seawater (or lakes) would break up and release the soluble organics into the water. The insoluble matter would have been disaggregated, particularly in shallow water environments subjected to storms and strong tidal action. Sedimented with the detrital fragments of the volcanic early Earth, the IOM would have been reprocessed by the very active hydrothermal systems at the surface and by hydrothermal fluids percolating through the upper crust (Martins and Pasek 2024).

Other sources of organic molecules to the early Earth would have included those formed in the atmosphere by lightening in CO_2-rich volcanic clouds (Cleaves et al. 2008) and those formed endogenously in the crust by Fischer Tropsch reactions during the serpentinisation of the ultramafic materials in the presence of hydrothermal fluids (Konn et al. 2009, 2015; McCollom et al. 2015).

The other critical ingredient for life is liquid water. This is because a liquid environment permits a variety of chemical interactions. On the one hand, certain molecules may be dissolved and transported while, on the other hand, others are hydrophobic and, therefore, stable. Thus, hydrophobic molecules, such as lipids, can form a micellar membrane that protects the internal machinery of a cell from a potentially hostile exterior environment while, at the same time, the porous hydrophobic membrane permits certain dissolved molecules to be transported in and out of the cell. Water is a particularly good combination with organic molecules (Brack 2002) because it is stable under a wide range of temperature and pressure conditions. Its density permits it to maintain critical concentrations of reactants, such as salts. It has high heat retention and surface tension, and a dielectric constant. Moreover, the fact that it is a polar molecule facilitates covalent bonding and therefore polymer building. Finally, water contributes to the structural organisation of molecules.

Finally, no chemistry can take place without an energy source. A large number of potential sources of energy for prebiotic reactions on the early Earth has been suggested. They range from highly energetic sources, such as UV (Pascal et al. 2013) and lightening (Pascal et al. 2013; Cleaves et al. 2008), through shock impact of meteorites (Martins et al. 2013), chemical redox reactions between water and highly reactive minerals, such as sulphides (Wächtershäuser 1998) and ferrous oxides (Russell and Hall 1997; Martin et al. 2008), to piezoelectric energy produced in deep-seated fault systems (Schreiber et al. 2012). Certain scenarios for the emergence of life suggest a nuclear energy source for the prebiotic reactions (Adam 2007; Adam et al. 2018) but there is no evidence for such an energy source on the early Earth (Westall et al. 2023a).

Within the constraints of the ingredients of life as we know it and the potential energy sources available, many hypotheses have been expounded as to where and how life emerged (see reviews in Westall et al. 2018, 2023a; Colón-Santos et al. 2024). Very briefly, the scenarios envisaged fall into two main groups, those implicating some form of hydrothermal activity (the majority) and those predicated on other sources of energy. Hydrothermal environments are popular for a number of reasons. They provide a source of energy in the form of heat and redox reactions and may be a conduit for organic molecules in hydrothermal effluent (e.g. Kelley et al. 2001; Konn et al. 2009, 2015; McCollom et al. 2015; Ménez et al. 2018). Hydrothermal environments are characterised by porous materials, either the vent constructs themselves or the sediments that surround them. These materials comprise reactive materials, either metal sulphides or mafic (magnesium and iron rich) minerals that produce energy through redox reactions at their surfaces. The surfaces of the reactive minerals and their alteration products (phyllosilicates, for instance), facilitate the concentration and conformation of organic minerals. Moreover, gradients in temperature, concentrations, pH form within the porous hydrothermal materials and gradients mean disequilibria, an essential element for abiogenesis and prebiotic chemistry.

Hydrothermal environments on the present Earth and on the early Earth occur in a wide range of settings. They are most commonly associated with deep sea rifting and plume volcanism, the higher temperature black smoker vents occurring closer to the volcanic sources with lower temperature white smokers or diffuse venting occurring further away. However, hydrothermal activity is not only a deep-sea phenomenon but also occurs in shallow water environments. This was particularly the case on the volcanically more active and hotter early Earth, as demonstrated by evidence from Palaeoarchaean rocks (Hofmann and Harris 2008; Westall et al. 2015a, 2023a). Moreover, there is some evidence for subaerial hydrothermal activity (at the edges of exposed landmasses) on the early Earth (Djorkic et al. 2017; Van Kranendonk et al. 2021). Furthermore, hydrothermal activity at the edges of impact craters may have been of importance (Osinski et al. 2013, 2020; Sasselov et al. 2020).

Hydrothermal environments offer a variety of characteristics that would have been potentially useful for prebiotic chemical reactions (cf. Russell and Hall 1997; Martin et al. 2008; Westall et al. 2018, 2023a; Van Kranendonk et al. 2021). Their generally porous structures comprise mafic (Fe and Mg-rich) minerals either as precipitates of hydrothermal effluent (metal oxides in black smokers), as well as detrital volcanic sediments, or the products of aqueous alteration of the mineral surfaces (phyllosilicate minerals, such as clays). All of these minerals have reactive surfaces that are conducive for trapping or chelating organics and that can contribute to their structural organisation and complexification. The porous hydrothermal structures and sediments permit the circulation of fluids and the formation of gradients in pH, temperature and concentration of reactants, thus creating chemical disequilibria.

Even though no deep-sea black smoker from early Earth environments have been preserved, given our knowledge of the geology of the early Earth, they must have existed. Thus, Westall et al. (2023a) conclude that subaqueous or subaerial hydrothermal environments, some exposed to UV radiation, some not, could have hosted abiogenesis. However, there is little evidence for other environments suggested for abiogenesis, pumice rafts (Brasier 2011), nuclear placer sands (Adam 2007; Adam et al. 2018), although deep seated faults would have been present (Schreiber et al. 2012).

Many scenarios for the potential emergence of life on other planetary bodies have been hypothesised for example hydrothermal activity at the edges of impact craters on Mars (Osinski et al. 2013, 2020; Sasselov et al. 2020)), as well as subaqueous hydrothermal activity (Westall et al. 2015b). In the case of an early (Pre-Noachian) land-locked planet such as Mars, environments exposed to UV radiation would have abounded. This may or may not have been the case on Venus (Westall et al. 2023b). However, such UV-bathed environments would not have been available on ocean planets. The question is, if UV energy was a prerequisite for the emergence of life on Earth, does this mean that life could not have emerged on ocean planets? To date, we do not know how life emerged on Earth. Moreover, if it did emerge in only one particular scenario, does this mean that it could not emerge in another on another planet? The finding of life on another rocky planet or an icy satellite in the Solar System will be highly significant in terms of confirming de Duve's (1995) hypothesis that, in the an appropriate environment, biology should emerge naturally from chemistry. Finding life on an icy satellite will tell us that life can appear without UV radiation, but not that this would be the only way in which it could occur. Nor will finding life on Mars constrain the environments for abiogenesis. Based on these essential requirements, life likely emerged on Earth during the Hadeon Eon (4.5–4.0 Ga), after the Moon-forming impact, variously dated between 4.51 Ga (Barboni et al. 2017) and 4.36 Ga (Borg and Carlson 2023).

How did life evolve on Earth and how did the geological evolution of the Earth influence biological evolution?

Eo-Palaeoarchaean eras: earliest traces of life

Some of the most ancient rocks and minerals on Earth may contain indications that life was present. 4.1 Ga old zircon crystals contain graphitic inclusions, whose $\delta^{13}C$ isotopic composition is consistent with that produced by life (Bell et al. 2015), but is also consistent with carbon isotope signatures from abiotic matter (cf. Summons et al. 2011). The oldest known sediments are preserved in 3.7-3.8 Ga rocks from the Isua Greenstone Belt in Greenland. They comprise highly metamorphosed rocks, including a banded iron formation (BIF). Again, evidence for life in these rocks was interpreted from carbon isotopes (Schidlowsky et al. 1988; Rosing et al. 1999). More recently, iron isotope measurements were used to interpret the possible presence of anaerobic phototrophs (Czaja et al. 2013), while Hassenkam et al. (2017) analysed carbonaceous inclusions in garnet crystals in the metasediments (metamorphosed sediments), concluding that the carbonaceous molecules were likely remnants of life. Metasediments of controversial age (4.3-3.7 Ga, Cates et al. 2013) from the Nuvvuagittuq terrane in northern Canada contain an apparent BIF in which large filamentous structures have been interpreted as iron oxidising microbes from hydrothermal vents (Dodd et al. 2017; Papineau et al. 2022). However, the filaments may be of abiotic origin (McMahon 2019; Lan et al. 2022). Moreover, the sediments are now believed to be Eoarchaean (4.0-.06 Ga) rather than Hadean (4.56-4.0 Ga) in age (Darling et al. 2013).

These are the oldest traces of life known to date. It could be argued that they are "tenuous" in the sense that, normally, reliable identification of traces of life would require a number of concurring lines of evidence, for example, physical remains of cells, colonies or microbial constructions; organic, isotopic or geochemical signature, amongst others (Summons et al. 2011; Neveu et al. 2018). However, in light of the very clear evidence of life in well-preserved metasediments in the Palaeoarchaean (3.6-3.2 Ga) (reviewed in Westall and Xiao 2025), the existence of life in Eoarchaean sediments is tenable. As of the Paleaoarchean, the fossilised evidence for life becomes stronger. Importantly, the oldest traces of life in these sediments document not only its widespread distribution (to be clear, its occurrence is limited to the few, preserved Palaeoarchaean terranes in the Pilbara Greenstone Belt, Australia and the Barberton Greenstone Belt in South Africa), but also its rapid evolution. The first cells are believed to have been chemolithotrophic autotrophs, *i.e.* organisms obtaining energy from oxidation of inorganic substances, such as hydrogen or sulphur, or redox reactions on mineral surfaces (e.g. (Russell and Hall 1997; Martin et al. 2008)). However, by about 3.5 Ga there is evidence of chemorganotrophs, *i.e.* organisms obtaining energy by oxidation of carbonaceous matter (Westall et al. 2006a, 2011, 2015a, 2025; Hickman-Lewis et al. 2020a; Clodoré et al. 2024), as well as phototrophs, *i.e.* organisms obtaining energy from sunlight (Noffke et al. 2008, 2013; Byerly et al. 1986; Walsh 1992; Hofmann et al. 1999; Tice and Lowe 2004; Allwood et al. 2006; Westall et al. 2006b), to name just a few of the many studies). All of these organisms were obligate anaerobes because oxygen was not present on the early Earth (although very small quantities were produced by a variety of abiotic processes, Westall et al. 2023a). Thus, the earliest, well-preserved biosignatures at 3.5 Ga document an already flourishing and evolved biome. This does not mean that life "started out with a bang", it is simply a reflection of the fact that any earlier traces of life were not preserved.

The Eo-Palaeoarchaean Earth (4.0-3.2 Ga) as an exoplanet

At this point, considering the search for life elsewhere in the Solar System and on exoplanets, it is useful to step back and observe the early Earth from the eyes (or space telescopes) of an extraterrestrial spectator. First of all, spectral data of the atmosphere of the early Earth would reveal the presence of N_2, CO_2, and possibly traces of CH_4 and H_20, depending upon the sensitivity of the instruments. The James Webb Space Telescope, for example, recently detected CH_4 and CO_2 (but no NH_3, H_2O or CO, Madhusudhan et al. 2021, 2023) in the atmosphere of planet K2-18 b, a "Hycean" world. These gases were interpreted as a potential biosignature (*op.cit.*), although Wogan et al. (2024) consider the planet to be a mini Neptune with CH_4 and CO_2 produced thermochemically. "Hycean" worlds are hypothetical planets that would resemble H_2O-rich sub-Neptunes with habitable surface oceans (Madhusudhan et al. 2021). There have been a number of recent reviews on potential exoplanet biosignatures (e.g. Catling et al. 2018; Schwieterman et al. 2018; Schwieterman and Leung 2024; Quanz et al. 2021). Quanz et al. (2021, after Catling et al. 2018) that list a number of atmospheric gases that could potentially be detected in exoplanet atmospheres. Some of these are considered to be biosignatures if both reduced and oxidised gases occur simultaneously, e.g. CH_4 and O_3, although both, individually, can be produced abiotically. A variety of additional potential gases of biological (and sometimes abiotic) origin has been proposed (Table 1), including the phosphine that was detected in the atmosphere of Venus (Greaves et al. 2021; Bains 2024; Schwieterman and Leung 2024). Importantly, the presence of gases in disequilibria would suggest some form of permanent production, for example O_2/O_3 and CH_4, CO_2 and CH_4, N_2 with O_2 and H_20, or N_20 and O_2/O_3 (Schwieterman and Leung 2024).

Other signatures that may signify the presence of life on an exoplanet include: pigments of various kinds, a vegetation red edge, molecular chirality, fluorescence and bioluminescence, and seasonality owing to temporal changes in gas levels due to seasonal productivity (Schwieterman and Leung 2024; see Table 2). Habitability may be indicated by "ocean glint" (Sagan et al. 1993) or hydrogen exospheres (Kameda et al. 2017). Recently Triaud et al. (2024) have proposed an alternative combination of signatures that might indicate life on an exoplanet that includes liquid water: "A low carbon abundance in the atmosphere of a temperate rocky planet, relative to other planets of the same system, traces the presence of a substantial amount of liquid water, plate tectonics and/or biomass."

Table 1. Potential biosignature and non biosignature gases in exoplanet atmospheres (after Quanz et al. 2021).

	Molecule	Origin
Potential biosignature gases	CH_4	Abiotic: volcanic outgassing Biogenic: methanogenesis
	O_3	From O2 via photochemical reactions
	N_2O	Abiotic: lightning, photochemical reactions, chemodenitrification Biogenic: denitrification
	NH_3	Abiotic production in hydrothermal vents Biogenic: volatilisation of organic waste
	CH_3Cl	Biogenic: from algae and tropical forests
	$(CH_3)_2S$	Biogenic: from plankton
	CH_3SH	Biogenic: mercaptogenesis
Other possible atmospheric gases	CO_2	Abiogenic: photodegradation, thermal degradation, oxidation of reactive oxidative species, inorganic chemisrty Biogenic: microbial degradation, extracellular oxidative metabolism
	H_2O	Abiogenic: from serpentinisation in crust, free molecules and ice
	N_2	Abiotic: redox reactions of Fe, Mn, photochemical and thermochemical reactions Biogenic: enzymer-mediated biological processes
	CO_2	Abiotic: CO2 photolysis and volcanoes Biogenic: photoxidation of organic matter
	C_2H_6	Abiotic: photochemically produced from CH4 and other C-bearing species
	H_2	Abiogenic: volcanic outgassing
	H_2S	Abiogenic: Volcanic gas Biogenic: microbial degradation
	SO_2	Abiogenic: volcanic gas
	H_2SO_4	Abiogenic: volcanic gas
	HCN	Abiogenic: hydrothermal environments, electrical discharge

Table 2. Variety of potential biosignatures (after Schwieterman and Leung 2024).

Surface and temporal biosignatures		
Biosignatures	**Origin**	**False positives**
Photosynthetic pigments	Clorophyll, bacteriochlorophyll	None
Non-photosynthetic pigments	Microorganism reflectance from pigments	None
Retinal pigments	Halaeoarchaea, bacteriorhodopsin	None
Vegetation red edge	Plant leaves, phototrophic cells	Surface minerals
Chirality	Nearly all biological molecules	weak enantiomeric excess in carbonaceous chondrites
Fluorescence and bioluminescence	Reprocessing or direct production of light from biological organisms	Production by non biological minerals
Seasonality	Temproal changes due to orbitally-constrained biological productivity	None

Chirality

Another biosignature noted by Schwieterman and Leung (2024) is chirality. Chirality refers to the structural configuration of molecules that can be mirror images of each other (L-left-, R-right handed). In biology, organic molecules are characterised by a distinct chirality, either left handed or right handed. For example, most terrestrial organisms synthesise and use α-amino acids exclusively in the L-form (homochirality), although some bacteria make and use D-amino acids (Summons et al. 2008). Degradation of the organic molecules with time tends to obliterate this effect. However, in order to determine the chirality of any organic matter on another planetary body, it is necessary to be able to sample the surface of the planet. We will see below that we have enough difficulty in sampling and analysing the surfaces of other planets in the Solar System that it is difficult to imagine a space faring life form from another stellar system being able to sample the Earth at any period during its geological history.

The appearance of phototrophy on Earth and its importance for the production of space-visible biosignatures

Considering this variety of atmospheric, surface and temporal phenomena that could indicate life on the surface of a planet, what could have been detected on an early Earth upon which life, in the form of anaerobic chemotrophs and phototrophs, flourished? In terms of gases, the only potential biogas might have been CH_4 (if it had been produced in sufficient quantities) and this would have been associated CO_2, but CH_4 can also be produced by volcanic outgassing and serpentinisation reactions, both of which would have been prevalent on the early Earth. In terms of any surface or temporal signatures, we could consider photosynthetic and nonphotosynthetic pigments and chirality from the list in Table 1. We have seen that the early, anaerobic terrestrial biome comprised both non photosynthetic microbes as well as photosynthetic ones, the former producing a variety of pigments and the latter characterised by chlorophyll and bacteriochlorophyll, all of which could have spectral signatures mimicked by minerals. Anaerobic chemotrophs on the early Earth inhabited benthic environments, generally away from the photic zone (Westall et al. 2006a, 2015a,b; Hickman-Lewis et al. 2020a; Clodoré et al. 2024; Westall and Xiao 2025). Whether or not they had pigments, their existence would not have been visible from space. Phototrophs, however, abounded in shallow water environments. The question is, what biomass did they represent and would they have been visible?

Des Marais (2000) calculated that the ability to capture photons as a source of energy that vastly increased the productivity of microorganisms: while chemolithotrophs obtain about (0.2 to 2.0) $\times$ 10^{12} mol C per year from inorganic (hydrothermal) energy sources, oxygenic phototrophs produce 9000 $\times$ 10^{12} mol C per year, *i.e.* two to three orders of magnitude greater carbon production. However, oxygenic phototrophy appears to have originated on Earth during the Mesoarchean (3.2-2.8 Ga), around 3.0 Ga (see review in Westall and Xiao 2025). We noted above the evidence for phototrophic mat development by 3.5 Ga (Noffke et al. 2013), as well as geochemical suggestions of possible (anoxygenic) phototrophy by ca. 3.8 Ga (Czaja et al. 2013). The lack of evidence for the production of oxygen associated with the Palaeoarchaean phototrophic mats indicates that the earliest phototrophs were anoxygenic. The evolution of anoxygenic phototrophy predated the more complicated but more efficient energy producing process of oxygenic phototrophy, whereby sunlight is used to split the water molecule (in abundant supply on Earth) to obtain hydrogen, which is then oxidised, while oxygen is liberated as a waste product. The early appearance of anoxygenic phototrophs was confirmed by gene sequencing analysis of the photosynthetic green sulphur bacterium *Chlorobium tepidum* and the green non-sulphur bacterium *Chloroflexus aurantiacus* (Xiong et al. 2000), which documented that purple bacteria were the earliest emerging photosynthetic lineage and that bacteriochlorophyll biosynthesis definitely evolved before chlorophyll biosynthesis. Recent estimations of an earlier appearance of oxygenic photosynthesis combining comparative structural biology, phylogenetic analyses and molecular clocks with experimentation and ecological studies suggests that oxygenic photosynthesis appeared already in the Palaeoarchaean (Cardona et al. 2019; Sánchez-Baracaldo and Cardona 2020). However, there is no direct fossil evidence for this and the geological underpinning of these trees is erroneous (Westall 2025).

Let us now consider the environments in which the early anoxygenic phototrophs flourished. The available phototrophic habitats comprised the shallow water environments around exposed landmasses. Today, planktonic phototrophy in the oceans is dominant but, on the early Earth, tenuous evidence for planktonic phototrophs only starts appearing after about 3.2 Ga (Javaux et al. 2010). Additionally, some large (tens of microns), enigmatic fossils from 3.43 Ga sediments in the Pilbara, Australia, have been interpreted as possible planktonic microorganisms of unknown affinity (Sugitani et al. 2015). These interpretations are still debated (e.g. Kapanarthy et al. 2024). Some studies have suggested that oxygenic planktonic phototrophs existed already by 3.8 Ga on the basis of bulk carbon isotope signatures and the presence of uranium measured in the metamorphosed sediments from the Isua Greenstone Belt, Greenland (Rosing et al. 1999; Rosing and Frei 2004). This date was also suggested by the phylogenetic studies of Sánchez-Baracaldo et al. (2022). However, modern contamination of the metasediments by endolithic oxygenic phototrophs (FW, personal observation, see also Westall and Folk 2003) will influence carbon isotopes measured on the bulk sample. Moreover, small amounts of uranium, an element most easily transported in its oxidised form and, hence, a frequent proxy for aerobic conditions, were common on the early Earth. At that time, very small and local amounts of oxygen were produced by abiotic processes, such as photolysis at the surface of water or dissociation of water vapour exiting shallow hydrothermal vents at high pressure (cf. Westall et al. 2023a). We can therefore exclude the planktonic realm from the Eo-Palaeoarchaean eras.

Landmasses on the early Earth comprised exposed land around volcanoes formed above mantle hot spots (similar to the Hawai'i Islands or the Réunion archipelago), or they formed on oceanic plateaus topped by volcanoes (similar to the Kerguelen Islands) that represented the protocontinents (Westall et al. 2023a; Westall 2024). Most of these low relief volcanic island landmasses formed in relatively shallow waters, similar to semi-enclosed basins around the volcanoes (cf. Nijman et al. 2017). The photic zone around the exposed landmasses would have been rather turbid during volcanic eruptions but in between eruptions because of the volcanic airfall would have been clearer. This is documented in sediments in the Palaeoarchaean terranes of the Pilbara and Barberton Greenstone Belts, where there is widespread evidence for flourishing phototrophic microbial mats (Noffke et al. 2008, 2013; Tice and Lowe 2004; Westall et al. 2006b; Hickman-Lewis et al. 2018, 2020b), and rare stromatolites (Byerly et al. 1986; Walsh 1992; Hofmann et al. 1999; Allwood et al. 2006). However, these are just two locations of probably many more that were not preserved.

To date, it is difficult to estimate the extent of shallow water biomes that would have been available to the early phototrophs. Estimates of the rate and amount of continental crust formation do not provide information about the aerial extent of the shallow water regions, because most of the early continents were effectively submerged (Fig. 1). The semi-circular water basins (collapse basins, Nijman et al. 2017) formed on top of them were not necessarily very deep, perhaps 50 to several hundreds of metres deep. Analysis of the sediments deposited in these basins documents shoreface to foreshore environments that were periodically influenced by storms with their deep reaching waves (Lowe 1999; de Vries et al. 2010; Westall et al. 2015a; Nijman et al. 2017). These shallow water environments were the recipients of nutrients sourced from the eroded, exposed landmasses and transported by rivers into the littoral areas around the landmasses, nutrients that contributed to the nourishment of the early microbial colonies (cf. Hickman-Lewis et al. 2020b). In the quieter periods in between volcanic eruptions, phototrophic mat formation could be quite extensive. For example, Westall et al. (2015a) document the formation of microbial mats in a period of quiet sedimentation (possibly lasting for hundreds of thousands of years, maybe even up to a million years; note that anaerobic phototrophs develop

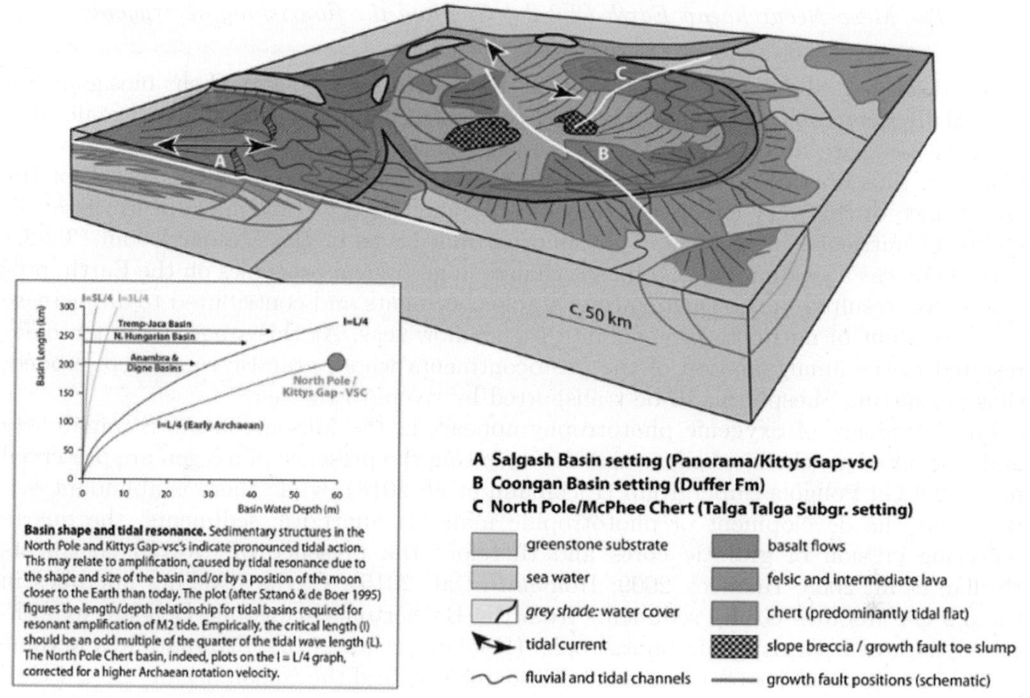

Figure 1. 3D schematic showing the development of broad, shallow semi-circular basins at the edges of exposed volcanic edifices, formed above oceanic plateaus (the protocontinents). The inset indicates the higher tidal resonance for the Palaeoarchaean compared to today (Nijman et al. 2017).

much more slowly than aerobic ones), in a basin that was probably many kilometres in diameter. In comparison, the modern Laguna Figuero on the Pacific coast of Mexico supports a vibrant (aerobic) phototrophic microbial mat community over an aerial extent of about 8 km^{2}that was compared to microbial mats of the Palaeoarchaean (Margulis et al. 1980). Given the degree of UV penetration through the early atmosphere because of the lack of attenuating oxygen and ozone (Sagan and Mullen 1972; Tehrany et al. 2002; for example, Cockell 2000 calculated DNA-weighted UVB and UVC values of 10 W/m^2, and up to 1000W/m^2 in a worse case scenario, compared to 1W/m^2 today), the early phototrophs would necessarily have had recourse to protective pigments.

In geological terms, the existence of Palaeoarchaean anaerobic phototrophic microbial mats having an aerial extent of several square kilometres and persisting over a period of several hundreds of thousands of years up to a million years is not significant. Moreover, this would be one isolated example on a planet that may have hosted several contemporaneously isolated examples. At best, the situation could be described as spatially and temporarily ephemeral and it is unlikely that any pigments could have been detected from space.

With the evolution of oxygenic photosynthesis and oxygenic photosynthesisers during the Mesoarchaean (3.2-2.8 Ga), microbial productivity increased by orders of magnitude, as noted by Des Marais (2000). Larger, domal stromatolites started to appear in the Neoarchaean (2.8-2.4 Ga) on well-developed continental platforms, at the same time as gradual changes in the geological evolution of the Earth led of the stabilisation of continents and the availability of greater extents of shallow water environments for colonisation by benthic phototrophic microbial mats (Westall and Xiao 2025).

The Meso-Neoarchaean Earth (3.2-2.4 Ga) and the flourishing of oxygenic photosynthesis

We discussed above the importance of photosynthesisers and their biosignatures potentially detectable from space, namely pigments. We noted also that oxygenic photosynthesisers are able to obtain orders more energy through reduction of the water molecule, thus liberating oxygen (Des Marais 2000). One of the limiting factors for the appearance of this very effective means of obtaining energy could have been the availability of nutrients, in particular phosphorus. Indeed, as of the Mesoarchaean (3.2-2.8 Ga), there was a gradual but significant change in geological processes on the Earth, processes that resulted in the emergence of stable continents and contributed to an increase in the amount of nutrients brought into the shallow seas. At this stage, plate tectonics resulted in the amalgamation of the protocontinents whose granitic roots were eroded, thus permitting phosphorus to be transported by rivers to the sea.

Fossil evidence of oxygenic phototrophy appears in the Mesoarchaean. Stromatolites and associated geochemical signatures documenting the presence of oxygen are preserved in the 2.9 Ga Pongola Supergroup (Eickmann et al. 2018), while there is abundant evidence for the development of phototrophic mats on quartzitic sediments, the quartz reflecting erosion of granitic cores and therefore the presence of evolved continents (Noffke et al. 2008; Heubeck 2009; Homann et al. 2015). Cavity-dwelling microbes in the 3.2 Ga Moodies Group sediments from the Barberton Greenstone Belt are hypothesised to have been oxygenic phototrophs (Homann et al. 2015). Stromatolites become increasingly more important in the late Mesoarchaean and the Neoarchaean (2.8-2.4 Ga), occurring on terranes as far flung as North America, Australia and Southern Africa.

The Great Oxidation Event and further evolution during the Proterozoic

The oxygen produced by the flourishing oxygenic microorganisms (cyanobacteria) of the Neoarchaean did not immediately appear in the Earth's atmosphere. The large amounts of reduced terrestrial rocks and minerals in the oceans, especially the shallow seas, acted as sinks for oxygen. The reductants, H_2, C, S, Fe^{2+}, that were oxidised were those previously utilised as energy sources for anoxygenic microorganisms. Their removal was a stimulus for the exploitation of another reductant, *i.e.* water, through oxygenic photosynthesis.

When most of the surficial, reduced materials had been oxidised, there is very strong evidence for a sudden rise in atmospheric oxygen at the end of the Neoarchaean/beginning of the Palaeoproterozoic, after 2.4 Ga, termed the Great Oxidation Event (GOE, Holland 2002; Lyons et al. 2014). The GOE coincided with major mantle overturn, the production of large igneous provinces, and the amalgamation of supercontinents and their subsequent erosion and breakup, as illustrated in Figure 2 (Gumsley et al. 2017; Chen et al. 2022). The change from a reducing atmosphere to an oxidising one is strongly marked in the mass dependent/mass independent sulphur isotope record (Fig. 3) (Farquhar et al. 2007). The significant rise in atmospheric oxygen and the consequent massive increase in biomass development led to the strongest known carbon isotope excursion in the geological record, the Lomagundi Excursion event that lasted from 2.3 to 2.1 Gyr ago (Karhu and Holland 1996; Bekker et al. 2006). These events brought the oxygen content of the atmosphere up to present day levels and lasted about 200 million years. Such an event would have been recognisable from space, but for a relatively short period of time (from the geological point of view). At this point, an extraterrestrial observer would have identified the presence of H_2O, O_2, O_3, and CO_2 in the atmosphere. Depending upon the observational capacities, perhaps land distribution would have been visible as well. Thus, evidence of life would have been detectable for the short window of time between 2.3-2.1 Ga.

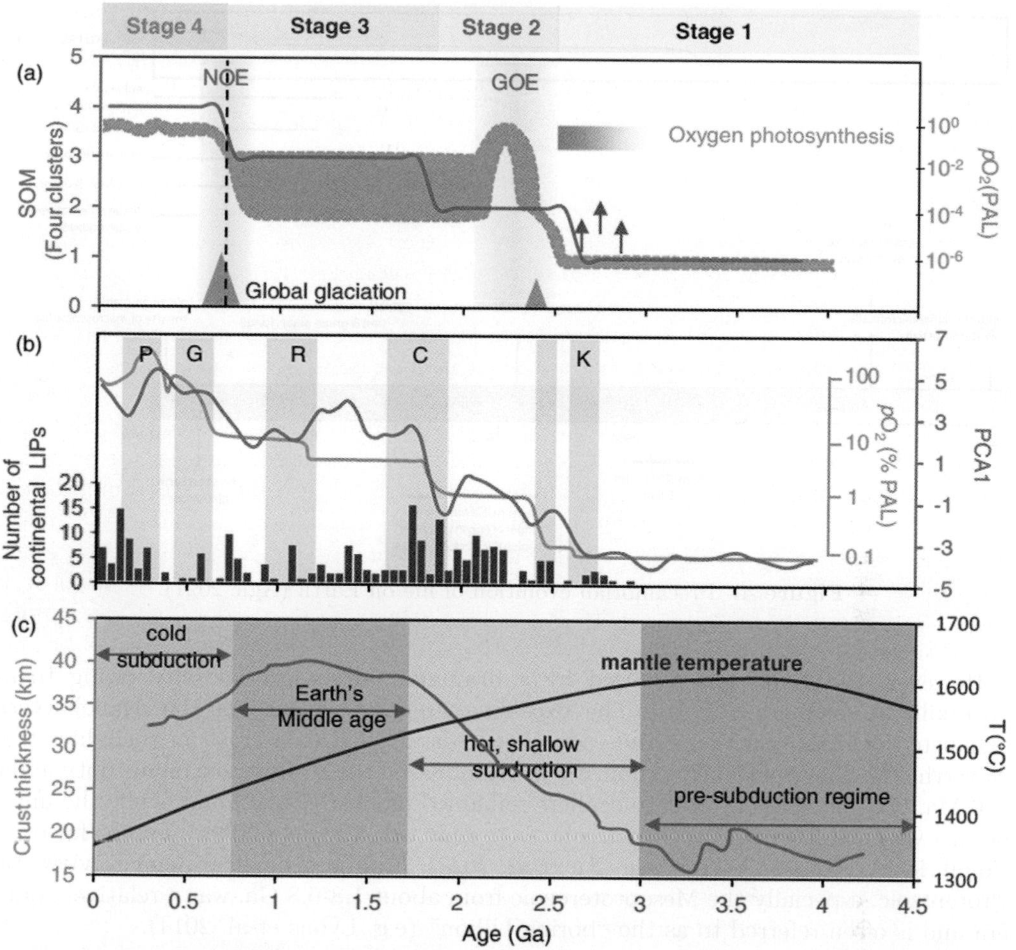

Figure 2. Co-evolution of atmospheric oxygen with crustal processes, tectonics and LIPs through geological time. (Chen et al. 2022).

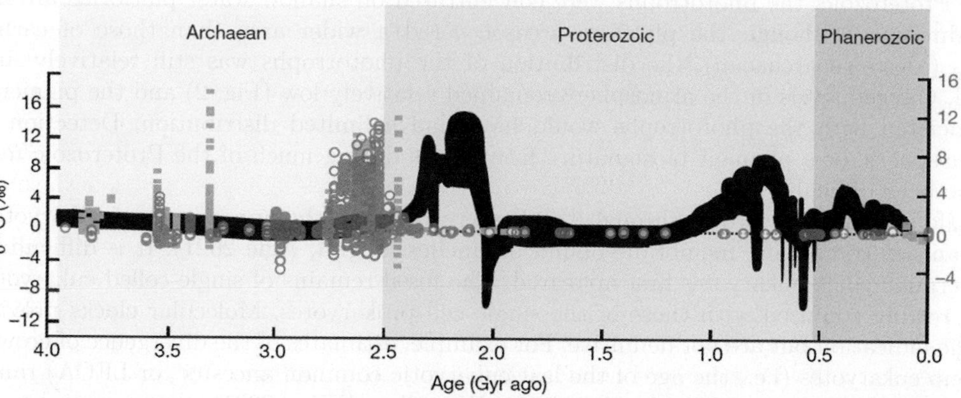

Figure 3. Sulphur and carbon (carbonate) isotope variations throughout geological time (Lyons et al. 2014).

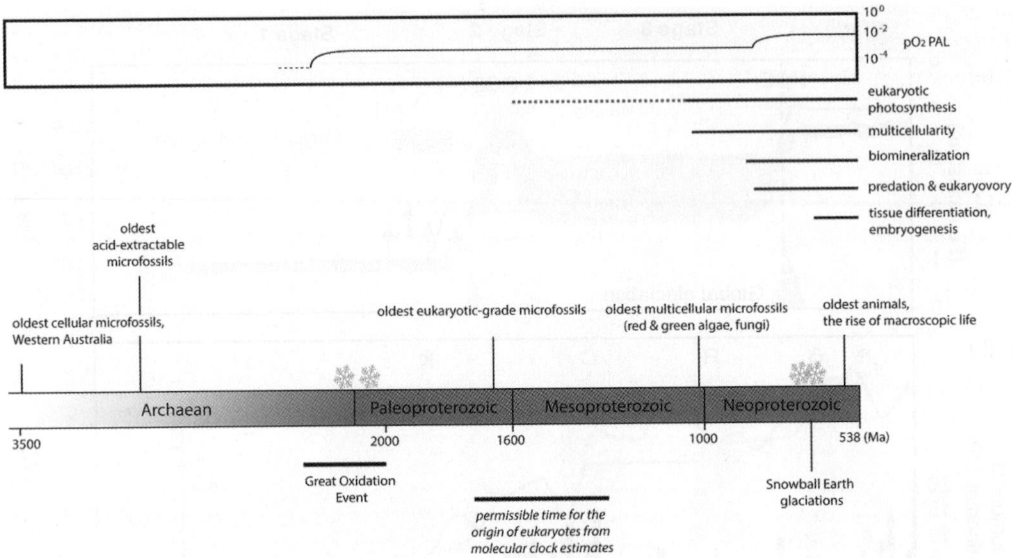

Figure 4. Precambrian evolution of life on Earth (Agič 2021).

However, the GOE was followed by a dramatic fall in productivity owing to an "overkill" in biomass activity. The excessive, frenetic activity of the cyanobacteria exhausted available nutrients and carbon sources and led to a crash in their life style. Nevertheless, they continued to flourish and dominated the Proterozoic biome until about 800 Ma. Oxygen levels in the atmosphere remained modest (Fig. 2) and, while the upper oceans were oxidised, the deeper oceans were reduced (Canfield, 1998), except for occasional, local oxidation events (e.g. Xu et al. 2023). From a geological point of view, the Proterozoic, especially the Mesoproterozoic from about 1.8-0.8 Ga, was a relatively quiet era and is often referred to as the "boring billion" (e.g. Lyons et al. 2014).

The Proterozoic Eon (2.4-0.56 Ga) was dominated by very large expanses of stromatolitic bioconstructions covered well-developed carbonate platforms around the margins of the stabilised continents (Westall and Xiao 2025). We note that throughout most of the Proterozoic, the phototrophs were concentrated on shallow water platforms around landmasses. Although the platform areas covered a wider area than those of earlier eras (Meso-Neoarchaean), the distribution of the phototrophs was still relatively limited. Oxygen levels in the atmosphere remained relatively low (Fig. 2) and the pigments associated with the phototrophs would have had a limited distribution. Detection of atmospheric or a pigment biosignature from space during much of the Proterozoic may have been difficult.

Life continued to evolve through the Proterozoic with the appearance of eukaryotes, organisms comprising membrane-bound organelles (Fig. 4, Agič 2021). It is difficult to determine when eukaryotes first appeared. The fossil remains of single-celled eukaryotes are readily confused with those of the single cell prokaryotes. Molecular clocks provide some indication but are not definitive. For example, estimates of the divergence of crown-group eukaryotes (i.e., the age of the last eukaryotic common ancestor, or LECA) range from >2,300 Ma to ~1,000 Ma (Agič 2021; Westall and Xiao 2025).

What is important for our discussion is the emergence of planktonic algae. As noted above, stromatolites and their cyanobacterial-dominated communities were responsible for the production of large amounts of oxygen in the Proterozoic, but these oxygenic phototrophic microorganisms develop best under conditions where there is a significant amount of CO_2. A decrease in atmospheric CO_2 would have inhibited their development and contributed to the flourishing of eukaryotic animals, such as green and red algae. Moreover, as Monty (1974) noted, blooms of algae produce toxins that have negative effects on potential eukaryotic grazers. This may have had an inhibiting effect on the rise of the latter in the Neoproterozoic. There is convincing evidence for multicellularity (and eukaryotes) in the Meso-Neoproterozoic, 1.8-0.56 Ga Agič 2021; Westall and Xiao 2025) and recently potentially multicellular microfossils (Qingshania magnifica) have been described from the ∼1635- million-year-old Chuanlinggou Formation, North China (Miao et al. 2024).

Although planktonic algae and cyanobacteria became widespread during the Neoproterozoic, molecular clock analyses indicate that phototrophic algae appeared earlier in the Mesoproterozoic (Sánchez-Baracaldo et al. 2022). The Neoproterozoic was characterised by the breakup of the supercontinent Rodinia and global glaciation, after which the spike in nutrient delivery to the oceans stimulated further evolution of life with the appearance of planktonic algae. Planktonic phototrophs, therefore, spread over the world's oceans and contributed to the further rise in atmospheric oxygen. This is called the Neoproterozoic Oxygenation Event (NEO). It is the appearance of the planktonic algae and the rise in oxygen that really placed the Neoproterozoic world on the galactic map of habitable planets. O_2 and O_3, together with smaller amounts of CO_2 and CH_4, would have been detectable in the atmosphere, as well as pigments from the ubiquitous algae (red and green) at the surface of the oceans (Schwieterman and Leung 2024). It is, therefore, at this point in time, about 3.7 billion years after the emergence of life, assuming that life emerged after 4.36 Ga (Westall et al. 2023b), that life would have been detectable from space – at least detectable with the kind of space telescopes operating today. Another, technologically capable civilisation may well have (had) more sensitive instrumentation and, thus, been able to detect life on Earth at an earlier stage.

At this point, we come back to the search for extraterrestrial life using the knowledge and instrumentation that we have at our disposal today. Apart from a brief moment in geological time, the GOE between about 2.3 and 2.1 Ga, when oxygen levels rose significantly in the atmosphere, it would be impossible to detect from orbit any forms of life "as we know" it on another planet until the extraterrestrial equivalent of phototrophic life forms would have become widespread on the surface of the planet. If we take the analogy of the Earth, this means significant co-evolution of that planet and its life. The evolution of life on Earth may have been retarded during the Proterozoic because of the absence of significant geological events that could have influenced it. An exoplanet hosting life may not have had to wait 3.5 billion years before being able to manifest its existence to outside observers. Nevertheless, evolution to the extraterrestrial "algal" stage may have required, at a minimum, more than 2 billion years.

What are the consequences of these reflections for the search for extraterrestrial life?

Our ability to detect traces of extraterrestrial life on exoplanets from orbit is limited to forms of life having signatures equivalent to, at a minimum, phototrophic planktonic algae, organisms that appeared on Earth only about 800 Ma. For 3.5 billion years, life flourished on Earth without demonstrating its existence. To external observers, that is an extremely long period of time. Moreover, any evolution of life during the first 3.5 billion years of its existence was predicated on co-evolution of the geology of the Earth. There are thus many constraining factors to take into account.

Let us go back to the beginning, the emergence of life. Life emerged on Earth in an, as yet, unknown location by an, as yet, unknown process after the Moon-forming impact (about 4.36 Ga; Borg and Carlson 2023). Nevertheless, we are able to constraint the general characteristics of the early Earth and find them to be similar to our concept of habitable rocky planets, *i.e.* a rocky planet with liquid water at moderate temperatures conducive to prebiotic chemistry (< ca. 120°C), an anaerobic atmosphere comprising CO_2, a source of organic molecules, and sources of energy. Rocky planets and, especially, rocky planets with liquid water, are the holy grail of exoplanet research (e.g. Foley 2024; Kaltenegger 2017). Providing that the parameters of the planetary systems around the star(s) permit the existence of liquid water at their surfaces (e.g. Kaltenegger 2017), we expect such planets to exist. Organic molecules abound in space (Guélin and Cernicharo 2022); they are found in the planetary building blocks on early Earth, and continue to bombard the surface of the Earth. Carbon being one of the most common elements in the Universe, organic molecules will be commonplace on any physical body in the Universe.

Water is also a common element in the Universe and the presence of liquid water on a planet is one of the constraints for defining the habitable zone around a star (Kasting et al. 1993; Seager et al. 2013). Water on Earth was likely sourced from a number of complementary sources, all formed in the outer, cold zones of the protoplanetary disk, including carbonaceous chondrites, comets, and asteroids, as could be common in other stellar systems (Genda 2016 and references therein; Triaud et al. 2024). Water is essential for the emergence of life, as we know it. It is also essential for the maintenance of an equable climate – but only in association with exposed landmasses where a complex feedback system of terrestrial weathering depletes the atmosphere of carbon (which is then replaced by volcanic effusions). Thus, according to Triaud et al. (2024) depletion of CO_2 in the atmosphere of a temperate rocky planet is considered to be a tracer of water oceans (a habsignature). The sources of energy evoked above for prebiotic chemistry on Earth would be among those available for the emergence of life on habitable exoplanets. These energy sources include UV radiation, redox chemistry, heat, and piezoelectric energy (in faults).

Thus, following the considerations of De Duve (1995), life should be a common occurrence on habitable rocky planets. The earliest life on Earth is believed to have used a chemolithotrophic metabolism (Russell and Hall 1997; Martin et al. 2008). Although life may have appeared as early as 4.2-4.3 Ga (or even earlier, depending upon when the surface of the early Earth became habitable; Westall et al. 2023a), the oldest, clear fossil evidence of anoxygenic phototrophy occurs at 3.5 Ga (Noffke et al. 2013) (although there may be geochemical evidence for its existence already at 3.7 Ga, Czaja et al. 2013). At a minimum, this is about 500–600 million years after the emergence of life. Moreover, we have seen that the early phototrophs would not have provided a signature visible from space, at least using present day space telescopes. In fact, although phototrophs continued to develop significantly during the Archaean, it is only for a very brief period during the GOE between 2.3-2.1 Ga that oxygen appeared in the atmosphere in sufficient quantities to be visible. All of this took place on an Earth that was continuously habitable and continued to evolve geologically with the initiation of horizontal plate tectonics, continent amalgamation, an increase in nutrient flux from the eroded continents to the seas, and the formation and breakup of supercontinents. Permanent indications of life in the form of a melange of oxidised and reduced atmospheric gases and pigments only arrived with the advent of planktonic algae about 800 Ma. Evidence from space of life on Earth depended upon many complementary contingencies and took a long time, even though we noted above that the very slow period of evolution during the Proterozoic was probably linked to a slowdown in mantle to surface tectonic processes.

The implications of this are that, in order to be detectable, life on another planet would need to have reached a relatively elevated degree of evolution, at least up to the equivalent of the terrestrial planktonic photosynthesis. Thus, for about 80% of its history (bar the GOE), it would not have been possible to detect evidence of life of Earth (with the equivalent of present day technology).

Life in the Solar System?

Let us consider the evolution of life on Earth with respect to life on other planetary bodies in the Solar System. Mars and Venus are rocky planets and the icy satellites may be considered as rocky planets covered with frozen oceans. The lack of data from Venus makes analogies difficult but we have much more data from Mars. Mars did not experience a Moon-forming impact that reset timing for the appearance of a volatile envelope having temperatures conducive to prebiotic chemistry. However, it lost its early reducing atmosphere due to the very high EUV radiation from the early Sun (Lammer et al. 2013). As the Sun's activity decreased, a secondary CO_2-rich atmosphere was outgassed and the volatile envelope, including water, was complemented by exogenous meteorites and micrometeorites. This occurred towards the end of the Pre-Noachian period, ca. 4.2 Ga (Lammer et al. 2013; Wordsworth 2016)). Compiling geological evidence with that obtained from measurements on martian meteorites and the *in situ* measurements from the MAVEN orbiter, Jakosky (2021) concludes that Mars may have lost as much as 80-99 % of its water inventory to space through geological time. However, although much of the planet (the Southern Highlands) apparently retains its ancient morphology dating back to the Noachian/Hesperion period, the history of water on the surface of the planet is complex. There was overall loss of surface volatiles to space or sequestered underground (in the form of ice or, potentially, carbonates). However, the strong obliquity experienced by the planet contributed to the transfer of water from the poles to the equator at various times in the history of the planet (Figs. 5 and 6, Wordsworth 2016; Jakosky 2021).

Mars never experienced the geological evolution that characterised the Earth. Being smaller (a dwarf planet), the internal dynamo producing a protective magnetic field faded away about 4.2 Ga (Acuna et al. 1998) leaving the secondary atmosphere exposed to erosion by the solar wind and, thus, the volatile envelope unprotected (Wordsworth 2016; Jakosky 2021). Thereafter, periodic warming influenced by volcanism and meteorite impacts (Wordsworth 2016), and possibly by changes in obliquity led to water activity, left permanent traces in the form of valley networks and fluvial features (Carr and Head 2015). Although it had previously been suggested that Mars had had a northern ocean (Clifford and Parker 2001), this possibility seems increasingly unlikely for geochemical and climatological reasons (Wordsworth 2016). However, recent geomorphological studies have, again, suggested a northern ocean during the Hesperion to early Amazonian period (3.6-2.5 Ga; Wang and Huang 2024).

Given the (very vague) geological history of Mars, can we consider the planet to have been habitable from the point of view of abiogenesis? Did Mars have the necessary ingredients of life: water, organic molecules, and energy? We have seen from the above that the volatiles, water and organic molecules would have been imported by meteorites and micrometeorites (*N.B.* exogenous organics have been detected in the Gale Crater sediments (Freissinet et al. 2015; Eigenbrode et al. 2018; Szopa et al. 2020). They would also have been degassed from the mantle and organic molecules could have been formed in volcanic clouds (cf. Cleaves et al. 2008). The same energy sources as those on the early Earth would have been available, namely UV radiation, redox reactions, heat, and piezoelectric energy in fault systems.

The variety of environmental scenarios available for the emergence of life would have differed, however. With no ocean and no plate tectonics, no deep ocean hydrothermal vents could be invoked. Standing water lakes existed but would have been fed by groundwater seepage (Kite et al. 2021) or rain or snow melt. However, the numerous impacts, filled by lakes, would have created associated hydrothermal environments (Osinski et al.

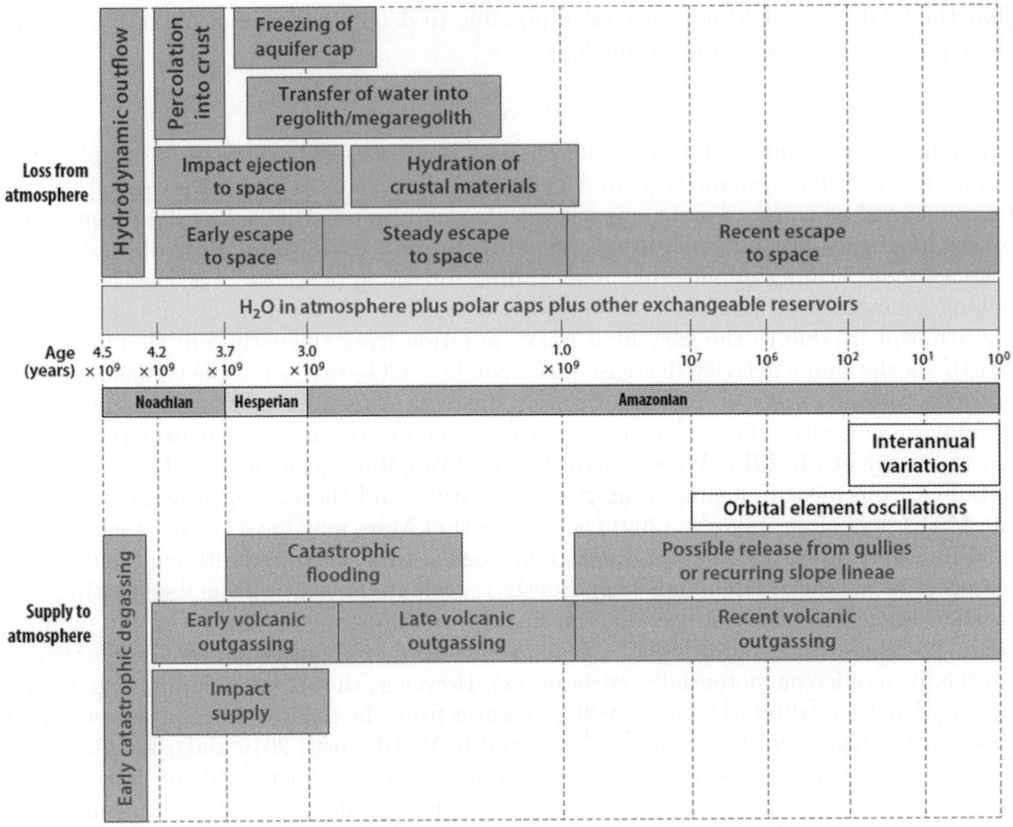

Figure 5. Summary of processes that have supplied water to the atmosphere or surface or have removed water from them (Jakosky 2021).

2013; Sasselov et al. 2020), a favoured environment for prebiotic chemistry. The abundance of exposed land on Mars would have also favoured reactions involving UV radiation (cf. Pascal et al. 2013).

While water was a permanent phenomenon on the early Earth, this was not the case on Mars. The aqueous environments in which life may have appeared on Mars would have been widely distributed in time and in space; they were not connected (Cockell et al. 2012; Westall et al. 2015b). We have no idea how much time it took for life to appear on Earth. For subsea scenarios, Westall et al. (2023a) argued that abiogenesis had to have taken place on a shorter rather than longer time scale because of the obturation of the porous materials hosting the prebiotic reactions caused by chemical precipitation of silica at the rock/water interface. But how short a time? We have, as yet, no answers from prebiotic chemistry. The time scale, however, is more likely to have been of the order of some 10^4 up to 10^6 years. If suitable environments existed on Mars, life may have appeared in more than one location – and may have disappeared, either extinguished or retreated through fracture systems into the subsurface (Friedmann and Koriem 1989; Michalski et al. 2018)).

We assume that primitive cellular life on Mars would have resembled that on the early Earth, but it would not have been genetically identical. Given the starting ingredients, reduced rocks and minerals enriched in Fe and Mg, as well as transition metals, and hydrothermal sources of energy, early martian life forms were likely to be based on a chemotrophic metabolism, *i.e.* obtaining their energy from inorganic sources in

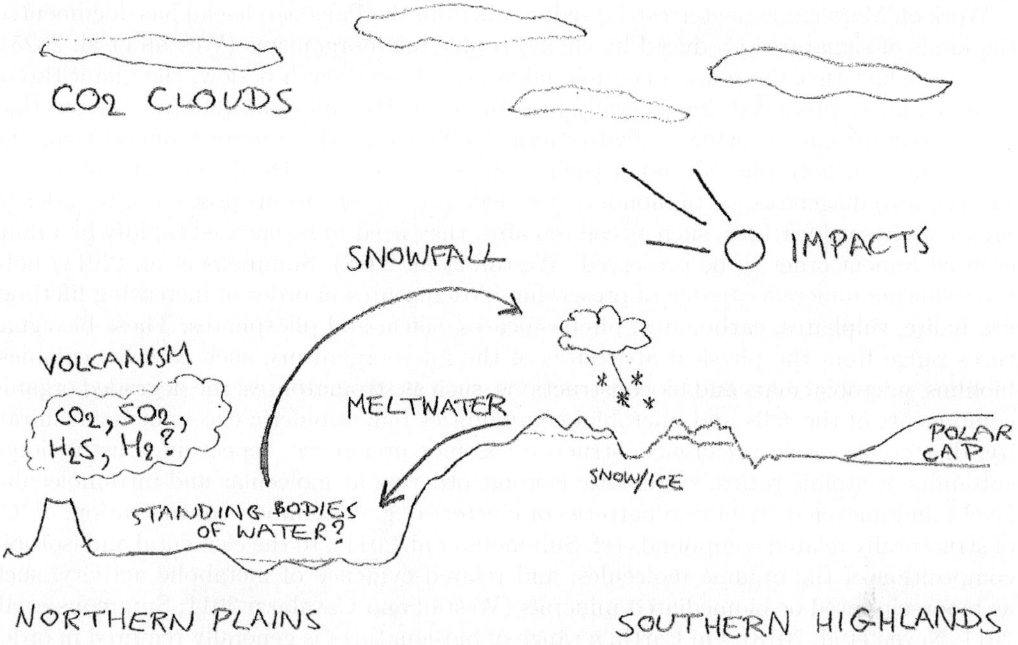

Figure 6. Schematic of the major climate processes on Mars in the Noachian and early Hesperian periods. This cartoon assumes an episodically warm scenario for the early climate with long-term transport of snow to the southern highlands interrupted by episodic melting events. Wordsworth 2016.

the case of chemolithotrophy, and from organic sources in the case of chemoorganotrophy (Westall et al. 2015b, 2025). As we noted above, organisms using these kinds of metabolisms do not produce traces of life that could be detected in an atmosphere, although CH_4 would have been produced by organisms similar to methanogens, albeit in very small quantities that would have been rapidly lost in the very localised biomes.

Could the chemotrophs have evolved some form of photosynthesis on Mars? The abundant shallow water habitats would have been conducive to this form of metabolism although we do not know exactly how long it took for phototrophs to evolve on the early Earth. As noted above, from the fossil record, we estimate that several hundreds of millions of years may have been needed. Could martian biomes have been stable sufficiently long for such a mutation to have taken place? The geological record indicates that the groundwater- or rain/melt water-fed lakes would have been transient rather than long-lasting. It is not possible to exclude the possibility of photosynthetic evolution on Mars, but the probability is low, based on our present understanding of the planet (Westall et al. 2015b). Again, on the global scale, detection of pigments associated with anoxygenic phototrophs on Mars would be arduous, given their very localised distribution, if they had evolved.

Having appeared in one, isolated location, we may envision the transport of living cells to another habitat via impact ejection, and possibly through subsurface aquifers, although these are likely to be frozen. Cockell et al. (2012) envisage the situation on Mars whereby potentially habitable habitats remain uninhabited, while Westall et al. (2015b) consider the spatial and temporal distribution of habitats, inhabited or uninhabited, as representing punctuated habitability. This punctuated habitability would have been a limiting factor for the evolution of life on Mars.

Work on Mars-analogue terrestrial sediments from the Palaeoarchaean has documented the kinds of signatures produced by chemotrophic microorganisms (Westall et al. 2025). Given the fact that these are anaerobic microbes whose growth is slow, the quantities of organic matter produced are generally low. However, Westall et al. (2015a,b) noted that chemoorganotrophs growing in hydrothermal effluent produce more biomass than the more elusive lithotrophs, although their biomass is still low. Dead microorganisms are subjected to diagenesis, or taphonomy, *i.e.* degradation by various processes. In order to preserve physical features, such as cell remains, they need to be encased rapidly in a mineralised cement order to be preserved (Westall et al. 2018). Summons et al. (2011) note the following minerals capable of preserving biosignatures in order of increasing lifetime: ice, halite, sulphates, carbonates, phyllosilicates, silica, and phosphates. These biosignatures range from the physical attributes of the microorganisms, such as cells, colonies, biofilms, microbial mats and bioconstructions, such as stromatolites; the degraded organic components of the cells and microbial communities (enantiomeric excess or homochirality, diastereoisomeric preference, structural isomer preference, repeating constitutional sub-units or atomic ratios, systematic isotopic ordering at molecular and intramolecular levels, and uneven distribution patterns or clusters (e.g., C number, concentration, $\delta^{13}C$) of structurally related compounds (cf. Summons et al. 2011); to the elemental and isotopic composition of the organic molecules; and related evidence of metabolic activity, such as bioprecipitated or biomediated minerals (Westall and Cavalazzi 2011; Summons et al. 2011; Neveu et al. 2018). On Earth, a range of biosignatures is generally required in order to identify the remains of microorganisms in the oldest rocks with any degree of certainty (Westall and Cavalazzi 2011) and this is the case also for Mars (Summons et al. 2011; Neveu et al. 2018). The reliability of a biogenic interpretation is increased by the complementarity of the biosignature assemblage, together with occurrence in a geologically plausible environment.

In situ detection of biosignatures on Mars is all the more arduous, given the limitations of instrumentation. This situation is underlined by the difficulties encountered in reliably identifying biosignatures in the oldest, well-preserved sediments on Earth. One study of fossilised (silicified) chemolithotrophs in Palaeoarchaean, coastal, volcanic sediments from the Pilbara, the 3.45 Ga Kitty's Gap Chert, documented features exhibiting the characteristics of cells, colonies and biofilms on the surfaces of volcanic particles and in the fine, volcanic dusty matrix of the sediments (Fig. 7a) (Westall et al. 2006a; Westall and Cavalazzi 2011). These would be the oldest fossil cells known to date. However, despite the combined physical organo-chemical and isotopic biosignatures presented, the features described were dismissed as abiotic artefacts (Wacey 2009) and ignored by the scientific community. At the time of this study, the polemic surrounding the possible identification of traces of life in the martian meteorite ALH84001 (McKay et al. 1996) rendered the scientific psychological environment hostile to ancient microfossil identification, especially that related to very small microfossils (individual cells in the Kitty's Gap Chert were 0.5 and 0.8 μm in diameter, too small to be visible with a light microscope). Large structures, such as stromatolites, were more readily identifiable (Fig. 7b) (e.g. Allwood et al. 2006). Recently, Clodoré et al. (2024) made an in depth investigation of the carbonaceous matter in the Kitty's Gap Chert using a wide variety of techniques, including synchrotron radiation, that demonstrated the presence of organic matter of mixed biogenic and abiotic origin in the sediments. This very broad investigation was complemented by a state of the art study (Westall et al. 2025) using a novel Cluster SIMS technique that definitively demonstrated the biogenicity of the cellular structures observed by Westall et al. (2006b). Such studies are crucial for demonstrating the level of investigation that will be necessary to determine the potential biogenicity of organic features in samples returned from Mars (cf. Bosak et al. 2024).

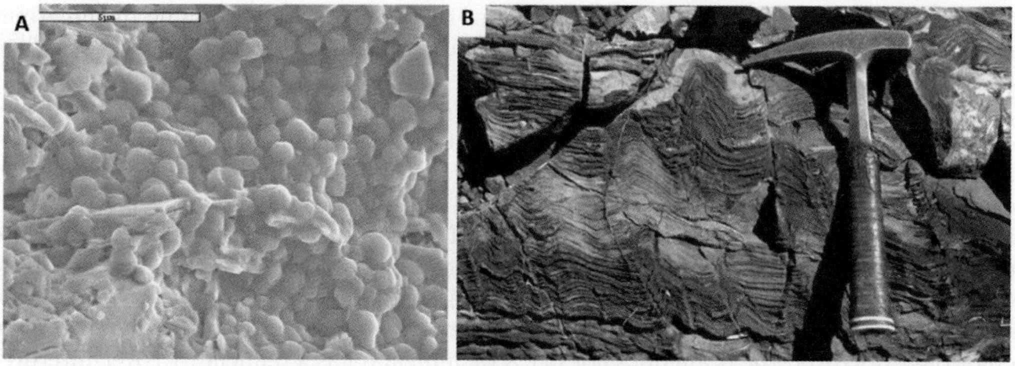

Figure 7. Fossil structures from the Palaeoarchaean. (A) Colony of small chemolithotrophs from the 3.45 Ga Kitty's Gap Chert, Pilbara (@ Westall). (B) stromatolites from the 3.4 Ga Trendall locality, Pilbara (@Marchesini).

The holy grail of the missions to Mars and the search for traces of martian life remains the larger scale biosignatures, microbial mats and stromatolites (e.g. Fig. 7b) (Noffke 2015; Hickman-Lewis et al. 2023). However, as we noted above, phototrophic microorganisms, if they ever evolved on Mars, are likely to be extremely rare.

The search for life in the Solar System is not only limited to Mars, the icy satellites of Saturn (Enceladus) and Jupiter (Europa) are objects of interest for space missions, such as the Europa Clipper mission (Pappalardo et al. 2024). If there is life on these icy worlds, that life will be similar to the terrestrial chemotrophs and, consequently, very difficult to identify. Recently, phosphine was identified in the atmosphere of Venus (Greaves et al. 2021). Phosphine is produced by a variety of biogenic and abiotic processes (Bains 2024). Its discovery has led to the Venus Life Finder Mission project (Seager et al. 2022).

The finding of traces of extraterrestrial life will lead to enormous polemic, as demonstrated by the furore following the NASA announcement of the finding of life in the martian meteorite ALH84001 (McKay et al. 1996). The NASA team made, for the time, a very careful study, documenting a number of features that contributed to their interpretation of traces of life: the presence of carbonaceous molecules in the meteorite (polycyclic aromatic hydrocarbons); the presence of structures that, to the team, had the appearance of fossilised microbial cells; the presence of tiny magnetite crystals that are very similar to those formed by magnetotactic microorganisms on Earth; and the presence of calcareous deposits thought to be biomediated precipitations, all formed in cracks in the meteorite through which water had flowed. At that time, understanding of the fossilised signatures of life was limited. We now know that carbonaceous matter of abiotic origin – from the mantle and from the infall of meteorites and micrometeorites – is common in martian materials (Steele et al. 2012; Freissinet et al. 2015; Eigenbrode et al. 2018). The fossilised cells, originally suggested to be "nanobacteria", were simply corrosion features of the carbonate deposits. The magnetite crystals are similar to those produced by one particular kind of magnetotactic bacterium, but other types of magnetotactic bacteria produce crystals of different shapes and sizes, and experiments show that they can also be formed by abiotic reactions (Golden et al. 2004). The calcareous deposits do appear to have been formed from low temperature water but can also be deposited by abiotic means. Thus, the balance of evidence is against a biogenic interpretation for the individual features described by McKay et al. (1996). However, as noted by Westall et al. (2021), the data are correct, although the interpretation erroneous but not surprising given the state of knowledge at that time.

A similar polemic followed the publication of the discovery of phosphine in the atmosphere of Mars (Greaves et al. 2021), but later investigations has confirmed its presence (Bains 2024; Schwieterman and Leung 2024).

The examples listed above, ranging from the oldest fossil cells on Earth, through traces of life in the meteorite ALH84001, and the phosphine signal in the Venusian clouds illustrate the heightened emotions surrounding such findings. It is not just a question of scientific veracity, but there is also communal and personal element to reactions that include simple denial and professional jealousy. This will happen repeatedly when future discoveries of extraterrestrial life are made, whether it be of simple organisms on Mars, or a biosignature gas on an exoplanet indicating that some kind of photosynthetic life form exists. Humans are fallible creatures, governed not just by the brain but also by emotions. Moreover, the brain does not like a vacuum and has a tendency to over-interpret data, especially the less the data there are. This does not help the scientific endeavour. Fortunately, we learn from our mistakes: the furore surrounding the ALH84001 meteorite stimulated the field of astrobiology and a plethora of original research is helping us to understand life and its traces better.

Conclusion

The evolution of life on Earth was not serendipitous. The composition of planet Earth and its position in the Solar System disk already placed it in an advantageous position for the creation of habitable conditions and the emergence of life. The continuous link between mantle processes, the surface of the Earth and the atmosphere has been maintained throughout geological history, albeit with variations. Life appeared sometime, somehow, somewhere in the Hadean, but by at least 3.5 Ga, if not before, it was widespread and relatively evolved, having diversified into chemotrophs and phototrophs. Certainly, the initiation of horizontal plate tectonics and creation of amalgamated continents with increased runoff of nutrients promoted the flourishing of oxygenic phototrophs, once they had appeared in the Mesoarchaean. Their rapid rise in the Neoarchaean resulted in the Great Oxygenation Event (GOE) in the Palaeoproterozoic, a geologically brief period of about a couple of hundred million years during which atmospheric oxygen levels rose to present day levels and would have been briefly visible to a space observer.

However, although oxygenic phototrophs dominated the biome since the Neoarachaean and throughout most of the Proterozoic, it was only with the evolution of eukaryotes, and the rapid rise of planktonic red and green algae during the Neoproterozoic, that an atmospheric signature of life would have become permanently visible to a space observer. Both the GOE and the Neoproterozoic Oxygenation Event (NOE) were the result of plate tectonic activity: the breakup of supercontinents, global glaciation, and an increase in nutrient supply. Without these stimuli to evolution, the rapid diversification of life that occurred at the cusp of the Neoproterozoic/beginning of the Phanerozoic would not have occurred.

Thus, for much of Earth's history, life flourished but would have been invisible from space, if it were observed using the kind of instrumentation we have today. However, more advanced technologies may have been able to detect the weak traces of biogases in the terrestrial atmosphere earlier – but this point is purely hypothetical. Moreover, the evolution of life on Earth was intimately interlinked with tectonic processes – and Earth has the right size for maintaining mantle overturn and a magnetic dynamo and maintaining habitable conditions.

Considering Christian de Duve's philosophy (1996) that biology should naturally evolve from chemistry, habitable rocky planets should be inhabited, at a minimum by organisms resembling terrestrial chemotrophs. These would be the kinds of organisms whose remains could be found on Mars or the icy satellites of the Solar System. Unfortunately for the red planet, its disconnected habitats and punctuated habitability argue strongly against the appearance of phototrophs, although we cannot conclude that such a metabolic strategy never evolved.

We have arrived at the era of sample return from asteroidal and planetary bodies in the Solar System. A range of instrumentation including highly sophisticated technology will need to be adopted to investigate the returned samples. As demonstrated by various polemical discussions concerning the oldest traces of life or of extraterrestrial life, very strong evidence of diverse nature is necessary to arrive at a degree of probability – weak, moderate, or strong. Unless the extraterrestrial organisms are living and can be analysed using the terrestrial tools at hand, we may only be able to arrive at the conclusion of "very strong probability" rather than absolute certainty.

In particular, we need to be prepared for a polemical reaction from scientists (with very human reactions) and from the general public, many of whom may fear that the foundations of their philosophy of life are threatened.

References

Acuna M, Connerney J, Wasilewski P, Lin R, Anderson K, et al. 1998. Magnetic field and plasma observations at mars: Initial results of the mars global surveyor mission. Science 279:1676–1680

Adam, Z. (2007). Actinides and life's origins. Astrobiology, 7(6), 852–872.

Adam, Z. R., Hongo, Y., Cleaves, H. J., Yi, R., Fahrenbach, A. C., Yoda, I., & Aono, M. (2018). Estimating the capacity for production of formamide by radioactive minerals on the prebiotic Earth. Scientific reports, 8(1), 265.

Agić, H. (2021). Origin and Early Evolution of the Eukaryotes: Perspectives from the Fossil Record. In: Neubeck, A., McMahon, S. (eds) Prebiotic Chemistry and the Origin of Life. Advances in Astrobiology and Biogeophysics. Springer, Cham. https://doi.org/10.1007/978-3-030-81039-9_11

Allwood, A. C., Walter, M. R., Kamber, B. S., Marshall, C. P., and Burch, I. W. (2006). Stromatolite reef from the early archaean era of Australia. Nature 441, 714–718. doi:10.1038/nature04764

Bains, W., Seager, S., Clements, D. L., Greaves, J. S., Rimmer, P. B., & Petkowski, J. J. (2024). Source of phosphine on Venus—An unsolved problem. Frontiers in Astronomy and Space Sciences, 11, 1372057.

Barboni, M., et al. (2017). Science advances, 3(1), e1602365.

Bekker, A., Karhu, J. A., & Kaufman, A. J. (2006). Carbon isotope record for the onset of the Lomagundi carbon isotope excursion in the Great Lakes area, North America. Precambrian Research, 148(1-2), 145–180.

Bell, E. A., Boehnke, P., Harrison, T. M., and Mao, W. L. (2015). Potentially biogenic carbon preserved in a 4.1 billion-year-old zircon. Proc. Nat. Acad. Sci. 112, 14518–14521. doi:10.1073/pnas.1517557112

Borg, L. E., & Carlson, R. W. (2023). Annual Review of Earth and Planetary Sciences, 51(1), 25–52. https://doi.org/10.1146/annurev-earth-031621-060538

Bosak, T., Shuster, D. L., Scheller, E. L., Siljeström, S., Zawaski, M. J., Mandon, L., et al. (2024). Astrobiological potential of rocks acquired by the Perseverance rover at a sedimentary fan front in Jezero crater, Mars. AGU Advances, 5, e2024AV001241. https://doi.org/10.1029/2024AV001241

Brack, A. (2002). Water, the spring of life. In Astrobiology: The Quest for the conditions of life (pp. 79-88). Berlin, Heidelberg: Springer Berlin Heidelberg.

Brasier, M. D., Matthewman, R., McMahon, S., and Wacey, D. (2011). Pumice as a remarkable substrate for the origin of life. Astrobiology 11, 725–735. doi:10.1089/ast.2010.0546

Byerly, G. R., Lower, D. R., & Walsh, M. M. (1986). Stromatolites from the 3,300–3,500-Myr Swaziland Supergroup, Barberton Mountain Land, South Africa. Nature, 319(6053), 489–491.

Canfield, D. E. (1998). A new model for Proterozoic ocean chemistry. Nature, 396(6710), 450–453.

Cardona, T., Sánchez-Baracaldo, P., Rutherford, A. W., Larkum, A. W., (2019). Early Archean origin of photosystem II. Geobiology, 17(2), 127–150.

Carr, M & Head, J. 2015. Martian surface/near-surface water inventory: Sources, sinks, and changes with time. Geophysical Research Letters 42:726–732.

Cates, N. L., Ziegler, K., Schmitt, A. K., & Mojzsis, S. J. (2013). Reduced, reused and recycled: detrital zircons define a maximum age for the Eoarchean (ca. 3750–3780 Ma) Nuvvuagittuq Supracrustal Belt, Québec (Canada). Earth and Planetary Science Letters, 362, 283–293.

Catling, D, C., KiangNancy, Y., RobinsonTyler, D., RushbyAndrew, J., & GenioAnthony, D. (2018). Exoplanet biosignatures: a framework for their assessment. Astrobiology, 18, 709–738.

Champion, D.C. and Smithies, R.H., 2018. Geochemistry of Paleoarchean Granites of the East Pilbara Terrane, Pilbara Craton, Western Australia: Implications for Early Archean Crustal Growth. In van Kranendonk et al., (Eds), Earth's Oldest Rocks, Elsevier, pp. 487–517.

Chen, X., Zhou, Y., & Shields, G. A. (2022). Progress towards an improved Precambrian seawater 87Sr/86Sr curve. Earth-Science Reviews, 224, 103869. https://doi.org/10.1016/j.earscirev.2021.103869

Cleaves, H. J., Chalmers, J. H., Lazcano, A., Miller, S. L., and Bada, J. L. (2008). A reassessment of prebiotic organic synthesis in neutral planetary atmospheres. Orig. Life Evol. Biosph. 38(2), 105–115. doi:10.1007/s11084-007-9120-3

Clifford SM, Parker TJ. 2001. The evolution of the martian hydrosphere: Implications for the fate of a primordial ocean and the current state of the northern plains. Icarus 154:40–79

Clodoré, L., Foucher, F., Hickman-Lewis, K., Sorieul, S., Jouve, J., Réfrégiers, M., ... & Westall, F. (2024). Multi-Technique Characterization of 3.45 Ga Microfossils on Earth: A Key Approach to Detect Possible Traces of Life in Returned Samples from Mars. Astrobiology, 24(2), 190–226. https://doi.org/10.1089/ast.2023.0089

Cockell, C. (2000) The ultraviolet history of the terrestrial planets: implications for biological evolution. Planet Space Sci 48:203–214. DOI: 10.1016/S0032-0633(99)00087-2

Cockell, C. S., Balme, M., Bridges, J. C., Davila, A., & Schwenzer, S. P. (2012). Uninhabited habitats on Mars. Icarus, 217(1), 184–193.

Czaja, A.D., C.M. Johnson, B.L. Beard, E.E. Roden, W. Li, and S. Moorbath. 2013. Biological Fe oxidation controlled deposition of banded iron formation in the ca. 3770Ma Isua Supracrustal Belt (West Greenland). Earth and Planetary Science Letters 363: 192–203. https://doi.org/10.1016/j.epsl.2012.12.025.

Colón-Santos, S., Vázquez-Salazar, A., Adams, A., Campillo-Balderas, J. A., Hernández-Morales, R., Jácome, R., ... & Trubl, G. (2024). Chapter 2: What Is Life?. Astrobiology, 24(S1), S-40.

Darling, J. R., Moser, D. E., Heaman, L. M., Davis, W. J., O'Neil, J., Carlson, R. (2013). Eoarchean to Neoarchean evolution of the Nuvvuagittuq supracrustal belt: new insights from UePb zircon geochronology. American Journal of Science 313, 844e876.

Dartois, E., Engrand, C., Brunetto, R., Duprat, J., Pino, T., Quirico, E., ... & Dumas, P. (2013). UltraCarbonaceous Antarctic micrometeorites, probing the Solar System beyond the nitrogen snow-line. Icarus, 224(1), 243–252.

De Duve, C. (1995). Vital dust: life as a cosmic imperative. Basic Book, 390pp.

de Vries, S. T., Nijman, W., & de Boer, P. L. (2010). Sedimentary geology of the palaeoarchaean Buck Ridge (South Africa) and Kittys gap (Western Australia) volcano-sedimentary complexes. Precambrian Research, 183(4), 749–769. https://doi.org/10.1016/j.precamres.2010.09.005

Des Marais, D. J. (2000). When did photosynthesis emerge on Earth?. Science, 289(5485), 1703–1705.

Djorkic, T., Van Kranendonk, M. J., Campbell, K. A., Walter, M. R., and Ward, C. R. (2017). Earliest signs of life on land preserved in ca. 3.5 Ga hot spring deposits. Nat. Commun. 8, 15263. doi:10.1038/ncomms15263

Dodd, M. S., Papineau, D., Grenne, T., Slack, J. F., Rittner, M., Pirajno, F., ... & Little, C. T. (2017). Evidence for early life in Earth's oldest hydrothermal vent precipitates. Nature, 543(7643), 60–64. https://doi.org/10.1038/nature21377

Eickmann, B., Hofmann, A., Wille, M., Bui, T. H., Wing, B. A., & Schoenberg, R. (2018). Isotopic evidence for oxygenated Mesoarchaean shallow oceans. Nature Geoscience, 11(2), 133–138. DOI: 10.1038/s41561-017-0036-x

Eigenbrode, J. L., Summons, R. E., Steele, A., Freissinet, C., Millan, M., Navarro-González, R., ... & Coll, P. (2018). Organic matter preserved in 3-billion-year-old mudstones at Gale crater, Mars. Science, 360(6393), 1096–1101.

Friedmann, E. I., & Koriem, A. M. (1989). Life on Mars: how it disappeared (if it was ever there). Advances in Space Research, 9(6), 167–172.

Freissinet, C., Glavin, D. P., Mahaffy, P. R., Miller, K. E., Eigenbrode, J. L., Summons, R. E., ... & MSL Science Team. (2015). Organic molecules in the sheepbed mudstone, gale crater, mars. Journal of Geophysical Research: Planets, 120(3), 495–514.

Farquhar J, Peters M, Johnston DT, Strauss H, Masterson A, et al. 2007. Isotopic evidence for Mesoarchaean anoxia and changing atmospheric sulphur chemistry. Nature 449:706–9. DOI: 10.1038/nature06202

Foley, B. J. (2024). Exoplanet Geology: What can we learn from current and future observations?. Reviews in Mineralogy and Geochemistry, 90(1), 559–594.

Genda, H. (2016). Origin of Earth's oceans: An assessment of the total amount, history and supply of water. Geochemical Journal, 50(1), 27–42.

Golden, D. C., Ming, D. W., Morris, R. V., Brearley, A. J., Lauer, Jr. H. V., Treiman, A. H., Zolensky, M. E., Schwandt, C. S., Lofgren, G. E. and McKay, G. A. (2004). Evidence for exclusively inorganic formation of magnetite in Martian meteorite ALH84001. American Mineralogist, 89, 681–695.

Greaves, J. S., Richards, A., Bains, W., Rimmer, P. B., Sagawa, H., Clements, D. L., ... & Hoge, J. (2021). Phosphine gas in the cloud decks of Venus. Nature Astronomy, 5(7), 655–664.

Guélin, M., & Cernicharo, J. (2022). Organic molecules in interstellar space: Latest advances. Frontiers in Astronomy and Space Sciences, 9, 787567.

Gumsley, A. P., Chamberlain, K. R., Bleeker, W., Söderlund, U., De Kock, M. O., Larsson, E. R., & Bekker, A. (2017). Timing and tempo of the Great Oxidation Event. Proceedings of the National Academy of Sciences, 114(8), 1811–1816.

Hassenkam, T., Andersson, M., Dalby, K., Mackenzie, D., and Rosing, M. (2017). Elements of Eoarchean life trapped in mineral inclusions. Nature 548, 78–81. doi:10. 1038/nature23261

Heubeck, C (2009) An early ecosystem of Archean tidal microbial mats (Moodies Group, South Africa, ca. 3.2 Ga). Geology 37:931–934. https://doi.org/10.1130/G30101A.1

Hickman-Lewis, K., Cavalazzi, B., Foucher, F., Westall, F., 2018. Most Ancient Evidence for Life in the Barberton Greenstone Belt: Microbial Mats and Biofabrics of the ~3.47 Ga Middle Marker Horizon. Precambrian Research 312, 45–67. https://doi.org/10.1016/j.precamres.2018.04.007

Hickman-Lewis, K., Cavalazzi, B., Sorieul, S. Gautret, P., Foucher, F, Whitehouse, M.J., Jeon, H., Georgelin, T., Cockell, C.S., Westall, F., 2020a. Metallomics in deep time and the influence of ocean chemistry on the metabolic landscapes of Earth's earliest ecosystems. Sci Rep 10, 4965. DOI: 10.1038/s41598-020-61774-w

Hickman-Lewis, K., Gourcerol, B., Westall, F., Manzini, D., Cavalazzi, B., 2020b. Reconstructing Palaeoarchaean microbial biomes flourishing in the presence of emergent landmasses using trace and rare earth element systematics. Precambrian Research 342, 105689.

Hickman-Lewis, K., Cavalazzi, B., Giannoukos, K., d'Amico, L., Vrbaski, S., Saccomano, G., ... & Westall, F. (2023). Advanced two-and three-dimensional insights into Earth's oldest stromatolites (ca. 3.5 Ga): Prospects for the search for life on Mars. Geology, 51(1), 33–38.

Hofmann, A., and Harris, C. (2008). Silica alteration zones in the Barberton greenstone belt: A window into subseafloor processes 3.5–3.3 Ga ago. Chel Geol. 257, 221–239. doi:10.1016/j.chemgeo.2008.09.015

Hofmann, H. J., Grey, K., Hickman, A. H., and Thorpe, R. (1999). Origin of 3.45 Ga coniform stromatolites in warrawoona group, western Australia. Geol. Soc. Am. Bull. 111, 1256–1262. doi:10.1130/0016-7606(1999)111<1256:oogcsi>2.3.co;2

Holland, H. D. (2002). Volcanic gases, black smokers, and the Great Oxidation Event. Geochimica et Cosmochimica acta, 66(21), 3811–3826.

Homann M, Heubeck C, Airo A, Tice MM (2015) Morphological adaptations of 3.22 Ga-old microbial communities to Archean coastal habitats (Moodies Group, Barberton Greenstone Belt, South Africa). Precambrian Res 266:47–64. https://doi.org/10.1016/j.precamres.2015.04.018

Jakosky, B. M. (2021). Atmospheric loss to space and the history of water on Mars. Annual Review of Earth and Planetary Sciences, 49(1), 71–93.

Javaux EJ, Marshall CP, Bekker A (2010) Organic-walled microfossils in 3.2-billion-year-old shallow-marine siliciclastic deposits. Nature 463, 934–938.

Kaltenegger, L. (2017). How to characterize habitable worlds and signs of life. Annual Review of Astronomy and Astrophysics, 55(1), 433-485.

Kameda, S. et al. Ecliptic north–south symmetry of hydrogen geocorona. Geophys. Res. Lett. 44, 11,706–11,712 (2017).

Kapanarthy, D. Westall, F., Lampe, M., Zhu, B., Boesen, T., Scheu, B., Klingl, A., Schwille, P., and Lueders, T., 2024. On the nature of the earliest known lifeforms. eLife. https://doi.org/10.7554/eLife.98637.1

Karhu, J.A., Holland, H.D., 1996. Carbon isotopes and the rise of atmospheric oxygen. Geology 24 (10), 867–870

J.F. Kasting, D.P. Whitmire, R.T. Reynolds, Habitable Zones around Main Sequence Stars. Icarus 101, 108–128 (1993). doi:10.1006/icar.1993.1010

Kelley, D. S., Karson, J. A., Blackman, D. K., FruÈh-Green, G. L., Butterfield, D. A., Lilley, M. D., ... & AT3-60 Shipboard Party. (2001). An off-axis hydrothermal vent field near the Mid-Atlantic Ridge at 30 N. Nature, 412(6843), 145–149.

Kite, E. S., Steele, L. J., Mischna, M. A., & Richardson, M. I. (2021). Warm early Mars surface enabled by high-altitude water ice clouds. Proceedings of the National Academy of Sciences, 118(18), e2101959118.

Konn, C., Charlou, J.L., Donval, J.P., Holm, N.G., Dehairs, F., and Bouillon, S. (2009) Hydrocarbons and oxidised organic compounds in hydrothermal fluids from Rainbow and Lost City ultramafic-hosted vents. Chem Geol 258:299–314.

Konn, C., Charlou, J. L., Holm, N. G., & Mousis, O. (2015). The production of methane, hydrogen, and organic compounds in ultramafic-hosted hydrothermal vents of the Mid-Atlantic Ridge. Astrobiology, 15(5), 381–399.

Lammer, H., Chassefière, E., Karatekin, Ö., Morschhauser, A., Niles, P. B., Mousis, O., ... & Pham, L. B. S. (2013). Outgassing history and escape of the Martian atmosphere and water inventory. Space Science Reviews, 174, 113–154.

Lan, Z., Kamo, S.L., Roberts, N.M.W., Sano, Y., Li, X.-H., 2022. A Neoarchean (ca. 2500 Ma) age for jaspilite-carbonate BIF hosting purported micro-fossils from the Eoarchean (=3750 Ma) Nuvvuagittuq supracrustal belt (Quėbec, Canada). Precambrian Research 377, 106728. https://doi.org/10.1016/j.precamres.2022.106728.

Lowe, D. R. (1999). "Petrology and sedimentology of cherts and related silicified sedimentary rocks in the Swaziland Supergroup," in Geologic Evolution of the Barberton Greenstone Belt, South Africa: Geol Soc America Special Paper 29, 83–114. https://doi.org/10.1016/j.earscirev.2020.103296

Lyons, Timothy W.; Reinhard, Christopher T.; Planavsky, Noah J. (February 2014). "The rise of oxygen in Earth's early ocean and atmosphere". Nature. 506 (7488): 307–315. doi:10.1038/nature13068

Madhusudhan, N., Piette, A. A., & Constantinou, S. (2021). Habitability and biosignatures of Hycean worlds. The Astrophysical Journal, 918(1), 1.

Madhusudhan, N., Sarkar, S., Constantinou, S., Holmberg, M., Piette, A. A., & Moses, J. I. (2023). Carbon-bearing molecules in a possible Hycean atmosphere. The Astrophysical Journal Letters, 956(1), L13.

Margulis, L., Barghoorn, E. S., Ashendorf, D., Banerjee, S., Chase, D., Francis, S., ... & Stolz, J. (1980). The microbial community in the layered sediments at Laguna Figueroa, Baja California, Mexico: Does it have Precambrian analogues?. Precambrian Research, 11(2), 93–123.

Martin, W., Baross, J., Kelley, D., and Russell, M. J. (2008). Hydrothermal vents and the origin of life. Nat. Rev. Microbio 6, 805–814. doi:10.1038/nrmicro1991

Martins, Z., Price, M. C., Goldman, N., Sephton, M. A., & Burchell, M. J. (2013). Shock synthesis of amino acids from impacting cometary and icy planet surface analogues. Nature Geoscience, 6(12), 1045–1049.

Martins, Z., & Pasek, M. A. (2024). Delivery of organic matter to the early Earth. Elements, 20(1), 19–23.

Maurette, M., and Brack, A. (2006). Cometary petroleum in Hadean time? Meteoritics Planet. Science 41, 5247

McCollom, T.M., Seewald, J.S., and German, C.R. (2015) Investigation of extractable organic compounds in deep-sea hydrothermal vent fluids along the Mid-Atlantic Ridge. Geochim Cosmochim Acta 156:122–144.

McKay, D. S., Gibson Jr, E. K., Thomas-Keprta, K. L., Vali, H., Romanek, C. S., Clemett, S. J., ... & Zare, R. N. (1996). Search for past life on Mars: possible relic biogenic activity in Martian meteorite ALH84001. Science, 273(5277), 924–930.

McMahon, S. (2019). Earth's earliest and deepest purported fossils may be iron-mineralized chemical gardens. Proceedings of the Royal Society B, 286(1916), 20192410.

Ménez, B., Pisapia, C., Andreani, M., Jamme, F., Vanbellingen, Q. P., Brunelle, A., ... & Réfrégiers, M. (2018). Abiotic synthesis of amino acids in the recesses of the oceanic lithosphere. Nature, 564(7734), 59–63.

Miao, L., Yin, Z., Knoll, A. H., Qu, Y., & Zhu, M. (2024). 1.63-billion-year-old multicellular eukaryotes from the Chuanlinggou Formation in North China. Science Advances, 10(4), eadk3208.

Michalski, J. R., Onstott, T. C., Mojzsis, S. J., Mustard, J., Chan, Q. H., Niles, P. B., & Johnson, S. S. (2018). The Martian subsurface as a potential window into the origin of life. Nature Geoscience, 11(1), 21–26.

Monty, C. (1974). Precambrian background and Phanerozoic history of stromatolitic communities, an overview. Annales de la Société géologique de Belgique. 90: 55–99.

Neveu, M., Hays, L. E., Voytek, M. A., New, M. H., & Schulte, M. D. (2018). The ladder of life detection. Astrobiology, 18(11), 1375-1402. https://doi.org/10.1089/ast.2017.1773

Nijman, W., Kloppenburg, A., and de Vries, S. T. (2017). Archaean basin margin geology and crustal evolution: An east Pilbara traverse. J. Geol. Soc. 174, 1090–1112. doi:10.1144/jgs2016-127

Noffke, N. (2015). Ancient sedimentary structures in the< 3.7 Ga Gillespie Lake Member, Mars, that resemble macroscopic morphology, spatial associations, and temporal succession in terrestrial microbialites. Astrobiology, 15(2), 169–192.

Noffke, N., Beukes, N., Hazen, R.M., and Swift, D. (2008) An actualistic perspective into Archean worlds—(cyano-)bacterially induced sedimentary structures in the siliciclastic Nhlazatse Section, 2.9 Ga Pongola Supergroup, South Africa. Geobiology 6:5–20.

Noffke, N., Christian, D., Wacey, D., & Hazen, R. M. (2013). Microbially induced sedimentary structures recording an ancient ecosystem in the ca. 3.48 billion-year-old Dresser Formation, Pilbara, Western Australia. Astrobiology, 13(12), 1103-1124. DOI: 10.1089/ast.2013.1030

Osinski, G. R., Tornabene, L. L., Banerjee, N. R., Cockell, C. S., Flemming, R., Izawa, M. R., ... & Southam, G. (2013). Impact-generated hydrothermal systems on Earth and Mars. Icarus, 224(2), 347–363.

Osinski, G. R., Cockell, C. S., Pontefract, A., and SapersHM (2020). The role of meteorite impacts in the origin of life. Astrobiology 20, 1121–1149. doi:10.1089/ast.2019.2203

Papineau, D., She, Z., Dodd,M. S., Iacoviello, F., Slack, J. F., Hauri, E., et al. (2022). Metabolically diverse primordial microbial communities in Earth's oldest seafloor hydrothermal jasper. Scie Adv. 8, eabm2296. doi:10.1126/sciadv.abm2296

Pappalardo, R. T., Buratti, B. J., Korth, H., Senske, D. A., Blaney, D. L., Blankenship, D. D., ... & Niebur, C. (2024). Science Overview of the Europa Clipper Mission. Space Science Reviews, 220(4), 1–58.

Pascal, R., Pross, A., & Sutherland, J. D. (2013). Towards an evolutionary theory of the origin of life based on kinetics and thermodynamics. Open biology, 3(11), 130156.

Quanz, S. P., Absil, O., Benz, W., Bonfils, X., Berger, J. P., Defrère, D., ... & Wyatt, M. (2021). Atmospheric characterization of terrestrial exoplanets in the mid-infrared: biosignatures, habitability, and diversity. Experimental astronomy, 1–25.

Rosing, M. T. (1999). 13C-depleted carbon microparticles in> 3700-Ma sea-floor sedimentary rocks from West Greenland. Science, 283(5402), 674–676.

Rosing, M. T., & Frei, R. (2004). U-rich Archaean sea-floor sediments from Greenland–indications of> 3700 Ma oxygenic photosynthesis. Earth and Planetary Science Letters, 217(3-4), 237–244. https://doi.org/10.1016/S0012-821X(03)00609-5

Russell, M. J., and Hall, A. J. (1997). The emergence of life from iron monosulphide bubbles at a submarine hydrothermalredox and pH front. J. Geol. Soc. Lond 154, 377–402. doi:10.1144/gsjgs.154.3.0377

Sánchez-Baracaldo, P. & Cardona, T. (2020). On the origin of oxygenic photosynthesis and Cyanobacteria. New Phytologist, 225(4), 1440–1446.

Sánchez-Baracaldo, P., Bianchini, G., Wilson, J. D., & Knoll, A. H. (2022). Cyanobacteria and biogeochemical cycles through Earth history. Trends in Microbiology, 30(2), 143–157. DOI: 10.1016/j.tim.2021.05.008

Sagan, C., & Mullen, G. (1972). Earth and Mars: Evolution of atmospheres and surface temperatures. Science, 177(4043), 52–56. DOI: 10.1126/science.177.4043.52

Sagan, C., Thompson, W. R., Carlson, R., Gurnett, D. & Hord, C. A search for life on Earth from the Galileo spacecraft. Nature 365, 715–721 (1993).

Sasselov, D. D., Grotzinger, J. P., and Sutherland, J. D. (2020). The origin of life as a planetary phenomenon. Sci. Adv. 6, eaax3419. doi:10.1126/sciadv.aax3419

Schidlowski, M. (1988). A 3,800-million-year isotopic record of life from carbon in sedimentary rocks. Nature, 333(6171), 313–318.

Schwieterman, E. W., Kiang, N. Y., Parenteau, M. N., Harman, C. E., DasSarma, S., Fisher, T. M., ... & Lyons, T. W. (2018). Exoplanet biosignatures: a review of remotely detectable signs of life. Astrobiology, 18(6), 663–708.

Schwieterman, E. W., & Leung, M. (2024). An Overview of Exoplanet Biosignatures. Reviews in Mineralogy and Geochemistry, 90(1), 465–514.

Schmitt-Kopplin, P., Gabelica, Z., Gougeon, R. D., Fekete, A., Kanawati, B., Harir, M., et al. (2010). High molecular diversity of extraterrestrial organic matter in Murchison meteorite revealed 40 years after its fall. PNAS 107, 2763–2768. doi:10.1073/pnas.0912157107

Schreiber, U., Locker-Grütjen, O., & Mayer, C. (2012). Hypothesis: Origin of life in deep-reaching tectonic faults. Origins of Life and Evolution of Biospheres, 42, 47–54.

S. Seager, Exoplanet Habitability. Science 340, 577–581 (2013).

Seager, S., Petkowski, J. J., Carr, C. E., Grinspoon, D. H., Ehlmann, B. L., Saikia, S. J., ... & Venus Life Finder Mission Team. (2022). Venus life finder missions motivation and summary. Aerospace, 9(7), 385.

Schmitt-Kopplin, P., Gabelica, Z., Gougeon, R. D., et al. (2010). High molecular diversity of extraterrestrial organic matter in Murchison meteorite revealed 40 years after its fall. Proceedings of the National Academy of Sciences, 107(7), 2763–2768.

Sephton, M. A., & Botta, O. (2008). Extraterrestrial organic matter and the detection of life. Space Science Reviews, 135, 25–35.

Steele, A., McCubbin, F. M., Fries, M., Kater, L., Boctor, N. Z., Fogel, M. L., ... & Jull, A. J. T. (2012). A reduced organic carbon component in martian basalts. Science, 337(6091), 212–215.

Sugitani, K., Mimura, K., Takeuchi, M., Yamaguchi, T., Suzuki, K., Senda, R., ... & Van Kranendonk, M. J. (2015). A Paleoarchean coastal hydrothermal field inhabited by diverse microbial communities: the Strelley Pool Formation, Pilbara Craton, Western Australia. Geobiology, 13(6), 522–545.

Summons, R. E., Albrecht, P., McDonald, G., & Moldowan, J. M. (2008). Molecular biosignatures. Strategies of Life Detection, 133–159.

Summons, R. E., Amend, J. P., Bish, D., Buick, R., Cody, G. D., Des Marais, D. J., ... & Sumner, D. Y. (2011). Preservation of martian organic and environmental records: final report of the Mars Biosignature Working Group. Astrobiology, 11(2), 157–181. DOI: 10.1089/ast.2010.0506

Szopa, C., Freissinet, C., Glavin, D. P., et al. (2020) First detections of dichlorobenzene isomers and trichloromethylpropane from organic matter indigenous to Mars mudstone in Gale Crater, Mars: results from the Sample Analysis at Mars instrument onboard the Curiosity rover. Astrobiology, 20(2), 292–306.

Tehrany, M. G., Lammer, H., Selsis, F., Ribas, I., Guinan, E. F., & Hanslmeier, A. (2002, December). The particle and radiation environment of the early Sun. In *Solar Variability: From Core to Outer Frontiers* (Vol. 506, pp. 209–212). http://adsabs.harvard.edu/full/2002ESASP.506..209T

Tice, M. M., & Lowe, D. R. (2004). Photosynthetic microbial mats in the 3,416-Myr-old ocean. Nature, 431(7008), 549–552. DOI: 10.1038/nature02888

Triaud, A. H., de Wit, J., Klein, F., Turbet, M., Rackham, B. V., Niraula, P., ... & Selsis, F. (2024). Atmospheric carbon depletion as a tracer of water oceans and biomass on temperate terrestrial exoplanets. Nature Astronomy, 8(1), 17–29.

Van Kranendonk, M. J., Baumgartner, R., Djokic, T., Ota, T., Steller, L., Garbe, U., et al. (2021). Elements for the origin of life on land: A deep-time perspective from the Pilbara Craton of Western Australia. Astrobiology 21, 39–59. doi:10.1089/ast.2019.2107

Vitas, M., & Dobovišek, A. (2019). Towards a general definition of life. Origins of Life and Evolution of Biospheres, 49, 77–88.

Wächtershäuser, G. (1998). Origin of life in an iron-sulfur world. The molecular origins of life, 206–218.

Wacey, D. (2009). Early life on earth: a practical guide. Dordrecht: Springer Netherlands.

Walsh, M. W. (1992). Microfossils and possible microfossils from the early archean onverwacht group, Barberton mountain land, South Africa. Precambrian research, 54(2-4), 271–293. DOI: 10.1016/0301-9268(92)90074-x

Wang, L. and Huang, J. (2024) Hypothesis of an ancient northern ocean on Mars and insights from the Zhurong rover. Nature Astronomy, 8(10), 1220–1229.

Westall, F. 2024. The early Earth: creating the conditions for the emergence of life. Oxford Encyclopedia of Science, in review.

Westall, F. and Folk, R.L., 2003. Exogenous carbonaceous microstructures in Early Archaean cherts and BIFs from the Isua greenstone belt: Implications for the search for life in ancient rocks. Precambrian Research 126, 313–330. DOI : 10.1016/S0301-9268(03)00102-5

Westall, F. ., de Vries, S.T., Nijman, W., Rouchon, V., Orberger, B. Pearson, V., Watson, J., Verchovsky, A., Wright, I., Rouzaud, J.-N., Marchesini, D.,& Anne, S. 2006a. The 3.466 Ga Kitty's Gap Chert, an Early Archaean microbial ecosystem. In Processes on the Early Earth (W.U. Reimold & R. Gibson, Eds.), Geol. Soc. Amer. Spec Pub., 405, 105–131. https://doi.org/10.1130/2006.2405(07)

Westall F, de Ronde C. E. J., Southam G., Grassineau N., Colas M., Cockell C., Lammer H. (2006b) Implications of a 3.472-3.333 Ga-old subaerial microbial mat from the Barberton greenstone belt, South Africa for the UV environmental conditions on the early Earth. Phil. Trans. Roy. Soc. Lond. Series B., 361, 1857–1875. DOI: 10.1098/rstb.2006.1896

Westall, F., Cavalazzi, B., 2011. Biosignatures in rocks. Encyclopedia of Geobiology (Eds.) V. Thiel, J. Reitner, Springer, Berlin, 189–201. https://doi.org/10.1007/978-1-4020-9212-1_36

Westall, F., Foucher, F., Cavalazzi, B., de Vries, S. T., Nijman, W., Pearson, V., ... & Anne, S. (2011). Volcaniclastic habitats for early life on Earth and Mars: a case study from ~3.5 Ga-old rocks from the Pilbara, Australia. Planetary and Space Science, 59(10), 1093–1106. https://doi.org/10.1016/j.pss.2010.09.006

Westall, F., Campbell, K. A., Bréhéret, J. G., Foucher, F., Gautret, P., Hubert, A., et al. (2015a). Archean (3.33 Ga) microbe-sediment systems were diverse and flourished in a hydrothermal context. Geology 43, 615–618. doi:10.1130/g36646.1

Westall, F., Foucher, F., Bost, N., Bertrand, M., Loizeau, D., Vago, J. L., et al. (2015b). Biosignatures on Mars: What, where and how? Implications for the search for martian life. Astrobiology 15, 998–1029. doi:10.1089/ast.2015.1374

Westall, F., Hickman-Lewis, K., 2018. Fossilisation of bacteria and implications for the search for early life on Earth and astrobiology missions to Mars. In Handbook of Astrobiology (Ed. Kolb, V.). CSC Press, pp. 609–631. https://doi.org/10.1201/b22230

Westall, F., Hickman-Lewis, K., Hinman, N., Gautret, P., Campbell, K.A., Bréhéret, J.-G., Foucher, F., Hubert, A., Sorieul, S., Kee, T.P., Dass, A.V., Georgelin, T., Brack, A., 2018. A hydrothermal-sedimentary context for the origin for life. Astrobiology 18, 259–293. DOI: 10.1089/ast.2017.1680

Westall, F., Hickman-Lewis, K., Cavalazzi, B., Foucher, F., Clodoré, L., & Vago, J. L. (2021). On biosignatures for Mars. International Journal of Astrobiology, 20(6), 377–393.

Westall, F., Brack, A., Fairén, A. G., & Schulte, M. D. (2023a). Setting the geological scene for the origin of life and continuing open questions about its emergence. Frontiers in astronomy and space sciences, 9, 1095701.

Westall, F., Höning, D., Avice, G., Gentry, D., Gerya, T., et al. (2023b). The habitability of Venus. Space Science Reviews, 219(2), 17.

Westall, F. and Xiao, S., 2025. Precambrian evolution of life on Earth and biogeodynamics. Precambrian Research, 414, 107589.

Westall, F. (2025). What the earliest evidence for life tells us about the early evolution of the biosphere. Phil. Trans. Royal Society B, 380:20240106. https://doi.org/10.1098/rstb.2024.0106

Westall, F., Purvis, G., Sano N., Sherrif, J., Clodoré, L., Foucher, F., Milojevic, T, 2025. The frequency of (elusive) life in the Universe. Nature Astronomy. https://doi-org.inc.bib.cnrs.fr/10.1038/s41550-025-02661-0.

Wogan, N. F., Batalha, N. E., Zahnle, K. J., Krissansen-Totton, J., Tsai, S. M., & Hu, R. (2024). JWST observations of K2-18b can be explained by a gas-rich mini-Neptune with no habitable surface. The Astrophysical Journal Letters, 963(1), L7.

Wordsworth, R. D. (2016). The climate of early Mars. Annual Review of Earth and Planetary Sciences, 44(1), 381–408.

Xiong, J., Fischer, W. M., Inoue, K., Nakahara, M., & Bauer, C. E. (2000). Molecular evidence for the early evolution of photosynthesis. science, 289(5485), 1724–1730.

Xu, D., Qin, Z., Wang, X., Li, J., Shi, X., Tang, D., & Liu, J. (2023). Extensive sea-floor oxygenation during the early Mesoproterozoic. Geochimica et Cosmochimica Acta, 354, 186–196.

Zerkle AL. Biogeochemical signatures through time as inferred from whole microbial genomes. Am J Sci 2005;305(6–8): 467–502; doi: 10.2475/ajs.305.6-8.467

(Toward) Discovery of Life beyond Earth and its Impact
Proceedings IAU Symposium No. 387, 2026
H. Landt, M. Dominik & C. Oliver, eds.
doi:10.1017/S1743921325000365

Using Linguistics and Information Theory to Assess the Diversity, Complexity, and Decoding of Interstellar Message

Saeed Jafari[1,2], Shirin Shaterzadeh Yazdi[3], Yadgar Karimi[4] and Parsa Bigdeli[5]

[1]Department of Linguistics, University of Kurdistan, Iran

[2]UNESCO's International Decade of Indigenous Languages 2022 - 2032;
email: Saeed.Jafari@spacegeneration.org

[3]Department of Physics, Shahid Beheshti University;
email: Shirin.Shaterzadeh@worldspaceweek.ir

[4]Department of English and Linguistics, Faculty of Language and Literature, University of Kurdistan;
email: Y.Karimi@uok.ac.ir

[5]Center for Complex Networks and Social Data Science, Shahid Beheshti University;
email: ParsaBigdeli77@gmail.com

Abstract. This abstract explores the application of quantitative information theory measures and linguistic features to analyze animal communication systems and extends this methodology to contemplate the possibilities of interstellar communication as a part of CETI practice. We will assess some early findings by using information theory on social species with sophisticated acoustic communication abilities, such as bottlenose dolphins and humpback whales (Hanser, Sean F., et al.) as well as birds, as examples of how the complex interplay between notions, data, and misinterpretations can become established as reliable knowledge and cognition about and understanding of an ETI civilization and culture. The study also looks into the potential implications of extraterrestrial contact. Given the diversity of language instances in Earth's evolutionary history, we discuss the choice of syntactic complexity of an "intelligent message" for potential alien civilizations. The central point of this paper is to examine the advantages and limitations of acoustic and visual communication, considering linguistical and mathematical constraints that may apply universally.

Keywords. Information theory and SETI; decoding interspecies communication; syntactic complexity; intelligent message; animal communication system complexity

---

1. Unique communication characteristics in animal communication systems

The claim that language is unique to humans is controversial, as numerous frameworks have been devised to differentiate between human language and animal communication. Two primary views exist regarding the relationship between language and animal communication. One group of scholars sees language as exclusively human, emphasizing a uniquely human trait Chomsky (1969), while another suggests a gradual evolutionary continuum Bickerton (1990). There is increasing evidence that some animal communication systems possess language-like features. Putty-nosed monkey pyow-hack sequences have been analyzed linguistically to demonstrate that their semantics are linked to non-threat-related group movement, even though their acoustic components are also separately used in alarm calls Schlenker et al. (2016). Honeybees demonstrate a limited form of displacement in their "dance language."

Animals produce a remarkable range of sounds, each with distinct acoustic characteristics. Using phylogenetic comparative analyses allows us to investigate the origins and implications of these variations across different taxa. A broad array of mechanisms, including vocalization, stridulation, and percussion, allows animals to produce sequences of sounds. To encode information, six paradigms have been proposed: Repetition, Diversity, Combination, Ordering, Overlapping, and Timing Kershenbaum et al. (2016). These paradigms facilitate a better understanding of information sequencing in species communication. Codas in are acoustic unit clicks where echolocation click timing moderates response by Sperm whale provide an example of this.

1.1. *Learning from Songbirds*

Birds are considered to have some of the most complex vocal systems among different species because, unlike nonhuman primates, they are vocal learners Marler (1967). For example, European starlings (Sturnus vulgaris) have the ability to recognize and classify acoustic patterns using recursive, context-free grammar, previously thought to be a uniquely human trait. They distinguish grammatical from ungrammatical sequences, indicating that recursive grammar understanding is not exclusive to humans; Engesser et al. (2016) studies on Japanese tits and southern pied babblers claim these songbirds combine vocalizations to create complex meanings, similar to human language. Furthermore, swamp sparrow songs vary by geographic dialect, not species, suggesting birds possess complex communication skills akin to human linguistic abilities Balaban (1988). These findings are incredibly fascinating, revealing how comparative linguistic and behavioural research in avian species can illuminate what aspects of language are uniquely human while also providing insights into the evolution of speech and communication in extraterrestrial species.

1.2. *How Dolphines Communicate*

The large brains and complex acoustic repertoire of dolphins have sparked considerable speculation and led to an ongoing scientific debate about the communicative function of dolphin whistles, particularly regarding whether a "dolphin language" exists akin to human language Caldwell and Caldwell (1968). While dolphins' vocalizations-squeaks and whistles-might not possess the linguistic complexity found in human speech, they can convey abstract information. This has led researchers to draw parallels between interspecies and interstellar communication, as both scenarios involve vastly different organisms attempting to exchange information. In this light, real-time contact with intelligent beings could model the interpretation of signals from alien species. Dolphin whistles are categorized as graded signals at the population level rather than individual communication, especially among individuals in close proximity. This graded signaling might serve a significant communicative role in such contexts. A captive playback study showed that dolphins can learn and easily use vocal labels to address each other King et al. (2013) while field studies demonstrated their use among familiar individuals in the wild Quick and Janik (2012). These experimental techniques raise questions about distinguishing stellar anomalies from natural or artificial signals, and how dolphins' spontaneous displays, mimicry, and imitation might facilitate the intention of initiating interaction between interstellar species.

1.3. *Humpback Whale as a Reflection of Aliens on Earth*

Since the discovery of whale songs nearly sixty years ago, scientists have been engaged in the intricate task of deciphering these fascinating vocalizations. Among animal species, the communicative abilities of humpback whales present significant challenges for interdisciplinary researchers studying animal communication McCowan et al. (2023). This complex image of the humpback whale emerges from an understanding that these intelligent and social mammals exhibit intricate behaviors and demonstrate sophisticated, elegant, and expansive migratory patterns. Researchers have established that humpback whales engage in extensive auditory communication, characterized by their unique 'songs.' It is hypothesized that the vocalizations produced by experienced whales migrating northward may serve to guide other members of their population Janik (2014). Such insights portray an animal capable of traversing diverse oceanographic systems, equipped with physiological and behavioral adaptations that allow them to respond effectively to changing environmental conditions. How likely is it that the social behaviors which cetaceans exhibit with their own species, or the empathy they show towards humans, would be similar for beings on exoplanets, assuming that they have developed a fusion of acoustic systems like whales in the form of interplanetary technology, the precondition for societies like those on Earth?

2. Application of Information Theory

Information theory was first introduced by Claude Shannon Shannon (1948) to treat the concept of information quantitatively. Since then, information theory has been applied as a mathematical framework for understanding communication process, including the encoding, transmission and decoding the information. In the past decades, Shannon's information theory and its elements has been a key measure for assessing a variety of communication systems, including the communication systems of animals Hanser et al. (2000) Hanser et al. (2004).

In this framework, zero order entropy (H_0) is a measure of repertoire diversity of a system, also known as the highest degree of freedom of the system (for example, for English alphabet $H_0 - 4.75$). Next is the first order entropy (H_1) which represents the distribution of each signal in the system (using the English alphabet example, letter A is significantly more frequent compared to letter Q). The next step is second order Entropy (H_2) that implies conditional probabilities of two sequential signals (if letter A occurs, what is the probability of the occurrence of letter B?). One can carry on and define third and fourth order entropies to measure the higher order conditional probabilities of a system Hanser et al. (2004)

2.1. *Intelligence filters*

Information theory along with all its benefits, still can not tell us about what one would say, but what one could say Doyle et al. (2011). Receiving a signal and analyzing it with information theory is more about the structure and complexity of that signal rather than its meaning and concept. Therefore, one of the most remarkable challenges in searching for extraterrestrial intelligence (ETI) is distinguishing between natural-random and artificial signals. In the context of information theory, some quantities could act as the intelligence filters for finding level of intelligence. This process is done through putting constraints on the degree of the optimization of the signals and/or the randomness of it, wether this signal is the song of a bird, signals of a pulsar or a technosignature.

2.1.1. Zipf's law

Zipf's law is an empirical law and a special case of Shannon's first-order entropy. Known also as Zipf's principle of least effort, this law plots the base-10 logarithm of the frequency of occurrence of each signal against the rank of that signal. The more frequently a signal appears, the higher its rank. When applying this mathematical relationship to most human languages, one can observe that these languages exhibit a -1 slope. Thus, a communication system with a -1 Zipf slope is often interpreted as being optimized. This law has been applied to several animal communication systems, and among them, Humpback Whales and bottlenose dolphins have a Zipf slope close to -1. Additionally, changes in the Zipf slope of bottlenose dolphins from infancy to adulthood resemble those observed in humans Doyle et al. (2011).

One serious criticism of Zipf's law is that it can arise from natural phenomena Hanser et al. (2004). Furthermore, it is not sufficient for estimating the complexity of a signaling system Kershenbaum et al. (2014). For these reasons, even if Zipf's law is retained as a potential filter for assessing intelligence in analyzing technosignatures, it cannot be the sole filter applied.

2.1.2. N-th order entropy and entropic slope

Another proposed intelligence filter that focuses on signal complexity is the entropy of signals in different orders and the change in higher-order entropies, known as the entropic slope. Entropy measures the uncertainty or unpredictability of intelligent signals, regardless of their source. Lower entropy indicates more predictable patterns in a signal, which may suggest some underlying structure.

In a structured and extended set of signals, entropy is expected to decrease as higher orders are calculated. If the entropic slope shows a gradual, shallow decrease at higher orders, it suggests low complexity. If the slope is around zero, it indicates a random signal. Finally, if the slope reaches a point where additional conditional entropies do not significantly change the entropy, it suggests that the system's complexity has approached a structural limit Doyle et al. (2011). This interpretation of entropic slope is particularly useful in distinguishing random stellar signals from potential technosignatures. As Doyle et al. (2011) suggest, considering elements of information theory and entropic slope could be crucial when crafting messages intended for extraterrestrial civilizations.

3. Technosignature and Network Science

Traditional approaches to detecting technosignatures have primarily relied on statistical methods like Zipf's law analysis Elliott et al. (2019) and information entropy measurements Doyle et al. (2011). While these methods have provided valuable insights, they often fail to capture the complex, dynamic nature of intelligent communication systems, such as cultural evolution Garland et al. (2013) Cantor et al. (2015), rapid song adoption or synchronization Garland et al. (2011). Recent studies in animal communication have demonstrated that complexity science, particularly network science, offers a more comprehensive framework for analyzing sophisticated communication patterns. For example, Allen et al. (2019) demonstrated that network analysis can reveal underlying syntactic rules in humpback whale songs that would be impossible to detect through traditional statistical methods, showing that these complex vocal displays follow specific structural constraints despite their apparent variability. Network science excels in revealing emergent properties, collective behaviors and complex structures that simple statistical measures might miss. Studies on bottlenose dolphins have shown how their vocal signatures support the stability of their complex multi-layer alliance network, which are invisible to traditional statistical approaches King et al. (2018). This suggests that technosignature detection might benefit from similar complexity-based analyses, as intelligent communication systems likely exhibit emergent properties that transcend simple power-law distributions.

Network construction in communication analysis typically follows the temporal networks approach that capture the sequential structure of signals. In animal vocalization studies, nodes often represent distinct vocal units, while edges represent temporal succession. Allen et al. (2019) Weiss et al. (2014) demonstrated that constructing networks where nodes represent song units and edges represent transitions between units can reveal previously unknown syntactic rules governing animal communication. Their analysis found small-world network properties in humpback whale and bird songs, suggesting that complex vocal learning follows universal organizational principles. Network analysis methods have revealed crucial features of complex communication systems: motif analysis can identify recurring patterns Allen et al. (2019), degree distribution analysis can examine the patterns that vocal unit frequencies follow such as Zipf's law, and the observation of small-world architecture Allen et al. (2019) Weiss et al. (2014) indicates a clustered structure with short paths between vocal units. Network entropy measurements have proven particularly valuable in quantifying the complexity of these communication systems, offering a more nuanced view than traditional entropy measures Anand and Bianconi (2009).

Complex communication systems exhibit specific network metrics and transition motifs that provide much stronger evidence of organized complexity than simple frequency distributions. Network-based approaches offer more robust alternatives by examining the structural and dynamic properties of communication systems. These features are less likely to arise from random processes and better reflect the organized complexity characteristic of intelligent communication. To implement these insights in technosignature detection, we propose a machine learning framework that combines network science with information theory. Deep learning architectures, particularly graph neural networks (GNNs), could be trained to recognize complex network features characteristic of intelligent communication Khemani et al. (2024) Farzaneh (2024). This approach could be enhanced by incorporating information-theoretic measures as additional features, creating a multi-modal detection system. Transfer learning from animal communication networks could provide valuable training data, as these systems share fundamental properties with potential technological signals. The framework could utilize network approaches (motifs, clusters, etc) alongside traditional information theory measures (entropy, mutual information) to create a more sophisticated detection filter. This combined approach would be more robust against false positives while remaining sensitive to the complex patterns that might characterize extraterrestrial technosignatures.

4. What Linguistics Can Tell Us about Talking to ETIs

4.1. *Language evolution*

The precise origins of human language remain a subject of considerable debate. Paleoanthropological evidence suggests that our capacity for language emerged within the last 100,000 years Tattersall (2009), the exact evolutionary trajectory remains elusive. Theories diverge on whether the capacity for hierarchical syntax arose abruptly during hominin evolution Berwick and Chomsky (2019) or gradually from early rudiments Martins and Boeckx (2020). While semantic and phonological systems exhibit clear evolutionary continuity with other species, Hauser et al. (2002) and Berwick and Chomsky (2015) posit that the syntactic computational system and the lexicon are uniquely human. Therefore, if we are to understand language in extraterrestrial species as something that evolved from pre-existing capacities of human and nonhuman species, this kind of species-specificity can provide us with a new angle for distinguishing received signals. Another way to think about the coevolution of language and civilization is to look at the effect of different informational structures into the evolution and emergence of communication patterns.

4.2. *Would ETIs Communicate Like Us?*

A key step in understanding the evolution of human language involves uncovering the origins of syntactic structure, the mechanism that allows us to combine meaningful units (e.g., words) into higher-order structures (e.g., compound words and phrases). Despite the central role of syntax in language, we must consider what elements and processes were involved in its evolutionary development. A growing body of evidence suggests that animals can combine meaningful vocalizations into larger structures with emergent or transparent meanings, but few species have been identified that combine meaningful calls into larger signals. However, Bolhuis et al. (2018) challenge these findings. For example, white-handed gibbons can manipulate the sequence of vocalizations to convey different meanings, such as alerting their group to the presence of predators. Honeybees employ a complex dance to communicate information about food sources, encoding both direction and distance in their movements Riley et al. (2005). Another factor that largely sets human language apart from other systems is Merge, which has led to the hypothesis that the generative power of language stems from the cognitive capacity for this mental operation. The most basic form of merge involves combining two words to form a new unit (e.g., eat + apple = eat apple, an + idea = an idea). Some argue that this operation is uniquely human and that cross-species comparisons are problematic without the presence of similar syntactic complexities in animals. Here, Zipf's law can be employed to reveal the rules and grammar of a language by plotting linguistic data. This implies that if a signal, whether from an animal species or an extraterrestrial source, adheres to Zipf's law, the hypothesis that it constitutes a language becomes tenable. Just as complex species have evolved and adapted to diverse environmental conditions on Earth, there could just as well be extraterrestrials from other planets who have followed evolutionary paths either similar to ours or entirely different.

5. Conclusion

How structured is the ETIs communication system? A complex society cannot exist without a complex communication system. We can better understand this by considering what evidence of the stages necessary for life on other planets would be observable on Earth to answer the question. While the nature of consciousness and cognition on exoplanets remains speculative, both the structure of ETI communication and its relationship to the intelligence of the communicating species are of paramount importance. Understanding the parallels between human language and animal vocalizations offers a promising avenue for developing intelligent filters capable of detecting and decoding extraterrestrial signals. Some of the mentioned results provide new evidence of an important commonality between human language and animal vocal communication, with respect to the basic structural patterns underpinning how signals are combined into larger structures. Moreover, we must evaluate the efficacy of tools, such as information theory and Zipf's law, and the extent of information they yield regarding the complicated nature of an extraterrestrial civilization's communication system. This would imply that the songs of birds, humpback whales, and other species with complex vocal patterns might best serve as animal models for studying syntactic-like structures and vocal diversity for conveying meaningful information. By comparing acoustic communication patterns through linguistic and mathematical models, astronomers could enable distinguishing astrophysical phenomena from intelligent signals by observing the variation in the structural dependencies of communication systems

References

Allen, J. A., Garland, E. C., Dunlop, R. A., & Noad, M. J. 2019, Network analysis reveals underlying syntactic features in a vocally learnt mammalian display, humpback whale song. *Proceedings of the Royal Society B*, 286(1916), 20192014.

Anand, K. & Bianconi, G. 2009, Entropy measures for networks: Toward an information theory of complex topologies. *Phys. Rev. E*, 80, 045102.

Balaban, E. 1988, Bird song syntax: learned intraspecific variation is meaningful. *Proceedings of the National Academy of Sciences*, 85(10), 3657–3660.

Berwick, R. C. & Chomsky, N., 2015, Why Only Us: Language and Evolution, MIT Press.

Berwick, R. C. & Chomsky, N. 2019, All or nothing: No half-merge and the evolution of syntax. *PLOS Biology*, 17(11), 1–5.

Bickerton, D. 1990, *Language and Species*. University of Chicago Press.

Bolhuis, J., Beckers, G., Huybregts, M., Berwick, R., & Everaert, M. 2018, Meaningful syntactic structure in songbird vocalizations? *PLOS Biology*, 16, e2005157.

Caldwell, M. C. & Caldwell, D. K. 1968, Vocalization of naive captive dolphins in small groups. *Science*, 159(3819), 1121–1123.

Cantor, M., Shoemaker, L. G., Cabral, R. B., Flores, C. O., Varga, M., & Whitehead, H. 2015, Multilevel animal societies can emerge from cultural transmission. *Nature Communications*, 6(1), 1–10.

Chomsky, N. 1969, *Aspects of the Theory of Syntax*. The MIT Press. MIT Press.

Doyle, L. R., McCowan, B., Johnston, S., & Hanser, S. F. 2011, Information theory, animal communication, and the search for extraterrestrial intelligence. *Acta Astronautica*, 68(3-4), 406–417.

Elliott, J., Atwell, E., & Whyte, B. 2019, Linguistic and mathematical properties of zipf's law for animal communication and human language. *Journal of Quantitative Linguistics*, 26(1), 1–21.

Engesser, S., Ridley, A., & Townsend, S. 2016, Meaningful call combinations and compositional processing in the southern pied babbler. *Proceedings of the National Academy of Sciences of the United States of America*, 113.

Farzaneh, A. 2024,. An information-theoretic analysis of temporal gnns.

Garland, E. C., Goldizen, A. W., Rekdahl, M. L., Constantine, R., Garrigue, C., Hauser, N. D., Poole, M. M., Robbins, J., & Noad, M. J. 2011, Dynamic horizontal cultural transmission of humpback whale song at the ocean basin scale. *Current Biology*, 21(8), 687–691.

Garland, E. C., Noad, M. J., Goldizen, A. W., Lilley, M. S., Rekdahl, M. L., Garrigue, C., Constantine, R., & Hauser, N. D. 2013, Cultural transmission of humpback whale song: Learning, memory, and revolution. *Advances in the Study of Behavior*, 45, 207–253.

Hanser, S., Doyle, L., McCowan, B., & Jenkins, J. 2004, Information theory applied to animal communication systems and its possible application to seti. *Proceedings of the International Astronomical Union*, 213, 514.

Hanser, S., McCowan, B., & Doyle, L. 2000, Information theory as a comparative measure of animal communication, 213, 613.

Hauser, M., Chomsky, N., & Fitch, W. 2002, The faculty of language: What is it, who has it, and how did it evolve? *Science (New York, N.Y.)*, 298, 1569–79.

Janik, V. 2014, Cetacean vocal learning and communication. *Current Opinion in Neurobiology*, 28, 60–65.

Kershenbaum, A., Blumstein, D., Roch, M., Akçay, Ç., Backus, G., Bee, M., Bohn, K., Cao, Y., Carter, G., Cäsar, C., Coen, M., Deruiter, S., Doyle, L., Edelman, S., i Cancho, R. F., Freeberg, T., Garland, E., Gustison, M., Harley, H., Huetz, C., Hughes, M., Hyland Bruno, J., Ilany, A., Jin, D., Johnson, M., Ju, C., Karnowski, J., Lohr, B., Manser, M., Mccowan, B., Mercado, E., Narins, P., Piel, A., Rice, M., Salmi, R., Sasahara, K., Sayigh, L., Shiu,

Y., Taylor, C., Vallejo, E., Waller, S., & Zamora-Gutierrez, V. 2016, Acoustic sequences in non-human animals: A tutorial review and prospectus. *Biological Reviews*, 91(1), 13–52. Publisher Copyright: © 2016 Cambridge Philosophical Society.

Kershenbaum, A., Bowles, A., Freeberg, T., Jin, D., Lameira, A., & Bohn, K. 2014, Animal vocal sequences: Not the markov chains we thought they were. *Proceedings. Biological sciences / The Royal Society*, 281.

Khemani, B., Patil, S., Kotecha, K., & Tanwar, S. 2024, A review of graph neural networks: concepts, architectures, techniques, challenges, datasets, applications, and future directions. *Journal of Big Data*, 11(1).

King, S., Sayigh, L., Wells, R., Fellner, W., & Janik, V. 2013, Vocal copying of individually distinctive signature whistles in bottlenose dolphins. *Proceedings. Biological sciences / The Royal Society*, 280, 20130053.

King, S. L., Friedman, W. R., Allen, S. J., Gerber, L., Jensen, F. H., Madsen, P. T., Manser, M. B., McGregor, P. K., Prosser, C., & Wilson, M. 2018, Bottlenose dolphins retain individual vocal labels in multi-level alliances. *Current Biology*, 28(12), 1993–1999.

Marler, P. 1967, Animal communication signals. *Science*, 157(3790), 769–774.

Martins, P. & Boeckx, C. 2020, Vocal learning: Beyond the continuum. *PLOS Biology*, 18, e3000672.

McCowan, B., Hubbard, J., Walker, L., Sharpe, F., Frediani, J., & Doyle, L. 2023, Interactive bioacoustic playback as a tool for detecting and exploring nonhuman intelligence: "conversing" with an alaskan humpback whale. *bioRxiv*.

Quick, N. & Janik, V. 2012, Bottlenose dolphins exchange signature whistles when meeting at sea. *Proceedings. Biological sciences / The Royal Society*, 279, 2539–45.

Riley, J., Greggers, U., Smith, A., Reynolds, D., & Menzel, R. 2005, The flight paths of honeybees recruited by the waggle dance. *Nature*, 435, 205–7.

Schlenker, P., Chemla, E., & Zuberbühler, K. 2016, What do monkey calls mean? *Trends in Cognitive Sciences*, 20.

Shannon, C. E. 1948, A mathematical theory of communication. *The Bell System Technical Journal*, 27, 379–423.

Tattersall, I. 2009, Out of africa: modern human origins special feature: human origins: out of africa. *Proceedings of the National Academy of Sciences of the United States of America*, 106, 16018–21.

Weiss, M., Hultsch, H., Adam, I., Scharff, C., & Kipper, S. 2014, Network analysis reveals ecological and evolutionary patterns in birdsong complexity. *Proceedings of the Royal Society B*, 281(1785), 20140287.

(Toward) Discovery of Life beyond Earth and its Impact
Proceedings IAU Symposium No. 387, 2026
H. Landt, M. Dominik & C. Oliver, eds.
doi:10.1017/S1743921324001509

Venus Phosphine: Updates and lessons learned

David L. Clements

Imperial College London
email: d.clements@imperial.ac.uk

Abstract. The discovery of phosphine in Venus's atmosphere provides lessons for the search for life. The detection has survived all challenges and has acquired independent support from archival data from PVP. The presence of phosphine in Venus' oxidising environment is perplexing, and comprehensive studies rule out all known abiotic sources. More data is needed to understand the origin of phosphine, leading to JCMT-Venus, a long term atmospheric monitoring programme. This can find how phosphine varies in relation to other species providing clues to its origin. We present the latest JCMT-Venus results. The discovery and subsequent papers were explicit that they did not constitute evidence for life, only of phosphine. Media and public reaction to the discovery and its implications provide lessons for future life searches, as does the reaction of the scientific community. How this was handled by the team, media, and general public will be reviewed.

Keywords. Astrobiology, Venus, Mm/sub astronomy

---

1. Introduction

While the surface of Venus is well known to be thoroughly hostile to the presence of life, with temperatures in excess of 450 C and atmospheric pressure nearly a hundred times that of sea level on Earth, there has long been speculation about the possibility of life in the clouds that make up a 'temperate zone' of the atmosphere at an altitude of $\sim$ 55 km. At these heights the atmospheric pressure is around 1 bar, and temperatures are around 60 C (Morowitz & Sagan 1967). These ideas persist to the present day, fuelled by a number of well established anomalies in various aspects of Venus' atmosphere that remain unexplained (Bains et al. 2021). These include an unidentified UV absorber, the presence of oxygen at $\sim$10 parts per million (ppm) levels in the clouds, the nature of large, 7μm sized, particles of unknown composition and possibly aspheric shape, and the vertical abundance variations of SO_2 and H_2O (see (Clements 2022) and references therein). Recent modelling of Venus' history, suggesting that it might (Way et al. 2016) (or might not (Turbet et al. 2021)) have had a warm wet past until as recently as 700 Myr ago, have added to this speculation, with the possibility that pre-existing surface life might have evolved to survive in the cloud decks as the surface became increasingly hostile.

In 2021, a new anomaly was added to the list of issues with our understanding of the atmosphere of Venus with the discovery of phosphine, PH_3, above the cloud decks (Greaves et al. 2021). This sparked considerable interest for a number of reasons. Firstly, phosphine, as a reduced chemical, should not have been present at all in the oxidising atmosphere of Venus. This is confirmed by detailed investigation of possible chemical pathways for phosphine's formation and destruction on Venus (Bains et al. 2021b; Bains et al. 2024). Secondly, phosphine has been proposed as an exoplanet biomarker gas

(Sous-Silva et al. 2020), suggesting that its presence on Venus might suggest the presence of life. Needless to say, this discovery led to considerable discussion and further work, some of which is detailed in Clements (2022).

We here present several updates on our observations of phosphine in the atmosphere of Venus, and especially the results of the JCMT-Venus long term monitoring programme. We also, in the spirit of this meeting preparing us for the discovery of life elsewhere, present a discussion on how the initial phosphine results were received by the scientific community and the general public, and what this might mean for the reception of future discoveries in astrobiology.

2. Current Status of the Phosphine Detection

The first announcement of the discovery of phosphine in Venus' atmosphere (Greaves et al. 2021) was a surprise and, understandably, was challenged on a number of fronts. These include the suggestions that the weak absorption line that indicated the presence of phosphine was a data processing artefact and not in fact real (Villanueva et al. 2021; Thompson 2021; Sous-Silva et al. 2020). These suggestions were subsequently shown to be incorrect (Greaves et al. 2021) since the chances of an artefactual line appearing at the exact frequency of a known absorption line in these datasets is $\sim 1\%$, and the chances of a line appearing at the same frequency in *both* datasets is far smaller. It was also suggested that the detected line, while real, was not phosphine but was in fact a misidentified SO_2 line (Lincowski et al. 2021). This cannot be the case for two reasons: firstly, the central line frequency of the SO_2 is inconsistent with the measured line centre by $\sim 3\sigma$, and, secondly, because contemporaneous or near-contemporaneous observations of a different SO_2 line limit the maximum contamination of phosphine to be $< 10\%$. Meanwhile, reanalysis of the Pioneer Venus Probe mass spectrometer *in situ* data from 1978 (Mogul et al. 2021) suggest the presence of phosphine at part-per-million (ppm) levels within the clouds, while the ALMA and JCMT data indicate 10-20 part-per-billion (ppb) levels above the clouds.

Other observations in search of phosphine have also been conducted, producing a mixed set of upper limits and detections (Clements 2022; Greaves et al. 2023), but a picture may be emerging with detections of phosphine appearing as the atmosphere as observed moves from night into day, while upper limits or weak detections are seen as it moves from day into night, suggestive of phosphine being destroyed by solar photolysis (Bains et al. 2021b) during daylight.

3. The Next Step: JCMT-Venus

One thing that is clear from the discussion of the Greaves et al. (2021) discovery of phosphine is that more and better data are needed to both confirm the detection and to begin investigations of how phosphine abundances might vary in relation to other properties, such as time of day or abundances of other chemical species such as H_2O or SO_2. The JCMT-Venus project was thus conceived (Clements et al. 2024). This is a long term JCMT programme to monitor the atmosphere of Venus at mm wavelengths using the new 'Ū'ū receiver on the Namakanui instrument (Mizuno et al. 2020). The previous receiver used, RxA3, was retired from the telescope in 2020, and the new instrument has a number of significant improvements relevant to this project. These include improved sensitivity and, more importantly, a significantly increased bandwidth, which allows phosphine, HDO, as surrogate for H_2O, and SO_2 lines at mm wavelengths to all be observed simultaneously. JCMT-Venus was awarded a total of 200 hours of telescope time to monitor Venus during three campaigns in Feb 2022, July 2023 and September 2023. So far about 100 hours of this time has been used.

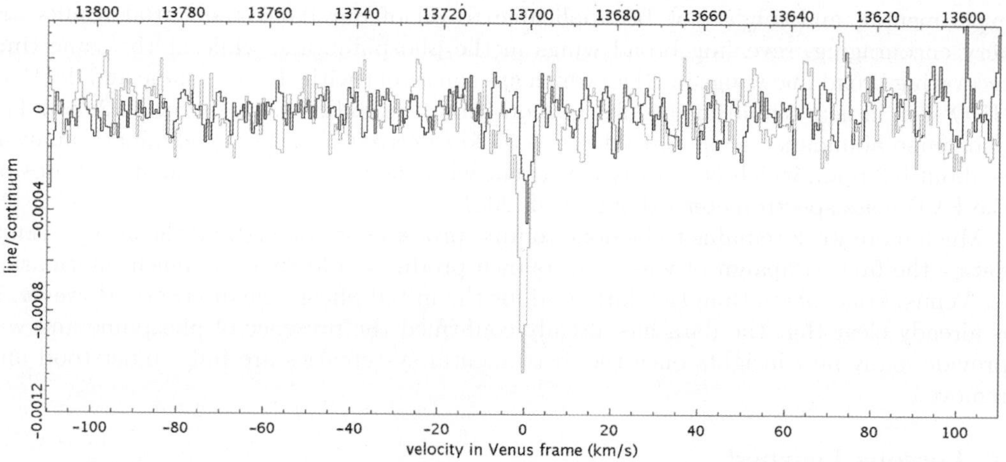

Figure 1. The result of applying the fourier method to remove ripples to the first few days of the first JCMT-Venus campaign. As can be seen both the HDO line (shown in blue) and the phosphine line (shown in black) are recovered.

The results of these observations are currently being analysed by the JCMT-Venus team. The data are not simple to deal with since the extreme brightness of Venus as a millimetre source, compared to the usual targets of the JCMT, leads to numerous systematic effects becoming significant. These are principally due to reflections of Venus from the telescope, telescope dome and support structure, and within the receiver cabin, entering the beam of the instrument. The digital autocorrelation analysis used by the instrument then leads these time delayed signals to appear in the resulting spectrum as various sinusoidal waves which contaminate the data in the form of large scale, time-varying baseline drifts and sinusoidal ripples with amplitudes exceeding those of the expected absorption lines. The large scale drifts can be dealt with by the subtraction of a running baseline average, though this comes at the cost of suppressing the broad wings of any absorption lines, which would originate from the clouds decks themselves thanks to pressure broadening. The remaining baseline ripples would be dealt with in narrow band instruments by fitting and removing a polynomial baseline from the data, with the positions of any expected absorption lines masked and interpolated over (see eg. Greaves et al. (2021)). For the broader spectral range in this data, which include multiple peaks and troughs from these sinusoidal ripples, such an approach would require a very high order polynomial, and so a different, Fourier based, approach may be needed (Greaves 2024), whereby the data is Fourier transformed, all but the most significant frequency components, representing the baseline ripples, are removed. This results in a 'ripple spectrum' which is transformed back and removed from the original data, suppressing the reflection-induced systematics. The result of applying this technique to the first few days of JCMT-Venus data for both the HDO and phosphine lines are shown in Figure 1. As can be seen both lines are recovered with good signal to noise.

While this approach allows us to retrieve the narrow components at the line cores, as can be seen in Figure 1, the broad line wings, coming from deeper in the atmosphere, remain suppressed. These broad wings are important as they allow us to assess the abundances of species deeper into the atmosphere. To investigate these broad line wings a new approach has been devised which relies on the repeatability of the large baseline ripples across our spectra to perform what is effectively *post-hoc* frequency switching. This removes the repeatable element of the large scale baseline ripples while retaining the broad wings of any absorption lines that might be present. While this method remains experimental, and might not be applicable to all of our data, the initial results are very encouraging, revealing broad wings in the phosphine line while at the same time recovering broad line wings in HDO which are consistent with the line shape and depth of HDO seen in near-simultaneous observations from the IRTF (Encrenaz et al. 2023). The phosphine abundance at at altitude of $\sim$ 55 km derived from this preliminary analysis is about 0.3 ppm, which is broadly consistent with the levels seen at similar altitudes by the PVP mass spectrometer (Mogul et al. 2021).

Much more work remains to be done to fully process and understand the JCMT-Venus data - the first campaign of which on its own produced 140 times as much information on Venus' atmosphere than the data used for the initial phosphine discovery. However, it is already clear that the data has already confirmed the presence of phosphine and will provide many new insights once the instrumental systematics are fully understood and removed.

4. Lessons Learned

While the origin of phosphine on Venus remains unclear, a biological origin cannot be discounted since no satisfactory standard chemical origin has yet been found (Bains et al. 2021). Our original discovery of phosphine in 2020 thus had a significant impact as a potential sign that life might exist elsewhere. As such, its reception, both scientifically and more broadly, can serve as a case study for what might happen when there is a clear detection of life elsewhere, potentially through the examination of the chemical composition of an exoplanet's atmosphere. We here discuss the scientific and public reaction to this result, and the lessons that might be learned.

4.1. *Scientific Reception*

As is appropriate for any potentially significant result, the scientific community rigorously tested our work (see discussion in Section 2). This was facilitated by making our data publicly available to any who wished to examine it. This is the first lesson that should apply to future astrobiology results: **the data used to derive any significant results should be made available to the astronomical community so that the results can be independently verified.**

Acceptance of our result is still far from universal, and this is likely to, at least initially, be the case for any detection of signs of life elsewhere. Additional observations, in our case from JCMT-Venus, will be critical to improve the robustness and understanding of any claimed life detections. There is a risk that the reluctance of the community to accept a radically new result may be an impediment to getting the additional telescope time to confirm and better understand such a result. We have certainly had this difficulty at some telescopes (though clearly not the JCMT). The second lesson learned is thus: **there need to be clear routes to obtaining confirmatory observations for claimed astrobiological detections even in the presence of skepticism by the community**. Given the highly oversubscribed nature of world ranking telescopes, and the resulting extreme conservatism of time assignment committees, this may mean that directors discretionary time is needed. Good communication with relevant telescope directors in advance of any detection will help this process.

In the case of Venusian phosphine we have the added advantage that ground-truth observations from flyby, orbiting, or descent probe missions that can provide fundamentally better and less ambiguous data than is possible for observations from Earth. In the long term, sample return missions will allow detailed biochemical analysis of any life found in the Solar System, answering fundamental questions about the origin and evolution of life. Such detailed biochemical analysis will never be possible for observations of exoplanets. This leads to the possibility that alternative, non-biological explanations for any claimed detection can never be fully excluded, and our third lesson learned: that **some people will never be convinced.**

4.2. *Public Reception*

We made detailed plans for public outreach in advance of the publication of the Greaves et al. (2021) paper, which was accompanied by a press conference and various media releases, some of which were recorded months before. We also, with the help of the EAO, arranged for the media impact to be monitored so that we could quantitatively assess the reception. The whole team was briefed that at all times we should make it clear that biological activity was only one possible explanation for the presence of phosphine in the atmosphere of Venus. This was, in contrast to the ALH84001 meteorite result, to avoid over-hyping the result and hopefully prevent any extreme reactions to the result. This was largely successful, though at least one journalist thought we were suggesting that penguins (whose guano is a known source of phosphine on Earth) were aliens originating from Venus (they were rapidly disabused of this notion when contacting team members).

There was considerable media interest, with a total aggregate readership of over 15 billion, making it one of the most impactful scientific announcements that year. The clear lesson to learn from this is to: **prepare for substantial media interest, but to emphasise that uncertainties in the conclusions remain, and to not engage with those who have extreme or unreasonable reactions.** It might also be useful to reflect on the political involvement with the ALH84001 result and how that led to the over-hyping of a result that had not yet been fully refereed in the standard publication process. We thus also conclude that: **publication by press release is a bad idea.**

The overall impact of the detection of phosphine on Venus is analysed from a rather different perspective, including social and psychological as well as scientific factors, elsewhere in this volume (Vidal & Bibas 2025).

5. Conclusions

Our discovery of phosphine in the atmosphere of Venus has so far survived detailed scientific scrutiny, and new observations from the JCMT-Venus project have confirmed its presence and have the potential to help us understand its origin and significance. New techniques are necessary to deal with systematic problems with the new data, and application of these leads to not only confirmation of phosphine, but detection of its pressure broadened line wings indicating amounts of phosphine in the clouds consistent with the ppm levels seen in PVP *in situ* mass spectrometer data. The reception of our original phosphine results provides lessons for the release of future exobiological results, including the need to make the data accessible, the need to be able to acquire followup observations, and the need to prepare an appropriately calm and consistent response to press and public interest.

Acknowledgements. Many thanks to Jane Greaves and Wei Tang for numerous contributions and discussions, and to the whole JCMT-Venus team without whom this work would not have been possible. The JCMT is operated by the East Asian Observatory on behalf of The National Astronomical Observatory of Japan; Academia Sinica Institute of Astronomy and Astrophysics; the Korea Astronomy and Space Science Institute; the National Astronomical Research Institute of Thailand; Center for Astronomical Mega-Science (as well as the National Key RD Program of China with No. 2017YFA0402700). Additional funding support is provided by the Science and Technology Facilities Council of the United Kingdom and participating universities and organizations in the United Kingdom and Canada.

References

Bains W., Petkowski J. J., Rimmer P. B., Seager S., 2021, PNAS, 118, e2110889118. doi:10.1073/pnas.2110889118.

Bains W., Petkowski J. J., Seager S., Ranjan S., Sousa-Silva C., Rimmer P. B., Zhan Z., et al., 2021, AsBio, 21, 1277. doi:10.1089/ast.2020.2352.

Bains W., Seager, S., Clements, DL., Greaves, JS., Rimmer, PB., Petkowski, JJ., 2024, Front. Astron. Space Sci., 11, doi:10.3389/fspas.2024.1372057.

Clements D. L., 2022, ConPh, 63, 180. doi:10.1080/00107514.2023.2184932.

Clements D. L., 2024, in prep.

Encrenaz T., Greathouse T. K., Giles R., Widemann T., Bézard B., Lefèvre M., Shao W., 2023, A&A, 674, A199. doi:10.1051/0004-6361/202245831.

Greaves J. S., Richards A. M. S., Bains W., Rimmer P. B., Sagawa H., Clements D. L., Seager S., et al., 2021, NatAs, 5, 655. doi:10.1038/s41550-020-1174-4.

Greaves J. S., Richards A. M. S., Bains W., Rimmer P. B., Clements D. L., Seager S., Petkowski J. J., et al., 2021, NatAs, 5, 636. doi:10.1038/s41550-021-01424-x.

Greaves J. S., Petkowski J. J., Richards A. M. S., Sousa-Silva C., Seager S., Clements D. L., 2023, GeoRL, 50, e2023GL103539. doi:10.1029/2023GL10353910.1002/essoar.10512932.1.

Greaves J.S., in prep.

Lincowski A. P., Meadows V. S., Crisp D., Akins A. B., Schwieterman E. W., Arney G. N., Wong M. L., et al., 2021, ApJL, 908, L44. doi:10.3847/2041-8213/abde47.

Mizuno I., Friberg P., Berthold R., Parsons H., Han C.-C., Acohido A., Bell G., et al., 2020, SPIE, 11453, 114533T. doi:10.1117/12.2561742.

Mogul R., Limaye S. S., Way M. J., Cordova J. A., 2021, GeoRL, 48, e91327 doi:10.1029/2020GL091327.

Morowitz, H., & Sagan, C., 1967, *Nature*, 215, 1259.

Snellen I. A. G., Guzman-Ramirez L., Hogerheijde M. R., Hygate A. P. S., van der Tak F. F. S., 2020, A&A, 644, L2. doi:10.1051/0004-6361/202039717.

Sousa-Silva C., Seager S., Ranjan S., Petkowski J. J., Zhan Z., Hu R., Bains W., 2020, AsBio, 20, 235. doi:10.1089/ast.2018.1954.

Turbet M., Bolmont E., Chaverot G., Ehrenreich D., Leconte J., Marcq E., 2021, Natur, 598, 276. doi:10.1038/s41586-021-03873-w.

Thompson M. A., 2021, MNRAS, 501, L18. doi:10.1093/mnrasl/slaa187.

Vidal, C. & Bibas, D., 2025, Proceedings IAU Symposium No. 387, H. Landt, M. Dominik, C. Oliver (eds.) doi:10.1017/S1743921324001996.

Villanueva G. L., Cordiner M., Irwin P. G. J., de Pater I., Butler B., Gurwell M., Milam S. N., et al., 2021, NatAs, 5, 631. doi:10.1038/s41550-021-01422-z.

Way M. J., Del Genio A. D., Kiang N. Y., Sohl L. E., Grinspoon D. H., Aleinov I., Kelley M., et al., 2016, GeoRL, 43, 8376. doi:10.1002/2016GL069790.

(Toward) Discovery of Life beyond Earth and its Impact
Proceedings IAU Symposium No. 387, 2026
H. Landt, M. Dominik & C. Oliver, eds.
doi:10.1017/S1743921324002072

Extending the Habitable Zone due to subglacial water

Amri Wandel

The Hebrew Univ. of Jerusalem
email: amri@huji.ac.il

Abstract. A long-standing issue in astrobiology is whether planets orbiting the most abundant type of stars, M-dwarfs, can support liquid water and eventually life. Often previous studies have raised doubts for the habitability of planets orbting M-stars, due to the intense stellar activity during the early evolution. Those include solar-wind-like eruptions of the host star that could erode the planetary atmosphere, as well as the intense XUV radiation from the host. A new study shows that subglacial liquid water that accumulates on the night-side of tidally locked planets may provide an answer, significantly extending the habitability region, in particular around M-dwarf stars, which are also the most promising for biosignature detection with the present and near-future technology.

Keywords. Habitable zone, exoplanets, M dwarf stars, Greenhouse effect, planetary climates, habitable planets, astrobiology, exoplanet atmospheres, tidal radius

1. Beyond the Habitable Zone

The Conservative Habitable Zone (CHZ) is the circumstellar region where a planet can maintain liquid water on the surface temperature given the stellar irradiation, or instellation, and the planetary atmospheric conditions, Albedo etc. The inner boundary of the CHZ is determined by the point at which water would evaporate due to excessive heat even before the boiling point is reached, due to processes such as the moist greenhouse effect (Kasting et al. 1993), while the outer boundary is determined by water freezing. These boundaries are consistent with the 1D climate models (Kopparapu et al. 2013; Yang et al. 2014). Alternatively, somewhat wider empirical HZ boundaries may be drawn by assuming that early Mars and recent Venus had liquid water (Figure 1).

These boundaries also depend (although weakly) on the stellar type (Shields et al. 2016), due to the different surface temperatures (and hence stellar spectra). In this work, following Wandel (2023a,b) we show that the CHZ may be significantly extended by considering subglacial liquid water (Fig. 2). While the outward extension is obvious, the extension inwards is enabled for tidally locked planets, on which ice can accumulate on the night side (Menou 2013).

Such tidally locked planets exist mostly around M-dwarf stars (the red ellipse in Fig. 2). Their night side is at the same time sheltered from the intense XUV radiation from the eruptions in the early evolutionary stages of those stars (see section 2 below), and the planetary atmosphere can be protected against erosion by stellar winds and giant mass ejections at the early phase by a magnetosphere, if the planet is not too small (Wandel and Gale 2020). On the other hand, if the surviving atmosphere is too dense, it can induce an efficient heat transport from the day side to the night side, causing an evaporation of the night side ice. This may reduce the inner boundary of habitability as shown by the percentile on the blue dashed lines in Fig. 3.

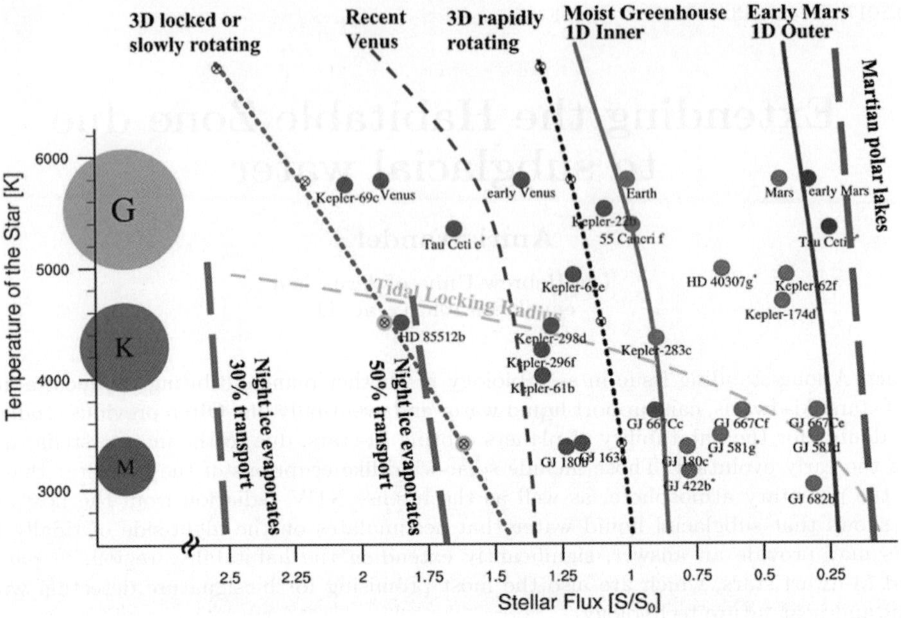

Figure 1. Boundaries of the Habitable Zone as a function of radiative flux received from the host star, relative to Earth vs. surface temperature of the host star (Wandel 2023a, adapted from Yang et al. 2014). Solid lines denote the CHZ 1D-boundaries (inner boundary green and the outer boundary violet). Dashed black and red lines represent 3D-simulations for inner HZ boundaries. Wide dashed lines represent the extended boundaries considering subglacial liquid water: blue at the inner edge (night side ice on tidally locked planets) and brown on the outer edge (Martian polar lake analog). The dashed grey curve marks the tidal locking radius. Colored circles denote exoplanets and Venus and Mars in the present and past.

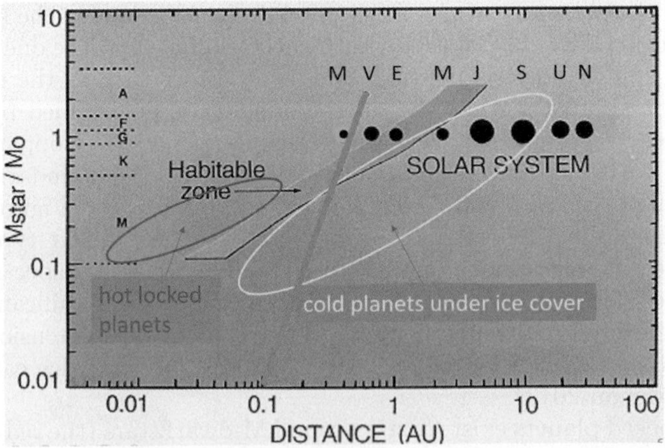

Figure 2. Extended Boundaries of the Conservative Habitable Zone (CHZ, in orange) as a function of distance from the host star and the mass of the Main Sequence host star (vertical axis). The purple diagonal line is the tidal locking radius. The light-blue regime represents the extension outwards of the CHZ, while the red one - extension inwards of the CHZ due to night-side ice on tidally locked planets.

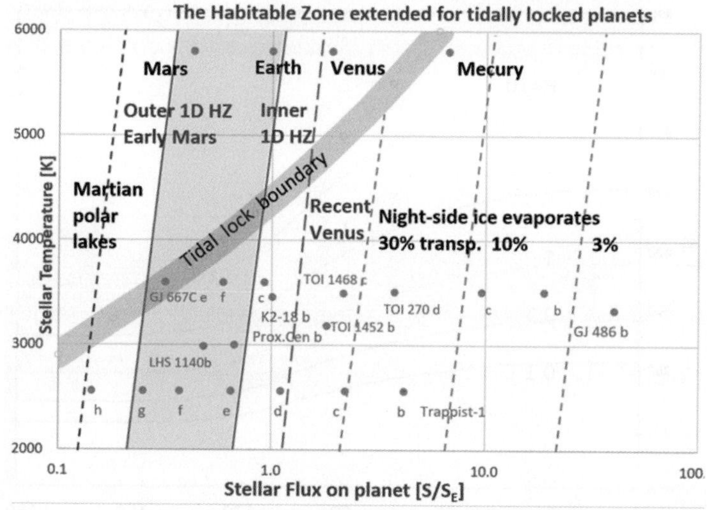

Figure 3. Boundaries of the Habitable Zone as a function of radiative flux received from the host star, relative to Earth (Wandel 2023b). Dashed blue lines represent the extended boundaries considering subglacial liquid water beyond the inner edge of the CHZ (see text). The dashed violet line is the outer HZ extended boundary (Martian polar lake analog). Solid green lines denote the conservative HZ boundaries. The orange curve marks the tidal locking radius. Blue dots denote the solar system and a few exoplanets orbiting M-dwarfs.

A better presentation of the enhanced habitability beyond the inner edge of the CHZ is shown in Fig. 3, where the instellation (horizontal axis) is shown logarithmically. The blue dashed lines correspond to subglacial liquid water on the night side ice on locked planets. The percentage marks the amount of heat transported from the day side, which when high enough would evaporate potential ice. This percentage depends on the instellation, and is calculated using the 1D inner boundary of the CHZ (Wandel 2023a). Similarly to subglacial water, also surface liquid water may exist on the night side of locked planets of M-dwarf exoplanets with high instellation (Wandel 2025), which were considered to be located way too close to their parent star for supporting liquid water.

2. Habitability of locked planets

While liquid water is essential, life as we know it requires in addition other factors—such as nutrient source, contact with the silicate sea floor, and stable energy sources. In this work, however, we only concentrate on liquid water. The energy required to keep the water liquid can come from the host star, or, in the case of planets further away from their host star than the outer boundary of the HZ, from geothermal heat. Another factor is protection from too much energetic radiation from the host star. It is often stated that planets orbiting M stars may be effected by the host's energetic outbursts in the early evolutionary stages, producing intense UV and X-ray radiation and stellar eruptions that may disintegrate water molecules (e.g. Luger and Barnes 2015). While super-Earth and larger planets are likely to have a thick H/He atmosphere that may survive the early active evolutionary phase of their M-dwarf host, the survival of water and atmosphere on smaller, Earth-sized planets orbiting M-dwarf stars was considered less certain. However, other works argue that atmosphere and water may survive (Wandel and Gale 2020 and refs. therein). An additional path to habitability of such planets is proposed by Wandel (2023a): sub-glacial liquid water could survive and replenish the water in the atmosphere

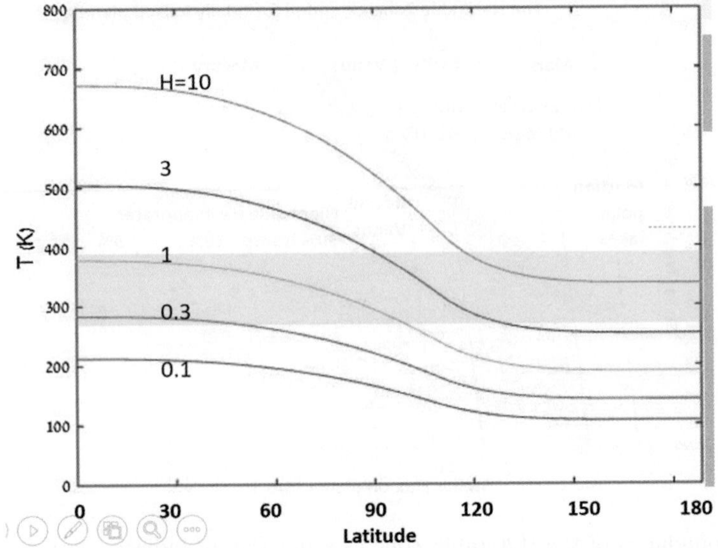

Figure 4. Surface temperature distribution on a locked planet vs. latitude (measured from the sub-stellar point). Curves are marked by their value of the atmospheric + stellar heating parameter H_{atm}. Global heat transport $f = 0.2$ is assumed. The light blue stripe marks the liquid warer regime (for a 1 bar atmosphere). Wandel 2018.

once the M-star host has calmed down. At that stage, surface liquid water could exist on some latitudes. This can be shown using the 1-D analytical model of Wandel (2018).

Planets in the CHZ of M-dwarf stars are likely to be tidally locked (Fig. 2) and may have temperatures adequate for liquid water, at least on part of their surface, under a broad range of atmospheric conditions (Fig. 4–6 and Wandel 2025). The atmospheric impact on the surface temperature distribution of a locked planet can be described by mainly two parameters: the atmospheric heating factor H_{atm} and the heat transport parameter f Wandel (2018). H_{atm} is the effective surface heating of the planet by the radiation from the host star, given by

$$H_{atm} = (1\text{-}A)\Gamma S/S_E.$$

Here A is the planet's average Albedo, Γ is the greenhouse factor and S/S_E is the radiative flux from the host star (instellation) relative to the flux on Earth. The heat transport parameter f is the fraction of the instellation transported by advection (mainly atmospheric) and distributed over the planet. For example, setting $f = 0.1$ we assume that at any point on the planet surface, 10% of the stellar energy radiated onto the planet is advected and finally distributed equally all over the planet, while $f = 1$ mean that all the radiated energy is equally distributed over the whole planet, leading to an isothermal surface. Fig. 4 shows the temperature distribution as a function of "latitude" (defined as the angle from the sub stellar point of a locked planet) for various values of the heating factor, assuming a heat transport parameter $f = 0.2$. Note that for a wide range of atmospheric heating there can be liquid water somewhere on the the planet.

Fig. 5 shows the range of temperatures on the surface on locked planets, as a function of distance from the host and for several values of the transport parameter f. The host star is assumed to be an M-type star of surface temperature 3,400 K. For example, the green vertical rectangle show the temperature range on the surface of a planet orbiting at 0.2 pc. Assuming the transport parameter is 0.5 (50% of the heat due to the stellar radiation is advected and equally distributed over the entire planet) The temperature at

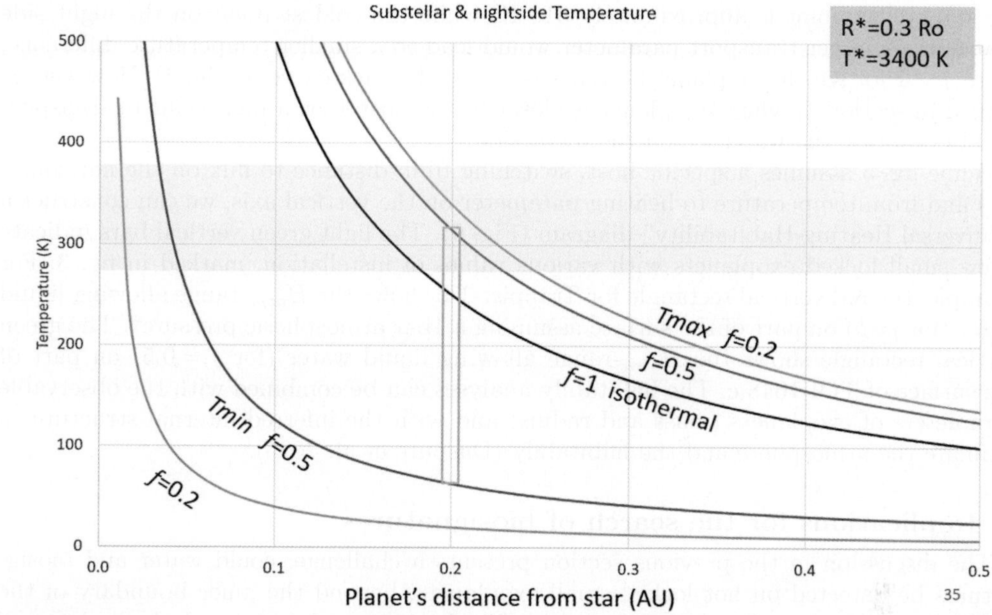

Figure 5. Maximum (sub-stellar point) and minimum (night-side off-stellar point) temperatures for locked planets orbiting a T=3400K M-star. Curves are marked by their value of the global heat transport f. The light blue stripe marks the liquid water regime (for a 1 bar atmosphere).

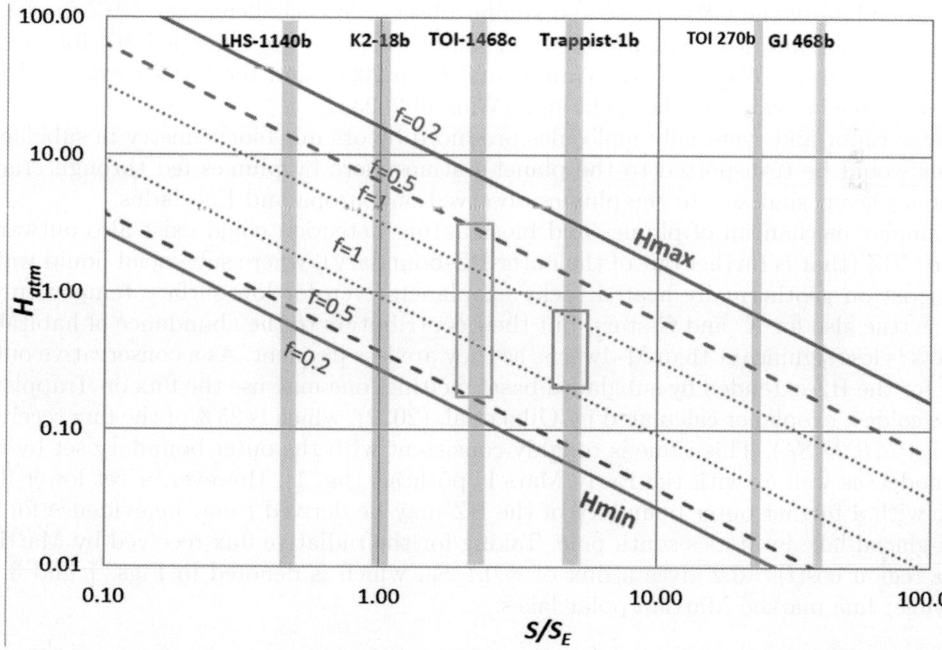

Figure 6. Universal Heating-Habitability diagram. Liquid water range of the stellar + atmospheric heating factor H_{atm} vs. stellar irradiation of the planet (instellation) relative to Earth. Curves are marked by their value of the global heat transport f. The lines indicate maximal (red) and minimal (blue) values of H_{atm} allowing surface liquid water for a given instellation. (Adapted from Wandel and Tal-Or 2019).

the sub stellar point is approximately 310 K, while the coldest point on the night side is at 60K. A larger transport parameter would lead to a smaller temperature difference, up to $f = 1$ for which the planet becomes isothermal (black curve in Fig. 4). However, as argued in section 1, when the planet is closer to the host star, a high value of transport could evaporate the water ice on the night side.

While fig. 5 assumes a specific host, switching from distance to flux on the horizontal axis and from temperature to heating parameter on the vertical axis, we can construct a "Universal Heating-Habitability" diagram (Fig. 6). The light green vertical bars indicate a few small locked exoplanets with various values of instellation, marked in fig. 3. For example, the red vertical rectangle for Trappist-1 b shows the H_{atm}-range allowing liquid water (for *f=1*) on part of the surface assuming a 1bar atmospheric pressure†. The green vertical rectangle shows the H_{atm}-range allowing liquid water (for $f = 0.5$) on part of the surface of TOI-1648 c. The habitabily-analyses can be combined with the observable parameters of exoplanets (mass and radius) and with the inferred internal structure to estimate the atmosphere and the habitabily (Daspute et al. 2025).

3. Implications for the search of biosignatures

The discussion in the previous section presents a challenge: could water and biosignatures be detected on hot locked small exoplanets, beyond the inner boundary of the CHZ? Actually, signs for water vapor on such planets may have already been detected by JWST, e.g. GJ 486 b (Moran et al. 2023) and TOI 270 b, perhaps also TOI 270 d (Holmberg and Madhusudhan 2024). These detections could be challenging, as the stellar flux on these planets is much higher than on Earth and their calculated average surface temperature is hundreds of degrees C. Being much closer to their host than the inner boundary of the CHZ, these and similar observations challenge the CHZ model. If confirmed, the empirical evidence of water vapor supports the extended HZ for locked planets of M-stars (Wandel 2018; Wandel and Gale 2020) and the further extended HZ due to subglacial water on the night side (Wandel 2023a).

Water vapor and eventually molecules produced by organic biochemistry in subglacial oceans would be transported to the planet's atmosphere by plumes fed through cracks in the ice layer, similarly to the plumes observed on Europa and Enceladus.

A similar mechanism of plumes and biosignature detection could exist also outwards of the CHZ (that is further out of the outer HZ-boundary), where subglacial liquid water may exist on geothermally heated rocky exoplanets even for low surface temperatures. This is true also for K- and G-stars, but their contribution to the abundance of habitable planets is less significant than M-dwarfs, as they are less frequent. As a conservative outer limit for the HZ extended by subglacial basal melting, one may use the flux on Trappist-1 g, the coldest exoplanet calculated by Ojha et al. (2022), which is 25% of the flux received by Earth ($0.25\ S_E$). This value is roughly consistent with the outer boundary set by the 1D model, as well as with the Early Mars hypothesis (fig. 1). However, a yet lower flux value with a further outer boundary of the HZ may be derived from the evidence for an intra-glacial lake in Mars's south pole. Taking for the radiative flux received by Martian polar region $\cos(i) = 0.2$ gives a flux of $\sim 0.1\ S_0$, which is denoted in Figs. 1 and 3 by the violet line marked Martian polar lakes.

Acknowledgements: this research has been supported by the Minerva Foundation, at the Center for Studying the Planetary Emergence of Life.

† Note that a recent work (Greene et al. 2023) suggests that Trappist-1 b may not have an atmosphere or a very thin one.

References

Daspute, M., Wandel, A., Kopparapu, R. K., et al. 2025, ApJ, 979, 158.
Greene, T.P., Bell, T.J., Ducrot, E., et al. 2023, Nature 618, 39.
Holmberg, M. and Madhusudhan, N. 2024 Astron. & Astrophys. 683, L2.
Kasting J. F., Whitmire D. P. and Reynolds R. T. 1993 Icarus 101, 108.
Kopparapu R. K., Ramirez R., Kasting J. F. et al. 2013 Astrophys. J. 770, 82.
Luger, R. and Barnes, R. 2015, Astrobiology, 15, 2.
Menou, K. 2013 Astrophys. J. 774, 51.
Moran, S. E., Stevenson, K. B., Sing, D. K., et al. 2023, Astrophys. J. 948, L11
Ojha L., Troncone B., Buffo J. et al. 2022 Nature Comm. 13, 7521.
Shields A. L., Ballard S. and Johnson J. A. 2016 Phys. Rev. 663, 1.
Wandel A. 2018 Astrophys. J. 858, 1.
Wandel A. 2023a Nature Comm. 14, 2125.
Wandel A. 2023b Astron. J. 166, 222.
Wandel, A., 2025, Exoplanets beyond the Conservative Habitable Zone. I. Habitability, ApJ, https://arxiv.org/pdf/2510.00736
Wandel A. and Gale J. 2020 IJAsB 19, 126W.
Wandel, A. and Tal-Or, L. 2019, Astrophys. J. Lett. 880, L21.
Yang, J., Boué, G., Fabrycky, D. C. & Abbot, D. S. 2014 Astrophys. J. Lett. 787, L2.

(Toward) Discovery of Life beyond Earth and its Impact
Proceedings IAU Symposium No. 387, 2026
H. Landt, M. Dominik & C. Oliver, eds.
doi:10.1017/S1743921324002084

Searching for Life in Space: Communicating the Uncertainties

David Whitehouse

writer and broadcaster
email: david@davidwhitehouse.com

All of us gathered at this conference share an interest in communicating science. This is my trade using words, written and spoken and here are a few of my own about it. There are many ways to tell people about science using various techniques and strategies, often aligned to a certain audience. I am here to tell you about addressing mass audiences which is one of the roles of journalists. We journalists have a large following reached daily or even hourly, so we must be swift if not always sure.

Journalism has been called literature in a hurry. Sometimes I think it's in too much of a hurry. We write the first draft of history and it's a special responsibility. Journalism has its own history, its own language and culture, and it understands the public. Because of this science journalism is not the same as science communication.

I have met many scientists who do not realise this, for many reasons. But I tell them, in defence of my profession, that it's just as hard, just as specialised and is a similar intellectual challenge as the work they do. It requires as much dedication to become an effective journalist as it is to become an effective scientist. Perhaps because some see journalism in all its various forms before them every day they do not appreciate it for what it is.

Before I transitioned from scientist to journalist I spoke to one of my media heroes - well-known broadcaster and author James Burke. He was encouraging but warned me about the clash of cultures fearing that I would give up in exasperation in short order. He also made me think about the special role of a science journalist when he said I should never ask the audience what they want but give them what they need. This wise advice would come back to haunt me as my BBC career progressed.

Racing forward a little, one of my conclusions is that you must understand journalism to communicate SETI effectively. Far too many scientists think that because they've been interviewed by the media or written a few articles, they understand the media, how it works and what it wants. This is not so. If you really understand journalism you will see that any plan to control the news of a SETI discovery will not work. I'm reminded of the words of the boxer and philosopher Mike Tyson when he said that everyone has a plan, until they are punched in the head! But if you know the system you can adapt.

Let me give you an example of a SETI plan that will not work. There is the RIO scale of SETI discovery significance which is based on the Torino scale previously devised to classify the threat from near earth-asteroids. It's said to be for the benefit of non-specialist organisations and audiences to aid their comprehension. The RIO version was proposed in 2000 as a means of quantifying the significance of a SETI discovery. Its purpose was to assist policy makers by telling them the likelihood, on a scale from 1 -10 that an extraterrestrial signal was produced by an alien civilisation. It was later modified to consider the finding of an artifact or technosignatures.

Level 0 is obviously nil; level 10 is extraordinary evidence. Level 4 is moderate importance; level 5 is intermediate confidence. An interesting intellectual exercise it might be, but it is useless in the real world of politicians and journalists. How would my story be different between a level 5 or 6? All I require is yes, no or maybe. Either option is a good story.

What Is a Story?

Which brings me on to stories, the fundamental unit of journalism. What is a news story? What indeed is news?

In a way news is like art. I can't really define it, but I think I know it when I see it. If you look at the world in the right way it can be surprisingly easy to find. But when news is news is a trickier question. It depends upon many things. Something that might be known to the scientific community might be unknown to the public so to them it will be news. Taking this approach can be fraught with problems as it can, if you are not careful, lead journalists to proclaim as news things that are just interesting. This is happening more and more these days. So, is it new, is it a discovery, is it important, and why would anyone else be interested? Give them what they need, not what they want, remember.

Let me give you an example of how to manufacture news. Late one hot Tuesday afternoon I was at a press conference in which a scientist was describing what he called memory metals that changed their shape depending upon temperature. For the journalists, suffering in a stuffy room, the story was going nowhere. Thern someone asked an obvious question, what could it be used for, and the answer given by the scientist was brilliant and turned a no-hoper into a must-use story. He said the metal would make a perfect underwire bra. I'll leave the rest to you!

Writing for news is not writing for scientists. You write to be effective, not to impress other scientists. Sometimes scientists do not like what we write or the approach we take, but then again when we write about them the politicians don't like it either.

One form of agony and culture clash I put up with for a few years when I was at the BBC was the annual 'scientists meet the media' bash organised by the Daily Telegraph. It usually took place at the Science Museum and was a throng of journalists, scientists and wannabees of both types, clutching drinks as well as a few speeches about the importance of getting good science in the media.

Walking into the gathering I used to think it was good of the Daily Telegraph to enable me to meet scientists, as I wouldn't normally do this in my day job as a science correspondent! More importantly I thought that if half the assembled scientists didn't hate more than half of the assembled journalists then we journalists were not doing our jobs properly. There should be a tension between the two communities. We are not automatically on the side of science despite the Telegraph's, and later the BBC's, desire to be part of the establishment. Journalists should shake the tree, not climb it.

So be aware that journalists, unless they are courting the establishment, are on the look-out for the awkward, the controversial, the gossip and the backstory. Despite their background, or perhaps because of it, they owe no allegiance to anything other than the story and the facts, and it might not be the facts you think of. Journalists are not your friends and tame science journalists make poor ones.

Involving The Media.

There are, of course, many reasons why scientists should be involved with the media. Science advocacy is vital for society, as are the interests of your university or institute. For some though personal advocacy is paramount. In my time as a journalist I have seen this rise to absurd levels.

Although it is rarely acknowledged there is often a cost in becoming involved in the media. Some scientists do not like it and think it demeans in some way the Olympian status of science. This was certainly the case when I was a scientist involved in the media. Only when I started working for the BBC did I find out that many of those in academia that disparaged my communication efforts were the very ones courting the media!

How do you write an effective press release? The measure of success here is that it must get used and result in media coverage. Pity the poor media officer who produces a press release that does not light a fire. There are ways to ensure it will.

One could say that one must figure out what doesn't need to be said, and then don't say it. You would be surprised by how much irrelevance there is in some press releases. Be aware that the public, even the scientifically interested public, have many draws on their attention and usually can't wait to stop reading you. Because of this the first sentence is the most important, and the second is the second-most important.

It's not an academic article. It's designed to pass in front of the eyes of a journalist or editor and make them pause to take a second glance. So, the headline is paramount. It must sum it up and have, well, the X-factor! This is not easy, there are in many publications professional headline writers, and they will tell you it's not easy!

Writing a news release is like producing a TV report which is the anthesis of a science lecture. You cannot lead-up to the discovery, there is no setting the scene. You must come out with the discovery or the news item straight away and fill in the background much later. As I have said, the key objective of a press release is to be noticed and to be used. Journalists see many releases every day and must make snap decisions to read beyond the headline or move on. You get one chance and one headline. So, what is the big thing of your story? If you can't sum that up in a headline then you haven't got one. One big idea per story is best.

Also, one thing that kills an article or press release is self-importance and pomposity. Cut it out and be aware that nobody has to read what you write and be aware that knowing nothing about a subject is not the same as being stupid or inferior. Don't patronise without realising it. Avoid cliches like the plague except when it works. Don't mix metaphors. Also use plain English and avoid jargon, and it has been said that the road to hell is paved with adverbs. Swiftly deliver the substance.

Helping the media this way is mutually beneficial. You want to get your message across, we want stories. If you do not take your place in the daily flood of content it will be taken by others.

If I have any advice to improve your science writing it would be not to read too many science books. I tell you that because many of them are agony to plough through. They are information slapped onto the page with no finesse. I will give you representative titles in private. You must do more than this, you have to help, tease and cajole the reader into making them want to read on. There is a saying in the publishing trade that if the public don't want to read your book or article then nothing will stop them. So read wide, take in and learn from the best writers on all subjects, novels, poetry especially. I have no proof, but I firmly believe that doing this enables you to communicate more smoothly and that it aids logical thought.

The Detection.

What happens when it comes to the big story, possible detection? When it comes to a SETI discovery expect it to leak immediately. When SETI searches used to take place from Arecibo I almost always got a call from someone involved, not the chief scientist, giving me their contact details and letting me know they would keep me informed if there were developments with an unspoken understanding that if there was a hint of a discovery they would be the one to be interviewed and become the face of SETI, and all that it implied. No matter what you do, or what procedures you put in place it will happen this way, remember Mike Tyson. So be quick, be open and tell all from the start even as you work towards verification, remember all we want is yes, no or maybe.

Which leads me to the case of EQ Peg. This was a possible detection flap that I started when I was BBC News Online Science Editor. It happened in 1998, and it is still instructive.

It was late on a drab Thursday evening and I hadn't had a good story for a few days, so like most journalists in that position my guard was a little lower than usual. I took a few phone calls and had a few emails about online discussion about a possible detection of a signal from EQ Peg. I wasn't that impressed as it has been picked up by a very small radio dish and not much was known about who found it or how to contact them. Still, the data was online.

It was picked up by small dishes like that used by the amateur radio enthusiasts of the SETI League, a US-based group. They were looking at the data and I thought that if they ever thought they had a chance of picking up a SETI signal then they should investigate this seriously. Which they did but failed to provide any useful comment we could use.

I contacted former colleagues at Jodrell Bank and the Max Planck Institute for Radio Astronomy. They said it was very unlikely, but worth a look. So, there was the story. Probably not a detection but an event that astronomers were checking out. I produced a brief story which accurately laid out what was happening emphasising the testing of the process of verification.

The story went around the world, was amplified and distorted. A few prominent American SETI scientists heavily criticised the BBC for being irresponsible. Listening to what they said it soon became clear that what we reported was exactly what they were saying to the American media and that their only serious gripe was that we at the BBC didn't call them!

A week later we ran a second story saying scientists were dismayed at an apparent hoax. The head of the SETI League said the affair had set back SETI a thousand years. This was absurd. What it had done was to show how little many in SETI understood the media, how they were too slow and didn't react quickly enough or even engage with the media. Part of the problem was that some prominent SETI personalities waited for the media to call them instead of calling the media and were miffed when they weren't contacted.

There was no plan, no adaptation to the peculiar circumstances. Just a defensive attitude that couldn't wait to criticise the media for not doing the story the way some scientists thought it should be done. If such an event happened today SETI scientists should have more of a plan, be more active and beware that if they don't get involved then the one of the media's first ports of call will be high-profile scientists like Avi Loeb!

I said at the start that science journalism is not science communication, and if you want to be effective in dealing with the media in all its forms, be seen and perhaps appreciated then you probably need a journalist to help you.

In my day-to-day life, whilst staring at a blank screen, I often reflect on the words of the Jack Kerouac when he wrote that, one day I will find the right words and they will be simple. That's the key. Get the message across as quickly and simply as possible. It's like a physicist looking for the simplest most elegant solution. I conclude my talk with the wise words of Joseph Pulitzer who does just this.

"Put it before them briefly so they will read it, clearly so they will appreciate it, picturesquely so they will remember it, and above all, accurately so they will be guided by its light."

(Toward) Discovery of Life beyond Earth and its Impact
Proceedings IAU Symposium No. 387, 2026
H. Landt, M. Dominik & C. Oliver, eds.
doi:10.1017/S1743921324001248

The definition of life and its role in astrobiology and the search for evidence of life in the universe

Nigel J. Mason[1], Dirk Schulze-Makuch[2,3], Terence Kee[4], Sohan Jheeta[5] and Juan Pérez-Mercader[6]

[1]School of Physics and Astronomy, University of Kent, Canterbury, CT2 7NH, United Kingdom
email: N.J.Mason@kent.ac.uk

[2]ZAA (Zentrum für Astronomie und Astrophysik), Technische Universität Berlin, 10623 Berlin, Germany
email: schulze-makuch@tu-berlin.de

[3]German Research Centre for Geosciences (GFZ), Section Geomicrobiology, 14473 Potsdam, Germany

[4]School of Chemistry, University of Leeds, Leeds, LS2 9JT, United Kingdom

[5]Network of Researchers on the Chemical Emergence of Life (NoRCEL), Leeds, United Kingdom

[6]Department of Earth & Planetary Sciences and Origins of Life Initiative Harvard University, Cambridge, MA, 02142-1204, United States of America

Abstract. How may life be defined? This is a question that has growing importance as humanity develops the ability to explore beyond the Earth and searches for evidence of life being present on other astronomical bodies from the planets and moons in our own solar system to the thousands of exoplanets we are detecting around other stars. Such a definition is non-trivial and, as we discover life in ever more diverse and 'extreme' environments on Earth broadening the regions of 'habitability', increasingly complex. In this short article, we review the different definitions of 'life', the need for such a definition and the implications that such a definition may have on the future development of astrobiology and the search for evidence of life in the universe.

Keywords. Astrobiology, biosignatures, extraterrestrial life, origins of life, technosignatures

1. Introduction

The need for a more scientific definition of life is a relatively recent concept since, for many, it was 'obvious' what was alive (humans, animals, plants) and what was not (rocks, materials (tables, chairs, cars, the elements). Indeed, several scientists and philosophers have stated that it is not possible to define life [Cleland (2019); Koonin (2012)]. However, as humanity has developed the technology to explore extraterrestrial worlds, from the other planets and moons in our own solar system to those 'exoplanets' around other stars, it has become necessary to identify 'signatures' that may indicate the existence of life elsewhere in the Universe. Furthermore, in defining life, we may also influence the development of technologies, creating 'life in a test tube, understanding the processes that are necessary for the origin of life, and making critical distinctions between artificial intelligence and machine life, robots and 'living' forms [Clark (2019)].

However, despite its importance, there is no widely accepted definition of life, and the debate continues. This paper reviews at least some of the definitions, the rationale for each, and the consequences of adopting such a definition for our development of the field of astrobiology and the search for evidence of life in the universe†. Through the definition of the characteristics of life we may identify biosignatures that may allow us to search for it, indeed this is already necessary for current searches for life in our own solar system, both extant and extinct. The derivation of 'biosignatures' is also highlighted in the study of exoplanets, where through atmospheric observation, we might determine whether life can have developed elsewhere in the Universe, addressing the most fundamental of questions; 'Are we alone?'. Such a definition may also be critical to studies of 'The Origins of Life' particularly here on Earth. It is also necessary to explore how such a definition may change as we appreciate the role of Artificial Intelligence (AI) and Machine learning in more advanced scientific and technologically developed civilisations whose legacy may be technosignatures.

2. Definitions of Life

The space agencies have been a major leader in establishing a definition of life. The NASA definition is deceptively simple; Life is "A self-sustaining chemical system capable of Darwinian evolution." This definition was derived in the 1990s by a NASA exobiology working group and has been widely adopted. It supports the concept that to be alive, it is necessary to replicate the genetic data being recorded and translated into a succeeding generation, which, in our case, is inscribed in our DNA, making DNA the 'molecule of life' and the subject of searches when we wish to identify life on Earth. Hence the search for DNA or its cousin, RNA and in a sample from Mars, or amongst the "ocean worlds" in the outer solar system – Jupiter's moon, Europa, Saturn's Enceladus or other moons of gas giants that are believed to have vast oceans beneath shells of ices in which life can develop.

Such a simple 'definition' is immediately open to criticism since using the reproduction criteria as a characteristic of life excludes some undoubtedly living creatures. The most common example being the mule. Mules are born when a male horse mates with a female donkey. This produces a genetically distinct creature but one that happens to be sterile as a species and so incapable of reproduction. Hence, by accepting the NASA definition of life, we should not classify a mule as alive. In addition, the NASA definition of life is highly unpractical. If we land a space probe on an unknown planet, would we really wait for an assessment whether we found life until we can observe in a putative organism Dawinian evolution? And what if the entity undergoes Lamarckian evolution?

Another complication of adopting the NASA definition is the virus. Should a virus be considered living or not? This has engendered a lot of debate [Farnsworth (2021); Dupré & Guttinge (2016); Juhas (2023)] both for and against their classification of a life form. Viruses have genetic information coded in DNA or RNA, so does that mean they can be considered 'alive'? However, viruses can only reproduce after 'infecting' a host – a process called viral replication. Once inside a living cell, viruses re-program the cell's machinery to produce viral proteins and genetic material to make new copies of themselves, these new virus particles then infect other cells. Viruses can also 'evolve'. Viral infection is a highly dynamic process, which leads to constant evolutionary changes

† This article is based on presentations and discussions at the KAVLI-IAU SYMPOSIUM (IAUS 387):(Toward) Discovery of Life Beyond Earth and its Impact Proceedings held In Durham, UK, April 14–19, 2024

diatoms consist entirely of silicon dioxide) [Irwin & Schulze-Makuch (2020)]. Arsenic has also been proposed as a substitute for phosphorous in alien RNA/DNA worlds [Wolfe-Simon et al. (2009)]. Therefore, we must be cognisant that other chemistries could be active elsewhere and that we should not be blinded by our current knowledge is based on only one biosphere and one biochemistry of life or will risk underestimating the forms and functions life can take.

Identifying the transition from prebiotic to biochemistry is essential in understanding the origins and evolution of life. Perhaps the most essential step in such evolution is the creation and development of an encapsulated structure, like cells within which the chemical processes that underpin life and drive metabolic and reproductive procedures occur. Such transitions may be modelled in an abiotic way, as demonstrated by Pérez-Mercader and colleagues, who created a micron scale self-assembled molecular system capable of embodying and mimicking some aspects of "simple" life, including self-assembly from a homogeneous but active chemical medium, membrane formation, metabolism, and a primitive form of self-replication [Albertsen et al. (2021), Pearce & Pérez-Mercader (2021)]. Such work may provide means to explore how life may evolve elsewhere in the universe from prebiotic chemistry and indicate extinct pathways on life on Earth pre-LUCA (Last Universal Common Ancestor). LUCA may be seen as a theoretical construct—it might or might not have been a particular organism [Weiss (2018)], and LUCA on another world may be distinct from those on Earth, leading to different evolutionary pathways. Indeed, we may ultimately come around to see how life is less about the state of any given system, LUCA or some other organism and more fundamentally about how such systems change *via* various pathways [Moody et al. (2024)].

Elsewhere in this volume Kee & McCrum (2024) propose replacing the three concepts that they identify with life; life as a property of a system, life as a purely biological phenomenon, and life as a binary concept with three operational principles; synonymity of 'life' with 'processes of change'; the foundation of change upon interaction; the recursion and integration of change over boundary conditions.

Therefore we must recognise the inherent uncertainty in defining life. Until we have (i) bona fide examples of extraterrestrial life and/or (ii) some more Universal model upon which to base our life definitions, we cannot say whether life as we know it is the only form of life possible or if a much broader definition is required to match observation. Life's definition, like life itself, is therefore expected to evolve and adapt to the scientific and technological environments in which the search for life is conducted.

Acknowledgments

This article arises from the discussion session at the KAVLI-IAU symposium:(Toward) Discovery of Life Beyond Earth and its Impact held in Durham UK April 15–19, 2024. The panel members were Professor Juan Pérez-Mercader Harvard University, USA; Professor Dirk Schulze-Makuch, Technical University Berlin, Germany; Dr Terence Kee, University of Leeds, UK and Dr Sohan Jheeta, Norcel, UK The Panel was chaired by Professor Nigel Mason, University of Kent the author of this report and discussion, any omissions in the discussion are his.

References

Albertsen, A. N., Szymański, J. K. & Pérez-Mercader, J. 2017, *Scientific Reports*, **7**, 41534

Brooks, H. & Schulze-Makuch, D. 2010, *Int. J. Astrobiol.*, **9**, 89–99

Clark, B. 2019, In: Kolb V M *Handbook of Astrobiology*, Boca Ratón: CRC Press

Cleland, C. E. 2019, *The quest for a universal theory of life*. Cambridge: Cambridge University Press

Dupré, J. & Guttinge, S. 2016, *Stud. Hist. Philos. Biol. Biomed. Sci.*, **59**, 109–116

Farnsworth, K. D., Nelson, J. & Gershenson, C. 2013, *Acta Biotheor.*, **61**, 61
Farnsworth, K. D. 2021, *Biosystems*, **200**, 104324
Irwin, L. N. & Schulze-Makuch, D. 2020, *Universe*, **6**, 130
Juhas, M. 2023, In: *Brief Lessons in Microbiology.*, Springer Cham
Kee, T. P. & McCrum, J. 2024, *Proceedings IAU Symposium No. 387*
Koonin, E. V. 2012, *J. Biomol. Struct. Dyn.*, **29**, 603–605
McCrum, J. & Kee, T. P. 2024, *Proceedings IAU Symposium No. 387*
Moody, E. R. R., Álvarez-Carretero, S., Mahendrarajah, T. A., Clark, J. W., Betts, H. C., Dombrowski, N., Szánthó, L. L., Boyle, R. A., Daines, S., Chen, X., Lane, N., Yang, Z., Shields, G. A., Szöllősi, G. J., Spang, A., Pisani, D., Williams, T. A., Lenton, T. M., & Donoghue, P. C. J. 2024, *Nat. Ecol. Evol.*
Onstott, T.C., Ehlmann, B.L., Sapers, H., Coleman, M., Ivarsson, M., Marlow, J.J. Neubeck, A. & Niles, P. 2019, *Astrobiology*, **19**, 1230–1262
Pearce, S. & Pérez-Mercader, J. 2021, *Polym. Chem.*, **12**, 29–49
Petkowski, J. J., Bains, W. & Seager, S. 2020, *Life*, **10**, 84
Rizzo, V. 2020, *Int. J. Astrobiol.*, **19**, 283–294
Schulze-Makuch, D. & Irwin, N. L. 2018, In: *Life in the Universe: Expectations and Constraints*, Springer, Berlin
Schwieterman, E. W., Kiang, N. Y., Parenteau, M. N., Harman, C. E., DasSarma, S., Fisher, T. M., Arney, G. N., Hartnett, H. E., Reinhard, C. T., Olson, S. L., Meadows, V. S., Cockell, C. S., Walker, S. I., Grenfell, J. L., Hegde, S., Rugheimer, S., Hu, R. & Lyons, T. W. 2018, *Astrobiology*, **18**, 663–708
Stern, A. & Andino, R. 2016, In: *Viral Pathogenesis*, Academic Press
Suazo, M., Zackrisson, E., Mahto, P. K., Lundell, F., Nettelblad, C., Korn, A. J., Wright, J. T., & Majumdar S. 2024, *MNRAS*, **531**, 695–707
Walker, S. I., Bains, w., Cronin, l., DasSarma, S., Danielache, S., Domagal-Goldman, S., Kacar, B., Kiang, N. Y., Lenardic, A., Reinhard, C. T., Moore, W., Schwieterman, E. W., Shkolnik, E. L. & Smith, H. B. 2018, *Astrobiology*, **18**, 779–824
Weiss, M. C., Preiner, M., Xavier, J. C., Zimorski, V. & Martin, W. F. 2018, *PLoS Genet.*, **14**, e1007518
Westall, F., Brack, A., Fairén, A. G. & Schulte, M. D. 2023, *Front. Astron. Space Sci.*, **9**, 1095701
Wilson, D. F. & Matschinsky, F. M. 2021, *Front. Physiol.*, **12**, 658997
Wolfe-Simon, F., Davies, P. C. W. & Anbar, A. D. 2009, *Int. J. Astrobiol.*, **8**, 69–74
Wright, J.T, Haqq Misra, J., Frank, A., Kopparapu, R., Lingam, M. & Sheikh, S. Z. 2022, *ApJL*, **927**, L30
Yolles, M. & Frieden, R. 2022, *Systems*, **10**, 70

(Toward) Discovery of Life beyond Earth and its Impact
Proceedings IAU Symposium No. 387, 2026
H. Landt, M. Dominik & C. Oliver, eds.
doi:10.1017/S1743921324002047

Re-evaluating Boundary Conditions of the Concept of Life

Terence P. Kee[1] and James McCrum[2]

[1]School of Chemistry, University of Leeds, Leeds, UK
email: t.p.kee@leeds.ac.uk

[2]School of Physics and Astronomy, University of St Andrews, St Andrews, UK
email: jm441@st-andrews.ac.uk

Abstract. It is argued that a central challenge to the task of developing a foundational model for life lies within the implicit propositions of the Western scientific view. These propositions constrain thinking about the concept of life. Three implicit boundary conditions in particular - life as a property of a system, life as a purely biological phenomenon, and life as a binary concept - are identified, and it is suggested to replace them with three operational principles - synonymity of 'life' with 'processes of change'; the foundation of change upon interaction; the recursion and integration of change over boundary conditions.

Keywords. Life, Recursion, Boundary Condition

---

1. Introduction

I attribute to Nature neither beauty nor ugliness, neither order nor confusion. For things can only be called beautiful or ugly, orderly or confused, in relation to our imagination. By the coherence of parts, I understand nothing but that the laws or nature of the one part so adapt themselves to the laws or nature of the other part that they are opposed to each other as little as possible. [...] From this it follows that every body, insofar as it exists modified in a certain way, must be considered as a part of the whole universe, must agree with the whole to which it belongs, and must cohere with the remaining bodies.

- Baruch de Spinoza 1665

Eight decades ago this year, Erwin Schrödinger published What is Life?, a seminal contribution in the quest to understand the concept of "life" (Schrödinger 1944). Whilst Schrödinger did not answer the question, precisely, in his book, he did lay a foundation of work building on and further developing an exploration of it. (Cleland & Chyba 2010; Cornish-Bowden & Cárdenas 2020; Trifonov 2011; Tsokolov 2009; Weber 2010; McKay 2004; Benner 2010; Jeuken 1975). Nonetheless, the question remains elusive.

There may, crucially, be a possible problem with the question *itself*, 'what is life?'. The word "is" relates to the stative verb "to be", which relates in turn to the condition of a given state at any given snap-shot in time. The implication then is that life is a quality or phenomenon which relates to "being", in the sense that a given system can be defined (in some way) as existing in a certain state (Musacchio 2005). This seems to run counter to all of our experiences of the phenomenon of life which have as their root, processes of change rather than states of being (Gómez-Márquez 2023). All biological organisms as we know it engage in the activity of change as a common theme and such changes are most commonly recognised through certain processes of change which we call variously; birth, death, ageing, evolution, mutation, selection, adaptation *etc.*

We must consider the way that we frame any such system. These frames are more commonly called boundary conditions (Paksi 2014; Polanyi 1968) and it seems logical to re-consider whether or not we have these frames set appropriately for the question(s) at hand; this work will reconsider such boundary conditions, to try and come to, at least, a better formulation of the question asked by Schrodinger.

2. Life as fundamentally a Boundary Condition Problem

We may re-express the problem of life as follows:

'Is there a meaningful, scientifically valid, and binary way to separate living things from non-living things, and to talk about why it is that life exists, and how it fits into the broader context of the world?'

This is clearly a two-part problem, and the two parts are not separable. It is of course possible to simply systematise the opinions of various observers on what is or isn't life, as the recent paper that distinguishes biosignatures using a neural network, ultimately trained on human judgement calls of such shows (Cleaves et al. 2023). But, in the first instance this approach is not completely consistent, as certain edge cases demonstrate (Cleaves et al. 2023). It doesn't really tell us what life is, independent of human perception. It just tells us what people, or rather some people, think life is.

Previous models of life have been incredibly useful, valuable, logical and well-constructed, and again, we by no means wish to cast aspersions on these works. But we are not the first to observe (Wolfram 1984; Persig 2014) that a binary definition of life may not be useful, or even possible. So, in un-asking the question of life, what are the assumptions, or boundary conditions, that we're objecting to? We identify three primary ones, which have proven particularly troublesome in trying to formulate a unified definition of life within the Western scientific cannon:

(i) Livingness is a quality, or property, inherent to a 'thing'; an object, a system, or an organism. (ii) Those things in proposition (i) are "biological things". (iii) There exists a binary distinction between livingness and non-livingness, such that 'things' can be identified as belonging in a category of either one or the other

Let us consider each of these in more detail.

2.1. *Implicit Proposition #1*

It is fundamental that livingness be associated with "things" (objects, systems, organisms etc. . .) according to this proposition.

If this is true, how do we know this? On one level it seems perfectly reasonable to consider many material forms as "things" or "systems". Some of these things appear to be inanimate in that they do not seem to change, in any human perceptual sense, in response to some action taken by that thing. They do not appear to have agency (Moss 2024) associated with the thing itself but respond only to their environment. Inanimate things then are usually classified as non-living. Other things do possess what we call agency and appear to direct any response to their environment, towards certain outcomes and not to others. When such things are "biological" in nature (whatever that may mean), they are usually classified as animate and living. In the case of both animate and inanimate things though, we perceive a physical differentiation between the thing itself (the system) and their environments.

But note the wording here - we *perceive.* To view something as a singular, differentiated form is to see it as dis-attached in some way from its wider context. Consequently, connections that are definitely physically apparent and relevant are ignored. In categorising things, humans first break a node out of what is an otherwise connected and integrated whole. Once separated, we attach a specific meaning to that separated node; a meaning that is different to what is not the thing. Finally, we attempt to reconstruct those connections that were necessarily present before thing and environment were conceptually separated; without a full understanding of what connectivity has been disconnected.

2.2. *Implicit Propositions #2 and #3*

Now that we have reasonable grounds to doubt proposition (i), let's move on to proposition (ii) and proposition (iii), which are closely interwoven. The inherent assumption that biology is the sole province of life is baked into the etymology of the field's name. The word biology is derived from the Greek, bíos, life, and logía, field of study. By implication then, something non-biological, that does not operate in the same way as a blood-and-flesh owl, is not alive by definition.

Humans have come to define a set of systems as having livingness, and systems that operate in a different way to be perceived as non-living. Sagan spoke of 'carbon chauvinism' or 'water chauvinism' to describe the assumptions made about how the chemical makeup of terrestrial biology must be mirrored in extraterrestrial life (Sagan 1973). These ideas are encoded within proposition (ii). Life is that which functions like we do. That which does not function as we do is not life. Is this a warranted belief? Well, we have something of an easy time saying that all life on Earth is of one phenomenon, one class, as they all share certain characteristics as a result of being descended from a single common lineage. Every single biological organism on Earth shares a common ancestor; in the theory of panspermia, this may even be true of all life everywhere! But what if an organism was found, say an extraterrestrial, or even a terrestrial one, of different descent? That formed from, evidently, a completely separate abiogenesis? It has no reason to use the same processes as we do (Davis et al. 2009).

The fact that we cannot even agree that some apparent offshoots of this common lineage - viruses, prions, etc - are 'alive' displays a further flaw in this argument. What set of characteristics, exactly, makes something that is biological alive? There is a conflation here of the concepts *biological* and *living*. Now, this conflation quite often has value, but what is it fundamentally based upon? There are definitions that would enclose things that do not have our lineage, and ones that would exclude things that don't! From a historical perspective, there was a time when both animate and inanimate substances were thought to be composed of the same set of fundamental "elements" [earth, air, fire & water as laid out by Empedocles (Russell 1991) expanding to include the concept of space (vide infra) in the Pañca-Mahā-Bhūta of Hindu philosophy (Gopal 1990). As our knowledge of the natural world increased however, it became clear that the operational principles, the mechanisms, of animate and inanimate systems differed. These differences then, helped lay the philosophical foundations for a separation of what is considered living from what is not, on mechanistic grounds (Bucci et al. 2010). This in turn influenced our collective, conceptual thinking about our world, which lies at the heart of implicit proposition (iii).

Is there some foundational, Universal principle that maintains such a binary distinction between livingness and non-livingness; and if so, on what is such a principle based? Whenever some natural phenomenon is assigned a "name", we are constraining that phenomenon. How is this so? In effect, we are placing that phenomenon within some conceptual framework or category. Consequently, what is outside of that conceptual categorization is de facto, not the same as "name", but part of a different category variously called "environment", "society", "colony", "world" etc...

In truth, we are not making categorization decisions of life (or anything else for that matter) based on universal, foundational principles, but rather using heuristic concepts.

(Ippolitti 2015). Such decision-making has served humans well in many scenarios of life; but its value is in providing something that is functional and expedient for the human rather than a solution to a problem based on foundational principles.

To conclude this section, we contend that a central problem with answering the question 'what is life' is that it relies on assumptions that are not themselves objectively true. Life exists in a place, has a shape, has a mechanism and process of existing and being and continuing, has a name and an idea. But these are problematic assumptions which cannot be reasoned to be fundamental or objective. The universe doesn't seem to have a differentiated concept of life; but we humans have created one and in so doing have created those differences accordingly. So then, if we not at all sure that the three implicit propositions (i)-(iii) listed above represent foundational, universal principles, through which we may view the concept of livingness; what happens when we reject them from our concept of life?

3. Interactions Lie at the Heart of Conceptual Boundaries.

Once the above implicit propositions within the western concept of life have been identified as being subjective, we can then consider how to probe more fundamental features. Note we say, 'more fundamental', as we are speaking only of models here, not underlying reality itself.

As a starting point, there is the view within physics that a more fundamental representation for the fabric of existence is a distributed and universal field than singular objects such as particles (De Angelis & Pimenta 2018). Thus, the particulate electron for example only has meaning because of the presence of distortions within a field of subjective interactions (Kuhlmann 2013). Note this means both that the idea of a singular and localised form of 'an electron' is not fundamental, and that it only exists as a consequence of interactions with other parts of the universe; a result which emerges from many experimental observations, such as the suite of double slit experiments (Chang 2024).

Parallel examples of this kind of thinking can be taken from human experience. We can say that a game of tennis cannot exist in the absence of interactions between players. The concept, game of tennis, has meaning only when players are mutually engaged in an interaction field which connects them in some way. One player alone cannot create game of tennis, even if they have the balls, the racquets, the nets, an umpire, and an audience. Similarly, two tennis players, all set to play a match, but without tennis balls, or racquets for example, cannot create game of tennis, because the interaction between players and necessary tools (the ball and racquets) are not present.

The point we want to emphasize here is that the concept of something we might call an electron or a game of tennis exists due to the operation of fundamental interactions, interconnections between various "actors" or distortions with a conceptual interaction field (such as electrostatic, magnetic fields in the case of the former and tennis balls and tennis players in the latter scenario). This collection of interaction space, whether electrostatic, magnetic, probabilistic, gravitational, or human, represents an interaction field. One may then think of this interconnectivity between actors (whatever they may be) or distortions within an interaction field as what creates the "concept" or "experience" (charged particle, game of tennis). How then might this fundamental focus on interactions influence our idea of the concept "life"?

All of biology is built around cellular interactions; interactions which occur either within each biological cell and interactions between differentiated cells. So, right at the outset we may delineate two distinct concepts to biological life. One concerns what happens within a biological cell, in which the cell itself is the conceptual system. The second conceptual system exists at a level higher of abstraction in which individual biological cells are interacting components within a concept called cell population. These two concepts, biological cell and cell population are not the same, but they are deeply and recursively connected (Liu et al. 2018). Each of these two concepts have their own internal sub-sets of structural, mechanistic and conceptual boundary conditions. For example, individuated bacterial and archaeal cells have different cell wall and lipid membrane components (structural boundary conditions) and perform different suites of metabolic chemical reactions (mechanistic boundary conditions). Consequently, the use of different names to symbolize these two families of organism provides a conceptual boundary condition between them.

Cell populations also possess structural, mechanistic and conceptual boundaries, but they are not necessarily the same as those of individuated cells. From a chemical perspective, similar classes of chemical entity (molecules, macromolecules, ions etc...) and mechanism (messenger-receptor, signalling cascades etc...) are used across such boundaries, but the ways in which they manifest, and their integration are different (Gerdes & Pepperkok 2013). The result of these variations in boundary conditions is that cell populations can do things that individuated cells cannot, such as build structures we call organs, tissues, organ systems and organisms. They display what is called *emergent behaviour* (Cuenca et al. 2022).

What we have here is an integrated network of mutually connected, recursive subsystems, which function at different levels across the system as a whole. Therefore, at each level of the recursive structure, we can identify concepts (cell, cell aggregations, organs, tissues, organism etc...) which have operational meaning as a result of the interactions present within that concept and between that concept and adjacent conceptual structures. In some cases, the operational meaning of a concept will be closely associate with either structural or functional (mechanistic) features of that concept (eg: a human heart is fundamentally associated with its function to pump blood around the body). In other cases, that meaning will be more closely associated with its connections to other concepts eg: the system of Linnaean taxonomy that runs from the more specific concept, species, to the more general concept of kingdom (Kuntner & Agnarsson 2006).

It's worth noting that we are far from the first researchers to identify recursion, especially in the geometric context of fractal shapes, as a key idea in the structure and nature of life(Azua-Bustos & Vega-Martínez 2013; Grosu et al. 2023).

4. Interactions Lie at the Heart of how Concepts Change. This Dynamic Process of Conceptual Change is Livingness.

At this point, it is worth pausing to reflect on what we have tried to lay out above; that concepts result from interactions, connections or distortions within some interaction field (in whatever form that may be envisioned). How does this then relate to the concept of life or livingness?

Most scientific models of life focus on the concept itself as possessing the quality we call life. For example, the physical concept or system called, dolphin is said to possess the concept of livingness, whereas the physical concept paving stone is thought of as non-living, in a conceptual way (Amilburu et al. 2021). However, what we wish to emphasize here is that the cognitive concept life is perhaps better viewed as a variable which can be framed in different ways according to the question being asked or the level of complexity with which one wishes to explore this concept. Thus, the concept livingness reflects the process by which the cognitive concept life changes. In essence then, livingness can perhaps be viewed more fundamentally as the conceptual, mechanistic and physical processes of change; or put another way, the processes by which a system (however that is defined) changes against some variables (eg: time, informational input etc...). This is neither a new nor original idea as the dynamic nature of livingness resulting in change is embedded in thinking from classical Greek-Persian philosophy (Müller-Merbach 2006) to current scientific models such as natural selection (Gregory 2009), the constructal law (Bejan & Marden 2009); assembly theory (Sharma et al. 2023); dynamic kinetic stability (Pross & Pascal 2013) and more recently models of functional information (Wong et al. 2023).

All these wonderful approaches (and many other besides) have the concept of change at their heart; changes of form, changes of mechanism of operation, changes in energy or information content. They are, essentially, techniques of modelling or parametrizing, the processes and mechanisms of change and, as such, lie seamlessly within the overarching picture presented here. Interestingly, many of the above models go to significant lengths to draw out the connections between animate and inanimate systems.

Of course, when one talks about the concept of change, a logical question to ask is: what is actually doing the changing? One way to think of this is to adopt a systems-type of approach. So, the question what is changing can be answered by, the system is changing. Well then, a further question raises its head: what is the system? To which one may respond that a system is some collection of entities or components which are mutually linked together in some way and which operate in such a way as to influence the collective of that system (Wilkinson 2011). Processes of change, at their simplest, are frequently represented as mechanistic elements of an input-manipulation(or processing)-output-feedback cycle.

These mechanistic or non-material components may be connected to some material parts of a system, such as a biological form (eg: hermit crab), non-biological form (eg: pulsar) or a human-designed object such as a TV or suspension bridge. Together the material and non-material or operational elements comprise what we might call a system. An important point is that a system needs (by its very definition) to incorporate mechanisms by which that system can change. Conceptually, a system must integrate those mechanisms of change; they must therefore be part of the system. Now they can be external to the structural objects within the system, but they must be a part of the system as a whole. Anything that undergoes change by mechanisms which are not considered to be part of that system is, in reality, not a system at all but an object which is acted upon.

If we imagine a single wave crashing against a surf-line, billions of grains of sand will be disturbed by the force of the wave. The retreating of this wave creates patterns of sandbanks called bed-forms; patterns which are unique to each wave cycle and never repeated exactly in form, size or duration. The concept *beach* has undergone a physical change in form as a result of the wave dissipating itself on the shore. In systems language, the system *beach* has transitioned from one state to another state as a result of an energy-input event. In this view, the concept *beach* remains unchanged, but the physical changes to the system that have occurred are technical in terms of how the systems components (the grains of sand in this case) are distributed. This would be in line with the western scientific mechanistic or reductionist approach to change.

An alternative view to this would be to consider the concept *beach* after the wave-crash to be a newly created concept to that of *beach* before the wave-crash. This new concept is created out of the interaction between two other concepts, wave-crash and *beach* before the wave-crash. So now, just like the concept car is an object but car-with-driver is a system (which must technically also include, car ignition mechanism, fuel and many other constituent, integrated actors), so the concept *beach* is better considered as an object but the combined concepts wave-crash and *beach* before the wave-crash comprise a system. Now this may seem like semantics, but the power of this flexible conceptualization, a willingness to consider conceptual boundary conditions as fluid, allows us to view processes of change as integrative not just transformative.

Thus, not only does the physical object *beach* after the wave crash emerge out of the interactions between other physical objects wave-crash and *beach* before the wave-crash, but the concept of *beach* after the wave crash only exists as a consequence of the integration of concepts wave-crash and beach before the wave-crash. The conceptual boundaries of our system have changed. Now, of course this example of waves and beaches applies not only to that single wave and that momentary structure of the object *beach*. It applies to each connected sequence of wave-beach interactions that occurs over any given period of time (a temporal boundary condition).

Not only that, but when we consider the concepts wave-crash and *beach*, where exactly are we drawing the structural and mechanistic boundaries in order to frame those individuated concepts? Does the concept wave-crash include the building of the wave? Does it include immediately preceding and proceeding waves into which our wave merges? Does it include the backwash once the wave has crashed? Once we recognise that structural, and mechanistic boundaries associated with concepts are often not clearly delineated, we can see that concepts too should also be considered as similarly fluid and evolvable. Clearly, the examples that we have given above are not ones that focus on biological concepts and hence some readers may find themselves wondering what the relationship are between complex biological systems and complex non-biological ones. This is a logical question and one that scientists are engaged with as outlined above.

We propose that the foundational behaviour of any form of livingness is therefore, this recursive integration of all processes of conceptual change across any given set of (user-)defined boundary conditions.

5. The Life of Samantha

As a jumping off point from the previous section, we used examples of dynamic processes of conceptual change which are not directly or identifiably biological. So, you may rightly ask, where is biology in all of this?

Let us start with an individual human being; we introduce her here as Samantha. Samantha is a biological organism (which is a concept of course), so we might logically conclude that it makes sense to view Samantha as a physical, cognitive and emotional system. Now, Samantha is not a closed system (Wilkinson 2011) but one who exchanges physical matter and energy with her surroundings. The system we have called Samantha functions as an operationally interconnected network, of chemical reactions, control mechanisms, thought management, processing and communication units. Collectively, we refer to such an interconnected set of units as biological mind and body.

We envisage that most of us would be happy in considering the system Samantha to be a living system so long as the collective biological interconnectivity network remains operational, at least physically. Now, if we look at some of the various units, or components, of this physical biological system, we might identify other physical concepts such as heart, lungs, liver, central nervous system etc... Each of these may be considered as a system as well as being a sub-system component of the larger system Samantha. As an individual system, the heart for example, is not a closed one either, it takes in matter, processes it according to a set of interconnected instructions and then moves it on to other systems.

Within the heart itself, various structurally identifiable objects (concepts) are present, such as a heart valve for example. This too can be considered as a system. Not only does the heart valve have identifiable structural boundaries and mechanistic boundaries in relation to its role as a component of the heart network system, but it also operates as such due to its own internal composition and interconnectedness. For example, mature valves are composed of extracellular matrix (ECM) material divided across three main layers, the fibrosa, spongiosa and atrialis (for the atrioventricular valves) or ventirularis for semilunar valves (Chester & Grande-Allen 2020). If we move to increasingly smaller scales and explore the ECM as smaller sub-system within the heart valve, then we recognize how ECM is composed of a dynamic network if interacting molecules and minerals, including collagen, glycoproteins, hydroxyapatite amongst others.

Indeed, each component macromolecule within ECM can itself be considered as network of interacting molecules, and beneath the molecules, atoms, sub-atomic particles, quarks, etc. . . .each a recursively smaller scale of concept comprising networks of dynamic, interconnecting concepts.

This picture is of course not a new one; the whole edifice of cellular and systems biology is built upon such models of systems-within-systems; rather like a set of matryoshka dolls (Miller 1972). The significance of this recursive relationship between systems is that each dynamic system can be considered as an example of livingness within it's own set of user-defined boundary conditions. Importantly, it is not so much that the system possesses a quality we call living, but the system engages in living processes. It is, collectively, one contiguous spectrum of living processes, each with different sets of boundaries. When one recursively integrates each and every interconnected chemical system of livingness up to and including the structural, mechanistic and conceptual boundary conditions that we call Samantha; we begin to frame what we might colloquially call "the life of this organism".

Of course, what we have described above is only a limited framing of the full livingness of Samantha; both biologically and otherwise. What is outlined above represents only the internal biological systems, presented through some hierarchical biological architecture, for the bounded concept, Samantha. Additional biological systems can be identified which, whilst being partially internal to Samantha, nevertheless have components which are external to the organism. For example, the system of visual stimulus, heat-regulation, skin pressure sensing etc. . . . In effect, all physical systems with which Samantha regulates and modifies herself as a biological organism. The dynamic operation of each such system, on a material physicochemical level contributes processes of livingness which incorporate the biological organism Samantha, but which have external actors or elements. Examples include reflected/refracted light from the Sun, objects which Samantha touches and the flow of air over skin through which radiant heat is exchanged. In effect, the concept Samantha is now an element or actor in living processes which transcend the structural boundaries of the Samantha organism herself. Samantha is therefore a component in the operation of a larger scale system - perhaps she could be thought of as a cell within the wider Gaia organism (Stolz 2017).

We can also move beyond these physicochemical processes to look at other forms of system that Samantha, the organism and concept, is a component of but whose mechanism of operation are not only physicochemical but emotive and/or cognitive. Samantha has parents, siblings, friends, work-colleagues with whom collectively, Samantha engages in cognitive exchange via mechanisms which include both the physical senses and other processes such as language and emotions. The integration of all such physicochemical, emotive and cognitive processes across all boundary conditions for which Samantha is an actor (or element) becomes a representation of the complete livingness of the concept Samantha. We tend to condense this rather cumbersome mouthful of a phrase as "the life of Samantha". Therefore, from the perspective of Samantha, any way in which the concept Samantha or any concept of which Samantha is a part, actor or involved with, changes or is created; becomes integrated into the concept "life of Samantha". Of course, this begs the question as to what it means to say Samantha is alive, primarily because the concept of livingness that we have outlined above is not object-centric but process-focused. In effect, it is more accurate to say that Samantha (the human organismic concept) engages in multiple processes of livingness simultaneously and contiguously. We can play games of classifying those processes of livingness according to, for example, structural, mechanistic or conceptual differences using all sorts of different techniques and mathematics; one of which, Conway's game of life, is discussed in more detail in another article on these proceedings.

Acknowledgements

The authors acknowledge many colleagues for their time and good offices in wider discussions on boundary conditions in the context of life which helped shape this contribution; specifically Andrew Kirton (Leeds), Clément Vidal (VUB, Brussels), Kuhan Chandru (NUM), Inge Loes ten Kate (Utrecht) and especially Dirk Schulze-Makuch (TUB) and Juan Perez-Mercader (Harvard) for helping crystallize what is contained herein at the Kavli-IAU (IAUS387) symposium.

References

Amilburu, A., Moreno, Á.& Ruiz-Mirazo, K. Definitions of life as epistemic tools that reflect and foster the advance of biological knowledge. 2021 Synthese, 198, 10565.

Azua-Bustos A., Vega-Martínez C. The potential for detecting 'life as we don't know it' by fractal complexity analysis. 2013 International Journal of Astrobiology. 12(4), 314.

Bejan, A., Marden, J.H. The Constructal Unification of Biological and Geophysical Design. 2009 Physics of Life Rev., 6, 85.

Benner, S.A. Defining life. 2010 Astrobiology 10(10), 1021.

Bucci, M., Goodman, C., Sheppard, T.L. A decade of chemical biology. 2010 Nat. Chem. Biol. 6(12), 847.

Chang, D.C. A New Understanding on Wave-Particle Duality: Comparing the Quantum Wave Model with the Copenhagen In-terpretation and Other Alternative Models. In: On the Wave Nature of Matter. Springer, Cham. 2024.

Chester, A.H., Grande-Allen K. J. Which Biological Properties of Heart Valves Are Relevant to Tissue Engineering? 2020 Front. Cardiovas. Med., 7, DOI=10.3389/fcvm.2020.00063.

Cleaves II H.J., Hystad, G., Prabhu, A., Wong, M.L., Cody, G.D., Economon, S. Hazen, R.M. A robust, agnostic molecular bi-osignature based on machine learning. 2023 PNAS, 120(41), e2307149120.

Cleland, C.; Chyba, C. Does 'life' have a definition?. Ch 26 in Planets and Life: The Emerging Science of Astrobiology. Cambridge University Press, Cambridge, 2010.

Cornish-Bowden, A.; Cárdenas, M. Contrasting Theories of Life: Historical Context, Current Theories. In search of an ideal theory. 2020 Biosystems. 188 104063.

Cuenca, M.B., Canedo, L., Perez-Castro, C., Grecco, H.E. An Integrative and Modular Framework to Recapitulate Emergent Be-havior in Cell Migration. 2022 Frontiers in Cell and Developmental Biology, 8, 10.3389/fcell.2020.615759.

Davis, P.C.W., Benner, S.A., Cleland, C.E., Lineweaver, C.H., McKay, C.P., Wolfe-Simon, F. Signatures of a shadow biosphere. 2009 Astrobiology, 9(2), 241-249.

De Angelis, A., Pimenta, M. Understanding the Universe: Cosmology, Astrophysics, Particles, and Their Interactions. In: Intro-duction to Particle and Astroparticle Physics. Undergraduate Lecture Notes in Physics. Springer, Cham. 2018.

de Spinoza, B. Letter 15, to Oldenburg: 'A Worm in the Blood'.

Gerdes, H.H., Pepperkok, R. Cell-to-cell communication: current views and future perspectives. 2013 Cell Tissue Res. 352, 1.

Gómez-Márquez, J. Reflections upon a new definition of life. 2023 Sci Nat 110 53.

Gopal, M. in K.S. Gautam (ed.). India Through the Ages. Publication Division, Ministry of Information and Broadcasting, Govern-ment of India, 1990.

Gregory, T.R. Understanding Natural Selection: Essential Concepts and Common Misconceptions. 2009 Evo. Edu. Outreach 2,156–175.

Grosu, G.F., Hopp, A.V., Moca, V.V., Bârzan, H., Ciuparu, A., Ercsey-Ravasz, M., Winkel, M., Linde, H., Murean, R.C. The fractal brain: scale-invariance in structure and dynamics. 2023 Cerebral Cortex, 33(8), 4574.

Ippolitti, E. Heuristic Reasoning, Springer, Switzerland 2015.

Jacobs, S.A. The Scientific Consensus on When a Human's Life Begins, 2021 Issues Law Med., 36(2) 221.

Jeuken, M. The biological and philosophical definitions of life. 1975 Acta Biotheor. 24(1-2), 14.

Kuhlmann, M. Physicists Debate Whether the World is Made of Particles or Fields - or Something Else Entirely, 2013 Sci. Am., 303, August 1st.

Kuntner, M., Agnarsson, I., Are the Linnean and Phylogenetic Nomenclatural Systems Combinable? Recommendations for Bio-logical Nomenclature. 2006 Systematic Biology, 55(5), 774.

Liu, Q., Herring, C.A., Sheng, Q., Ping, J., Simmons, A.J., Chen, B., et al. Quantitative assessment of cell population diversity in single cell landscapes. 2018 PLoS Biol. 16(10), e2006687.

McKay, C.P. What is life–and how do we search for it in other worlds? 2004 PLoS Biol. 2(9), E302.

Miller, J.G. Living systems: The organization. 1972 Behavioral Science, 17(1), 1.

Moss, L. Normativity, Autonomy, and Agency: A Critical Review of Three Essays on Agency in Nature, and a Modest Proposal for the Road Ahead. 2024 Biol Theory 19 73.

Müller-Merbach, H. Heraclitus: philosophy of change, a challenge for knowledge management? 2006 Knowl. Manage. Res. Pract. 4, 170.

Musacchio, J.M. Why Do Qualia and the Mind Seem Nonphysical? 2005 Synthese, 147(3) 425.

Paksi, D. The concept of boundary conditions. 2014 Polanyiana, 1-2(23), 5.

Persig, R.M. Zen and the Art of Motorcycle Maintenance, 2014, Vintage Books, London.

Polanyi, M. Life's irreducible structure. 1968 Science, 160(3834) 1308.

Pross, A., Pascal, R. The origin of life: what we know, what we can know and what we will never know. 2013 Open Biol. 3, 120190.

Russell, B., History of Western Philosophy (2nd ed.), Routledge, London, 1991.

Sagan, C. The Cosmic Connection, Anchor Press/Doubleday, New York, 1973.

Schrödinger, E. What Is Life? The Physical Aspect of the Living Cell. Cambridge University Press, Cambridge, 1944.

Sharma, A., Czégel, D., Lachmann, M., Kempes, C.P., Walker, S.I., Cronin, L. Assembly theory explains and quantifies selection and evolution. 2023 Nature 622, 321.

Stolz, J.F. Gaia and her microbiome. 2017 FEMS Microbiology Ecology, 93(2), fiw247, https://doi.org/10.1093/femsec/fiw247

Trifonov, E.N. Vocabulary of Definitions of Life Suggests a Definition, 2011 J. Biomol. Str. & Dynam., 29(2), 259.

Tsokolov, S.A. Why is the definition of life so elusive? Epistemological considerations. 2009 Astrobiology. 9(4), 401.

Weber, B.H. What is life? Defining life in the context of emergent complexity, 2010 Orig Life Evol Biosph. 40(2), 221.

Westland, J.C. Information loss and bias in Likert survey responses. 2022 PLoS ONE 17(7), e0271949.

Wilkinson, L.A. Systems Theory. In: Goldstein, S., Naglieri, J.A. (eds) Encyclopedia of Child Behavior and Development. Springer, Boston, MA. 2011.

Wolfram, S. Cellular automata as models of complexity. 1984 Nature 311, 419.

Wong, M.L., Cleland, C.E., Arend Jr., D., Bartlett, S., Cleaves II, H.J., Demarest, H., Prabhua, A., Lunine, J.I., Hazen, R.M. On the roles of function and selection in evolving systems. 2023 PNAS, 120(43), e2310223120.

(Toward) Discovery of Life beyond Earth and its Impact
Proceedings IAU Symposium No. 387, 2026
H. Landt, M. Dominik & C. Oliver, eds.
doi:10.1017/S1743921324001662

Conway's Game of Life as an Analogue to a Habitable World: Livingness beyond the Biological

James McCrum[1] **and Terence P. Kee**[2]

[1]School of Physics and Astronomy, University of St Andrews, St Andrews, UK
email: jm441@st-andrews.ac.uk

[2]School of Chemistry, University of Leeds, Leeds, UK
email: t.p.kee@leeds.ac.uk

Abstract. Conway's Game of Life is a cellular automaton noted for its rich, complex, and emergent behaviour, which seems qualitatively 'lifelike'. It exists within a wider space of different rulesets of cellular automata, none of which have been found to display behaviours that seem as rich as Conway's selected example. We present here a set of three quantitative tests for 'lifelike' behaviour, based on the critical brain theory, Shannon's theory of information entropy and integrated information theory, all of which are successfully able to select Conway's Game of Life as an outlier within this set, which is a non-biological analogue to the selection of a habitable planet or universe amongst a wider space of similar settings that cannot support the same kinds of living systems.

Keywords. Cellular Automaton, Ruleset, Criticality, Entropy, Life, Livingness, Integration, Information

1. Introduction

Life is an exceptionally difficult concept to pin down with specificity and validity. The exact parameters of what constitutes life and its distinctive behaviours and attributes are subject to immense debate and controversy, and represent one of the most important unsolved problems in the interdisciplinary sciences. (Cleland & Cheba 2010; Cornish-Bowden & Cárdenas 2019; McKay et al. 2004)

In light of this, those trying to solve this problem may wish to start with a simple representation of a lifelike system in order to try and build up from first principles what we mean when we say that a behaviour is lifelike, and therefore gain some insight into what we mean when we talk, in general, about life.

John Von Neumann and others created the most basic and influential way to represent a simplistic 'lifeform'; cellular automata, a broad set of variable and versatile programs. (von Neumann & Burks 1966)

'Automata' here refers to the idea that the program acts as a simulacrum of a biological organism, without itself being biological. Cellular refers to the fact that the particular feature of organic systems that the structure of the program emulates is modularity.

One may think of how a neural network creates a simplified automaton of a brain by selecting out of its many possible emulatable characteristics the specific characteristic of network structure and emulating this; a cellular automaton is similar in that it essentially reduces life to inter-cellular interactions, and emulates only this structure.

Each cell has some finite number of discrete states as well as some finite set of neighbours to which it connects, and evolves in discrete generations according to its own state and the states of its neighbours.

This is an absolutely bare-bones representation of how a living system functions, and yet they can be incredibly complex and display emergent behaviour that appears subjectively lifelike.

Even more striking than the initial cellular automata of the 1940s - which are relatively complicated in structure - is Conway's Game of Life (Gardner 1970), developed in the 1970s by Cambridge student John Conway (Izhikevich et al. 2015), who would go on to become one of the most influential mathematicians of the late 20th century, with many further accomplishments beyond this.

Conway's Game of Life (CGOL) is striking because it is simultaneously one of the simplest cellular automata in structure and yet, one of the most complex in behaviour. (Berkelamp et al. 2004) (Schiff 2011). As the name suggests, CGOL is exceptionally subjectively lifelike, and yet the basic essence of the program can be boiled down to two short binary strings.

Each cell has two states, 'dead' or 'alive', so functions as a bit. Each cell carries out a simple logical operation on its 8 neighbours on the simple 2D grid it occupies - referred to as its Moore Neighbourhood - and then reacts based on its own state. All it does is count how many of its neighbours are alive. If there are two or three live neighbours and it is alive, it remains alive. If there are three neighbours and it is dead, it becomes alive. In any other case it will be dead in the next generation.

This is relatively wordy spelled out in English, but if we write out two binary strings, one for a live cell and one for a dead cell, where the nth bit corresponds to n live neighbours and the state of that bit is 1 if this produces a live cell and 0 if it produces a dead cell, it reduces to:

$$(0, 0, 1, 1, 0, 0, 0, 0, 0)$$
$$(0, 0, 0, 1, 0, 0, 0, 0, 0)$$

In essence you have an incredibly concisely defined artificial 'genotype' that nonetheless produces a shockingly elaborate phenotype.

We can certainly say that in CGOL we have located a very simplistic representation of a life-like system. We can now ask what exactly it has in common with biological life - what quantitative traits produce its qualitative 'livingness'?

One core aspect of CGOL is that it is highly unpredictable. It is in general difficult to predict how patterns will evolve, which gives the system some sense of agency and independence. The first way to parametrise this unpredictability would be to say that the system is displaying **criticality**.

Criticality refers to the behaviour of a system very near a phase transition, such that small perturbations in the parameter(s) controlling the phase will shift the system to and from different phases, resulting in unpredictable behaviour. Familiar examples can be found in thermodynamics and in nuclear physics, but also in biology - the human brain is a critical system, and it has in fact been hypothesised that such criticality is essential to its proper functioning. (Tian et al. 2022) Furthermore, a key marker of criticality is fractality and self-similarity, which is also a key feature of many biological systems. (Azua-Bustos & Martínez 2013; Grosu et al. 2023)

Like living things, CGOL is critical, in that very small changes to an initial condition on the board can lead to wildly divergent results; it also displays the corresponding 'organic' fractal patterns. (Alstrøm & Leão 1994) (Bak et al. 1989)

Strikingly, CGOL's criticality directly parallels the brain. The brain alternates between phases where neural firing is amplified, leading to propagation and an explosion in the number of firing cells, and where it is ignored, leading to decay and a population collapse after a timelapse. (Tian et al. 2022) CGOL alternates between phases where an initial small pattern of living cells will grow to cover much of the board, or will disintegrate into nothingness, or will stabilise in size. (Wolfram 1984)

The second way to discuss CGOL's wild unpredictability is that it is **computationally irreducible**.

Computational irreducibility is a concept developed by Wolfram which, unlike criticality, was first derived from cellular automata like CGOL and then applied to biological systems like the brain. (Wolfram 2002) A computationally irreducible system is one that is unable to be approximated by a simpler system, so cannot be 'shortcut', but can nonetheless be deterministic in nature.

CGOL's computational irreducibility is a byproduct of its **Turing Completeness**.

The Turing Completeness property of CGOL is that one can build any computable program as a configuration in CGOL and then run said program by allowing the game to evolve from that configuration. If one could, given only the state of generation 1 of a game, predict the state of generation 10,000 one would be able in general to determine the outcome of such a program running. Turing showed that this - the Halting Problem - is undecidable. (Turing 1937) Therefore one cannot 'shortcut' the system - the only way to find the state of generation 10,000 from generation 1 is to evolve all of the intervening states in CGOL through.

Similarly, Wolfram argued that the brain is computationally irreducible, in that one could know the exact state of a brain at a given time, but be unable to predict its state some time later using any simple algorithm or rule. Even if the brain, he argues, evolves deterministically according to some exact computation like CGOL, the only way to find the state is to let it evolve through all the in-between states, much as one cannot 'skip' states in CGOL. He presents this as an argument for free will without non-determinism.

What's doubly striking here is that Conway located his specific eponymous game within a larger 'multiverse' of different games, not all of which are necessarily lifelike in this way, and that the reasoning he used to do was deeply tied to both of these models of its behaviour. Conway, as a Cambridge student, was interested in games and cellular automata, and wished to create a game that would be 'universal' - that is, Turing complete, able to simulate any program within its ruleset - and also unpredictable, in that it would not be trivial to determine how any given starting configuration will generally play out. He noted that he devised many prototype games over the course of 18 months, and discarded them all. He found them uninteresting, because they were highly predictable - initial patterns tended to either very predictably grow, or very predictably disintegrate. (Izhikevich et al. 2015)

Although he did not phrase it in these terms, what Conway was searching for here was a critical system - one that is not in the phase of 'population decay' or 'population growth', but at an unpredictable liminal point between these. He eventually found this, and also a Turing complete system - another related desire of his for his game - in 'Life'. (See fig 1a).

Although we cannot know what all of the rulesets Conway tested exactly were, we can easily define a space of rulesets that are similarly simple and adjacent to CGOL by simply modifying the population values in the Moore Neighbourhood that produce life or death. In practice that means that any two pair of nine-length binary strings can define a ruleset, which gives 2^{18} possible configurations. These are referred to as Lifelike Cellular Automata (LCA).

Now that we have defined a very simple representation of a living system, and a number of directly comparable representations that are not as strongly associated with life, we can

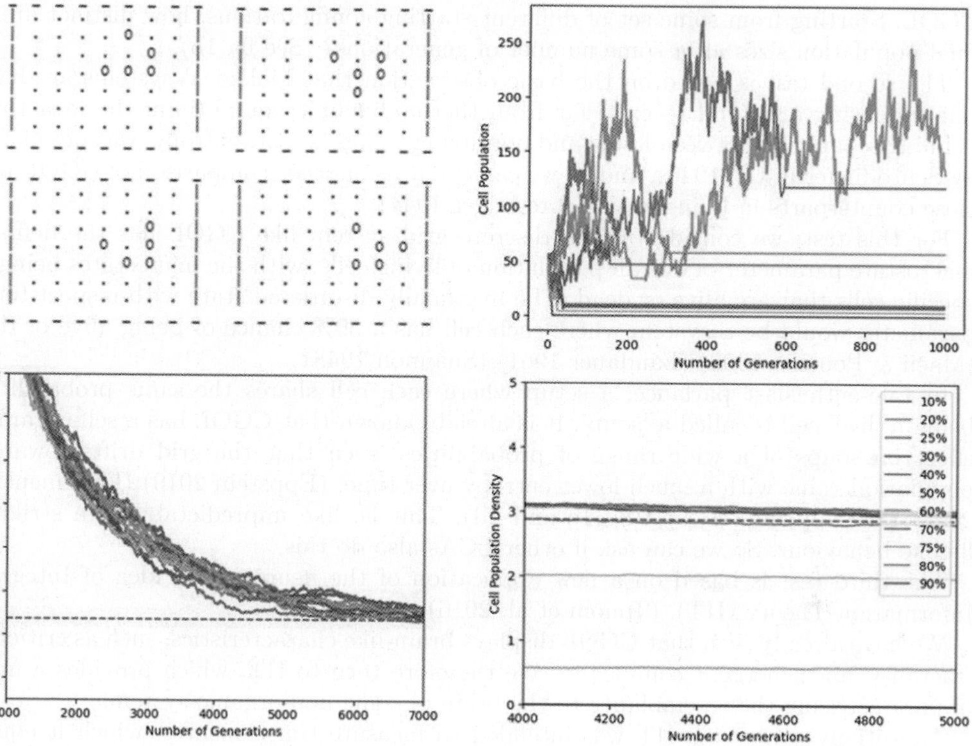

Figure 1. a) the 'Glider', a simple pattern in CGOL that carries information across the board, allowing for Turing Completeness. b) A set of random starting configurations in CGOL allowed to freely evolve; the difficulty in predicting when a given configuration will die out is a collorary of the Halting Problem. c) A set of maximally entropic 'soups' of 50% life probability in CGOL converging to an equilibrium; Flammenkamp's value of this is shown in red. d) The convergence of all but the most heavily populated soups to the Flammenkamp value.

ask how, exactly, CGOL objectively differs, in order to gain insight into how biological life differs from abiotic systems.

2. Methodology

We here define three objective tests to which we submit any LCA in order to compare it to CGOL.

Test 1) Given some set of initial starting conditions, what is the standard deviation of final population size compared to that of CGOL?

Test 2) Given a maximally disordered starting system, what is the disorder of the final system?

Test 3) What is the integration parameter, Φ, of this system?

We already know from Conway's own anecdotal account (Izhikevich et al. 2015) that many cellular automata - although not necessarily LCAs - have highly similar population values from a wide variety of different starting conditions, unlike CGOL, which is highly divergent from even minor variations in starting condition. So the first test serves as an easy way to compare the criticality/computational irreducibility of a system relative to

CGOL. Starting from some set of different starting configurations, how distinct are the final population sizes after some number of generations? (See fig 1b).

The second test is based on the basic observation that biological systems are highly self-organising and tend to exist far from thermodynamic equilibrium; the idea that a defining separation between living and non-living things is 'anti-entropy' was popularised by Schrodinger in the 1940s, and so we can use it as another comparison of CGOL to its close counterparts in livingness. (Schrodinger 1944)

For this test, we consider that a discrete grid system like CGOL has the definable macrostate parameter of overall population of living cells, with the microstates being the specific cells that are alive or dead. The maximally disordered state with respect to this parameter would be a system where each cell has a 50% chance of being alive or dead. (Macii & Poncino 1996) (Landauer 1961) (Shannon 1948)

In Life-enthusiast parlance, a setup where each cell shares the same probability of being a 'live' cell is called a 'soup'. It is already known that CGOL has a self-organising effect on soups of a wide range of probabilities, such that the grid drifts towards a population value with a much lower entropy over time. (Eppstein 2010) (Flammenkamp 2004) (See appendix 1, and fig 1c and 1d). This is, like unpredictability, a strikingly lifelike behaviour. So we can ask if other LCAs also do this.

The third test is based on a new application of the neuroscience idea of Integrated Information Theory (IIT). (Tononi et al. 2016)

We have already seen that CGOL displays brain-like characteristics, such as criticality, fractality, and emergent complexity. We therefore turn to IIT, which provides a means of parametrising the resemblance to the brain of some non-organic system.

As initially conceived, IIT was intended to measure consciousness, which it equates to the qualia of self-awareness. This is a highly controversial and convincingly disputed theory. (Doerig et al. 2019) (Merker et al. 2022)

It may, however, be possible to reinterpret Φ, the 'consciousness parameter', as a 'livingness parameter', if we reason that systems that are highly self-connected will tend to more closely resemble living systems, as these systems are able to self-regulate and to recover from damage more effectively in a manner selected for by evolution. This is by far the most speculative of the three tests. Φ can be seen as a measurement of the extent to which a system is indivisible, such that it cannot be defined merely as the sum of its parts. We test it here as a measurement of lifelike behaviour, using the PyPhi implementation. (Mayner et al. 2018)

Given that we already know that CGOL was selected by a human observer - Conway - as especially lifelike and interesting out of the set of different cellular automata, if these tests are able to correctly isolate CGOL as an outlier from the broader LCA set, then they can be said to be able to successfully parametrise the human intuition of 'livingness'.

3. Sampling

100 of the 2^{18} different LCAs were randomly selected by independently randomly generating binary strings in the range from 0 to 2^9 for the living and dead cell rulesets.

There are a number of different conventions in the literature used to refer to cellular automata of the lifelike class. Here, we use a novel one in which each cellular automaton is specified uniquely by a pair of decimal numbers between 0 and 2^9, where the two numbers are the decimal representations of their binary rules trings. For example, CGOL is written 96/32, which is just the decimal representation of 001100000 / 000100000.

The full list of 100 LCAs used in this study for the first two tests, named in this notation, is given in Appendix 2.

Each LCA was uniquely defined in two independent ways; on a box grid, where cells on, for example, the top left edge of the grid, have no neighbours above or to the left;

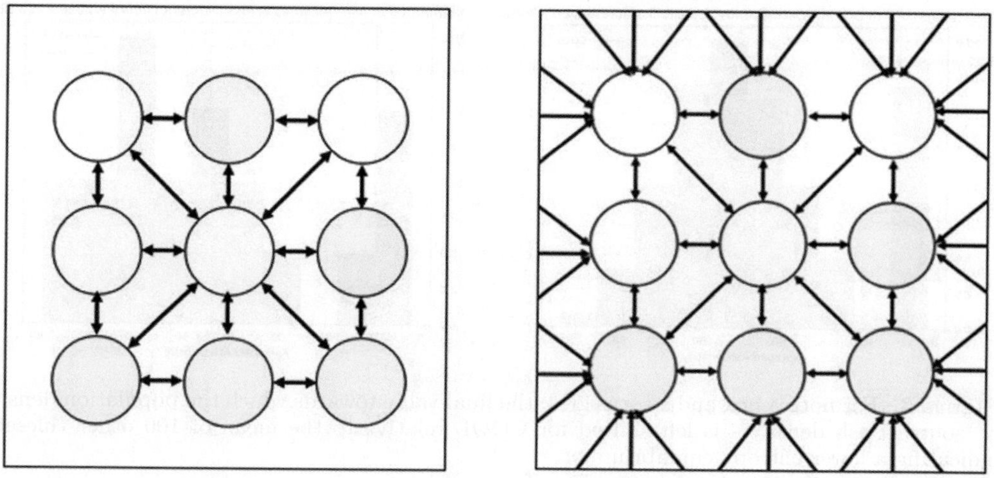

Figure 2. Network diagrams of a box grid (left) and a toroidal grid (right). Note that cells in the right image connect to cells on the opposite side of the 'grid', producing many more interconnections.

and on a toroidal grid, where cells on the top left edge of the grid neighbours cells on the bottom right edge of the grid. A toroidal grid wraps back on itself and maps to a torus - a box grid maps to a chess or Go board (the latter of which was used by Conway in his initial experiments).

For Test 1, five distinct 'soups', with each cell having 50% probability of being alive or dead, were randomly generated, evolved for 100 generations, then had their final population taken. This was carried out on both box and toroidal grids, both with 2048 by 2048 cells overall.

For Test 2, all of the 2^{16} different configurations of a binary 4 by 4 grid were tested as the initial central square of a 25 by 25 grid, and again evolved for 100 generations, both on box and torus.

For Test 3, only a 2 by 2 toroidal grid was used for each ruleset. This has 16 possible configurations, but only a few of these will give valid Φ values. At this stage it is necessary to justify the use of such a relatively small system for testing using IIT.

For the third test, a given ruleset first had to be conceptualised as a system of logic gates, which could then be converted into a Transition Probability Matrix (TPM). This TPM is the raw input of this ruleset into the IIT algorithm, and is an array of dimensions $n \times 2^n$, where n is the number of nodes in the system. For a square grid of side length s, n = s^2. This is inputted alongside the Connectivity Matrix (CM), which encodes the relevant grid structure, and has dimensions $n \times n$.

Just as these input values grow exponentially in size as the number of cells considered grows, the runtime of the PyPhi algorithm will also grow exponentially, going as $O(n \times 53^n)$. Furthermore, each distinct valid configuration of a set of cells - of which there are at maximum 2^n - will have its own distinct Φ value.

In practice, therefore, it was found that a maximum of n = 4, or a 2 by 2 grid, was the largest feasible sample that could be studied for each ruleset.

4. Results

All three tests were successfully able to select CGOL as an outlier in the LCA set, thus parametrising an intuitive sense of livingness, with very similar results on both box and torus grids for the first two tests.

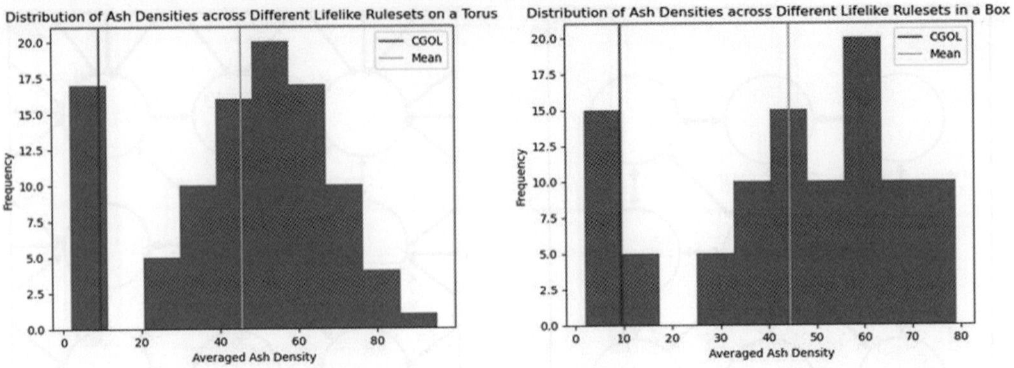

Figure 3. For both a box and a torus grid, the final value towards which the population density of 'soups' - ash density - is left-shifted for CGOL relative to the mean of 100 other rulesets, indicating a lower entropy equilibrium.

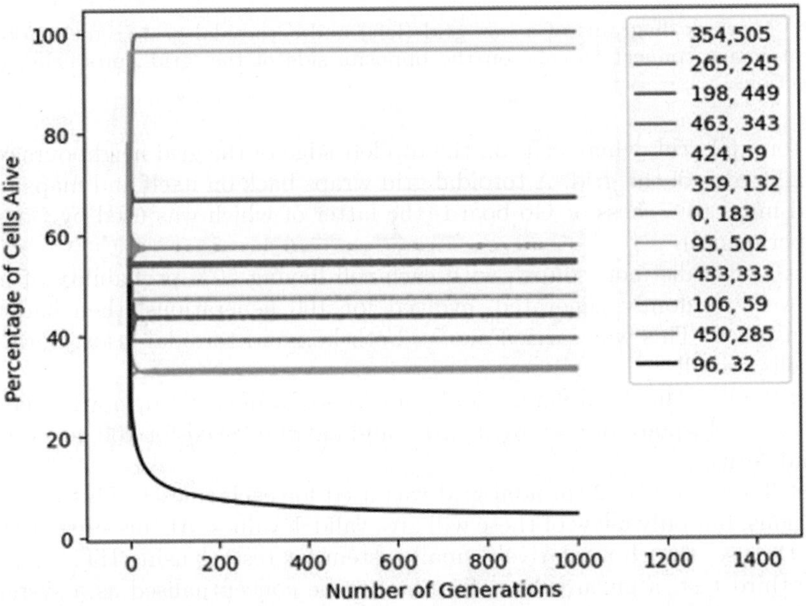

Figure 4. A set of different rulesets are allowed to evolve from soup to ash. The lowest line is CGOL, which is clearly again moving towards a lower entropy equilibrium than its counterparts.

CGOL was unusual in how low entropy its final state is after evolution from a maximum entropy 50% state, although most LCAs did tend to move away from thermodynamic equilibrium as they evolved, and it was not the lowest entropy of all states, which can be attributed partially to the fact that the minimum entropy state of a CA grid is empty, so that any predictably exploding or decaying ruleset will tend to produce a stronger anti-entropic effect by this metric. (Figure 3). This was also true on a much longer timescale. (Figure 4).

As expected, CGOL had a higher standard deviation of final population than the majority of other rulesets tested. (Figure 5). This means that given all the possible initial 'seeds' of different possible fillings of the 16 central cells, most rulesets evolved them in relatively predictable and convergent ways, unlike CGOL and a few other outliers, which evolved in divergent and chaotic fashion.

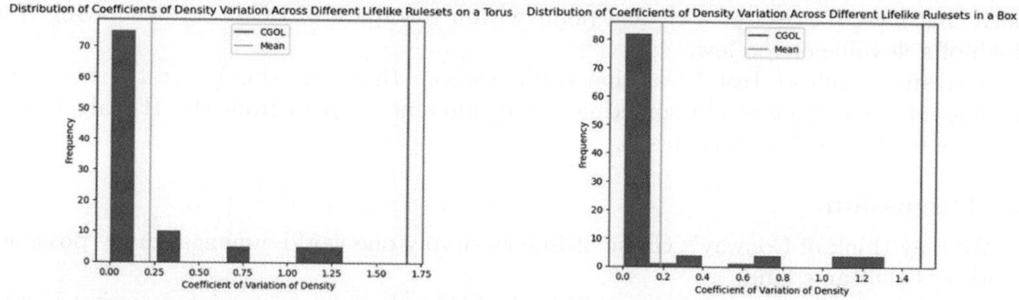

Figure 5. The divergence of different initial configurations is much higher - right-shifted - for CGOL than for the mean of other rulesets on both a torus and box.

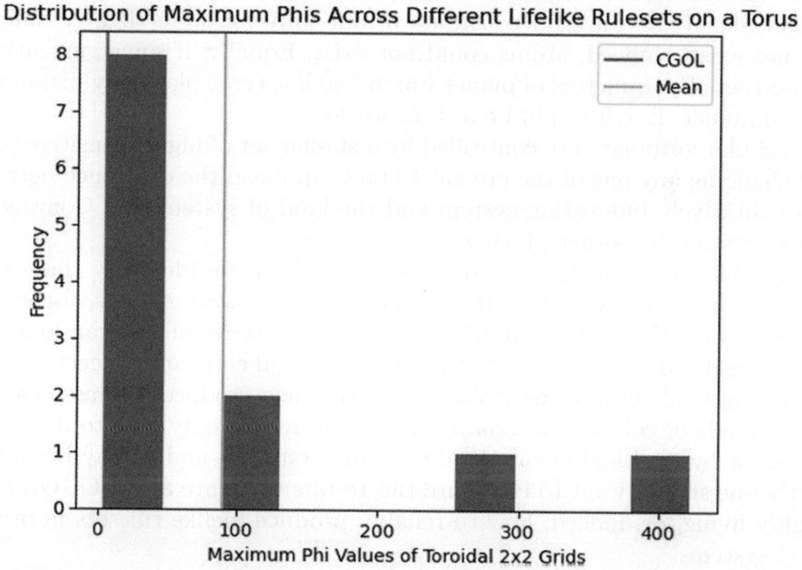

Figure 6. The a) minimum, b) mean, and c) maximum phi values of the different valid configurations of a 2 by 2 grid were all smaller for CGOL than for most other LCAs. d) The different configurations (white = dead, black = live) of a 2 by 2 grid and their outputs from the IIT algorithm for the CGOL ruleset.

It was also found that some rulesets displayed a surprising correspondence between ash and seed density. Of the 100 rulesets, the mean population density of soup after 100 generations - averaged over five runs - was within —0.3— % of the mean population density of a patch of 16 randomly generated cells - averaged over all its possible configurations - for 64. By way of comparison, this difference was around 7% for CGOL. It is unclear why this is the case.

Finally, the minimum, mean, and maximum Φ value of CGOL across its valid configurations was distinctly lower than the corresponding average value of the other LCAs tested (Figure 6).

This is interesting, as going by a naive implementation of the algorithm as originally conceived in IIT, higher Φ would correspond to a greater 'consciousness', or in this case livingness. However, connecting back to the idea of criticality, increasing the Φ of a real brain beyond a certain point would correspond to supercriticality, and in fact would represent a state of epileptic seizure. (Hesse & Gross 2014) This actually supports the idea that CGOL exists at a critical point within the space of LCA, neither having too high of a Φ value or too low.

A strange result of Test 3 was that configurations that were simply rotations or mirrorings of each other would sometimes give different outputs from the IIT algorithm (Figure 6). It's unclear why this is.

5. Discussion

We may think of Conway's Game of Life as simply one world, amongst many possible worlds of the same kind.

We can analogise its special place within the LCA ruleset to the special place that Earth occupies amongst known planets, or the special place that our 'fine-tuned' universe may theoretically occupy within a wider multiverse. (Goff 2024)

It is the configuration of a specific set of parameters that is most conductive to what we may call life. If the physical constants of our universe were slightly different, then life could not exist - indeed, atoms could not exist. Equally, if some currently not fully characterised set of parameters of planet Earth - stellar type, planetary distance, satellite size - were different, Earth would be a dead world.

Lifelike cellular automata are controlled by a similar set of highly sensitive parameters, such that changing any one of the crucial 18 bits can mean the difference between a very rich and qualitatively interesting system and the kind of system that Conway dismissed as too uninteresting to bother playing with.

Notably, unlike those analogue cases, we have very little idea how these parameters exactly relate to the behaviour of the system. We are aware of how, for example, the Earth's position relative to the Sun affects its living systems, and so can look for planets with similar positioning. We know that altering physical constants in certain ways would result in certain kinds of universes. While Wolfram has produced a classification system for different kinds of cellular automata, even Conway was only able to find the class he was looking for by trial and error. We do not understand - and it may be unknowable - how exactly one should want to configure the 18 bits to create a non-CGOL ruleset that seems highly living, or, indeed, how to reliably produce lifelike rulesets in other cellular automaton systems.

However, we can, in essence, identify 'biosignatures' that mark out habitable rulesets. Because we know that CGOL is a qualitatively lifelike system, we can hypothesise that it will have certain quantitatively lifelike traits, and then search for these traits across the space. In this study, we have identified three such abstract biosignatures - criticality, anti-entropic behaviour, and interconnectivity. The first two were definitively able to separate CGOL from similar rulesets; the third requires further study, but seems promising in this very early stage.

This is instructive both in the development of new ways to locate habitable, or inhabited, planets and worlds, and also in the use of non-biological systems as more than simply an analogue to a living system. CGOL is an excellent example of many different physical processes, and indeed, it is an excellent example of a kind of non-biological life in its own right. Its similarities are more than just surface deep; it has quantitative properties in common, not just a qualitative resemblance.

Further research is needed to characterise more of the lifelike ruleset of cellular automata, as well as to run similar tests on other cellular automata outside of this ruleset. Other rich and interesting rulesets may be found in this way, and the underlying markers of such systems understood more clearly. Furthermore, the tests for criticality and integration in addition to far-from-equilibrium behaviours can be applied to a wider variety of systems than relatively simple cellular automata, which stand here as a case in point to testify to their usefulness.

6. Acknowledgements

The authors would like to thank Gaelan Steele for her commentary and support during the early stages of this work, which had a great impact on inspiring and improving its path to its final form. The authors would like to thank the Life community for its tireless categorisation and study of the game and its variants for more than fifty years since its inception, especially Achim Flammenkamp, whose extensive experiments with it formed a core part of this publication. This paper would also of course be completely impossible without the work of John H Conway, and would be greatly lacking were it not for his detailed and insightful account of the process of creation of his work in cellular automata. May he rest in peace.

7. Appendix I

The Flammenkamp value of the attractor of soup population density for the CGOL ruleset - 2.87115% - is strangely close to Euler's number, e. This is not the only similar value for CGOL that is close to this constant.

If we start with a 50% population density grid and run one iteration of Conway's Game of Life, the population density will generally be around 27%. This experimental result can also be predicted using this equation:

$$P = \frac{1}{2}B\left(3, 8, \frac{1}{2}\right) + \frac{1}{2}B\left(2, 8, \frac{1}{2}\right) + \frac{1}{2}B\left(3, 8, \frac{1}{2}\right)$$

$$P = \frac{1}{2}(0.21875) + \frac{1}{2}(0.10938) + \frac{1}{2}(0.21875) = 0.27344$$

Recursively applying this equation with a starting value of 0.5 for CGOL gives a limiting value of approximately $\frac{1}{e}$.

8. Appendix II

Table 1. Rulesets used in this paper.

448/131	221/151	333/329	398/165	153/487	080/280	323/000	244/130	292/361	509/212
286/117	234/381	002/484	369/366	347/116	480/505	053/364	132/413	404/024	477/354
411/168	203/496	405/247	218/059	098/413	242/341	074/170	500/368	096/139	325/193
438/370	148/278	066/044	253/355	174/493	494/194	134/122	119/237	012/244	289/185
453/222	358/008	069/440	051/481	036/369	438/040	202/411	234/500	483/389	187/478
338/404	116/491	450/322	266/128	446/449	150/337	376/054	160/192	183/374	331/314
161/043	454/031	387/367	317/008	302/403	209/486	312/113	052/268	261/323	370/443
191/163	267/315	325/077	162/274	473/160	146/127	361/250	347/476	293/350	480/474
426/167	186/307	071/104	292/148	181/411	020/296	308/456	501/212	287/279	446/388
134/028	342/320	095/121	495/188	478/281	229/273	433/387	479/485	484/024	153/350

References

Alstrøm P. & Leão J., "Self-organized criticality in the "game of Life"," Physical Review E, vol. 49, no. 4, 1994.

Azua-Bustos A., Vega-Martínez C. The potential for detecting 'life as we don't know it' by fractal complexity analysis. 2013 International Journal of Astrobiology. 12(4), 314.

Bak, P., Chen K., & Creutz M., "Self-organized criticality in the 'Game of Life'," Nature, vol. 342, pp. 780–782, 1989.

Berkelamp, E., Conway, J.H., & Guy R., Winning Ways for your Mathematical Plays, 2nd ed., AK Peters Ltd, 2004.

Cleland, C. & Chyba, C. (2010). Does 'life' have a definition?. Planets and Life: The Emerging Science of Astrobiology. 119–131. 10.1017/CBO9780511730191.032.

Cornish-Bowden, A. & Cárdenas, M. (2019). Contrasting Theories of Life: Historical Context, Current Theories. In search of an ideal theory. Biosystems. 188. 104063. 10.1016/j.biosystems.2019.104063.

Doerig, A., Schurger, A., Hess, K., & Herzog M.H, "The unfolding argument: Why IIT and other causal structure theories cannot explain consciousness," Consciousness and Cognition, vol. 72, pp. 49–59, 2019.

Eppstein, D., "Growth and decay in life-like cellular automata," in Game of Life Cellular Automata, A. Adamatzky, Ed., Springer, 2010, pp. 71–98.

Flammenkamp, A., "Most seen naturally occurring ash objects in Game of Life," 2004. [Online]. Available: `https://wwwhomes.uni-bielefeld.de/achim/freq_top_life.html`. [Accessed 25th May 2024].

Gardner, M. "The fantastic combinations of John Conway's new solitaire game "life"," Scientific American, Mathematical Games, vol. 223, no. 4, pp. 120–123, 1970.

Goff, P. "Is the fine-tuning evidence for a multiverse?". Synthese 204, 3 (2024). https://doi.org/10.1007/s11229-024-04621-z

Grosu, G.F., Hopp, A.V., Moca, V.V., Bârzan, H., Ciuparu, A., Ercsey-Ravasz, M., Winkel, M., Linde, H., Mureşan, R.C The fractal brain: scale-invariance in structure and dynamics. 2023 Cerebral Cortex, 33(8), 4574.

Hesse J, Gross T. Self-organized criticality as a fundamental property of neural systems. Front Syst Neurosci. 2014 Sep 23;8:166.

Izhikevich, E. M. Conway, J. H. & Seth, A. "Game of Life," 2015. [Online]. Available: `http://www.scholarpedia.org/article/Game_of_Life`. [Accessed 26th May 2024].

Landauer, R. "Irreversibility and Heat Generation in the Computing Process," IBM Journal of Research and Development, vol. 5, no. 3, pp. 183–191, 1961.

Macii, E & Poncino, M. "Exact Computation of the Entropy of a Logic Circuit," Proceedings of the Sixth Great Lakes Symposium on VLSI, 1996.

Mayner, W., Marshall, W., Albantakis, L., Findlay, G., Marchman, R., & Tononi, G., "PyPhi: A toolbox for integrated information theory," PLOS Computational Biology, vol. 14, no. 7, 2018.

McKay, C.P. What is life–and how do we search for it in other worlds? 2004 PLoS Biol. 2(9), E302.

Merker, B., Williford, K., & Rudrauf, D., "The integrated information theory of consciousness: a case of mistaken identity," Behavioural and brain sciences, 2022.

Schiff J., Cellular Automata: A Discrete View of the World, Wiley & Sons, Inc, 2011.

Schrödinger, E. What Is Life? The Physical Aspect of the Living Cell, Cambridge University Press, 1944.

Shannon, C.E., "A Mathematical Theory of Communication," Bell Systems Technical Journal, vol. 27, pp. 379–423, 1948.

Tian, Y., Tan, Z., Hou, H., Li, G., Cheng, A., Qiu, Y., Weng, K., Chen, C., & Sun, P., "Theoretical foundations of studying criticality in the brain," Network Neuroscience, vol. 6, no. 4, pp. 1148–1185, 2022.

Tononi, G., M. Boly, M. Massimini and C. Koch, "Integrated Information Theory: From consciousness to its physical substrate," Nature Reviews Neuroscience, 2016.

Turing, A.M, "On Computable Numbers, with an Application to the Entscheidungsproblem," Proceedings of the London Mathematical Society, Vols. s2-42, no. 1, pp. 230–265, 1937.

von Neumann, J. & Burks, A. W., Theory of Self-Reproducing Automata., University of Illinois Press, 1966.

Wolfram, S., "Universality and complexity in cellular automata," Physica D: Nonlinear Phenomena, vol. 10, no. 1-2, pp. 1–35, 1984.

Wolfram, S., A New Kind of Science, Wolfram Media, 2002.

(Toward) Discovery of Life beyond Earth and its Impact
Proceedings IAU Symposium No. 387, 2026
H. Landt, M. Dominik & C. Oliver, eds.
doi:10.1017/S174392132400231X

Broadband linear polarimetry of exoplanet Upsilon Andromedae b: Constraints on the orbital and physical parameters

Yasir Abdul Qadir[1], Andrei V. Berdyugin[1], Vadim Kravtsov[1], Vilppu Piirola[1], Takeshi Sakanoi[2], Masato Kagitani[2] and Juri Poutanen[1]

[1]Tuorla Observatory, Department of Physics and Astronomy, FI-20014 University of Turku, Finland
e-mail: `yasir.abdulqadir@utu.fi`

[2]Graduate School of Sciences, Tohoku University, Aoba-ku, 980-8578 Sendai, Japan

Abstract. *Context.* Modern polarimeters enable precise measurements of exoplanetary systems, providing valuable constraints on orbital geometry and atmospheric scattering properties.

Aims. We aim to investigate the orbital parameters of the non-transiting Jupiter-like exoplanet υ And b, which orbits close to its host star, using polarized scattered light observed with the DiPol-2 polarimeter on a 60 cm telescope.

Methods. Over nearly three years, we collected high-precision polarimetric data of the υ And system. Using Lomb-Scargle periodograms, we identified variability close to the planetary orbital phase and modeled the signal with the Rayleigh-Lambert approximation.

Results. A weak polarimetric signal at half the orbital period of υ And b was detected. The V-band MCMC fits yielded an orbital inclination of $i = 28^{+88^\circ}_{-18}$, an argument of periastron $\omega = 102^{+139^\circ}_{-45}$, a longitude of ascending node $\Omega = 107^{+134^\circ}_{-43}$, and a geometric albedo of $p = 19^{+9}_{-8}\%$. Although uncertainties remain large, the inclination result agrees with earlier estimates, supporting a low-inclination, non-transiting geometry.

Conclusions. While the detection is not definitive, these results demonstrate the capability of high-precision polarimetry to probe the orbital parameters of non transiting exoplanets. Further observations with larger telescopes and improved sensitivity will be required to refine the constraints on the orbital parameters.

Keywords. planets and satellites: atmospheres – planets and satellites: individual: Upsilon Andromedae b – polarization – scattering– techniques: polarimetric

1. Introduction

Upsilon Andromedae b (υ And b) is the innermost exoplanet in the planetary system of Upsilon Andromedae. This exoplanet can be categorized as a close-in extrasolar giant planet (CEGP, or hot Jupiter), as it orbits its host star with an orbital period of just above 4.6 d and has a mass of about 1.7 M_J (Piskorz et al. 2017). It is one of the earliest discovered exoplanets detected using Doppler spectroscopy (Butler et al. 1997). The host star is an F8 dwarf and hosts another three exoplanets (McArthur et al. 2010). The υ And b is a non-transiting exoplanet and therefore by using polarimetry, we can potentially reveal orbital parameters, e.g., orbit inclination, which are not possible by the use of other observational techniques. The υ And b due to its size and location in the sky has been considered as one of the most promising targets that is expected to show high degree of polarization mainly caused by Rayleigh scattering.

Most exoplanets have been discovered by either the transit method or Doppler spectroscopy. Therefore, exoplanetary studies are mainly limited to transiting planets. However, some exceptions such as υ And b and a few other CEGPs have been directly detected by variable infrared flux (Harrington et al. 2006; Crossfield et al. 2010). The first exoplanet that was studied using optical polarimetry is HD189733b (Berdyugina et al. 2008, 2011). From the data analysis of polarimetric observations, we can obtain the orbital inclination and by using that we can derive the true mass, radius and hence density of an exoplanet. Furthermore, polarimetry of exoplanets can be used to obtain the geometrical albedo and by using this information, we can deduce the atmospheric composition of an exoplanet. (Seager & Whitney 2000; Stam et al. 2004; Fluri & Berdyugina 2010; Berdyugina et al. 2008, 2011).

An exoplanetary polarimetric signal even in the case of a CEGP such as υ And b is very weak, at the level of $\leq 10^{-5}$ at most, and thus requires highly sensitive polarimeters to detect the signal. The small degree of polarization arising due to scattering from the atmosphere of an exoplanet can get easily lost in either the noise due to the observational errors or irregular/semiregular polarization signal arising from the host star. Our polarimetric observations of υ And b has clearly demonstrated this difficulty. In this paper, we are reporting the methods and techniques that we used to model our data in order to deduce the orbital parameters and physical properties of υ And b.

2. Observations

We observed υ And b during 114 nights from 2016 October 25 to 2019 January 16 (Table 1) using DiPol-2 polarimeter installed at the 60 cm Tohoku 60 (T60) telescope at Haleakalā Observatory, Hawaii. The host star of the planetary system υ And is a bright target with an apparent magnitude of 4.60 and thus visible to human eye. At the Haleakalā summit, υ And is well visible from September through March. Skyflat images that are used for calibration purposes were regularly taken at the time of twilight hours, either at the beginning of observing night or at the dawn. Once per night, the series of dark and bias images were also taken.

DiPol-2 is a highly sensitive polarimeter capable of measuring polarization with precision better than 10-5. It splits in-coming light into B, V, and R passbands using dichroic beam-splitters, and the light is simultaneously recorded by three CCD cameras. A rotatable polarization modulator with a super-achromatic waveplate allows measurements across 400 to 800 nm. To minimize errors from atmospheric changes, the calcite plate generates two orthogonally polarized images, canceling sky polarization. The defocusing technique avoids pixel saturation, spreading the stellar image across more pixels. Remote operation is possible via a connected PC. For linear polarization, a typical measurement cycle includes 16 exposures, and for circular polarization, 4 exposures. Polarization accuracy reaches 0.002%–0.015% across different passbands, and standard stars are regularly observed for calibration.

Data reduction involves bias, dark frame subtraction, and flat fielding, followed by aperture photometry to determine intensities of orthogonally polarized images. Stokes parameters q and u are computed using Fortran scripts, with weights adjusted based on deviations from the expected noise. Instrumental polarization is corrected using zero-polarized standard stars, with accuracy around 0.002%–0.003%. Polarization for υ And is corrected for both instrumental polarization and angle zero-point, determined from highly polarized standard stars. Though inter-stellar polarization for υ And is negligible due to its proximity, time-dependent trends in the data were removed before analysis.

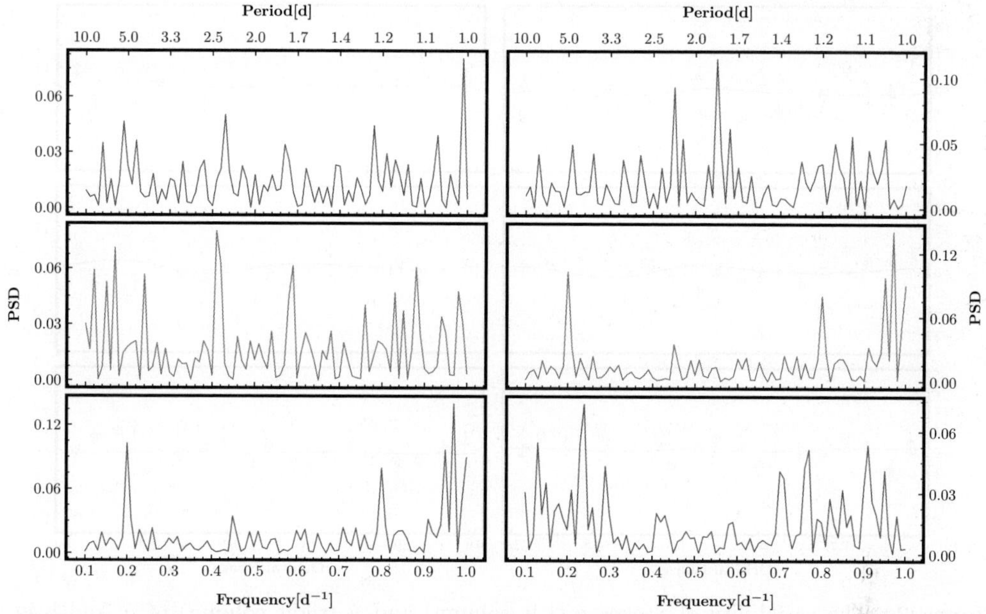

Figure 1. Lomb-Scargle periodograms for Stokes q (left column) and u (right column) of υ And in B, V, and R passbands (top, middle, and bottom panels respectively).

Table 1. Log of polarimetric observations for υ And b.

Dates	MJD	$N_{\rm obs}$
2016 October – 2016 November	57686 – 57701	14
2017 September – 2018 January	58025 – 58140	58
2018 October – 2019 January	58406 – 58499	42

3. Data Analysis

3.1. *Period Search*

To identify periodic signals in the polarimetric data of υ And, we applied the Lomb-Scargle algorithm (Lomb 1976; Scargle 1982) using the spectral packages in R programming language. The algorithm is effective for unevenly sampled data, making it well-suited for astronomical observations. Figure 1 shows Lomb-Scargle periodograms for Stokes q and u in the B, V, and R passbands.

While the periodograms appear noisy due to the sensitivity limitations of polarimeters, peaks around 2.3 days—half of the orbital period of υ And—are discernible in the B and V passbands, albeit with low power spectral density (PSD). Some peaks, particularly those around 1.0 day, may result from nightly observations conducted at consistent times. Alias peaks at around 1.8 days also appear due to the sampling frequency not meeting the Nyquist criterion Abdul Qadir et al. (2025).

3.2. *Orbital Parameters*

To model the observed Stokes q and u variations, we applied the Rayleigh–Lambert (RL) approximation (Wiktorowicz 2009; Fluri & Berdyugina 2010), which assumes scattering by condensates in a Lambertian atmosphere. This approach enables constraints on orbital inclination (i), argument of periastron (ω), longitude of the ascending node (Ω), and geometric albedo (p), while also accounting for a constant polarization offset Z.

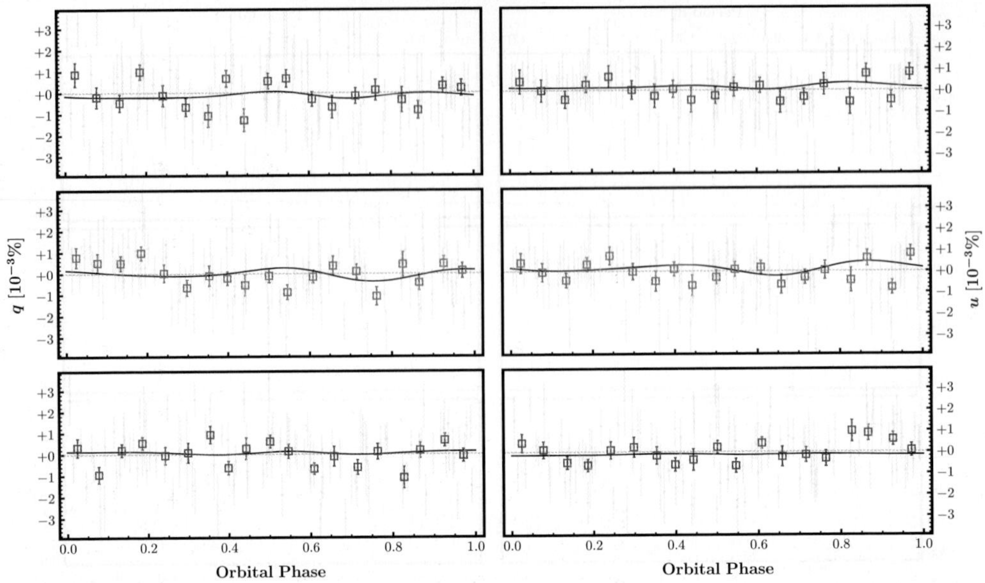

Figure 2. The variability of Stokes q (left column) and u (right column) of υ And b in B, V, and R passbands (from top to bottom), phase-folded at the orbital period of 4.62 d. The filled squares are the data points with error bars that correspond to the average values of the individual observations and the standard errors of the mean calculated for the orbital phase bin size of 0.053. The black solid line corresponds to the best-fit parameters given in Table 2.

Table 2. Best-fit orbital parameters for υ And b.

Passband	$R_{\mathrm{RL}}[R_{\mathrm{J}}]^a$	$i[\mathrm{deg}]^b$	$\omega[\mathrm{deg}]^b$	$\Omega[\mathrm{deg}]^b$	$p[\%]^b$	$Z_{\mathrm{q}}[\%]^b$	$Z_{\mathrm{u}}[\%]^b$	χ^2/d.o.f.
B		72^{+26}_{-38}	117^{+69}_{-80}	105^{+68}_{-85}	16^{+17}_{-8}	$0.000214^{+0.000129}_{-0.000218}$	$-0.000116^{+0.000164}_{-0.000092}$	466/222
V	1.6	28^{+88}_{-18}	102^{+139}_{-45}	107^{+134}_{-43}	19^{+9}_{-8}	$0.000003^{+0.000077}_{-0.000112}$	$-0.000123^{+0.000079}_{-0.000124}$	412/222
R		87^{+30}_{-46}	162^{+90}_{-104}	171^{+139}_{-126}	9^{+12}_{-5}	$-0.000137^{+0.000281}_{-0.000087}$	$0.000150^{+0.000076}_{-0.000252}$	423/222

Notes. $^{(a)}$ Fixed parameter. $^{(b)}$ Uncertainties represent 16th/84th percentiles ($\pm\sigma$ equivalent).

The data were phase-folded using $T0 = 2450034.058 \pm 0.3$ JD and $P = 4.61711 \pm 0.00018$ d (Deitrick et al. 2015), and binned at $\Delta\phi = 0.053$ with inverse-variance weighting to suppress noise. We adopted a planetary radius of 1.6 R_J, a compromise between earlier estimates of 1.3 R_J (Crossfield et al. 2010) and ~1.8 R_J (Deitrick et al. 2015). Separate fits were performed in the B, V, and R passbands, since Rayleigh scattering amplitudes are ex-pected to decrease with increasing wavelength. The resulting fits (Fig. 2) show weak signals, with poor agreement in R where the curves are nearly flat.

To mitigate overfitting, we included both a 1% relative error floor and an additional white noise term f in the likelihood. MCMC analysis with 50 walkers and 10,000 steps yielded mean acceptance fractions of 0.20 (B), 0.25 (V), and 0.15 (R). Priors were uniform in i [0°, 180°], ω, Ω[0°, 360°], and p [1, 100]%, with offsets Zq, Zu ~[-0.001%, +0.005%]. Initial guesses were i = 25°, ω = 44°, Ω= 60°, and p = 37%. Posterior distributions (Table 2) indicate very broad uncertainties, especially in B and R. Only the V band produced a plausible inclination of $28^{+88°}_{-18}$, consistent with the $24° \pm 4°$ value reported by Piskorz et al. (2017).

4. Discussion

Our period search confirms that polarimetric variability in υ And is consistent with the planet's orbital period, marking an important step toward orbital characterization in a non-transiting system. Polarimetry provides a unique probe of orbital inclination, which cannot be derived from radial velocity measurements alone (Seager & Whitney 2000; Fluri & Berdyugina 2010). The V-band detection stands out as the most credible result, with a posterior inclination in line with previous dynamical estimates.

Nevertheless, the uncertainties in our fits are substantial. The B- and R-band solutions yielded inclinations of $72^{+26°}_{-38}$ and $87^{+30°}_{-46}$, respectively, values inconsistent with the non-transiting nature of the system. The posterior distribution for i in the V band shows a primary peak around 28°but also exhibits secondary maxima near 85° and 170°. These secondary peaks, likely introduced by noise and degeneracies in the RL model, inflate the error margins and highlight the limits of the current dataset.

A key challenge arises from the low polarimetric amplitudes and possible contamination by stellar activity. The presence of starspots, which can produce polarization signatures at levels comparable to or larger than the planetary signal Kostogryz et al. (2015). This contamination likely dilutes the true reflected-light signature and drives the low values of p we report. Since no reliable prior albedo estimates exist for υ And, our results should be treated as lower limits rather than physical measurements.

Simulations using the RL model with our phase coverage and uncertainties confirmed that the high scatter masks much of the planetary signal. Importantly, they revealed no intrinsic low-inclination bias, supporting the plausibility of the V-band inclination result. However, the simulations also reinforced that our current data quality is insufficient to resolve degeneracies among orbital parameters.

Improving polarimetric constraints will require either significantly longer nightly integrations—on the order of six hours compared to the 1.5 hours obtained here—or the use of larger aperture telescopes. Previous attempts with the 1 m Calern telescope yielded limited gains due to weather conditions, but future prospects include deploying DiPol-2 on the 1.27 m Greenhill Observatory telescope. In the long term, space-based polarimetry may be essential for overcoming atmospheric systematics and achieving the precision required for robust orbital solutions.

5. Conclusions

This study demonstrates the potential of ground-based polarimetry to constrain orbital parameters of close-in giant exoplanets. Using DiPol-2 on a 60 cm telescope, we detected weak polarimetric variability from υ And consistent with the planetary orbital period:

- Lomb–Scargle periodograms revealed half-period peaks in Stokes q across all passbands, and in Stokes u for B and V, though with high FAPs.
- Bayesian MCMC fits indicated that only the V-band solution ($i = 28^{+88°}_{-18}$) is consistent with previous results (Piskorz et al. 2017); B and R bands gave implausible inclinations.
- Geometric albedo estimates were low, likely suppressed by stellar activity, preventing robust atmospheric conclusions.
- Future progress will require longer integrations or larger telescopes, with space-based polarimetry offering the most promising long-term path.

While the uncertainties remain large, these results provide valuable groundwork for future polarimetric studies of non-transiting hot Jupiters. With improved data quality, polarimetry could become a powerful complementary technique for probing orbital geometry and scattering properties in exoplanetary atmospheres.

References

Abdul Qadir, Y., Berdyugin, A. V., Piirola, V., et al. 2025, A&A, 697, A133

Berdyugina, S. V., Berdyugin, A. V., Fluri, D. M., & Piirola, V. 2008, ApJ, 673, L83

Berdyugina, S. V., Berdyugin, A. V., Fluri, D. M., & Piirola, V. 2011, ApJ, 728, L6
Butler, R. P., Marcy, G. W., Williams, E., Hauser, H., & Shirts, P. 1997, ApJ, 474, L115
Crossfield, I. J. M., Hansen, B. M. S., Harrington, J., et al. 2010, ApJ, 723, 1436
Deitrick, R., Barnes, R., McArthur, B., et al. 2015, ApJ, 798, 46
Fluri, D. M. & Berdyugina, S. V. 2010, A&A, 512, A59
Harrington, J., Hansen, B. M., Luszcz, S. H., et al. 2006, Science, 314, 623
Kostogryz, N. M., Yakobchuk, T. M., & Berdyugina, S. V. 2015, ApJ, 806, 97
Lomb, N. R. 1976, Ap&SS, 39, 447
McArthur, B. E., Benedict, G. F., Barnes, R., et al. 2010, ApJ, 715, 1203
Piskorz, D., Benneke, B., Crockett, N. R., et al. 2017, AJ, 154, 78
Scargle, J. D. 1982, ApJ, 263, 835
Seager, S. & Whitney, B. 2000, in ASP Conf. Ser., Vol. 212, From Giant Planets to Cool Stars, ed. C. A. Griffith & M. S. Marley (San Francisco: Astron. Soc. Pac.), 232
Stam, D. M., Hovenier, J. W., & Waters, L. B. F. M. 2004, A&A, 428, 663
Wiktorowicz, S. J. 2009, ApJ, 696, 1116

(Toward) Discovery of Life beyond Earth and its Impact
Proceedings IAU Symposium No. 387, 2026
H. Landt, M. Dominik & C. Oliver, eds.
doi:10.1017/S1743921324001637

JCMT-Venus: Monitoring HDO Absorption in Venus' Mesosphere

Wei Tang

Imperial College London
email: wei.tang16@imperial.ac.uk

Abstract. In the JCMT-Venus project, we conducted three multi-week disc-integrated observations of Venus' atmosphere using the JCMT 'Ū'ū receiver. This paper focuses on the HDO 2(2,0)-3(1,3) transition line at 266.1611 GHz, employed as a proxy for water abundance under the assumption of a constant D/H ratio. Based on the HDO line depth, we observed significant and rapid short-term variations, and have tentatively identified potential long-term effects. Water vapor abundance values were retrieved daily, and their variations will be displayed and discussed in our forthcoming publications.

Keywords. Venus, Mm/submm astronomy, Atmospheres

1. Introduction

Water is an essential component for life and plays a vital role in driving the photochemical processes within Venus' mesosphere (60-120 km altitude). Despite its significance, our understanding of water vapour's variability in Venus' mesospheric layers remains limited. Previous observations have been constrained by either brief durations or coarse sampling, which hinders the identification of long-term effects and the characterisation of short-term variations (Sandor and Clancy 2005; Chamberlain *et al.* 2020). The JCMT-Venus project enabled extensive observations of the integrated disc during three multi-week campaigns, which took place from 1–20 February 2022, 6–27 July 2023, and 24 August to 22 September 2023, yielding 191, 122, and 114 observations, respectively (Clements *et al.* 2024). In our research, we utilised deuterated water (HDO) as a proxy for assessing water vapor abundance, assuming a constant deuterium to hydrogen (D/H) ratio, which is over 100 times higher than Earth's. Specifically, our analysis focused on the HDO 2(2,0)-3(1,3) transition line at 266.1611 GHz. Here we present our results on the daily variation of HDO line depth across the three campaigns.

2. Data Processing

The intense brightness of Venus presented significant challenges in our spectral analysis, primarily due to the Venusian light reflecting within the instrument environment. This reflection caused extra time-delayed signals to re-enter the receiver, subsequently leading to sine-wave ripples in the resulting spectrum. Our spectrum displayed a mixture of short-period and long-period ripples, with the ripple period being inversely proportional to the travel time of the reflecting signals. To mitigate these effects, we used a mean filter to remove the long-period ripples. The more challenging short-period ripples were addressed using two independent techniques: polynomial fitting and Fourier transforms, as displayed in Figure 1. The spectrum from each observation was processed individually and then combined into daily spectra.

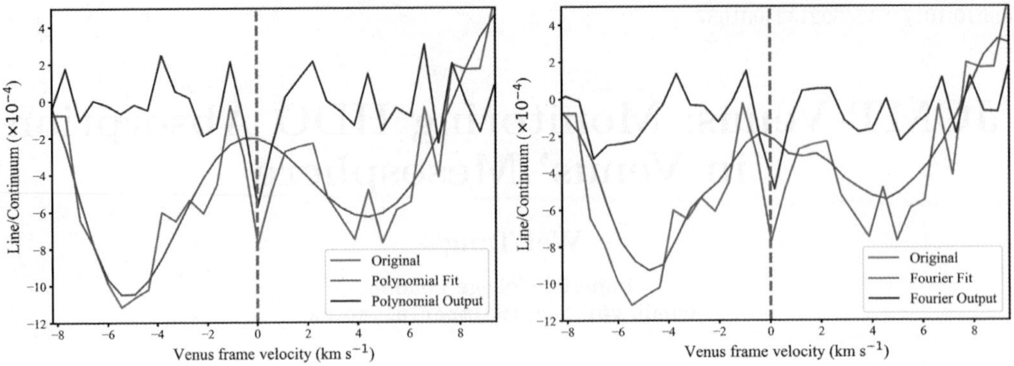

Figure 1. Illustrations of two methods for ripple removal applied to data from February 7, Campaign 1. The locations of the HDO lines are indicated by grey dashed lines. The original spectra are displayed in red, the identified ripples in green, and the results after cleaning in blue.

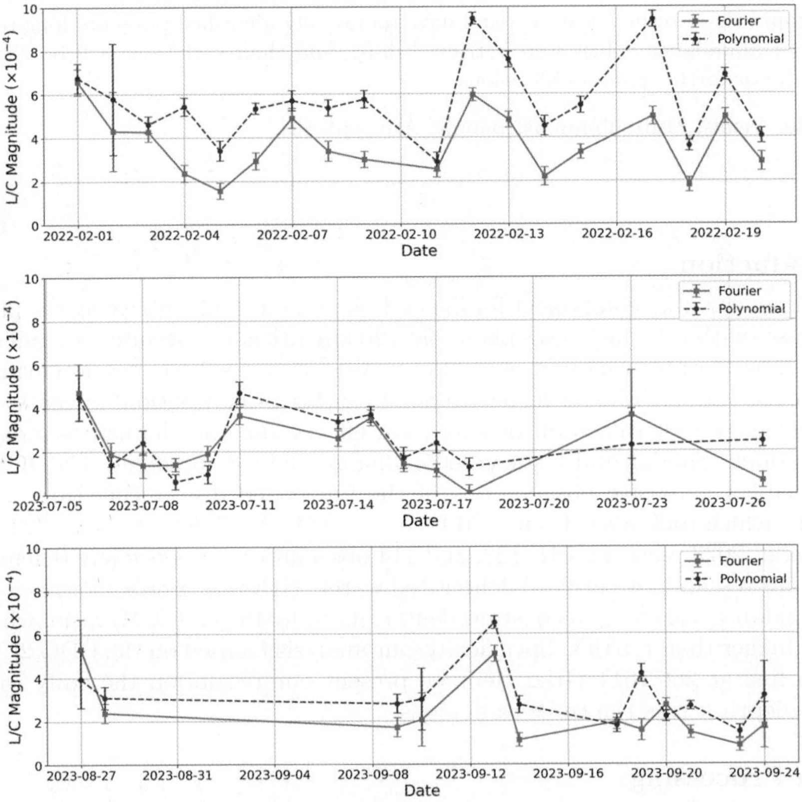

Figure 2. Daily variations in line depth across the three campaigns. Results from the polynomial method are depicted with dashed blue lines, while results from the Fourier method are shown in solid red lines. Error bars indicate the standard errors.

3. Variation in HDO

The depth of the HDO line was determined from each day's spectra, with variations across the three campaigns illustrated in Figure 2. Systematic offsets were observed between the two methods, with the Fourier method considered less reliable due to its suppression of line depths (for more details, see Tang *et al.* (2024)). Despite the offsets, the results from the two methods exhibit very similar trends. Significant day-to-day

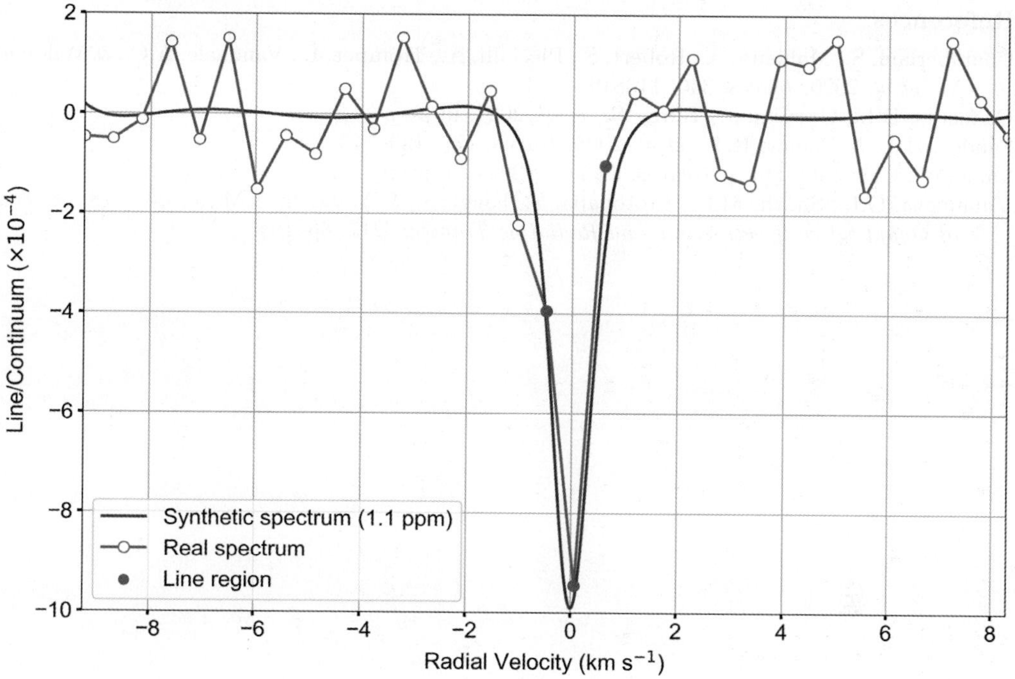

Figure 3. Example of abundance retrieved for February 12, Campaign 1: the model spectrum is depicted in blue, featuring an input profile of 1.1 ppm at 97-107 km, and the real spectrum is shown in green. The discrete spectral points constituting the real spectrum are marked by empty circles. Solid circles highlight the points within the line region used to calculate the chi-squared value.

variations are observed, notably the increase from February 11 to 12, 2022. Long-term effects also appear to be present, as indicated by the decreased line depths in the two campaigns of 2023 compared to Campaign 1 in 2022.

4. Water Abundance Retrieval

Based on the polynomial results, daily water vapour abundances were retrieved using synthetic spectra generated by the Planetary Spectrum Generator (Villanueva *et al.* 2018). These synthetic spectra underwent identical polynomial cleaning before comparison with the real spectra. The retrieved abundances were determined by identifying the best fit, which was established based on the minimum chi-squared value within the HDO line region (± 1 km s^{-1} around the line). An example is illustrated in Figure 3, where the synthetic spectrum includes an input profile of 1.1 ppm water vapor at altitudes between 97-107 km. The results of water vapor variations, including both short- and long-term effects, will be presented and discussed in Tang *et al.* (2024).

5. Future Work

We will explore all potential factors that could influence the observed variations, including vertical transport, temperature, local solar flux intensity, and interactions with cloud particles and aerosols through evaporation and condensation processes. Additionally, we need to address the systematic offset between the two processing methods and recover broad features corresponding to the deeper atmosphere, which were removed alongside the ripples. Ultimately, our goal is to correlate water vapor variations with phosphine (PH_3), which is the primary focus of the JCMT-Venus project.

References

Chamberlain, S., Mahieux, A., Robert, S., Piccialli, A., Trompet, L., Vandaele, A.C., & Wilquet, V., *et al.* 2020, *Icarus*, 346, 113819

Clements, D.L., Greaves, J., Tang, W., *et al.* 2024, in prep

Sandor, B.J., & Clancy, R.T., *et al.* 2005, *Icarus*, 177, 129–143

Tang, W., Clements, D.L., Greaves, J., *et al.* 2024, in prep

Villanueva, G.L., Smith, M.D., Protopapa, S., Faggi, S., & Mandell, A.M., *et al.* 2018, *Journal of Quantitative Spectroscopy and Radiative Transfer*, 217, 86–104

(Toward) Discovery of Life beyond Earth and its Impact
Proceedings IAU Symposium No. 387, 2026
H. Landt, M. Dominik & C. Oliver, eds.
doi:10.1017/S1743921324001625

The Exoplanet Citizen Science Pipeline: Human Factors and Machine Learning

Oisín Creaner, Anna Preis, Cormac Ryan and Nika Gorchakova

Dublin City University
email: oisin.creaner@dcu.ie

Abstract. We present the progress of work to streamline and simplify the process of exoplanet observation by citizen scientists. International collaborations such as ExoClock and Exoplanet Watch enable citizen scientists to use small telescopes to carry out transit observations. These studies provide essential supports for space missions Such as JWST and ARIEL. Contributions include maintenance or recovery of ephemerides, follow up confirmation and transit time variations. Ongoing observation programs benefit from a large pool of observers, with a wide variety of experience levels. Our projects work closely with these communities to streamline their observation pipelines and enable wider participation. Two complementary approaches are taken: Star Guide applies human-centric design and community consultation to identify points of friction within existing systems and provide complementary online tools and resources to reduce barriers to entry to the observing community. Machine Learning is used to accelerate data processing and automate steps which are currently manual, providing a streamlined tool for citizen science and a scalable solution for large-scale archival research.

Keywords. Exoplanet, Citizen Science, Machine Learning

---

1. Introduction

We work to remove barriers to entry for citizen scientists. Our work addresses both technical and human factors which impede participation. By removing these barriers, we permit a larger number of people to participate in the study of exoplanets.

Astronomy has long been a field where amateur observers can make a significant contribution. Even in the era of the James Webb Space Telescope (JWST) and forthcoming Atmospheric Remote-sensing Infrared Exoplanet Large-survey (ARIEL) missions, amateur observers remain indispensable. Amateur astronomers can enhance exoplanet ephemerides (Noguer et al. 2024), or visually inspect light curves with *Planet Patrol*, one of many Zooniverse-hosted projects (Kostov et al. 2022). By gathering large pools of observers, citizen science projects can bypass the bottleneck of limited time and availability of professional observatories (Peluso et al. 2023). The forthcoming Vera C. Rubin provides a pipeline directly from the Rubin Science Platform to the Zooniverse to support scientists in successfully running crowd science projects, including those in the field of Exoplanets Schwamb et al. (2021).

We collaborate formally with Exoplanet Watch (Zellem et al.) and ExoClock (Kokori et al. 2022), citizen science projects which enable members of the public to contribute observations to exoplanet science. These projects have identified areas which pose a challenge, especially to inexperienced observers. Our ongoing work will elucidate areas of difficulty and work to address those challenges.

Our team includes an academic, undergraduate and postgraduate students and a web and software development professional without a background in astronomy. This diverse

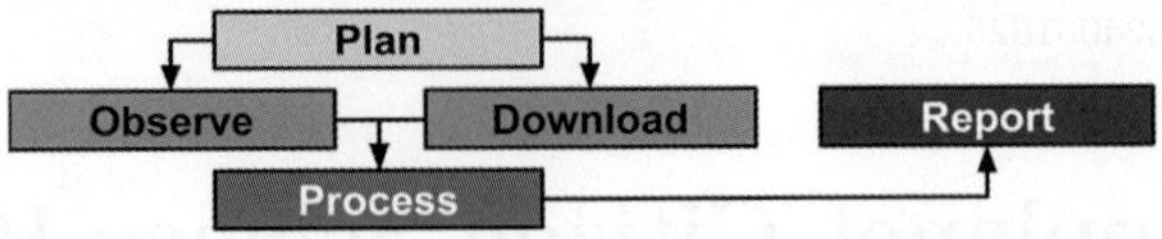

Figure 1. The workflow of a citizen scientist working on exoplanet transits. **Plan**: select a target. **Observe** a target directly or **Download** data from a previous observation. **Process** the raw images into a lightcurve. **Report** the results by uploading the lightcurve.

group has been essential in identifying areas for improvement which would not be obvious to a more traditional, purely academic group. For example, jargon terms familiar to a professional (e.g. ephemerides, limb darkening) may be unknown to citizen scientists. These terms can be off-putting and thus act as a barrier to entry.

2. Current Pipeline

The process by which citizen scientists observe exoplanets can be treated as a pipeline as illustrated in Figure 1. Observers must first plan their observation, selecting the star to observe determining when to make the observation. Observers must then gather data, either by carrying out the observation themselves or by use of remote facilities. They must then process that data at home, generating a light curve with software on their home computer. Finally, they submit their results to a central repository, where they can be collated with results from across the collaboration. The details of this process vary by project, but the overall principles are consistent (Zellem et al.; Kokori et al. 2022; Peluso et al. 2023).

Plan: In the planning phase, the observer must determine a target to observe. Tapir (Jensen 2013) (also known as the Swathmore Transit Finder) and ExoWorldsSpies Transit Scheduler (Tsiaras and Kokori 2017) are online tools for identifying observable transits. They both permit an observer to input parameters such as their location, equipment and time of day/year, and will return lists of transits which will be observable.

To mitigate against atmospheric effects which would otherwise dominate over the small effect of a transit, techniques such as differential photometry and temporal binning are used (Hartley and Wilson 2023). Careful selection of reference stars can mitigate against wavelength-dependent effects such as atmospheric reddening and techniques such as the Locus Algorithm (Creaner et al. 2022) can help identify optimum fields of view to observe.

Observe: Observers typically use Charge-Coupled Device (CCD) or other digital cameras to capture images, and computer-controlled telescope mounts to keep the object in frame. Many citizen science projects rely on highly heterogeneous equipment to gather their data(Zellem et al.; Kokori et al. 2022). The UniStellar project is an exception, relying on the Enhanced Vision Telescope (eVscope), potentially simplifying data processing (Peluso et al. 2023).

Download:An alternative approach is for citizen scientists without access to a telescope (and/or clear, dark skies) to make use of remotely controlled telescopes such as MicroObservatory (Fowler et al. 2021). This can broaden the pool of potential observers. Data gathered remotely is processed identically to data gathered locally.

Process: Gathered data is typically stored in Flexible Image Transport System (`FITS`) files, a lossless data storage system designed for astronomy (Wells et al. 1981). The data is processed using a variety of tools, such as EXOTIC (EXOplanet Transit Interpretation Code) (Zellem et al.) and HOPS (HOlomon Photometry Software, Kokori et al. (2022)). These tools typically require user input to identify the target and reference stars, and the selection of these is not always obvious. These software packages process the images into lightcurves.

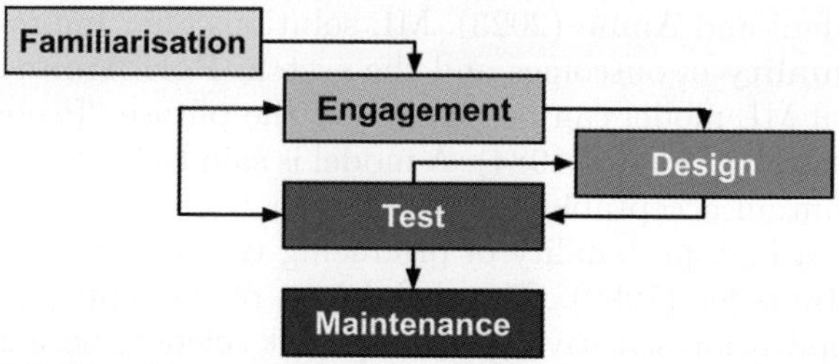

Figure 2. The workflow of the group in addressing human factors in exoplanet citizen science projects. First, we **familiarise** ourselves with the systems. Next, we **engage** with the community to identify their needs. We then **design** solutions to these problems in an iterative cycle which includes **testing** and continued engagement with observers. Finally, we intend to launch and **maintain** the solutions developed.

Report: These lightcurves can be visually inspected by the observer to approximate factors such as transit depth and signal-to-noise ratio (SNR). The underlying data, together with a report on the instrument and software used to process it, are uploaded to a remote server. There, they are combined with data from other observers to produce aggregate data with greater fidelity (Zellem et al.; Kokori et al. 2022; Peluso et al. 2023).

3. Human Factors

It can be seen from the above pipeline that there is significant direct participation by the citizen scientist in the observation process. While participation is the strongest form of public engagement, it can prove daunting for inexperienced observers. The make-up of our team as outlined in Section 1 allows us to readily identify these barriers. Our work builds on industry expertise and the principles of design affordance to streamline the pipeline wherever possible. Our approach has five key phases as shown in Figure 2.

Familiarisation: Internal use of existing systems by members of the project team gives us a basis to understand the processes currently followed. This also ensures sufficient familiarity with the systems to interpret comment and feedback from the community.

Community Engagement: We Work with the existing user community to identify points of friction and areas with potential for improvement. This takes the form of both informal interviews and anonymous community surveys carried out using online tools. The results of these surveys are expected to be available in Q3 of 2024.

Design Solutions: Iterative design processes will allow new solutions to be developed with a rapid feedback cycle. This includes work on user guides (Preis et al. 2024a), website content (Preis et al. 2024b) and software (Creaner et al. 2024), which are currently in progress.

Community Testing: Throughout the development of solutions, we remain engaged with community members to ensure their needs are met. This includes the recruitment of alpha- and beta-testers for software and discussions at community meetings.

Launch and Maintenance: Due to constrained timelines for this project, when we finalise and launch any content, it will be intended to function with limited maintenance. However, we will continue to monitor performance and seek feedback on the project while practicable.

4. Machine Learning

Machine Learning (ML) is an umbrella term for a category of computer algorithms and models which "learn" from their input data El Naqa and Murphy (2015), allowing computers to perform tasks such as data analysis and prediction with minimal explicit programming Sharifani and Amini (2023). ML solutions offer improvements in areas we categorise as the **Quality** of outcomes and the system **Performance**.

Quality: An ideal ML model can be defined by the phrase "Probably Approximately Correct" model, coined by Valiant (1984). A model is said to be "Approximately Correct" if its output is within an acceptable margin of error from the true result. A "Probably Correct" model has a high probability of producing the correct result over a large sample of predictions Haussler (1990). The definitions of "acceptable margin" and "high probability" are application-sensitive. For exoplanet science, because transits are rare, we value *precision* ($\frac{True\,transits}{Transits\,predicted}$) over *accuracy* ($\frac{Correct\,predicitons}{Sample\,size}$). One method to assess these quality factors is to divide input data into "Training" and "Validation" samples. Training data is used by the model to learn how to map inputs to outputs. The Validation data is used to assess the model outputs against known outputs by determining how *probable* it is that the model is *approximately* correct Choi et al. (2020).

Performance: ML has significant potential to accelerate a variety of workflows in exoplanet observation Jara-Maldonado et al. (2020). A key point in accelerating a process is to identify the bottlenecks in the process. In general, manual processes are slower than automated processes, and ML algorithms tend to outpace programmed solutions. Benchmarks such as CPU/GPU time per process, memory consumption and so on are used to assess performance Takamoto et al. (2022).

We have identified several areas of the exoplanet pipeline in which relatively little ML research has been focused, but which show potential for growth. Key among these is computer vision: using machine learning techniques to support extraction of light curves from data by minimising the effect of noise on the outcome. Another area of potential is observation planning: using archives and catalogues together with machine learning technology to plan efficient observing campaigns. Key to evaluating the success of these approaches will be the development of relevant metrics against which a ML solution can be evaluated.

Acknowledgements

Star Guide (OC, AP, CR) is funded by the National Open Research Forum (NORF) Open Research Fund 2023, Strand II: Open Research Stimulus.

NG and OC are funded by Science Foundation Ireland through the SFI Centre for Research Training in Machine Learning (Grant No. 18/CRT/6183) Call 2023. This funding is provided through the Higher Education Authority (HEA).

References

Choi, R. Y., Coyner, A. S., Kalpathy-Cramer, J., Chiang, M. F., & Campbell, J. P. 2020, Introduction to machine learning, neural networks, and deep learning. *Translational vision science & technology*, 9(2), 14–14.

Creaner, O., Nolan, K., Hickey, E., & Smith, N. 2022, The locus algorithm: A novel technique for identifying optimised pointings for differential photometry. *Astron. Comput.*, 38, 100537.

Creaner, O., Yoo, T., Venkatesan, V., & Anumba, K. 2024, Stellar Matchmaker.

El Naqa, I. & Murphy, M. J. 2015, *What is machine learning?* Springer.

Fowler, M., Sienkiewicz, F., Zellem, R., & Dussault, M. 2021, Observing exoplanets with the MicroObservatory: 43 new transit light curves of the hot jupiter HAT-P-32b. *J. Br. Astron. Assoc.*, 131, 359–368.

Hartley, K. E. & Wilson, R. 2023, Optimized temporal binning of comparison star measurements for differential photometry. *MNRAS*, 526(3), 3482–3494.

Haussler, D. 1990, Probably approximately correct learning.

Jara-Maldonado, M., Alarcon-Aquino, V., Rosas-Romero, R., Starostenko, O., & Ramirez-Cortes, J. M. 2020, Transiting exoplanet discovery using machine learning techniques: A survey. *ESIN*, 13, 573–600.

Jensen, E. 2013, Tapir: A web interface for transit/eclipse observability. ASCL.

Kokori, A., Tsiaras, A., Edwards, B., Rocchetto, M., Tinetti, G., Wünsche, A., Paschalis, N., Agnihotri, V. K., Bachschmidt, M., Bretton, M., *et al.* 2022, Exoclock project: an open platform for monitoring the ephemerides of ariel targets with contributions from the public. *Exp. Astron.*, 53(2), 547–588.

Kostov, V. B., Kuchner, M. J., Cacciapuoti, L., Acharya, S., Ahlers, J. P., Andres-Carcasona, M., Brande, J., de Lima, L. T., Di Fraia, M. Z., Fornear, A. U., *et al.* 2022, Planet patrol: Vetting transiting exoplanet candidates with citizen science. *PASP*, 134(1034), 044401.

Noguer, F. R., Corley, S., Pearson, K. A., Zellem, R. T., Simon, M. N., Burt, J. A., Huckabee, I., August, P. C., Weiner-Mansfield, M., Dalba, P. A., *et al.* 2024, Enhancing exoplanet ephemerides by leveraging professional and citizen science data: A test case with wasp-77a b. *arXiv preprint arXiv:2405.19615*.

Peluso, D. O., Esposito, T. M., Marchis, F., Dalba, P. A., Sgro, L., Megowan-Romanowicz, C., Pennypacker, C., Carter, B., Wright, D., Avsar, A. M., *et al.* 2023, The unistellar exoplanet campaign: Citizen science results and inherent education opportunities. *PASP*, 135(1043), 015001.

Preis, A., Ryan, C., & Creaner, O. 2024,a. Exoplanet watch users guide. in prep.

Preis, A., Ryan, C., & Creaner, O. What is exoplanet watch? In Brennan, P., Walbolt, K., & Schirner, T., editors, *Exoplanet Watch Website* 2024,b. NASA.

Schwamb, M., Lintott, C., Smith, K., Robinson, J., Williams, R., Fortson, L., Fulton, M., Miller, G., Omer, R., Smartt, S., *et al.* 2021, Enabling citizen science with rubin observatory alert stream brokers.

Sharifani, K. & Amini, M. 2023, Machine learning and deep learning: A review of methods and applications. *World IT and Eng. J*, 10(07), 3897–3904.

Takamoto, M., Praditia, T., Leiteritz, R., MacKinlay, D., Alesiani, F., Pflüger, D., & Niepert, M. 2022, Pdebench: An extensive benchmark for scientific machine learning. *Advances in Neural Information Processing Systems*, 35, 1596–1611.

Tsiaras, A. & Kokori, A. 2017, Transit scheduler.

Valiant, L. G. 1984, A theory of the learnable. *Comm. ACM*, 27(11), 1134–1142.

Wells, D., Greisen, E., & Harten, R. 1981, FITS: a Flexible Image transport System. *A&A Supp.*

Zellem, R. T., Pearson, K. A., Blaser, E., Fowler, M., Ciardi, D. R., Biferno, A., Massey, B., Marchis, F., Baer, R., Ball, C., *et al.* Utilizing small telescopes operated by citizen scientists for transiting exoplanet follow-up. *PASP*.

(Toward) Discovery of Life beyond Earth and its Impact
Proceedings IAU Symposium No. 387, 2026
H. Landt, M. Dominik & C. Oliver, eds.
doi:10.1017/S1743921324001327

The Fermi paradox in an age of discovery

Stephen Webb

University of Portsmouth, Hampshire, United Kingdom (retired)
email: stephen.webb@physics.org

Abstract. The title of the Kavli–IAU Symposium *(Toward) Discovery of Life Beyond Earth and its Impact* invites us to speculate upon the consequences for humanity of the detection of biological processes out there in the cosmos. The discovery of any form of life beyond Earth would be a momentous event for science, but for many people it is the discovery of *intelligent* life that would have the most profound impact—not just on science, but on philosophy, religion, and society in general. Equally profound, however, might be the impact of a continuing *non*-discovery of extraterrestrial intelligence. In this paper I outline some personal thoughts regarding the implications of continuing non-discovery.

Keywords. Fermi paradox, technosignatures, astrobiology

1. Introduction

In the summer of 1950, a lunchtime conversation between Enrico Fermi, Emil Konopinski, Edward Teller, and Herbert York, four colleagues at Los Alamos National Laboratory, turned to the subject of flying saucers, a topic of much public concern at the time (Jones 1985). Fermi famously asked: "Where *is* everybody?" His question referred, of course, to the lack of observational evidence for visitations from extraterrestrial intelligence.

Although Fermi did not publish on this topic, we can reasonably reconstruct the thinking that led to his question. We know that Fermi was exceptionally skilled at computing order-of-magnitude estimates of physical quantities, and he therefore likely generated a ballpark figure for the number of extraterrestrial civilisations in the galaxy. The spirit of his approach to estimation manifests itself in the well known equation, developed by Frank Drake, a decade after Fermi asked his question (Drake 1992). This equation cannot be "solved" nor even calculated with any accuracy. (As the SETI scientist Jill Tarter once remarked (see Achenbach 2000), the Drake equation is a "wonderful way to organize our ignorance".) Rather, it provides a framework for estimating N, the number of extraterrestrial civilisations with whom communication is possible. It does so by multiplying several relevant quantities:

$$N = R_\star \times f_\mathrm{p} \times n_\mathrm{e} \times f_\mathrm{l} \times f_\mathrm{i} \times f_\mathrm{c} \times L.$$

Here, the quantity $R_\star$, the rate of star formation, is the number of stars formed each year with conditions that are suitable for the development of intelligent life; f_p is the fraction of those stars that possess planetary systems; n_e is the number of planets per solar system that provide an environment that is suitable for life; f_l is the fraction of suitable planets upon which life in fact appears; f_i is the fraction of those life-bearing planets upon which *intelligent* life develops; f_c is the fraction of civilisations that go on to devise a technology that produces signs of their existence that other civilisations might detect; and L is the average number of years that civilisations produce those signs of their existence. In 1950, Fermi's estimate for most of these quantities would be no less reliable than an estimate made today. (The term f_i, for example, the fraction of planets with life that go on to develop intelligence that supports a civilisation, is entirely unknown. We have no data to guide us.) Fermi could therefore quite reasonably have estimated a large value for N. After all, the original meeting at Green Bank during which Drake developed his eponymous equation, concluded that "the number of communicative civilizations in the Galaxy might range from less than 1000 to one billion" (Billingham 2011).

Up to one billion civilisations in the Galaxy! If Fermi did indeed arrive at a large number for N, then when he asked his question his voice might well have held at least a note of puzzlement if not full-blown paradox. The reason is that one would expect many of those extraterrestrial civilisations to be older than our human civilisation and presumably, therefore, to be more technologically advanced. Fermi understood that technology can make *incredible* advances even over a single lifespan. He himself had experienced a rapid growth in aerospace technology. He was born in 1901, into a species that was literally terrestrial: this was before the Wright brothers made the first sustained flight by a crewed heavier-than-air powered and controlled aircraft. Fermi lived to see craft reach the Kármán line. In a technical sense, humanity was at that point a space-faring species. (Fermi died young. Had he enjoyed an average lifespan he would have seen the Voyager craft launch. Had he lived to be an old man, he would have seen Voyager 1 reach the heliopause. In a technical sense, humanity became at that point an interstellar species.) Instead of contemplating progress over a single lifespan, imagine what progress might be possible if a species persists in advancing its technology for a few centuries or a few millennia. Such a timescale is a mere eyeblink in cosmic terms. A species that develops a Kardashev type I civilisation (one that can access all the energy available on its planet) might quickly, on cosmic timescales, evolve into a Kardashev type II civilisation (one that can access all the energy available from its star). And one expects a Kardashev type II civilisation to be capable of sending probes to travel through the Galaxy and making its presence known across interstellar, even intergalactic, space (Armstrong and Sandberg 2013).

There is, then, a tension here. On the one hand, the universe is large. If one adopts the Copernican principle, which tells us there is nothing particularly special about Earth, then one would expect there to be many planets with life, and some of those life forms will have evolved intelligence. On the other hand, the universe is old. There has been time for intelligent life elsewhere to develop technological civilisations capable of disturbing the universe in a way we could detect. And yet (so far) we see and hear nothing. This tension typically goes by the name of the "Fermi paradox", although some would argue that it is not a paradox and, even if it were, Fermi's should not be the sole name attached to it since thinkers such as Konstantin Tsiolkovsky anticipated the concept in the 1930s (Lytkin, Finney and Alepko 1995). Regardless of the status and history of the paradox, Fermi's question surely demands some response. If the universe is indeed home to advanced technological civilisations, why do we observe no signs of them?

Where *is* everybody?

2. Approaches to the Fermi paradox

When we address Fermi's question we naturally find ourselves in an arena in which multi-disciplinary conversations can take place, and thinkers from a range of disciplines have, indeed, discussed a variety of ways of tacking the paradox. In 2002, I grouped 50 approaches to the paradox into three broad classes (Webb 2002). Class I approaches suggest that, in one way or another, extraterrestrial intelligence is currently within the solar system or else has visited in the past. Class II approaches adopt what is perhaps the consensus among the scientific community: extraterrestrial intelligence exists, but we have yet to find evidence. Class III approaches claim that Earth is unique in being home to an intelligent species capable of spaceflight or of communication over interstellar distances. In the decade following the publication of that book, researchers pointed out many more approaches. In 2014, I discussed 75 approaches to the paradox and again grouped them into those same three broad classes (Webb 2014). In the decade since the publication of that second edition, numerous new suggestions have been made. For each of those three classes, let me provide just two recent suggestions.

2.1. *Class I solutions*

Some recent Class I solutions resolve the paradox by pointing to terrestrial or solar system observations that are supposedly only explicable in terms of extraterrestrial technology.

The discovery (Meech et al. 2017) of 1I/2017 U1, commonly called 'Oumuamua, has puzzled the astronomical community. 'Oumuamua is the first interstellar object (ISO) to be detected traversing the Solar System, and its unusual constellation of properties led to the suggestion (Shmuel and Loeb 2018) that the object was the product of extraterrestrial technology. In particular, the technosignature claim explains the observations of 'Oumuamua in terms of a thin solar sail being accelerated by solar radiation pressure. The vast majority of the astronomical community, while acknowledging the unusual properties of this extraterrestrial visitor, are content to interpret the observations in terms of known populations of objects such as asteroids and comets.

Another recent suggestion belonging to Class I approaches points to unidentified aerial phenomena (UAP), which some might consider to be merely a rebrand of the term unidentified flying object (UFO). What perhaps distinguishes UAP from UFO is the number of sober and highly skilled individuals, and in particular military pilots, who claim to have observed them. US Defense, wisely in my opinion, broadened the definition of these sightings; the term UAP now refers to unidentified *anomalous* phenomena. The renaming takes into account that the observations might not be aerial in nature. They could, for example, be caused by software glitches or malfunctioning hardware. The All-domain Anomaly Resolution Office (AARO) was established in 2022, and in 2024 published the first comprehensive historical record report (AARO 2024). As Sean Kirkpatrick, the first AARO director has stated, there is no evidence "that any US government investigation, academic-sponsored research, or official review panel has confirmed that any sighting of a UAP represented extraterrestrial technology".

While undoubtedly interesting, the 'Oumuamua and UAP observations do not resolve the Fermi paradox in a way that has found general acceptance.

2.2. *Class II solutions*

Class II solutions typically accept the existence of extraterrestrial intelligence but argue that, for some reason, we have yet to receive signals from them or observe evidence of their technosignatures. The search for extraterrestrial intelligence is undeniably more difficult than the proverbial search for a needle in a haystack, and there remains a vast amount of phase space still to explore. The search is currently more than sixty years old; perhaps we need another sixty, or another six hundred, before we can expect to see success. But perhaps we should broaden the search to include different places or to use different search methods. Consider Dyson swarms, for example.

A Dyson swarm, a construction in which innumerable objects are put into orbit around a star, is often considered to be an activity in which a Kardashev Type II civilisation would engage. If humankind were able to put trillions of small objects in orbit around the Sun at 1 AU, then our civilisation would be able to harvest essentially all the Sun's output. We would have access to about 10^{26} W of power and to unimaginable amounts of real estate. The suggestion is that other civilisations will want to do the same around their star. Well, suppose Dyson swarms are common in the Galaxy. A Dyson swarm will block starlight, making a star appear fainter (and thus more distant) than it really is. However, the *Gaia* astrometry mission (ESA nd) is measuring the parallax, and thus the geometric distance, to more than a billion stars. We can therefore compare the true distances to stars as measured astrometrically with estimates from spectrophotometry. Stars displaying a discrepancy could be the target of follow-up observations for an infrared excess or for radio signals (Zackrisson, Korn, Wehrhahn and Reiter 2018). In other words, by piggy-backing on top of existing astronomical observations we might be able to discern a technosignature, or at least aspects of a possible technosignature.

Vidal (2019) provides a different example of how observations made in the course of traditional astronomical research might be used to search for technosignatures. Vidal points out that millisecond pulsars (MSPs) rotate with a stability comparable to atomic clocks. If one can compare the arrival times of four millisecond pulsars then one can determine one's position in space. The existence of MSPs thus provides us with ready-made GPS—a "galactic positioning system". If we can use MSPs in this way then extraterrestrial civilisations could do likewise. Perhaps advanced civilisations modify MSPs in order to maximise their utility as a positioning system? Perhaps truly advanced civilisations *move* MSPs or even create them for GPS purposes. Vidal suggests that we look, for example, for time correction signals sent towards MSPs. Or perhaps we could examine the spatial distribution of MSPs to check for signs of design in the network.

2.3. *Class III solutions*

Class III solutions argue that, for some reason, we are alone. We see no evidence of extraterrestrial intelligence because intelligence is a purely terrestrial phenomenon. Many reasons have been given for why Earth, or our own species, might be unique. Here are two recent suggestions for why life might be rare.

Phosphorus is one of the six key elements required for life *as we know it*. For example, the element plays a crucial role in the molecule ATP. In unpublished preliminary work, Greaves and co-workers have studied phosphorus in the supernova remnants Cassiopeia A and Crab Nebula. They detected much less phosphorus in the Crab Nebula than in Cassiopeia A (perhaps because these were different types of progenitor star). If planetary phosphorus is sourced from supernovae and then travels across space in meteoritic rocks that seed the planet then, for life to develop, the planet must be born around the "correct" sort of supernova. Is this yet another goldilocks condition that must be "just right" for a planet to develop and sustain life?

Another feature of life *as we know it* is the way in which ordered information is stored in DNA/RNA molecules. The DNA molecule in particular, however, is a complex structure. How did such a complicated feature arise from abiotic processes? The RNA-world hypothesis is one prominent suggestion for explaining the development of DNA.

RNA is capable of playing both genetic and catalytic roles, so the suggestion is that the simpler RNA molecule came into existence first and then evolution led to the more complicated world of DNA and proteins that we inhabit today. For self-replicating activity it appears that one requires an RNA molecule of length at least 40 nucleotide bases. Again, though, the question becomes: how did nature produce such a polymer from abiotic processes?

The answer to this question is not clear, and in the absence of a well understood mechanism it has been suggested that the first self-replicating RNA molcule arose by randomly adding monomers to a growing chain—a Poisson process. However, as Totani

(2020) points out, there are $4^{40} \approx 10^{24}$ unique 40-mers. The vast majority of those 40-mers will possess no functional properties. Since the observable universe contains of the order of 10^{22} Sun-like stars, the formation of a long polymer having the correct nucleotide sequence is statistically unlikely. However, Totani makes a further point relating to cosmology. A number of problems with the standard big bang model of cosmology led theorists to posit that the early universe underwent a brief period of exponential expansion, a phenomenon known as cosmic inflation. If the duration of inflation is just twice that needed to solve the problems for which the phenomenon was introduced, then the homogeneous universe possesses about 10^{78} times the volume of the observable universe and it contains about 10^{100} Sun-like stars. Totani points out that in this scenario the stochastic process leading to a self-replicating RNA molecule would be certain to occur many times in the broader homogeneous universe, but it would be statistically unlikely to occur again within our observable universe. Life is common. But in the *observable* universe we are alone.

2.4. *Signs of progress?*

The above suggestions are, of course, not intended to be taken as definitive solutions to the Fermi paradox. Rather, I present them here to illustrate the fact that, despite a decade since the publication of my collection of "solutions" in 2014, and more than two decades since the publication of my first collection in 2002, little progress appears to have been made. Researchers are *still* engaging with the puzzle on the same terms. Indeed, we seem no further forward than we were when Fermi first asked his question almost three quarters of a century ago.

This lack of progress might be regarded as curious because, in the 75 years or so since Fermi posed the question, humanity has seen striking improvements in its understanding of science and impressive developments in its technology. Consider, for example, the period from 1984 to 2024. (I pick this period because I first learned about the Fermi paradox in 1984 and I delivered this paper at the Kavli–IAU Symposium in 2024.) Almost any field of science or technology one cares to pick has seen substantial progress over those four decades.

For example, at the start of that period cosmologists were debating whether the Hubble parameter H_0 possesses the value 50 or 100 (in standard units of $\mathrm{km\,s^{-1}\,Mpc^{-1}}$). Now, at the end of this period, they debate whether H_0 is 67.5 ± 0.5 or 73 ± 1 (in the same units). The advent of precision cosmology means cosmologists have advanced from the "Hubble wars" to a dispute over the significance of the "Hubble tension". Or consider the exponential increase in various metrics associated with computing technology. In 1984, the IBM PC AT operated at 1.2 MegaFLOPS; the 2024 iPhone 15 Pro puts 2.15 TeraFLOPS into a person's back pocket. Or consider developments in genetics, which could be said to be faster-than-exponential. In 1984, any DNA sequencing that could take place happened manually; in 2024, whole genome sequencing is cheap.

Over the same period, those involved in the search for extraterrestrial intelligence have enjoyed equally significant improvements in the technologies required for the task. Observing instruments are better. (Those four decades have seen the introduction of the Keck telescopes; the Hubble and James Webb space telescopes; ALMA; FAST; and a host of other observatories operating across the entire electromagnetic spectrum.) Vastly more computational power can be thrown at the problem. (The 1985 Cray-2 supercomputer reached 1.9 GigaFLOPS; the 2023 HPE Cray EX235a reaches 1.2 ExaFLOPS.) And the supporting infrastructure is incomparably better, with improved storage, faster networks, and entirely new techniques such as machine learning. These developments have ushered in a golden age of observational astronomy, and they support the claim that we are living in an age of discovery. Drake surely foresaw this astonishing development in scientific capability, which led him in 1992 to write the following about the detection of extraterrestrial intelligence ((Drake 1992), my italics): "This discovery, *which I fully expect to witness before the year 2000*, will profoundly change the world".

And yet, despite all that technical and scientific advance, we continue to observe a silent universe.

The lack of observational evidence for extraterrestrial intelligence does not allow us to conclude we are alone. To repeat the point made above about Class II approaches: there remains a vast amount of phase space still to explore. And yet the continuing silence seems increasingly deafening. Indeed, two recent papers, (Uno et al. 2023) and (Choza et al. 2024), suggest that Kardashev type II civilisations, if they exist at all, are a rarity in the local universe.

None of this is to suggest that we should discontinue the search. The role that biology plays in the universe is one of the outstanding questions in science, and we should deploy all the tools at our disposal to continue the search. And our age of discovery will provide scientists with new tools for the quest. For example, the Vera Rubin Observatory will soon see first light, and within a few years it is likely to discover many more interstellar objects similar to 'Oumuamua. Proposals have been published (Stern et al. 2024) to send flyby probes to an ISO. We might visit Rama! Forthcoming terrestrial observatories such as the Square Kilometer Array will expand the capacity of radio SETI. New space-based telescopes such as the Nancy Grace Roman Space Telescope will enhance our ability to look for biosignatures. We have by no means exhausted our options in this age of discovery. All these new tools, allied to new techniques such as artificial intelligence, might prove successful. But they might not. After how many years of continuous silence will humanity conclude that, probably, we are alone? What would be the impact of that conclusion? And how should we act if we believe we are the only sapient beings in the galaxy?

3. The Intelligence Principle

The NASA historian Steven J. Dick has argued (Dick 2003) for something he calls the *Intelligence Principle*. The Principle states that "the maintenance, improvement and perpetuation of knowledge and intelligence is the central driving force of cultural evolution, and that to the extent intelligence can be improved, it will be improved". Dick argued that, in terms of knowledge, this principle can be seen in daily operation at both the individual and societal levels. However, Dick was more interested in applying this principle at the species level. He argued that intelligence, which he asserted is a function of brain size and structure, has permitted the species *Homo sapiens sapiens* to outcompete other life forms and to dominate the planet. Brain size and structure, however, has not changed in 100,000 years. The Intelligence Principle, Dick contended, implies that given the opportunity to increase intelligence—and one can readily appreciate how AI presents us with this opportunity; biotechnology and genetic engineering are two other relevant technologies—any society would do so. It would fail to do so at its peril.

One can criticise the underlying assumptions and implications of the Intelligence Principle, as applied to modern humans, on several grounds. For example, the Principle lacks a solid grounding in empirical evidence (Lineweaver (2008) points out that a non-biased review of the paleontological record, with cranial capacity acting as a proxy for intelligence, shows no trend towards increasing intelligence) and it could be seen to downplay the role of chance and contingency. The Principle relies on a concept, namely "intellligence", the definition of which remains fundamentally ambiguous. It has been argued (Bräuer et al. 2020), for example, that the related term "cognition" has often been used by applying an anthropocentric rather than biocentric viewpoint, and that there is not "one cognition". The Principle takes relatively little account of the critical role played by cultural transmission and imitation; aspects of the Principle adopt a reductionist view of the inherently complex, multifactorial phenomenon of cultural evolution, a phenomenon that is shaped by social dynamics, environmental influences, historical contingencies, and economic systems. Finally, the Principle could be seen to possess an apparent teleological aspect.

Dick himself was aware of these critiques, but to repeat: he was concerned more with long-term evolution than with the complexities of present-day cultural developments. He argued convincingly that, when one considers cosmic timescales, one should adopt a mindset popularised by Olaf Stapledon. This English philosopher was concerned with the long-term prospects of the human species, and therefore with questions such as: How might our species evolve? What technological futures are most likely? What might humanity be capable of achieving? The adoption of such a mindset led Dick to postulate that biological and cultural evolution would necessarily lead to a postbiological universe, one in which embodied intelligence cedes to artificial intelligence.

Dick's paper refers briefly to the Fermi paradox, and notes that the "roaming of intelligent probes might lead to an AI version" of the paradox. But the Intelligence Principle also opens further avenues for contemplating Fermi's question. The behaviour of postbiological beings is unknown (and perhaps from our vantage point unknowable; except that, perhaps, the evolution of their intelligence is likely to be more explicitly goal-oriented and driven by design) but presumably they would not operate under the same restrictions as biological beings. They would not be confined to the surfaces of planets, for example. Perhaps, then, we hear a silent universe because our search techniques have listened for biologicals rather than postbiologicals?

4. The spreading of intelligence

I believe the phenomenon of human intelligence arose through a myriad of chance events. Teleology played no role. No evolutionary drive led to humans sitting at the top of a ladder of increasing sapience. Notwithstanding Dick's Intelligence Principle, there is no guarantee that, in the short term, human intelligence (however defined) will improve. Indeed, it seems possible that at the species level a combination of high intelligence and opposable thumbs could lead in our case to disaster. Nevertheless, I concur with Dick regarding the importance of maintaining and perpetuating intelligence in the cosmos. Recent work (see, for example, Haba et al. 2021) suggests that around the time of the formation of our solar system, both type Ia and core-collapse supernovae seeded heavy elements into the cloud from which Earth and the other planets condensed. Some 4.5 billion years later, a combination of labyrinthine chemical processes and the press of evolution brought into being stable combinations of those elements possessing the quality of self-awareness. As Sagan (1980) noted, in what I would argue is one of the most profound statements with philosophical content made by a practising astronomer, "we are a way for the universe to know itself". There is something wonderful—in the word's original sense—in the notion that a mixture of inanimate chemical compounds, the building blocks of which were cooked in the interiors of ancient stars, can come to "know itself".

I would like to argue that a universe possessing a component capable of knowing itself is more interesting than a universe devoid of such components. Intelligence, however it is defined and whichever way it manifests itself, is more interesting than inanimate material. Sentience is more interesting than sand. The greater the depth of intelligence, the greater the variety of types of intelligence, then the more interesting and meaningful the universe becomes. Although it remains an article of faith, particularly when the maintenance of intelligence on planet Earth is far from guaranteed, I would argue that in our local corner of the universe intelligence—whether human; human-developed AI; or some merger of human with AI—will come to appreciate the wonder in Sagan's statement that *we are a way for the universe to know itself.* At that point, for reasons of altruism rather than selfishness or competition, Dick's Intelligence Principle would come to apply: to the extent intelligence can be improved, it will be improved.

I believe, however, that an implicit aspect of the Intelligence Principle should be made explicit. We now understand how the same astrophysical processes that seeded the primordial cloud from which our solar system developed, and provided the elements from which intelligence arose here, are equally capable of destroying intelligence (whether biological or postbiological). Supernovae possess the capacity to destroy life on nearby planets; as do gamma-ray bursters; as do magnetars. More local celestial threats come from meteor impact events; cometary impact events; coronal mass ejections. Even more locally, existential threats for biological life include climate change; pandemics; environmental degredation. Postbiological entities would likely not be immune from local existential threats, albeit those threats would likely be of a different nature. If intelligence of whatever form is localised to one planetary system then it is at risk of extinction. I therefore believe the Intelligence Principle requires a small extension: "... to the extent intelligence can be improved, it will be improved *and it will spread*" (my italics).

This extension to the Intelligence Principle, however, sharpens the Fermi paradox. If Earth-bound intelligence (human; human-developed AI; or some merger of human with AI) concludes that expansion through the cosmos is necessary to ensure that the universe can "know itself" then other civilisations would presumably also have reached that conclusion. Those civilisations should now be moving through the galaxy, increasing the quantity and quality of intelligence in our part of the cosmos. The same should have happened in other galaxies. And yet, as repeatedly noted above, despite us being in an age of discovery for astronomy, we have no observational evidence for the existence of life elsewhere in the cosmos.

As I argued earlier, we should intensify our search in order to understand the role biology plays in the universe. In the absence of evidence for extraterrestrial life, however, I believe it is prudent to assume we are alone. In that case, how should we respond? The phenomenon of intelligence, if it is confined to this one planet, could easily be removed from being whether from celestial perils or local hazards. The universe would then lose its wonderful ability to "know itself". The modified Intelligence Principle implies we should spread that ability. In order to do that, however, we must in the first instance *maintain* and *perpetuate* that ability. As the Italian ethicist Lo Sapio (2022) has argued, with our current level of understanding of intelligence in the universe we should adopt the principle of "survival at any cost" as a guide to our actions.

5. Conclusion

Recent technological developments are consistent with observations made by the philosopher David Chalmers (2010), namely that humanity might be at the cusp of an "intelligence explosion". Such an event would be one of the most significant events in the history of life on this planet, and the question of where this development might lead has numerous practical and philosophical aspects. In this paper I have focused on a single aspect: the speculation that terrestrial intelligence (whatever form it takes) will appreciate the value in ensuring the universe can continue to "know itself" and this in turn will lead to the spreading of intelligence through the cosmos.

This speculation then encounters the Fermi paradox. Would intelligence elsewhere not possess the same drive to spread? And yet, although we are in an age of discovery, thanks to vast improvements in telescopes, computation, and general technology, the more than six decades of SETI endeavour has failed to find evidence for the existence extraterrestrial intelligence. If we adopt a conservative solution to Fermi's question—namely, that we are alone—a large burden rests upon us. It becomes incumbent upon humanity to act to overcome current existential threats; to navigate an intelligence explosion in such a way that issues of value and morality are addressed so the outcome is good rather than bad; and then ... to reach for the stars.

References

AARO (2024) *Report on the Historical Record of U.S. Government Involvement with Unidentified Anomalous Phenomena, Volume 1.* (Washington, D.C., AARO)

Achenbach, J. (2000) Life beyond Earth. *National Geographic* **194**, 24.

Armstrong, S. and Sandberg, A. (2013) Eternity in six hours: intergalactic spreading of intelligent life and sharpening the Fermi paradox. *Acta Astronaut.* **88** 1–13.

Billingham, J. (2011) SETI: The NASA years. In: *Searching for Extraterrestrial Intelligence: SETI Past, Present, and Future* Ed. H. Paul Schuch. pp. 65–85. (Berlin: Springer).

Bräuer J., Hanus, D., Pika, S., Gray, R., and Uomini, N. (2020) Old and new approaches to animal cognition: there is not 'one cognition'. *J. Intell.* **8**(3) 28.

Chalmers, D. (2010) The singularity: a philosophical analysis. *J. Consciousness Studies* **17**(9–10) 7–65.

Choza, C. et al. (2024) The Breakthrough Listen search for intelligent life: technosignature search of 97 nearby galaxies. *Astro. J.* **167**(1) 10.

Dick, S.J. (2003) Cultural evolution, the post biological universe and SETI. *In. J. Astrobiol.* **2**(1) 65–74.

Drake, F. and Sobel, D. (1992) *Is Anyone Out There? The Search for Extraterrestrial Intelligence* (New York: Delacorte Press).

ESA (nd) *GAIA—ESA's billion star surveyor.* `www.esa.int/Science_Exploration/Space_Science/Gaia`

Haba, M.K., Lai, Y.-J., Wotzlaw, J.-F., Yamaguchi, A., Lugaro, M., and Schönbächler, M. (2021) Precise initial abundance of niobium-92 in the solar system and implications for *p*-process nucleosynthesis. *Proc. Natl Acad. Sci. USA* **118**(8):e2017750118.

Jones, E.M. (1985) Where is Everybody? An account of Fermi's question. *Phys. Today* (Aug) pp 11–3.

Lineweaver, C.H. (2008) Paleontological tests: human-like intelligence is not a convergent feature of evolution. In: Seckbach J. and Walsh M. (eds). *From Fossils to Astrobiology.* pp 355–68. (Berlin: Springer)

Lo Sapio, L. (2022) The ethics of astrobiology: humanity's place in the cosmos and the extinction problem. *Cosmos: A Personal Voyage.* Episode 1. PBS, New York.

Lytkin, V., Finney, B., and Alepko, L. (1995) Tsiolkovsky, Russian cosmism and extraterrestrial intelligence. *Frontiers Astro. Space Sci.* doi: 10.3389/fspas.2022.1008265.

Meech, K.J. et al. (2017) A brief visit from a red and extremely elongated interstellar asteroid. *Nature* **552**, 378–81.

Sagan, C. (1980) *Cosmos: A Personal Voyage.* Episode 1. PBS, New York.

Shmuel, B. and Loeb, A. (2018) Could solar radiation explain 'Oumuamua's peculiar acceleration? *Astrophys. J.* **868**(1), L1.

Stern, S. A. et al. (2024) A study of an interstellar object explorer (IOE) mission. *Planetary & Space Sci.* **241** 105850.

Totani, T. (2020) Emergence of life in an inflationary universe. *Nat. Sci. Rep.* **10**:1671.

Uno, Y. et al. (2023) Upper limits on transmitter rate of extragalactic civilizations placed by Breakthrough Listen observations. *Mon. Notes R. Astro. Soc.* **522**(3) 4649–53.

Vidal, C. (2019) Pulsar positioning system: a quest for evidence of extraterrestrial engineering. *Int. J. Astrobiol.* **18** 213–34.

Webb, S. (2002) *If the Universe is Teeming with Aliens... Where is Everybody? Fifty Solutions to the Fermi Paradox and the Problem of Extraterrestrial Life* (New York: Copernicus).

Webb, S. (2014) *If the Universe is Teeming with Aliens... Where is Everybody? Seventy-five Solutions to the Fermi Paradox and the Problem of Extraterrestrial Life* (Berlin: Springer). 2nd edn.

Zackrisson, E., Korn, A.J., Wehrhahn, A. and Reiter, J. (2018) SETI with Gaia: observational signatures of nearly complete Dyson spheres. *Astrophys. J.* **82**, 21.

(Toward) Discovery of Life beyond Earth and its Impact
Proceedings IAU Symposium No. 387, 2026
H. Landt, M. Dominik & C. Oliver, eds.
doi:10.1017/S1743921324002114

What Makes "Weird" Life Weird?

Philipp Spillmann

Department of History and Philosophy of Science, University of Cambridge Free School Ln, Cambridge CB2 3RH, United Kingdom
email: ps856@cam.ac.uk

Abstract. In this article, I analyze the claim that alien life could be "weird". I propose to distinguish between two types of astrobiological weirdness: a naturalistic and an epistemic one. Weirdness, naturalistically construed, is an attribute for astrobiological traits. Weirdness, epistemically construed, is an attribute for scientific hypotheses and theories about the nature of such traits.

Keywords. Weird Life, Life as We Do Not Know It, Life on Earth, Alien Biochemistry

1. Introduction

It is often said in astrobiology that extraterrestrial life could be very different from life on Earth. Alien organisms might use body plans, energy sources, or reproductive strategies unlike anything we know from Earth. Perhaps life could be based on solvents other than water. Perhaps life could exist on Jovian planets, in interstellar dust clouds, or in the interior of stars. Not all hypothetical, alternative ways of life are different in the same way. Some potential kinds of life seem more similar to life on Earth than others. Some seem quite plausible, while others sound outright fictitious.

This article discusses how to measure and classify ways in which potential, alternative forms of life may deviate from the kind of life we find on Earth. My focus will lie on the notion that life could be "weird". The scientific literature is considerably ambiguous in the way astrobiological weirdness is spelled out. I propose to distinguish between two different conceptions of weirdness: a naturalistic conception and an epistemic conception. Weirdness, naturalistically construed, is an attribute for traits possessed by organisms, taxa, ecosystems, or biospheres. Weirdness, epistemically construed, is an attribute for scientific knowledge claims that can be made about such traits. While I don't have the time to formulate an exact account for each kind of weirdness, I will sketch roughly what such accounts should have to look like in order to be scientifically useful.

2. Weird Life

The vocabulary used to describe weird kinds of life is notoriously messy. The NASA astrobiology strategy (Hays 2015) uses the label "weird life" to specifically refer to organisms and biospheres whose biochemistry is based on molecular scaffolds other than carbon and chemical solvents other than liquid water. Davies et al. (2004), on the other hand, speak of "weird life" to refer to any form of life that is biochemically different from known life on Earth, including carbon- and water-based organisms and biospheres. Worse even: In the scientific literature, we find a kaleidoscope of labels that have been used to refer to virtually any kind of way in which alien life might be biochemically or otherwise different from the kind of life we find on Earth: "life as we do not know it", "exotic life", "unfamiliar life", "life as it could be", etc.

In order to prevent getting lost in terminological quandaries, it is useful to specify what the notion of weirdness is supposed to do. If we consult the scientific literature, we find two contrasting roles that the vocabulary of weirdness performs in astrobiological theory and practice. First of all, the vocabulary of weirdness is used as a classificatory tool to separate forms of life that are similar to the kind of life we find on Earth from genuine "alternative" forms of life. Secondly, the vocabulary of weirdness is used as a measure for the degree of dissimilarity between known life on Earth and some genuine alternative form of life. In this regard, it is often proposed that differences in the traits that make life weird translate into differences in the ability to justify scientific beliefs about weird life. NASA (Hays Ibid., 147–148) writes:

"Astrobiology has involved exhaustive study of past and present life on Earth, providing lessons that can be extrapolated to environments on other worlds. This [...] is at the heart of how we plan to search those environments for signs of Earth-like life [...]. Those who propose to search for "weird life" must start by developing viable working models of such life."

In order to identify how to search for life outside of Earth, we need models for the kind of chemistry (of organisms as well as their abiotic habitats) we are looking for. Among many other things, this includes a model of the biological and abiotic chemical pathways that produce a given chemical substance, or pattern that can be recognized. The NASA report argues that the chemical models we have to search for the signatures of life we know are not informative in determining what kind of signatures to search for that might be produced by some form of life that does not share our own biochemistry. Arguably, in order to determine what phenomena to look for, where to look for them, etc. when searching for "weird life", we therefore first need to develop "viable working models" of such forms of life. Davies et al. propose a slightly different way to analyze weirdness. Rather than a catch-all distinction between forms of life that are ordinary life and forms of life that are weird, they (Ibid, 242) argue that weirdness comes in degrees, so that some forms of life might be weirder than others:

"A fundamental obstacle in searching for life as we do not know it is that we cannot be sure what to look for. It is therefore useful to construct a hierarchy of weirdness that progressively departs in its biochemistry from standard life. [. . .]The hierarchy of increasing weirdness is paralleled by the increasing difficulty in our ability to detect that life, even if it is located in our immediate vicinity."

According to Davies et al, some form of life A is weirder than another form of life B, if A is biochemically more different than B from the kind of biochemistry we know on Earth. The larger the difference in biochemistry, the more difficult it is arguably to predict what such life will look like. While Davies et al. and Hays seem to disagree about the extent to which some hypothetical biochemistry would have to differ in order to deserve to be labeled as "weird", they share some common ideas. Two of these ideas are relevant to spell out right now. (1) Astrobiological weirdness is a naturalistic attribute: Some form of life is weird if, and only if, it is biochemically different from the kind of life we are familiar with on Earth. (2) Biochemical difference corresponds to and/or correlates with predictive difficulty: The bigger the gap between the biochemistry of weird life and familiar Earth-life, the more difficult it is to make reliable predictions about the phenomena produced by weird life. Before we can put pressure on these claims, let's spell them out more clearly.

3. Naturalistic Weirdness

Davies et al. suggest to distinguish between "Earth-life", which supposedly refers to "all life on Earth, regardless of its biochemistry", and "standard life", which refers to "life that uses the general biochemical architecture for life on Earth as it is taught in standard textbooks". Their motivation for this distinction is that it could be the case that some yet undiscovered "shadow"biosphere exists on Earth that does not possess the general biochemical architecture taught in standard textbooks. This is sensible: If some non-standard life was able to originate on Earth, it would be odd not to call it Earth-life. As scientists may have not been able to discover all life that ever has evolved on Earth, it also makes sense to distinguish between Earth-life per se and"known" (or "familiar" or "extant") Earth-life. Usually, known life on Earth is characterized as an evolutionary tree or web of life that can be reconstructed from the three extant domains of life – Prokarya, Archaea, and Eukarya –, which share a last universal common ancestor (LUCA) (Weiss et al. 2016). Davies et al., as we have seen, think of astrobiological weirdness as some sort of a measure for how different some instance of life is from extant Earth-life, biochemically speaking. What weirdness measures is some sort of physicochemical distance in the space of biochemical possibility: The more the biochemistry of some hypothetical or actual form of life departs from standard life, the weirder it is.

The physicochemical perspective is compelling for a number of reasons. To begin with, it allows us to use the notion of weirdness across different levels of biological hierarchy. It also incorporates ecological considerations: Cellular biochemistry is causally, structurally, and functionally linked to the physics and chemistry of the abiotic environment. The structure and function of biomolecules can change drastically if environmental conditions (pH-levels, temperature, etc.) change. What kind of biochemical traits are possible or beneficial to have for some organism or population is constrained by the physics and chemistry of its environment. Finally, the physicochemical perspective provides us with a framework for identifying the kind of traits that some form of life must have to be considered weird: Biochemical or biophysical traits that life might have instead of standard traits.

In the astrobiological literature, we find a wide variety of proposals for such traits. Among other things, this includes alternative atomic constituents (bulk elements or trace elements), energy sources and energy-harvesting mechanisms, building blocks (other sets of sugars, lipids, nucleic acids, and amino acids, or entirely different classes of molecules), chemical solvents (non-aqueous liquids) or chemical milieus (fully or partially non-liquid), habitability constraints (temperature, pressure, salinity, UV-tolerance, etc.), molecular scaffolds (inorganic polymers and the chemical pathways to synthesize them), genetic codes (e.g., alternative substrates to encode hereditary information or different codification schemes), and morphotypes (e.g., cellular boundaries other than cell walls) (Grefenstette et al. 2024).

Davies et al.'s suggestion, as we have seen, is to "construct a hierarchy of weirdness that progressively departs in its biochemistry from standard life". It's not entirely clear what such a hierarchy should look like, considering the plurality of potential non-standard traits that have been suggested. For example, some hypothetical type of cell could use standard nucleic acids (RNA and DNA) but not for self-replication and still use amino acids to store hereditary information – some sort of "protein world". Another type of cell might use nucleic acids for self-replication and amino acids to build proteins but none of the standard building blocks to build these polymers, e.g., nucleotides other than A, T, C, G, and U. Which type of cell should be considered more different from standard cells?

Davies et al.'s suggestion seems to be that the astrobiological weirdness hierarchy is not strictly speaking a continuum but a ladder: Silicon-based life might not clearly be weirder than boron-based life or sulfur-based life. But it is clearly weirder than a carbon-based biochemistry with a proteome built from non-standard amino acids would be. A ladder of weirdness makes more sense, biochemically speaking. The molecular architecture of life is an integrated whole whose parts are highly specific in the way they fit to each other and work together. There are various developmental, chemical, and physical constraints that may prevent replacement of molecular constituents with potential analogues. Some constituents might be easy to replace without substantially changing the rest of the cell. But others will only be possible to replace along with an entire group. You can't replace phosphate without getting rid of (standard) nucleic acids, and you can't get rid of nucleic acids without losing the genetic code. Some potential ways to build a molecular architecture will unlikely overlap in any way at all. Water and ammonia, for example, have quite different melting and freezing points, so that they cannot co-exist in the liquid phase. Water-based life and ammonia-based life (if it exists), thus, will have a considerable gap between the physicochemical conditions (e.g., pressure and temperature) under which each of them can be found (Bains 2004)

A ladder of weirdness could be potentially modeled as a hierarchy of different degrees of biochemical incompatibility. According to this view, a weird form of life would be one that is biochemically incompatible with standard life. The more strongly incompatible it is, the weirder it is. A weak sense of incompatibility could be reproductive: Roughly speaking, we can say that some hypothetical or actual form of life is reproductively weird, if it cannot evolve from standard life. This could be due to developmental constraints, constraints in the architecture of the replication system, etc. A stronger sense of incompatibility could be ecological: Roughly speaking, life is ecologically weird, if it cannot exist in the same ecological niche as standard life. These suggestions are tentative. Further specification and differentiation will be necessary. What matters for scientific theorizing and practice is to provide a coherent frame-work for naturalistic weirdness that allows scientists to compare how distant different kinds of biochemical architectures are to each other in terms of their (in)compatibility, so that more compatible ones are more close to each other than less compatible ones.

4. Epistemic Weirdness

Is it more difficult to predict what some form of life might look like, the more its biochemical architecture departs from standard life? Davies et al. (Ibid.) argue that we will only be able to successfully detect alternative forms of life if they produce phenomena which we can reliably distinguish from the abiotic environment and artifacts on the basis of our current scientific knowledge of life. Accordingly, what explains predictive difficulties in hypothesizing about non-standard life is not that it possesses non-standard traits. Rather, it is that we possess much less reliable scientific knowledge about non-standard life than we possess about standard life. One of the reasons why silicon has been considered attractive as a potential carbon-alternative is that it can form molecular structures analogous to functional groups from organic chemistry (Petkowski, Bains & Seager 2020). A silicon-based metabolism would of course use a completely different, i.e., inorganic, set of chemical substances as standard life. However, this does not mean that silicon biochemistry and standard biochemistry cannot lead to various evolutionary convergences. This might include highly similar morphological traits on levels of biological hierarchy higher than the cellular level, such as camera-eyes of nervous systems. It also might include inorganic waste products of metabolic activity or other causal impacts on the abiotic environment produced by extant life. In other words, what makes it so difficult to determine how to search for silicon worlds is not that silicon-based biospheres cannot produce phenomena that standard life produces. Rather,the problem is that we can only constrain the possibilities very little on the basis of the evidence available to us. Davies et al. (Ibid.) high-light this point when they write that a "fundamental obstacle in searching for life as we do not know it is that we cannot be sure what to look for."

Some form life A that is biochemically more incompatible with standard life than some other form of life B will share less biochemical traits which we can hypothesize about on the grounds of currently available evidence – textbook knowledge, data from experimental studies, computer-simulations, theoretical investigations, etc. In this sense, it is accurate to consider predictions about silicon worlds more difficult to justify than predictions about RNA worlds. It is still unclear whether an RNA world ever existed on Earth. Likewise, it is unclear whether it is possible for silicon-based life to exist. However, there are more aspects of the molecular architecture of RNA-based biospheres that we can reliably hypothesize about than there are aspects of the molecular architecture of silicon-based life. It seems reasonable to expect that non-standard life forms closer to standard life than silicon-based life would be but more distant than RNA-worlds would fall somewhere in between.

Having said that, any such correlation is contingent on the evidence that is actually available to the scientific community at any given time. Consider a hypothetical scenario in which some future astrobiologists had successfully gained access to some biochemically very distant form of life. Imagine, for instance, that these future astrobiologists had been able to detect and exhaustively study a silicon-based biosphere S in the methane lakes of Titan. In that case, scientists may have been able to develop chemical models for the search for silicon-based life on extrasolar worlds. And yet, like astrobiologists today, these researchers would still lack a working model of some non-standard kind of carbon-based life C that does not use nucleic acids and amino acids. If it was more difficult to search for S than C because S is more distant than C from standard life, these hypothetical future astrobiologists should still find it more difficult to search for S than C.

As it does not have to be the case that predictions about alien life are less reliable, the more its biochemical architecture departs from standard life, it makes sense to analyze how unpredictable life is separately from how biochemically different it is. In both instances, we can think of "weirdness" as a form of incompatibility. We can use the term "naturalistic weirdness" to refer to forms of life that are biochemically incompatible with standard life. The more incompatible its biochemical architecture with standard life, the weirder it is, naturalistically speaking. Another sense of incompatibility is epistemic. Roughly speaking, two facts or claims are incompatible with each other, if they cannot both be true at the same time in light of the available evidence. "Epistemic weirdness", in this regard, can be thought of as a measure for how incompatible some hypothetical or actual form of life is with currently established, commonly accepted scientific knowledge.

Again, this suggestion is tentative. Further specification and differentiation will be necessary. What matters for scientific theorizing and practice is to provide a coherent framework for epistemic weirdness that allows scientists to compare how implausible or improbable it would be for life in our universe to have some traits rather than others.

5. Conclusion

In conclusion, it is fruitful to distinguish between ways in which alien life might be incompatible with the standard chemical architecture of extant life and ways in which alien life could be incompatible with commonly established scientific beliefs about life. The former type of incompatibility we can dub "naturalistic weirdness", the latter one "epistemic weirdness".

References

Bains, W. 2004, *Astrobiology*, 4, 2

Davies, P.C.W., Benner, S.A., Cleland, C.E., Lineweaver, C.H., McKay, C.P. & Wolfe-Simon, F. 2004, *Astrobiology*, 9, 2

Grefenstette, N., Chou L., Colon-Santos, S., Fisher, T.M., Mierzejewski, V., Nural, C., Sinhad, P., Vidaurri, M., Vincent, L. & Weng, M.M. 2024, *Astrobiology*, 24, S1

Hays, L.E. 2015, *NASA*

Petkowski, J.J., Bains, W. & Seager, S. 2020, *Life*, 10, 6

Weiss, M.C., Filipa L.S., Mrnjavac, N., Neukirchen, S., Roettger, M., Nelson-Sathi, S., & Martin, W.F. 2016, *Nat. Microb.*, 1, 9

(Toward) Discovery of Life beyond Earth and its Impact
Proceedings IAU Symposium No. 387, 2026
H. Landt, M. Dominik & C. Oliver, eds.
doi:10.1017/S1743921324001364

Equivocal encounters: alien visitation claims as a societal problem

Tony Milligan

School of History, Philosophy and Digital Humanities, University of Sheffield, UK
email: t.milligan@sheffield.ac.uk

Abstract. The contention of this paper is that alien visitation claims are a societal problem when they (a) move into the mainstream of discourse to the extent that government policy has to respond to them; (b) when they generate background noise which impedes science communication; and (c) when they become entangled with indigenous origin narratives, making it hard to recover the latter. Where this is the case, periodic debunking looks like a failed paradigm. Something closer to a scientific research program (SRP) might be called for, at some point. This is an idea which has already been advanced by Avi Loeb and Martin Elvis (albeit in significantly different ways and for different reasons). It is not clear that we are already at the stage where an SRP is required, but such a requirement does seem to be on the near horizon.) The paper concludes by setting out a number of framing requirements for any such SRP.

Keywords. alien visitations, Unidentified Aerial Phenomena, government policy, debunking, scientific research programme

1. Background noise

As a commonplace observation, the world could have been very different. It is possible to imagine a world identical to our own until around 1950 but which then diverged. In this alternative world, a small number of artefacts have since been discovered, giving plausible indications of the advanced engineering required for spaceflight. While not definitive, the artefacts are not easily attributable to any terrestrial culture or military program. In this alternative world, there are good reasons for belief that aliens have visited Earth. (For convenience, I will refer to this as belief in 'alien visitation.') The evidence is not absolutely conclusive, but reasonable standards of evidence have been met. Science in this alternative world is much the same as it is in our world, with the addition of research programs into possible alien technologies and the propulsion systems which might help to account for their presence.

This is a description of a possible world, but it is not our world. In our world, talk about alien visitation belongs mostly to counterculture and light entertainment. In our world, alien visitation claims are typically underwhelming. More particularly, they are regarded as underwhelming within the space community. But if we are underwhelmed by what has been offered as evidence, then we may owe a story about what would count as convincing evidence. A story that does not set up standards which are a great deal more demanding than those accepted in normal areas of research where a good deal is inferred rather than directly experienced. Science in our world often involves inference to the best explanation (IBE), with complex discipline-specific norms for what counts as 'best.'

So far, the best explanation for any natural phenomenon has never been alien visitation. Since around 1950, in our world, discussions about contact with other, similarly intelligent, alien beings have generally taken the form of a borderline scientific interest in SETI and METI together with philosophical speculation about the likelihood that any such beings exist, protocols for a hypothetical first contact, and speculation about whether or not evolution will have converged or taken some radically different path elsewhere. Beyond this, the dominant scholarly and scientific response has taken the form of periodically debunking visitation claims without become drawn into any ongoing research process that might accomplish little more than repeatedly pointing to the exceptionally low quality of the evidence presented and the poor argumentative framing. (Usually, some appeal to analogy rather than anything so robust as IBE.)

There are three reasons why this approach may not hold indefinitely. In one way or another, all of them involve the idea that widespread and unfounded alien visitation claims are a problem for society and hence a problem for science. Here, I am not interested in the peculiarities of individual agents many of whom, in any good society, are likely to believe all manner of strange things. Rather, I am strictly concerned with the societal role of widespread and unfounded belief in visitation.

The first reason why debunking is unlikely to be good enough response for much longer is a shift of belief in visitation from the counter-cultural fringes over into the political mainstream, evidenced by effective pressure in the US to engage in a mass release of Unidentified Aerial Phenomena related documentation. If beliefs of this sort (in conspiracy, concealment and collaboration) have made it into the mainstream, then periodic debunking has simply not worked. Moreover, if we hold that the practice of science in a democratic society requires the answerability of the science community to sustained public concerns then something more robust may be due. This will be the case even if the end story that is told ('no aliens, no coverup, no conspiracy') is likely to be the same.

The second reason for a different kind of response, beyond debunking, is that alien visitation narratives create a great deal of unhelpful background noise that gets in the way of science communication. Visitation claims, when they operate in this way, are again a societal problem. Moreover, the background noise of such claims has undergone exponential expansion with the arrival of social media. Much of the focus has been upon biology, through attempts to represent the discipline and its core concepts as little more than ideologically driven social constructs which are overly responsive to power elites. Effective science communication has become harder as the historical standing of science as one of our most reliable pathways to knowledge has come to count for very little online. If we value science communication (as we should do) then there well may come a time when we have to accept the need for a more robust response.

A third and related reason for looking beyond the paradigm of occasional debunking is that this same background noise has already compromised cultural astronomy, respect for which is a crucial aspect of good relations between indigenous peoples and an astronomy community whose ground infrastructure is often sited in remote locations and sometimes on indigenous lands. My own initial interest in visitation narratives stemmed from a growing sense of the problem of disentangling indigenous storytelling from counter-cultural narratives. One of the pathways for entangling is political activism where the authority of indigenous storytelling carries a tempting political cachet which can lead to an overriding of fidelity to the storytelling itself as narratives are reframed for the purposes of dissent. Again, this looks very much like a societal problem.

It may still be too early to call for an actual paradigm shift in the response of the science (and especially the astronomy) community. However, I do want to suggest that the possibility of such a shift sits somewhere on the horizon. Eventually, the background noise may become overwhelming. Whether or not we are 'there yet' is unclear, and I will remain officially neutral on the question. I will also concede that there could be a downside to such a change. A more engaged paradigm could initially increase the background noise in the hope of better long-term containment. Such long-term containment might never arrive. This is a significant concern. And so, I will restrict myself to the weaker claim that even if we are not yet ready for a more engaged response, we should already start to consider what such a response might look like.

2. Equivocal encounters

If this response takes the form of ongoing scientific research programs (SRPs) to evaluate alien visitation hypotheses, these need not go beyond the bounds of appeals to the natural, and hence they need not sit outside of the bounds of genuine science. However, the capacity of SRPs to command respect across a range of related and involved scientific disciplines would depend upon exactly what sort of hypotheses were being evaluated, as much as it would upon how the evaluation might proceed.

There is a lesson that might be learned here from the emergence of parapsychology departments. Even if the methodology happened to be rigorous, the subject matter itself puts the research beyond the bounds of anything likely to be taken seriously by peers in psychology or in the physical sciences. Similarly, anyone considering whether or not the pyramids were built using alien technology is unlikely to be taken seriously. Some filtering must take place to help differentiate between what is worth examining and what is beyond the pale.

But we need not think in terms of the outlandish. Rather, we can consider only possible artefacts of the sort brought into play in the scenario of the imagined world. In the absence of any response to METI, the artefact-focused approach has been the most credible game in town. It is the option pursued notably by Avi Loeb and by a number of Harvard researchers based around Loeb's Galileo Project. Rather than targeting the wilder horizons of dubious testimony about abduction, they have focused upon equivocal material evidence in forms such as possible derelict craft and possible physical residues. The interstellar object 'Oumuamua, in particular, has been claimed as a possible derelict spacecraft (Loeb 2021), and metallic spherules off the coast of Papua New Guinea have been claimed as possible residue which cannot readily be explained by the industrial revolution (Loeb et al. 2024). Claims which have been strongly disputed.

Of the two, the 'Oumuamua claim has been the more productive. While few astronomers have ever believed that it is an actual derelict alien craft, the exercise of showing that its trajectory can be explained in far simpler terms (e.g., by localised gas pocket formation followed by outgassing as a result of proximity to the Sun) has been a worthwhile exercise. It would also not be unfair to point out that there is a widespread view, across the space science community, that Loeb's overall approach is shaped too much by wanting to believe, that it is too entangled in the kinds of populist narratives about power relations within science that I allude to above.

In fairness to Loeb, it may simply be difficult to build any robust SRP program dedicated to evaluation of artifact claims under current conditions, without involving a disproportionate number of people who also want to believe, and who have a certain attitude towards the conservatism of more mainstream lines of scientific research. Furthermore, scientific expectations sometimes do shift over time. If are currently in an encounters-are-unlikely phase, this may reflect our current technologies rather than attitudes that we might on the other side of the revolution in robotics and AI. Expectations can shift in the light of technological change.

In line with this, I want to differentiate between two claims. With allowances for some slippage into more dramatic claims in public statements, what I take to be Loeb's view is that sooner or later, we will stumble upon remnants of, or trace residue from, an alien craft, and we may already have done so. Instead, I am sympathetic towards a position set out by Martin Elvis in a sympathetic but critical response to the Loeb claim about 'Oumuamua: a broad scientific research program can be built around evaluation of alien craft claims for a suitable range of objects from beyond the Solar System (Elvis 2022). This view is consistent with the idea that extraordinary claims require extraordinary evidence. Or, with a softer version, such that the evidence needs to meet a usual standard but we are entitled to set up a default position of rejecting extraordinary claims. In the case of alien visitation, we are entitled to assume that any such claims are far more likely to be false than true. A default assumption that Loeb seems, at times, to regard as closed minded. Rather, I take it that defaults of this sort are a familiar part of normal science, and they are often part of any disciplined form of enquiry.

At least three reasonable assumptions are shared across the the Loeb and Elvis approaches, in spite of their significant differences. First, when it comes to interstellar distances, ships are better at survival than crews. Any alien craft that we encounter would probably be a derelict and so it might not exhibit ongoing control. Second, an alien craft really could look like a naturally occurring object. A little like the papyri at Herculaenum which initially appeared to be charred pieces of wood rather than a precious treasury of text. Such an appearance could be accidental (a result of the ravages of time) or intentional (to avoid cultural impact or else to avoid any risks of detection and tracing back to a homeworld). Third, in line with the previous two points, any encounter could well be equivocal, i.e., what is encountered may be less than clear cut. The equivocal nature of such encounters with interstellar objects might leave open the possibility that one or more such encounters with a difficult to identify object, spread over a sufficient period of time, could eventually turn out to be the real thing. No laws of nature would be violated. Nothing odd need occur. And accepting this in no way does not automatically step outside of the normal bounds of science.

3. How can we structure SRPs for equivocal encounters?

The mere fact that I think that this will not happen is neither here nor there. Patterns of belief around these matters are social phenomena and as an ethicist it is how science deals with social phenomena that is my primary concern. Visitation beliefs and evaluation are socially significant even in the absence of visitation. To this end, I will suggest that, beyond focusing upon something tangible and within normal bounds (such as candidate artefacts), and beyond abiding by standard naturalistic constraints, any scientific research program in this area should meet at least three minimal requirements.

1. Independent value: a scientific research program geared to evaluating alien artifact and trace evidence claims should follow the astrobiological model in which the scientific value of the work does not depend upon actual discovery of alien life. Work on 'Oumuamua is a nice example of research which meets this independent value requirement. Research to explain 'Oumuamua's unusual trajectory did play off of the derelict craft claim, and upon a resulting search to find simpler ways to explain its acceleration. Simpler, in an ontological sense of positing a more economical range of entities.

2. Noise minimization: a good SRP should not feed the unhelpful background noise beyond its initial phases of operation. This requirement captures and makes explicit what is already done on a regular basis in science communication: keep reportage close to the research; avoid sensationalism but expect that it might be smuggled back in when bylines are written; and choose platforms for communication with a degree of caution. These are normal good practices. Of course, the price of any science communication at all comes at the risk of repackaging in more sensational terms and the initial stages of an SRP in this area would attract attention.

3. Evidential standards: the approach towards justification of claims adopted by a good SRP should be translatable into robust argument structures such as inference to the best explanation, rather than relying heavily upon analogy which has been a staple of alien visitation claims. This leads them to look suspiciously like creationist arguments in which intelligent guidance is repeatedly inserted into naturally occurring processes. The requirement for significantly higher evidential standards is not a special raising of the bar but a door to normalization of sufficiently well-formed and well-directed discourse.

4. Default option: the ontologically simplest option should normally be adopted. Having a default assumption in which we follow Ockham, and do not appeal to more entities than we need in order to explain phenomena, is not prejudice. It is normal science. All other things being equal, explanations in which appeal is made to aliens and explanations in which there is no need to appeal to aliens are not equal. Matters would be different if we already knew that there were aliens and that their traces could be discovered ready to hand. Such explanations would not then introduce an entity of a new sort. Until such times, the default holds. This also helps to explain why adoption of a default is not arbitrary: the default is indexed to our existing body of knowledge. Science does not begin anew with each arising problem.

References

Elvis, M. 2022, *Oumuamua Considered as an Alien Craft, International Journal of Astrobiology* 21, 63–77.

Loeb, A. 2021, *Extraterrestrial: The First Sign of Intelligent Life Beyond Earth Houghton Migglin Harcourt, New York, 2021.*

Loeb et al. 2024, *Recovery and Classification of Spherules from the Pacific Ocean Site of the CNEOS 2014 January 8 (IM1) Bolide, Research Notes of the AAS*, 8, 1, 39.

(Toward) Discovery of Life beyond Earth and its Impact
Proceedings IAU Symposium No. 387, 2026
H. Landt, M. Dominik & C. Oliver, eds.
doi:10.1017/S1743921324002060

Planetary Protection: Fact, Fantasy, and Its Future

John D. Rummel

Friday Harbor Partners LLC
email: rummelj@ecu.edu

Abstract. Planetary Protection is an effort to prevent inadvertent biological or organic contamination in solar system exploration. That effort cannot be perfect in actual exploration missions, but the price of not attempting it can be seen in numerous actual contamination events on Earth (e.g., kudzu and starlings in the US, Australia-released rabbits) and the more fanciful, but possibly valid, works of fiction (*The War of the Worlds, The Andromeda Strain*). Extraterrestrial contamination spread by Earth missions is known as forward contamination, while backward, or back contamination refers to contamination brought to Earth. Both robotic and human missions may be affected by planetary protection practices, which strive to apply the most current science to those efforts, but it is clear that those same missions may discover new information (the hope of most missions) that could argue for new and potentially altered requirements on the next mission, or on the ongoing missions themselves.

Keywords. Harmful Contamination, Life Detection, Planetary Science, COSPAR

1. Introduction

When we "boldly go" into near-Earth space our ignorance proceeds us. What we don't know can kill us! For example, we never expected the scope of the current space debris problem, but it is far worse than we thought possible. As we send more and more capable missions beyond Earth orbit, we need to keep in mind the limitations of our data and our understanding of what we see, adopting the reality that "ignorance is not bliss." We need to grow that understanding without causing irrecoverable harm to the space exploration enterprise or to our home planet. Here we will discuss planetary protection as both a practical and theoretical effort, and address its implications regarding emerging aspirations in space exploration and use.

In the mid-2020s, there are those who anticipate benefits – even riches – by moving into cislunar space and the asteroid belt (e.g., Kornuta et al. 2019, Stuart 2022), yet we don't really understand the physics of those locations (Metzger and Mantovani 2021), let alone the economics of working there. Our missions are continuing lunar exploration by robots while preparing to extend the hands-on knowledge gained by the human missions first undertaken by Project Apollo in the late 1960s and early 1970s. Missions to the lunar poles may discover accessible water ice of some quantity, or not. Mineable ice and minerals may be found on the Moon, and almost certainly in the asteroids (the key word is "mineable"). Given that such locations exist – and from an exploration perspective, some look very promising – we have to wonder, as stewards of these potentially rich locations, how far will we go before we regret the way that we have treated our new frontiers?

Some are in love with the concept of the frontier for its own sake (Service 1907), and those who promise a near-term economic return may depend on that love, rather

If an action or policy has a suspected risk of causing harm to the public or to the environment, in the absence of scientific consensus that the action or policy is harmful, the burden of proof that it is not harmful falls on those taking the action.

Figure 1. Narrow Precautionary Principle.

than on plans that include innovative mining technologies and proven ore deposits. Even then, the economics of space-dependent commodity markets are uncertain, given the need for a large up-front investment just to start such a business, and the problems of ending scarcity when space exploration yields large quantities of rare-on-Earth material. Commodity markets can be tricky.

But let us assume that economically productive locations do exist in outer space, and that missions to them are both achievable and supportable, and might even support long-range human habitation. If we learn the basics about what it takes to go there, stay there, and keep on going, we will find that those economics (and possible military concerns) will drive future resource development and habitation, along with concerns about harmful contamination in the exploration and use of outer space.

2. Ethical issues and the province of all mankind

Article I of the UN Outer Space Treaty (UN 1967) states that "the exploration and use of outer space, including the Moon and other celestial bodies, shall be carried out for the benefit and in the interests of all countries, irrespective of their degree of economic or scientific development, and shall be the province of all mankind." This may sound encouraging, and in the timeframe of the late 1960s reflected the dominance of the USSR and USA, as countries. Nonetheless, with current plans for the exploration and use of outer space trending more and more into the private sector, Article I has been almost completely overcome by events. There is no generally accepted way to apportion "benefit," let alone a way to capture and distribute it.

And of course, there are no guarantees in space exploration – it is not clear how one might apportion the various risks of space exploration and use, as part of the net benefit referred to (as an aspiration) in Article I.

Accordingly, a useful questions might be, "What are the costs (benefits vs. risks) of space exploration and human habitation off-planet? Who benefits? and Who pays? There is lots there for an accountant to do, on both sides of the ledger.

The basic concepts of ethical behavior are not complex: don't do things that you shouldn't do, and don't fail to do things that you should. Simple, but who decides what those things are? And how? When undertaking activities, such as space exploration, that have both obvious and hidden risks, an applicable principle to decide some of these questions is the Narrow Precautionary Principle (Fig. 1). This principle sounds good in terms of fairness, benefit to the actors vs. risk to the uninvolved public, but it is generally unworkable when applied to something as pervasive as space exploration and use. For example, the potential degradation of low Earth orbit caused by large satellite constellations has been balanced against the world-wide availability of internet services, while the development of huge, reusable launch vehicles balances the environment of South Texas or Florida against the overall survival of the human species in the face of dangers such as asteroid impacts and climate change on Earth (Musk 2017).

Perhaps such benefit vs. risk calculation could be better viewed with a nod to environmental ethics as understood for Earth environments, alone.

Environmental ethics, as applied to Earth environments, are concerned with the moral status of the environment and its non-human contents (individual organisms, species, communities, and ecosystems), and focuses on applying traditional ethical theories to the above concerns. It may be that all of environmental ethics can be fit within these

- How broad should our precautionary principle be for space exploration? Should it be applied to far-future human beings and Earth environments?
- Should we spend valuable resources that could be used on Earth to, *instead*, explore our solar system? What return could we expect from that investment?
- What non-human entities should have intrinsic moral status?
- What obligations do we have to living systems that are not part of the Earth's ecosystem?
- What benefits might the discovery of those systems bring?
- Where does the human ecosystem begin and end?
- What do we have to know before we let humans go beyond lunar orbit and be exposed to planetary environments?

Figure 2. Some Ethical Issues in Space Exploration.

Figure 3. The ice table beneath the 2008 Phoenix mission, cleared of dust by the landing plume.

theories, but this is not the place to debate that. Rather, a list of how such issues might be applied to space exploration is provided for consideration (Fig. 2).

3. Indications of our ignorance

The common characteristic tying together the ethical issues that populate this list is a reflection of our overall ignorance regarding other worlds, and to a great extent our lack of knowledge about some of the most basic facts about living systems – particularly microbial life in extraterrestrial environments. Because much of the Earth is relatively benign (to us), many of the adaptations available to extreme environments here, and potentially on planets such as Mars, are unfamiliar or unknown to us. One particularly interesting example is the situation at the landing site of the Phoenix mission which landed in the polar regions of Mars in 2008 (cf., Smith et al. 2009).

Fig. 3 shows a view of the Phoenix landing site below the lander itself. Of particular interest is the presence of an ice table with dust removed by the 12 landing thrusters. This ice table contains more water ice than can be held by the pore-space of the martian regolith at that location. Elsewhere, the mission excavated ice-laden dust, which sublimated over the course of a few days. The full depth of these icy deposits was not measured.

As interpreted in light of Fig. 4, the Phoenix site also provided an intriguing puzzle for those looking for a research project studying microbes, particularly hearty lichens,

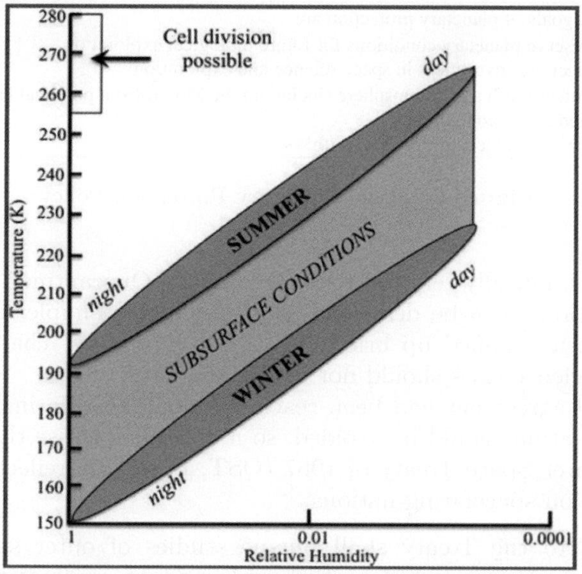

Figure 4. Subsurface conditions at the Phoenix site (Rummel et al. 2014). Note that cold temperatures (below 200 K/−73°C) are present whenever the relative humidity is above 60%, while summer temperatures of greater than 255K are associated with relative humidities well below 10%

in cold environments. For that project, it will be noted that there are lichens that can survive on the water provided by water vapor at relative humidities below 1, and it is also true that cell division by lichens can also occur when temperatures are approximately 25C below freezing.

At the Phoenix site, both sufficient water activity (as a vapor) and warmer temperatures may be present in the summer within the same 24-hour cycle, but not simultaneously. The question is whether there are lichens that can both divide and survive when water vapor and warm temperatures are not delivered at the same time. There appears to be no biophysical reason why they could not do that (Rummel et al. 2014)

Are there Earth microbes that might be pre-adapted for Northern Mars? Mars lichens? We just don't know! But we do know that such an opportunity exists, somewhere on Mars. Not only do we have ample evidence of past liquid water on Mars, but recent evidence (Wright et al. 2024) suggests a large volume of liquid water exists today, at depth, even in the mid-latitudes of Mars.

That is another example of how much we don't yet know about solar system environments – even though we have been to Mars numerous times. It's a big planet.

4. Wise heads and a measured approach

Long before space travel was a reality, the implications of biological contamination control were appreciated and illustrated in fiction—most notably in H. G. Wells (1898) novel, *War of the Worlds.* In that novel, pathogenic Earth organisms killed the Martian invaders and saved the day for the earthlings. (Note that such an outcome also provides a timely and instructive comment on one potential concern that might cause the human invasion of Mars to have a negative outcome.)

In the late 1950s, as the exploration of space became a reality, serious scientists and others saw the need to prevent the biological contamination of extraterrestrial environments, and to exercise caution when bringing extraterrestrial samples to Earth. Actions to deal

The basic goals of planetary protection are: • To preserve planetary conditions for future biological exploration – Protect our investment in space science and exploration • To protect Earth and its biosphere (including the Moon) from potential harmful extraterrestrial contamination – Simple prudence; protect the Earth!

Figure 5. Basic Planetary Protection Policy

with those concerns, initially referred to as "Planetary Quarantine," and now known as "Planetary Protection," can be described at-length, but a simple rendering is given in Fig. 5, and has been summed up briefly (by "The Simpsons" character, Bart, on the black board) as "Science class should not end in tragedy."

By 1967, general agreement had been reached among spacefaring nations that interplanetary contamination should be avoided, so it was no surprise that Article IX of the United Nations Outer Space Treaty of 1967 (OST, UN 1967) reflected that agreement, placing obligations on spacefaring nations:

> "...parties to the Treaty shall pursue studies of outer space including the Moon and other celestial bodies, and conduct exploration of them so as to avoid their harmful contamination and also adverse changes in the environment of the Earth resulting from the introduction of extraterrestrial matter..."

NASA had named its first Planetary Quarantine Officer a few years earlier under the authority of the Space Act of 1958, but the UN treaty made international cooperation and consultations about planetary protection a regular part information exchange regarding the search for life in space. This exchange was effected, in part, through COSPAR, the Committee on Space Research.

In 1958, COSPAR, was formed as an interdisciplinary committee of ICSU, the International Council of Scientific Unions (now International Council on Science), and event quite significant to the story of planetary protection. From the first, COSPAR had both an international scientific perspective and an interest in extraterrestrial contamination (Meltzer 2011). Following that interest (especially since 1999), COSPAR has promulgated a planetary protection policy that has both served international space cooperation on its own, and has regularly supported the work of the UN Committee on the Peaceful Uses of Outer Space (COPUOS). For example, COPUOS, in 2017, specifically recognized:

"the long-standing role of COSPAR in maintaining the planetary protection policy as a reference standard for spacefaring nations and in guiding compliance with Article IX of the Outer Space Treaty."

5. COSPAR's planetary protection policy

COSPAR, with a current membership of 46 National Scientific Institutions and 13 International Scientific Unions, has a number of roles in space science and exploration. COSPAR strives to promote the use of space science for the benefit of humankind by organizing biennial scientific assemblies, with strong contributions from most countries engaged in space research, and provides the means for rapid publication of results in its journals. Most importantly, here, COSPAR advises, as required, the UN and other intergovernmental organizations on space research matters or on the assessment of scientific issues in which space can play a role, and prepares scientific and technical standards related to space research.

COSPAR has had a policy regarding the potential risks for contamination of other solar system bodies since 1963, and further elaborated that policy through a number of decisions and resolutions thereafter. Those were separated and often hard-to-find, leading some OST signatories to plan missions outside of the COSPAR guidelines out of ignorance. To address that problem, in 1999 the COSPAR Panel on Planetary Protection was formed, and in October 2002 the COSPAR Bureau and Council adopted a new, complete, planetary protection policy for the use of space-faring nations. That 2002 version of the policy also specified that the policy should be supported by the best available science, and that it should be amended, as necessary, to meet the needs of entities participating in space exploration and use.

As of this writing, the COSPAR Policy on Planetary Protection (COSPAR 2024) has been added-to and modified, but the basic tenets of the 2002 version continue in the most recent version issued in 2024. As that policy states: "COSPAR maintains this Policy on Planetary Protection (hereafter referred to as the COSPAR PP Policy) for the reference of spacefaring nations as an international voluntary and non-legally binding standard for the avoidance of organic-constituent and biological contamination introduced by planetary missions." The 2024 version of the COSPAR policy has benefitted from the editing and additional explanation of the policy, and efforts "to highlight the key assumptions that form the basis for the technical guidelines. No new assumptions have been added at this time, but additional rationale and background references have been added to each assumption." (Ehrenfreund et al. 2024).

The COSPAR PP Policy "emphasizes . . . that all entities conducting space activities beyond Earth orbit must implement controls commensurate with the mission type and targeted body's significance for understanding the process of chemical evolution and/or the origin of life and the potential for adverse impacts to the Earth's biosphere." The controls applied are, of necessity, different for robotic missions than those for missions involving a human crew.

5.1. *COSPAR planetary protection categories for robotic missions*

For robotic missions, avoiding both the contamination of other space environments (forward contamination) can be a priority, while protecting Earth (from backward, or back contamination) must trump even that. Accordingly, there are a number of spacecraft and mission preparations that are available, depending on the exact nature of the target bodies to be encountered. In the COSPAR policy, such missions are sorted into four different categories for outbound missions, and a fifth category that applies to sample-return missions.

Category I missions comprise all types of missions to a target body which is not of direct interest for understanding the process of chemical evolution and/or the origin of life. No protection of such bodies is warranted, and no planetary protection guidelines are imposed by the COSPAR PP Policy.

Category II missions comprise all types of missions to those target bodies where there is significant interest relative to the process of chemical evolution and/or the origin of life, but where scientific opinion provides a remote chance of contamination by organic or biological materials which could compromise future investigations of the process of chemical evolution and/or the origin of life. [There are subcategories regarding specific missions to the Earth's Moon.]

Category III/IV missions comprise certain types of missions (mostly flyby and orbiter) to a target body of chemical evolution and/or origin of life interest and for which scientific opinion provides a significant chance of contamination by organic or biological materials which could compromise future investigations of the process of chemical evolution and/or the origin of life. [There are subcategories regarding specific missions to the planet Mars.]

Category V missions comprise all Earth-return missions and is given in addition to the outbound (I-IV) categorization. The concern for these missions is the protection of the terrestrial system, the Earth and the Moon. (The Moon should be protected from backward contamination by other celestial bodies to ensure unrestricted Earth-Moon travel.) For solar system bodies deemed by scientific opinion to have no indigenous life forms, a subcategory, "unrestricted Earth return," is defined. For all other Category V missions, a subcategory is defined as "restricted Earth return."

As noted, the categories assigned to the various bodies of the solar system depend on the most current science regarding those bodies, and to some extent the understanding of Earth organisms. As an example, if the current categorization scheme had existed for the Viking missions of the 1970s, the Viking landers would have been assigned to Category IV. That would not have been immediately done when Viking was first approved – rather, Viking's requirements were based on 1960s advice and 1970s research, and the eventual mission objectives to try to detect microbial metabolism and the presence of martian organic chemistry. Today that would result in a Category IVb classification and a stringent requirement for lander cleanliness. Viking took that very seriously, and showed that a full-system, terminal sterilization, can work well, when specified in advance.

For missions to elsewhere in the solar system, COSPAR has assembled a list showing the likely current assignment for missions to other solar system bodies:

5.2. *Category-specific listing of target body/mission types for robotic missions*

Category I: Flyby, Orbiter, Lander: Undifferentiated, metamorphosed asteroids; Io; others to-be-defined (TBD).

Category II: Flyby, Orbiter, Lander: Venus; Moon (Cat. II, IIa & IIb); Comets; Carbonaceous Chondrite Asteroids; Jupiter; Saturn; Uranus; Neptune; Ganymede*; Callisto; Titan*; Triton*; Pluto/Charon*; Ceres; Kuiper-belt objects $> 1/2$ the size of Pluto*; Kuiper-belt objects $< 1/2$ the size of Pluto; others TBD *The mission-specific assignment of these bodies to Category II should be supported by an analysis of the remote potential for contamination of the liquid-water environments that may exist beneath their surfaces.

Category III: Flyby, Orbiters: Mars; Europa; Enceladus; others TBD

Category IV: Lander Missions: Mars; Europa; Enceladus; others TBD

Category V: Any Earth-return mission

- "restricted Earth return": Mars; Europa; others TBD
- "unrestricted Earth return": Venus, Moon; others TBD

5.3. *COSPAR planetary protection policy for human missions*

Recognizing that a human crew would generate its own potentially contaminating community of microbes, and are generally unwilling to submit to "full-system, terminal sterilization," and other mitigating anti-microbial efforts, COSPAR uses a different approach in its PP Policy for human missions. Due to the nature of the lunar environment, the planetary protection challenges for human missions to the Moon are no longer thought to include a focus on the presence of contaminating microbes – neither those taken to the Moon, nor dangerous lunar microbes brought to Earth.

Humans can return to the Moon and stay without major efforts in the implementation of planetary protection measures, but the next target for human missions, and the applicable COSPAR policy, appears to be a trip to Mars. Thus:

> <u>Principles and Guidelines for Human Missions to Mars</u>
> The intent of this planetary protection policy is the same whether a mission to Mars is conducted robotically or with human explorers. Accordingly, planetary protection goals should not be relaxed to accommodate a human mission to Mars. Rather, they become even more directly relevant to such missions—even if specific implementation requirements should differ.

General principles include:

Safeguarding the Earth from potential back contamination is the highest planetary protection priority in Mars exploration.

The greater capability of human explorers can contribute to the astrobiological exploration of Mars only if human-associated contamination is controlled and understood.

For a landed mission conducting surface operations, it will not be possible for all human-associated processes and mission operations to be conducted within entirely closed systems.

Crewmembers exploring Mars, or their support systems, will inevitably be exposed to martian materials.

In accordance with these principles, specific implementation Guidelines for human missions to Mars include:

Human missions will carry microbial populations that will vary in both kind and quantity, and it will not be practicable to specify all aspects of an allowable microbial population or potential contaminants at launch. Once any baseline conditions for launch are established and met, continued monitoring and evaluation of microbes carried by human missions will be required to address both forward and backward contamination concerns.

A quarantine capability for both the entire crew and for individual crewmembers shall be provided during and after the mission, in case potential contact with a martian life-form occurs.

A comprehensive planetary protection protocol for human missions should be developed that encompasses both forward and backward contamination concerns, and addresses the combined human and robotic aspects of the mission, including subsurface exploration, sample handling, and the return of the samples and crew to Earth.

Neither robotic systems nor human activities should contaminate “Special Regions” on Mars, as defined by this COSPAR policy.

Any uncharacterized martian site should be evaluated by robotic precursors prior to crew access. Information may be obtained by either precursor robotic missions or a robotic component on a human mission.

Any pristine samples or sampling components from any uncharacterized sites or Special Regions on Mars should be treated according to current planetary protection category V, restricted Earth return, with the proper handling and testing protocols.

An onboard crewmember should be given primary responsibility for the implementation of planetary protection provisions affecting the crew during the mission.

> Planetary protection requirements for initial human missions should be based on a conservative approach consistent with a lack of knowledge of martian environments and possible life, as well as the performance of human support systems in those environments. Planetary protection requirements for later missions should not be relaxed without scientific review, justification, and consensus.

These are fully reasoned principles and guidelines for human missions, and COSPAR continues to address the knowledge gaps that must be filled to allow them to be part of a future, successful, Humans-to-Mars strategy (Spry et al. 2024). They are an essential part of that future that is often overlooked in programs that are trying to fit a crewed mission to Mars into a series of variable allocations in a governmental budget cycle. Doing Mars "on the cheap" is seen as a recipe for failure, immediate or long term.

Then how do space explorers escape the vagaries of government funding?

5.4. *A gold rush needs gold – commercial implications of the COSPAR policy*

As we learn more, but by no means all, of the secrets of the solar system it is not unusual to find technocrats and investors who are looking to make money – and even non-governmental money, in space. This may be a step in the right direction, although often times their business plans have subtle flaws that are due to a lack of understanding of the real solar system, versus the hoped-for solar system that meets their expectations. For example, where do you sell rocket fuel made from lunar polar water, and how much does it cost to get the fuel to a paying customer? And are you going in the same direction of your customer when you get there?

That's important!

In a perfect world, the question would be, "does your plan meet COSPAR's guidelines?" And the answer would be, "Thanks?" But for the most part, New Space companies seem to prefer to opt out, resisting even the simplest steps to understand enough about Mars (or elsewhere) to allow commerce-oriented Martians to realize that it's not only the red dust that they need to conquer. Rather than trying to kill off measures that address ethical concerns (and Article IX of the OST), a simple reading of the COSPAR PP Policy should suffice. Therein, we see that, for most of the solar system, there are no obvious conflicts between adherence to the COSPAR Planetary Protection Policy and the needs/desires of commercial entities. Under that policy, missions to the asteroids and the Moon would not have operational requirements imposed on them—only record keeping would be required. Beyond those missions, for the foreseeable future planetary protection conflicts would be limited to those generated by human activities on and around Mars, which has many conflicting usage issues (science, human exploration and colonization, and commercial use). Outside of national agencies, however, interest in Mars is limited to a few companies with the nascent capability of having their own missions.

COSPAR guidelines need not be a barrier to future commerce, even in the face of a rush to get rich in space. It would behoove nations wishing to cooperate in space to work together on the maintenance of those guidelines. Even if commercial space missions become dominant, robotic missions to intriguing destinations – like the Europa Clipper – should still attract funding and scientific inquiry (Fig. 6).

6. The Artemis Accords and the ILRS

Whereas it may be important to cooperate in space science and exploration as the core of future economic activity in space, generating a single treaty that can regulate such activity is not simple. It's not just that COPUOS operates by consensus, or that one person's enlightened regulation is another's unbearable "yoke." More likely it is because

Figure 6. The Europa Clipper aims to characterize Jupiter's moon Europa, and its potentially inhabited ocean. A COSPAR Category III mission. (Credit: NASA/JPL-Caltech)

there is currently no clear advantage for any nation to impose such regulations, despite the example of Earth-orbital debris and large satellite constellations, and their increasing harm to the space environment.

One effort to instill norms of behavior among space faring nations, and their commercial offspring, is known as "The Artemis Accords," because its focus has been to pave the way for cooperation in the further exploration of the Moon and cislunar space. Nonetheless, the full title to the Accords is, "*Principles for Cooperation in the Civil Exploration and Use of the Moon, Mars, Comets, and Asteroids for Peaceful Purposes.*" The Accords is a US-Government-based effort that is implemented through a series of bilateral agreements between the US and the other signatories. So far the Accords have been been signed by over 40 Nations, including many of the European Space Agency's 22 member states.

The principles set out in the Accords "are intended to apply to civil space activities conducted by the civil space agencies of each signatory," and "each signatory commits to taking appropriate steps to ensure that entities acting on its behalf comply with the principles" of the Accords. While the main thrust of the Accords follows the OST, the principles elaborated therein are listed as "Peaceful Purposes, Transparency, Interoperability, Emergency Assistance, Registration of Space Objects, Release of Scientific Data, Preserving Outer Space Heritage, Space Resources, Deconfliction of Space Activities, and Orbital Debris."

Only after one lists the principles of the Accords does it become apparent that no provision regarding harmful contamination exists in the Accords, and that only the sharing of scientific data is to be protected, but not the ability to conduct science, itself. A missed opportunity, to be sure.

Elsewhere, China and Russia are embarked towards building an International Lunar Research Station (ILRS), which has its own, undisclosed set of principles for participation. Its stated purpose is to support lunar research, exploration, and technology verification, as well as crewed lunar missions. Aside from the desire by the parties to show leadership and address their own priorities, there appear to be no specific technical/engineering incompatibilities between the ILRS and the Artemis Accords. One area that may need standardization could be the anticipated cabin pressures of any habitat/cabin, versus pressures for any EVA suit used to support human exploration, outside. This is an area where human limitations cry out for wise choices.

7. Things to watch in the future

Planetary protection protocols have been adopted to preserve science opportunities inherent or explicit in space exploration and use. To date, those science opportunities have been the prime motivation for all space missions beyond Earth orbit – scientists have been the customers, and they have been able to justify large infusions of public funds to pay their bills. Those funds have come with the understanding that the science community would respect its own, internally generated, rules and procedures in an effort to maximize scientific productivity and avoid tawdry disagreements within that community (above a certain level). One such set of rules and procedures is planetary protection, whereas another would be the general understanding that, in the US and Europe, approximately every 10 years a scientific consensus will be generated for the various science disciplines, and the community will support that "decadal survey" in as much of a single voice as is possible. Either the community chooses what's next, or the politicians do. (It could be noted that scientists can make fine politicians, and that the process of building a scientific consensus can, at times, be quite messy.)

At present, there are a number of factors driving a desire for privatization and commercialization of planetary space missions. One, clearly, is to short-cut the 10-year process noted above. Scientific data in many areas of planetary science could be quite valuable, if taken with an eye to supporting economical resource extraction or unlocking the potential in the possible discovery of extraterrestrial life. Governmental log-jams and open-data policies are not necessarily helpful under some scenarios, whereas private/public partnerships might be compelling, if properly structured. Either way, the quality of the data is essential to success, and planetary protection guidelines are essential to that quality. The discovery of microbes from Florida on the surface of Mars would likely not attract much venture capital.

Likewise, for those seeking to meet other, long-term goals on the Moon or on Mars – such as establishing a beach-head for humanity elsewhere in the solar system – then planetary protection measures are really about choosing your neighbors. Crop-dependent life support systems will not thrive if crop pathogens are also imported our new home, and we can all do without *Anthrax*. And given that there are Earth organisms that can seal cracks in poured concrete (Jonkers et al. 2010), we need to consider how well a martian aquifer would do with such tenants. Existing Martians may not be anticipated in our plans, but we need to know who we are bringing with us, regardless. Avoiding surprises in the space business is a good survival skill.

And if we intend to survive and thrive in outer space, planetary protection provisions are something we need to take with us.

Acknowledgements

I would like to thank Hermine Landt for her invitation to participate in an excellent symposium, and to thank the Kavli Foundation and the IAU for making it possible for me to be a part of it. I would also like to thank Vicky Hipkin for her reading of the paper and helpful suggestions.

References

COSPAR. 2024. COSPAR Planetary Protection Policy. Space Research Today 220, 14

Ehrenfreund, P., Worms, J-C., Coustenis, A. *et al.* 2024. Editorial to the new restructured and edited COSPAR Policy on Planetary Protection. Space Research Today 220, 10

Jonkers, H.M., Thijssen, A., Muyzer, G., *et al.* 2010. Application of bacteria as self-healing agent for the development of sustainable concrete. Ecol Eng 36:230

Kornuta, D., Abbud-Madrid, A., Atkinson, J., *et al.* 2019. Commercial lunar propellant architecture: a collaborative study of lunar propellant production. Reach, 13, 100026

Metzger, P.T., and Mantovani, J.G. 2021. Dust transport and its effects. In: The Impact of Lunar Dust on Human Exploration. (Levine JS ed.) Cambridge Scholars: Newcastle Upon Tyne; pp. 88–111

Musk, E. 2017. Making humans a multi-planetary species. New Space 5, 46

Rummel J. D., Beaty D., et al. 2014. A new analysis of Mars "Special Regions": Findings of the second MEPAG Special Regions Science Analysis Group (SR-SAG2). Astrobiology 14, 887

Service, R. W. 1907. The spell of the Yukon. In: Songs of a Sourdough, T. Fisher Unwin, London

Smith, P.H., Tamppari, L.K., Arvidson, R.E., *et al.* 2009. H_2O at the Phoenix landing site, Science 325, 58–61

Spry, J.A., Siegel, B., Bakermans, C., *et al.* 2024. Planetary protection knowledge gap closure enabling crewed missions to Mars. Astrobiology 24, 230 https://doi.org/10.1089/ast.2023.0092

Stuart, C. 2022. A race is afoot to make billions from the Moon's resources. Science Focus. https://www.sciencefocus.com/space/new-space-race-moon (Downloaded 6 September 2024)

United Nations. 1967. Treaty on Principles Governing the Activities of States in the Exploration and Use of Outer Space, Including the Moon and Other Celestial Bodies, A/RES/21/2222 (19 December 1966)

Wells, H. G. 1898. The War of the Worlds. Heinemann, London. 287 pp

Wright, V., Morzfeld, M., and Manga, M. 2024. Liquid water in the martian mid-crust. PNAS 121, e2409983121

(Toward) Discovery of Life beyond Earth and its Impact
Proceedings IAU Symposium No. 387, 2026
H. Landt, M. Dominik & C. Oliver, eds.
doi:10.1017/S174392132400187X

Making Habitable Worlds: Planets Versus Megastructures

Raghav Narasimha Ponnaganti, Margarita Safonova and Chandra Sivaram

Indian Institute of Astrophysics, Bangalore, India
email: narasimha.p@gmail.com

Abstract. With billions of planets in the galaxy, advanced civilizations could relocate planets within or into their planetary systems rather than destroy entire planetary systems to construct megastructures. Such shifts could create Strange Exoplanetary Architectures (SEA), with unusual planetary arrangements potentially indicating deliberate actions by extraterrestrial intelligence (ETI). Searching for biosignatures and technosignatures in these systems could be a promising method for detecting ETI activities.

Keywords. Megastructures, Habitable Zone, Free-Floating Planets, Service Worlds, Technosignatures

1. Introduction

Star KIC 8462852† caught the world's attention when the observers (Boyajian *et al.* 2016) struggled to explain the irregular dips in the brightness. These unusual dips could not be explained by stellar variability or planetary transits. Theories involving swarms of comets or asteroids proved inadequate, thereby propelling an alternative hypothesis into the forefront – that KIC 8462852 could be enveloped by a Dyson Sphere (DS) constructed by an advanced ETI to harness the star's energy output, and this is what causing these variations (Wright *et al.* 2016). This has put the world into an observational frenzy centred on this star. Finally in 2018, the light curve made in three colours showed chromaticity – the blue light was blocked much more than red light during the star's dimming, a characteristic of the dust (from comets or asteroids) rather than of something made from solid material (Boyajian *et al.* 2018; Deeg *et al.* 2018; Schaefer *et al.* 2018)), but it was to late to thwart ET hopes – the world was woken up to the idea of megastructures.

The possibility of building giant structures by advanced civilization to better utilize the energy output from their star was suggested very early in the 20th century, possibly first in a 1929 non-fiction book by J. D. Bernal, which describes the space habitats intended for permanent residence in case Earth becomes unlivable, able to host several hundred thousand people (Bernal 1929). The knowledge of the Solar System was well developed at that time (Pluto was discovered in 1930), and Bernal was suggesting using asteroids, Saturn's rings, etc., to construct his habitation globes: "*...the great bulk of the structure would be made out of the substance of one or more smaller asteroids, rings of Saturn or other planetary detritus.*". Olaf Stapledon further developed the concept of space habitats in his 1937 book *The Star Maker*, where he described that stars without natural planets can be made habitable by constructing artificial habitats around them (Stapledon 1937). These artificial worldlets – hollow globes of different sizes – were to

† Formally Boyajian's star, now known as Tabby's star after its discoverer Tabetha Boyajian.

be arranged in concentric rings at some particular distance from the star. All this was possibly the inspiration for Dyson to introduce his megastructure. Dyson acknowledged that Stapledon's conceptualization of an artificial biosphere served as an inspiration for his own work (Dyson 1979).

2. Dyson Sphere

The Search for Extraterrestrial Intelligence (SETI) enterprise originated from Morrison and Cocconi's suggestion to search for radio signals emitted by ETI (Cocconi and Morrison 1959). In 1960, Freeman Dyson proposed an alternative way of detecting ET technology – by looking for sources in the infrared (IR) region that are radiating at temperatures suitable for sustaining life, yet with a power output equivalent to that of a star (Dyson 1960). Dyson's conceptualization of his artificial biosphere (megastructure) was based on three key assumptions:

- *Malthusian theory of population growth.*
- *Our own Solar System as a model for other planetary systems.*
- *Jupiter as the resource for building the megastructure.*

Dyson hypothesized that if the total mass of Jupiter were evenly distributed within a spherical shell around the Sun at an orbital radius of approximately 1 astronomical units (AU), the resulting shell – with a thickness of 2 to 3 meters and a mass† of 200 g/cm^2. A shell of such thickness could be made comfortably habitable and can contain all machinery required for utilizing all solar radiation falling on it creating an artificial biosphere around the star. The temperature of such megastructure might be around 200 to 300 K, which would be radiating at a wavelength around 10 micron (Dyson 1960). Dyson further expounded his idea by saying that (Dyson 1966)‡:

- *It is possible to build* ***large*** *rigid structures in space.*
- *It is possible to build* ***light*** *rigid structures in space.*
- *It is possible to take planets apart.*

3. Drawbacks of Dyson's Idea

3.1. *Wrong assumptions*

A. Malthusian theory of population growth.

The Malthusian theory, which posits that population growth will inevitably outstrip the availability of resources, has faced considerable criticism due to its inability to account for technological advancements, shifts in societal norms, and other factors that have influenced population dynamics. It has been shown to be flawed in various respects (Kallis 2019). Dyson himself acknowledged that his assumption of population growth surpassing the Malthusian limit was "anthropomorphic" (Maddox 1960).

B. Our own Solar System as a model for other planetary systems:

Dyson wrote this article over three decades prior to the discovery of exoplanets; he must have thought that our Solar System is a typical planetary system. In fact, most known planetary systems do not have enough planets (rocky construction material) to build megastructures; the majority of known planets are in single-component systems; only one more system is known, apart from the solar system, to host eight planets (KOI-351).

C. Jupiter as the resource for building the megastructure:

Only about 13% of Jupiter's mass is practically usable for construction, as Jupiter is primarily composed of hydrogen and helium (Militzer *et al.* 2008). Even if we were to break up all the inner planets, we would only obtain 11.79×10^{24} kg of material, which is sufficient for a 1-AU shell with a mass of 42 kg/m^2, equating to a mere 1.5 cm in

† Expressed here in mass per unit area, because of unknown construction material.

‡ The emphasis is added by authors.

thickness. If we used all the rocky material available in the Solar System (approximately 1.82×10^{26} kg, including the gas giants' cores), it would allow for a 1-AU shell with an average thickness of 8 cm, resulting in a mass of 650 kg/m^2 (Harrop and Schulze-Makuch 2010). Dyson himself clarified that "a solid shell or ring surrounding a star is mechanically impossible." What Dyson actually envisioned was a "collection or swarm of objects traveling on independent orbits around the star" (Maddox 1960).

3.2. *Destruction of Heliosphere*

Solar System's heliosphere serves as a protecting shield for life on our planet, shielding more than 75% of cosmic rays from the Milky Way galaxy (Cummings *et al.* 2016). Constructing a DS around the Sun will trap the solar wind inside, resulting in the obliteration of the heliosphere. This, in turn, would allow the interstellar medium and cosmic rays to affect the DS. Whether the biosphere is inside or outside the DS, it will be affected. Even shrinking of the heliosphere can affect life on the Earth. Notably, even a reduction in the size of the heliosphere can exert profound effects on life on the Earth. A historical instance exemplifies this: approximately 2 to 3 million years ago, the heliosphere experienced a contraction, shrinking to a mere 0.22 AU in radius. Coinciding with this event was one of the episodes of mass extinction that occurred during the same period (Opher *et al.* 2022; Opher and Loeb 2022). This correlation underscores the intricate relationship between the state of the heliosphere and the sustenance of life on Earth, emphasizing the significance of its role as a protective boundary against cosmic hazards.

4. An Alternative: Moving Planets into the Habitable Zone

The motivation behind constructing megastructures stemmed from the aim of addressing the energy crisis faced by advanced overpopulated civilisation – with the eventual need for extra living space and resources. It is worth considering that we can potentially tackle the energy crisis by delving into aspects that Dyson might not have accounted for initially. Dyson's premise hinged on the idea of utilizing Jupiter as a resource, albeit without acknowledging a critical factor: only a fraction of Jupiter's mass, approximately 13%, is practically employable for construction purposes due to the predominance of hydrogen and helium in its composition.

If the energy requirement of some advanced civilisation increases, they even might have advanced commercial nuclear fusion reactors at their homes to fulfill their energy needs (like how we have solar panels right now) instead of building DS. Even if the energy sources run out on the Earth because of high energy demands, the gas giants are there for the rescue! If we convert Jupiter's hydrogen content to energy by thermonuclear reactions, the energy released is $\sim 10^{42}$ J. Even if we consume energy at the rate of solar luminosity, we can manage for more than 100 million years (Shklovskii and Sagan 1966). However, as the global population increases, the urgency for more habitable area and resources intensifies, necessitating an expansion beyond our planet. Basically, we may need more habitable planets!

It is theoretically possible to manipulate planetary trajectories using lasers. By deploying multiple lasers from different directions, it is feasible to manage the transfer of momentum, thereby enabling precise adjustments to a planet's trajectory while minimizing unintended effects on other planets. Any displacement of a planet should be done very gradually to minimize the disruptions to the orbital dynamics of the system. Sudden changes could have unpredictable effects, potentially destabilizing other orbits. The use of laser technology to reposition planets in the habitable zone (HZ) would facilitate the terraforming process. The notion of moving planets existed more than a century ago. In 1895, Konstantin Tsiolkovsky suggested that the intelligent beings of the future may

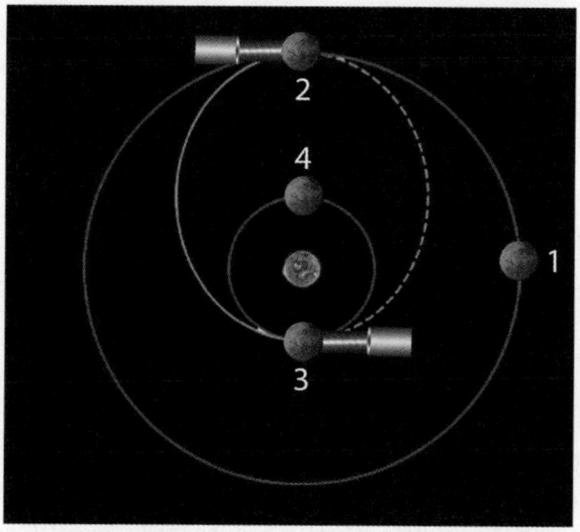

Figure 1. Hohmann transfer of Mars using lasers (the image is not to scale). The representation of a single large laser is merely illustrative; in practice, the configuration might involve multiple lasers operating from various orientations.

control the motion of the planets, saying "*They drove the planet as we drive horses*" (Tsiolkovsky 1895).

Shifting planets like Mars, Pluto, or free-floating planets into the Solar System's HZ could significantly alter their environments, potentially making them more suitable for life. For Mars, moving it closer to the Sun might warm the planet, melting polar ice caps and releasing water, which could aid in terraforming efforts. However, the increased solar radiation could also cause atmospheric changes, potentially leading to a greenhouse effect or atmospheric loss. Pluto, if relocated to the HZ, would experience a substantial rise in temperature, requiring atmospheric and surface transformations to support life. Free-floating planets brought into the HZ could also undergo similar changes, possibly developing habitable conditions if they possess a sufficient atmosphere and water. However, the challenges and risks associated with such endeavors would require careful planning and advanced technology. Even Trans Neptunian objects (TNO) and asteroids, when moved into the HZ, could become sources of water or other resources and possibly develop more dynamic environments.

ETI might intentionally relocate planets within their system for purposes like industrial exploitation, energy generation, or waste processing. These uninhabitable planets, termed "service worlds" (Wright *et al.* 2022), could serve various functions according to their characteristics. For example, gas giants might be moved closer to harness their hydrogen for nuclear fusion, while icy planets like Pluto could support aquatic life or aquaculture. Rocky planets could be used for large-scale agriculture or even as astronomical observatories.

5. Energetics involved in moving planets into the HZ

5.1. *Hohmann Transfer*

The Hohmann Transfer (HT) is an efficient method to move a smaller object between orbits around a larger body, like the Sun (See Fig. 1), using minimal energy by applying two velocity changes (Sivaram 2008). Shifting Mars 0.3 AU into the HZ requires slowing it by 1.3 km/s at 1.5 AU and 1.4 km/s at 1.2 AU, using $\sim 10^{32}$ J and taking ~ 10 months. Moving Pluto 37.8 AU into the HZ requires slowing it by 3.5 km/s at 39.5 AU and 10.7 km/s at 1.2 AU, using $\sim 10^{29}$ J and taking ~ 47 years. While HT is efficient, slower transfers may be preferable for maintaining planetary stability.

5.2. *Laser Power for Continuous Thrust*

We assume the laser used to shift a planet operates at constant power, positioned close enough to the planet that beam divergence is negligible. The equation of motion for a constant-power laser is as follows:

$$\frac{d^2r}{dt^2} = \frac{P}{M_p \cdot c},$$

where r is the distance to be moved, P is laser power, M_p is the mass of the planet and c is the speed of light. The required laser power is:

$$P = \frac{M_p \cdot r \cdot c}{t^2}.$$

For example, shifting a Mars-sized planet 0.3 AU into the HZ over 10^{11} seconds (few thousand years) requires $\sim 10^{21}$ W and $\sim 10^{32}$ J. Similar energy levels apply for moving Pluto- or Earth-sized planets, while shifting asteroids would require significantly less energy.

It is suggested that power-beaming technology can be used for launching spacecraft, raising satellite orbits, facilitating space-to-space transfers, and even for missions to the outer solar system and beyond (J.N. Benford and D.J. Benford 2016). Power beaming could potentially be extended to include the displacement of asteroids and planets, which would require significantly higher orders of power.

6. Detection methods

Searches for ETI in radio, infrared, and optical regions have formed the basis of modern SETI (Shuch 2011). However, despite decades of our efforts, concrete evidence of ETI remains elusive. The null result does not mean that there are no ETI. "*Our current search completeness is extremely low, similar to having searched for something like a large hot tub or a small pool's worth of water out of all of Earth's oceans*" (Wright *et al.* 2018).

Continuous improvement of search parameters and models is essential, along with exploration of novel detection ideas. If advanced civilizations use high-power lasers to shift planets, this would create detectable technosignatures. These sustained laser signals, or High-Power Laser Technosignatures (HPLT), could be sought, especially in systems with unusual planetary arrangements, which might indicate ETI manipulation.

Sofia Sheikh created a framework to rank technosignatures based on nine factors (Sheikh 2019). This framework allows people to critically evaluate technosignatures on the basis of detection probability. We considered the HPLT and ranked it by the following nine parameters. Here is the ranking and the rationale behind the ranking:

- **Observing Capability** We can detect shifting of planets using powerful lasers with our current observing capabilities. No need to build any new instruments for observation.
- **Cost** The cost of searching for laser signatures is relatively low as no new instrumentation is required.
- **Ancillary Benefits** HPLT detection has no present ancillary benefits.
- **Detectability** HPLT are detectable even over a kiloparsec.
- **Duration** HPLT are not transient in nature as they operate continuously over years. The time period of the high power laser signature depends upon the distance to which the planet need to be moved. We estimate that if the acceleration is low (like 10^{-10} m/s^2), then it may take around 1000 years to move a Earth-sized planet over 0.5 AU, and even more time is required to shift a FFP from interstellar space to the resident star system.
- **Ambiguity** The high-power narrow-band laser beam has no natural source. Therefore, these technosignatures are unambiguous.

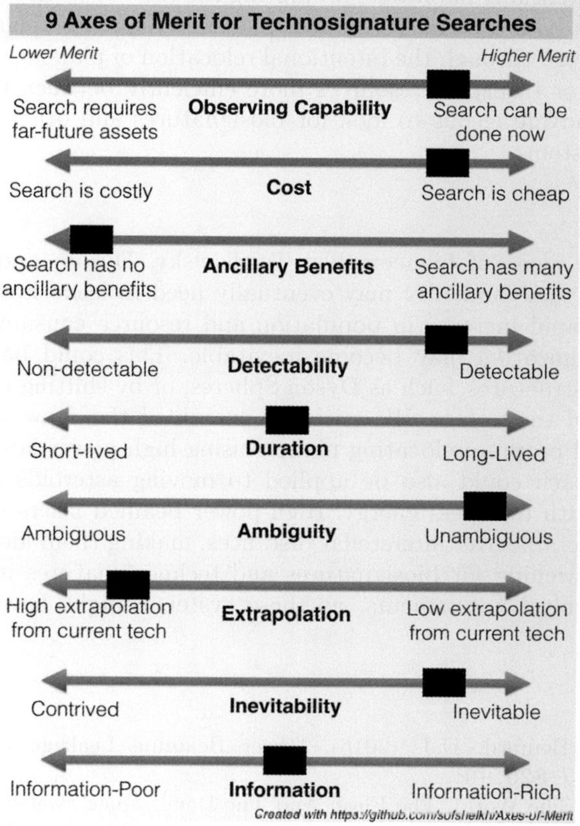

Figure 2. Axes of Merit for high-power laser technosignature.

- **Extrapolation** High-power lasers (at ZW scale) operating continuously are yet beyond our current technological capabilities, but we already have such powerful lasers operating at short pulses. Thus, building such high-power lasers operating in continuous mode is plausible in future.
- **Inevitability** The need for additional living space and resources is inevitable. Therefore, additional habitable planets could eventually be absolutely necessary for advancing and space-faring civilizations.
- **Information** ETI can also transmit information through their lasers and the information is proportional to the bandwidth of the laser.

These merits are visually represented in Fig. 2, with Low to High merits displayed from left to right.

If an advanced species had already relocated planets into their HZ, we might expect to observe an unusually high concentration of planets in close proximity and SEAs. For example, TRAPPIST-1 has four potentially habitable planets, while Kepler-20 shows strange orbital patterns that suggest possible artificial manipulation. A recent discovery of the HD 110067 planetary system (Luque *et al.* 2023) has reinforced our proposal to look for SEA systems. The star is an old K0-type star with mass and radius of $\sim 80\%$ of the Sun hosting 6 sub-Neptune planets orbiting in a stable perfect resonance, which means that all 6 of them align every few orbits, or 492 days. Since all planets are in the star's HZ and three of the planets have low-densities with atmospheres, they are already considered to be potentially habitable, prompting the recent search for radio technosignatures in this system (Choza *et al.* 2024). It is plausible that these planetary systems were artificially constructed through the intentional relocation of planets, allowing inhabitants to facilitate travel or transport resources more efficiently between them. Therefore, we suggest that it is advantageous to look for biosignatures and other technosignatures in such engineered systems.

7. Conclusion

Building on the ideas of pioneers like Tsiolkovsky, Dyson, and Sagan, there is a growing consensus that humanity may eventually need to leave Earth and expand into space. Given the rapid increase in population and resource consumption over the past century, space colonization may become inevitable. This could be achieved either by constructing megastructures, such as Dyson Spheres, or by shifting planets into the habitable zone (HZ) of their stars. We critically examined the drawbacks of constructing Dyson Spheres and propose relocating planets using high-power lasers as a viable alternative. This approach could also be applied to moving asteroids or Trans-Neptunian Objects (TNOs) with even less energy. High-power beamed lasers used in these operations could be detectable over interstellar distances, making them ideal technosignatures. We recommend searching for biosignatures and technosignatures in planetary systems with significant orbital adjustments, as these systems might have been intentionally engineered.

References

Benford, J.N. and Benford, D.J. (2016). Power Beaming Leakage Radiation as a SETI Observable. *ApJ.* **825**, 101.

Bernal, J.D. (1929). The World, The Flesh And The Devil. *Foyle Publishing.*

Boyajian, T.S., Alonso, R., Ammerman, A., et al. (2018). The First Post-Kepler Brightness Dips of KIC 8462852. *Astrophys. J. Lett.* **853**, L8. doi:10.3847/2041-8213/aaa405.

Boyajian, T.S., LaCourse, D.M., Rappaport, S.A., et al. (2016). Planet Hunters IX. KIC 8462852 - where's the flux? *Monthly Notices of the Royal Astron. Soc.* **457**, 3988–4004. doi:10.1093/mnras/stw218.

Choza C., Croft S., Siemion A. P. V., et al. (2024). A Radio Technosignature Search of Six Resonant Sub-Neptunes Orbiting HD 110067. *Research Notes of the American Astronomical Society* **8**. doi:10.3847/2515-5172/ad235f.

Cocconi, G. and Morrison, P. (1959). Searching for interstellar communications. *Nature* **184**, 844–846. doi:10.1038/184844a0.

Cummings, A.C., Stone, E.C., Heikkila, B.C., et al. (2016). Galactic Cosmic Rays in the Local Interstellar Medium: Voyager 1 Observations and Model Results. *Astrophys. J. Lett.* **831**, 18. doi:10.3847/0004-637X/831/1/18.

Deeg, H.J., Alonso, R., Nespral, D. and Boyajian, T.S. (2018). Non-grey dimming events of KIC 8462852 from GTC spectrophotometry. *Astron. Astrophys.* **610**, L12. doi:10.1051/0004-6361/201732453

Dyson, F.J. (1960). Search for artificial stellar sources of infrared radiation. *Science* 131, 1667–1668. doi:10.1126/science.131.3414.1667.

Dyson, F. (1966). The Search for Extraterrestrial Technology, in: *Perspectives in Modern Physics.* Essays in honor of Hans A. Bethe, pp. 641–655.

Dyson, F. (1979). Disturbing the Universe. *Basic Books, Inc., Publishers. New York.* 211.

Harrop, B.L. and Schulze-Makuch, D. (2010). The solar wind power satellite as an alternative to a traditional Dyson Sphere and its implications for remote detection. *Int. J. Astrobiology* **9**, 89–99. doi:10.1017/S1473550410000066.

Kallis, G. (2019). Limits: Why Malthus Was Wrong and Why Environmentalists Should Care. Stanford University Press.

Kane, S.R., Turnbull, M.C., Fulton, B.J., et al. (2020). Dynamical Packing in the Habitable Zone: The Case of Beta CVn. *Astrophys. J.* **160**, 81. doi:10.3847/1538-3881/ab9ffe.

Kardashev, N.S. (1964). Transmission of Information by Extraterrestrial Civilizations. *Soviet Astronomy* **8**, 217.

Luque, R., Osborn, H.P., Leleu, A. et al., (2023). A resonant sextuplet of sub-Neptunes transiting the bright star HD 110067. *Nature* **623**, 932–937. https://doi.org/10.1038/s41586-023-06692-3.

Maddox, J. (1960). Artificial biosphere. *Science* **132**, 250–251.

Militzer, B., Hubbard, W.B., Vorberger, J., et al. (2008). A massive core in jupiter predicted from first-principles simulations. *Astrophys. J.* **688**, L45. doi:10.1086/594364.

Niven, L. (1974). Bigger Than Worlds. A Hole in Space. New York: Ballantine Books. pp. 111126. ASIN B002B1MS6U.

Opher, M. and Loeb, A. (2022). Terrestrial Impact from the Passage of the Solar System through a Cold Cloud a Few Million Years Ago. *Bulletin of the AAS* **54**. https://baas.aas.org/pub/2022n6i227p06

Opher, M., Loeb, A. and Peek, J. (2022). Evidence that Earth was exposed to cold dense interstellar medium 2 MyrsaAgo as a result of the encounter with Local Lynx Cold Clouds, in: *AGU Fall Meeting 2022*, pp. SH22C-05.

Safonova, M. and Sivaram, C. (2021). Planetary Protection: Too Late. A chapter in *Planet Formation and Panspermia. New Prospects for the Movement of Life through Space*, V. Branislav, G. Richard and S. Joseph (eds.), pp. 55–92. Volume in the series *Astrobiology Perspectives on Life of the Universe*, Series Eds: R. Gordon & J. Seckbach (Wiley-Scrivener, Beverly, Massachusetts, USA). ISBN: 9781119640394

Schaefer, B.E., Bentley, R.O., Boyajian, T.S., et al. (2018). The KIC 8462852 light curve from 2015.75 to 2018.18 shows a variable secular decline. *Monthly Notices of the Royal Astron. Soc.* **481**, 2235–2248. doi:10.1093/mnras/sty1644.

Sheikh, S.Z. (2019). Nine axes of merit for technosignature searches. *Int. J. Astrobiology* **19**, 237–243. doi:10.1017/s1473550419000284.

Shklovskii, I.S. and Sagan, C. (1966). Intelligent Life in the Universe. *HoldenDay, Inc., San Francisco.*

Shuch, H. P. (2011). Searching for Extraterrestrial Intelligence: SETI Past, Present, and Future. The Frontiers Collection. Springer-Verlag Berlin Heidelberg. `https://books.google.co.in/books?id=jAk9bTm3Sj4C`.

Sivaram, C. (2008). Rocket Dynamics And Space Flight. *Ane Books India.*

Stapledon, O. (1937). Star Maker. *Methuen and Co. Ltd., London.*

Tsiolkovsky, K. (1895). Dreams of Earth and Sky. Athena Books, 2004.

UNOOSA, 2008. United Nations Treaties and Principles on Outer Space. United Nations Publication. UNITED NATIONS New York, ISBN 92-1-100900-6, 56 p.

Wright, J.T., Cartier, K.M.S., et al. (2016). The G Search for Extraterrestrial Civilizations with Large Energy Supplies. IV. The Signatures and Information Content of Transiting Megastructures. *Astrophys. J.* **816**, 17. doi:10.3847/0004-637X/816/1/17.

Wright, J.T., Kanodia, S. and Lubar, E. (2018). How much SETI has been done? Finding needles in the N-dimensional cosmic haystack. *Astrophys. J.* **156**, 260. doi:10.3847/1538-3881/aae099.

Wright, J.T. (2020). Dyson spheres. *Serb. Astron. J.* **200**, 1–18.

Wright, J.T., Haqq-Misra, J., Frank, A., et al. (2022). The case for technosignatures: Why they may be abundant, long-lived, highly detectable, and unambiguous. *Astrophys. J. Lett.* **927**, L30. doi:10.3847/2041-8213/ac5824.

(Toward) Discovery of Life beyond Earth and its Impact
Proceedings IAU Symposium No. 387, 2026
H. Landt, M. Dominik & C. Oliver, eds.
doi:10.1017/S1743921324001339

Simulating the characteristics of extra-terrestrial civilizations that encounter Earth

Kevin H. Knuth[1,2]

[1]University at Albany (SUNY), Albany NY USA

[2]Stellenbosch Institute for Advanced Study (STIAS), Stellenbosch, South Africa
email: kknuth@albany.edu

Abstract. It is generally believed that it is unlikely that our civilization is alone in this galaxy. This belief is central to the premise of the Search for Extraterrestrial Intelligence (SETI), which has focused mainly on searching for radio signals originating from extraterrestrial communications, since it is believed that extraterrestrial craft visiting Earth would be an extremely unlikely event. However, the fact that we ourselves are currently working on developing probes to send to the Alpha Centauri system by 2069, strongly suggests that other civilizations may make similar, or more ambitious, efforts. Therefore, it is reasonable to inform our expectations by considering what characteristics and capabilities would be required for an interstellar civilization to find and visit Earth. In this paper, a physics-based analytic model of expanding interstellar civilizations is developed. A million civilizations that encounter Earth are simulated and their statistics are studied to determine their characteristics.

Keywords. extra-terrestrial, interstellar travel, interstellar colonization

1. Introduction

Given what is known about the populations of planets (Traub 2011; Seager 2013; Batalha 2014; Kane *et al.* 2016), and the generation, prevalence and distribution of complex organic molecules in space (Ehrenfreund and Sephton 2006; Knuth *et al.* 2007; Salama 2008; Walsh *et al.* 2014; Cami *et al.* 2018; Ohishi 2019), it is generally believed that it is unlikely that our civilization is alone in this galaxy (Sagan 1963; Shklovsky and Sagan 1966; Drake 1980; Grosz 2013). That we are not alone is a central tenet of the Search for Extraterrestrial Intelligence (SETI) program (Tarter *et al.* 2010) and the more recent Breakthrough Listen program (Worden *et al.* 2017), which mainly focus on searching for extraterrestrial radio signals, since it is thought that extraterrestrial spacecraft visiting Earth would be an extremely unlikely event.

However, the fact that in the last decade we have observed three interstellar objects, the meteor CNEOS 2014-01-08 (Siraj and Loeb 2022), the asteroid 1I/2017 U1 ('Oumuamua) (Meech *et al.* 2017), and comet C/2019 Q4 (2I/Borisov) (Guzik *et al.* 2020), and that we are currently working on developing probes to send to the Alpha Centauri system by 2069 (Clery 2016; Daukantas 2017; Litchford and Sheehy 2020), a century after the first Moon landings, suggests that other civilizations may make similar, or even more ambitious, efforts (Freitas 1983). It is therefore reasonable to utilize our current understanding of physics to precisely delineate our prior expectations of the possibility that an extraterrestrial civilization could visit Earth. These expectations can be made more precise by identifying which characteristics and capabilities such an interstellar civilization would have to possess.

We are relatively knowledgeable about the physics involved in interstellar travel, even if we are still naive about the requisite technologies and whether or not they can be created. In this paper, we use our knowledge of physics to create models of interstellar civilizations that expand their domains over time by traveling at potentially relativistic speeds in order to explore and occupy nearby star systems. The model we adopt is similar to the model recently investigated by Carroll-Nellenback *et al.* (2019) where a civilization experiences resource-limited exponential growth resulting in an expanding domain front. By modeling each civilization with several general parameters quantifying the basic capabilities for interstellar travel, we can identify which civilizations would have been capable of finding and visiting Earth. From this subset of civilizations we compile statistics on their parameter values thus objectively quantifying our prior expectations on the nature of such civilizations. In addition, we will learn which qualities and capabilities are critical for a civilization to successfully engage in interstellar exploration and occupation. Once identified, these qualities and capabilities could be nurtured and developed to enable our civilization to embark on interstellar exploration.

Since these models are constructed based on our current knowledge of physics, the results of these simulations ought to conform to our expectations. In this sense, we do not expect any dramatic surprises beyond those that could have been carefully reasoned. Although by simulating interstellar colonization, we can expect to add a layer of precision to our expectations by quantifying civilization model parameters with mean values and uncertainties.

2. The Milky Way Galaxy

The Milky Way Galaxy is believed to be a barred spiral galaxy with a diameter estimated to be about 150 000 ly to 200 000 ly containing about 100 to 400 billion stars with at least as many planets (Villard and Sahu 2012). In addition to the spiral disk, the Milky Way has a central elliptical bulge and is surrounded by a spheroidal halo of stars. The Milky Way is dynamic and is in the process of evolving in a complex manner as small neighboring galaxies continue to merge with the Milky Way forming streams of stars within the galaxy (Newberg *et al.* 2002; Ivezić *et al.* 2008). The stellar density of the galaxy is neither uniform nor smooth with density structures ranging from the more familiar spiral arms to the more recently discovered corrugation of the galactic disk (Xu *et al.* 2015). The limits in our knowledge about the galactic structure precludes the utility of a detailed galactic model in this study.

For these reasons, we adopt a simplified picture of the galaxy as a 900 ly thick disk of stars with uniform stellar density equal to that of the local stellar density. Given that there are approximately 317 stellar neighbors within $10\,\text{pc} \approx 32.6\,\text{ly}$ of the Sun (Henry *et al.* 2018), the stellar density is approximately $\rho \approx 0.0022\,\text{systems/ly}^3$. The stars in the galaxy are assumed to be stationary over the times scales of a civilization. This neglects the rotation of the stars about the galactic center, which given proper velocities ranging from around 200 to 250 km/s (Mróz *et al.* 2019) result in displacements on the order of 10 000 ly over a period of 10 million years, which would be relevant for older civilizations. By not modeling stellar motions, we restrict this study to civilizations with lifetimes less than 10 million years.

Our stationary approximation also neglects random, or thermal, stellar motions, which have been modeled by Carroll-Nellenback *et al.* (2019) as a Maxwell-Boltzmann distribution with an average speed of 10^{-4} c. These thermal motions, which would introduce a small diffusive component, become relevant in the low stellar density limit where

Table 1. Civilization Parameters and their Prior Probabilities.

Model Parameters				
Symbol	**Parameter**	**Prior Range**	**Units**	**Prior Type**
x_o	galactic x-coordinate	within the GHZ	ly	Uniform
y_o	galactic y-coordinate	within the GHZ	ly	Uniform
z_o	galactic z-coordinate	$[-450, 450]$	ly	Uniform
$\mathscr{R}$	resilience	$[0, 1]$	unitless	Uniform
T	civilization lifetime	$[1, 10 \times 10^6]$	years	Jeffreys
t_{col}	colonization time	$[1, 5000]$	years	Uniform
τ_{max}	endurance	$[1, 1000]$	years	Jeffreys
a_{max}	max. acceleration	$[1, 25000]$	g	Jeffreys
v_{max}	max. velocity	$[0.001, 1)$	c	Jeffreys

Derived Parameters			
Symbol	**Parameter**	**Units**	**Constraints**
δ	interstellar reach	ly	$\delta > 7.37$ ly
R	characteristic domain radius	ly	
v_{dom}	domain expansion velocity	c	
d_e	distance to Earth	ly	
t_e	time of first contact	y	

Note: Jeffreys priors were handled by working with uniformly-distributed logarithms of the parameters.

some civilizations would have to wait for a settleable star systems to come into range (Carroll-Nellenback *et al.* 2019).

Limiting our study to situations we are familiar with, such as terrestrial, icy, or ocean-bearing worlds, places constraints on where in the galaxy such life might evolve. To form terrestrial planets the stellar environment must have a sufficient supply of heavy elements, or metallicity (Wang and Fischer 2014). In addition, the local environment must be relatively calm for the length of time required for the evolution of complex life. This means that both nearby supernovae and stellar interactions need to be relatively rare. This results in a Galactic Habitable Zone (GHZ) that is expected to be an annular disk-shaped region with an inner radius of $4\,\text{kpc} \approx 13,000$ ly and an outer radius of $11\,\text{kpc} \approx 36,000$ ly (Lineweaver *et al.* 2004). Stars close to the galactic center will be older generation stars with high metallicity and a commensurate tendency to have large close-orbiting massive planets (Wang and Fischer 2014), which may disrupt smaller terrestrial worlds (Raymond *et al.* 2006; Haghighipour 2007). More importantly, the greater density of stars near the galactic center will result in more frequent stellar interactions, which threaten the stability of planetary systems, as well as a significantly higher supernova rate resulting in more frequent sterilizing events. On the other hand, stars in the outer reaches of the galaxy tend to be earlier generation stars with low metallicity, which is expected to result in terrestrial planets being more rare (Lineweaver *et al.* 2004).

3. Model Parameters Describing a Civilization

A civilization's **origin** (x_o, y_o, z_o) is sampled uniformly from the GHZ, and its **lifetime** T is sampled from the range $[1, 10 \times 10^6]$ y. Table 1 lists the civilization's model parameters, ranges, and the prior probability distributions from which they are sampled. Parameters are either uniformly distributed, or distributed with a Jeffrey's prior, which is invariant with respect to the choice of units (and thus greatly favors smaller values).

Civilizations are modeled independently (no interactions) and it is assumed that the civilization's parameters do not change over their lifetime. This way we can more effectively model what one expects from a civilization with a given set of parameters. A civilization's dispersion across interstellar space will be highly species and culture dependent. It will also depend greatly on the type of occupation, which could range from setting up small bases or research stations to exoforming planets to support large populations (colonization). To model these possibilities, we employ a parameter called the **resilience**, $\mathscr{R}$, which is the probability that a given system will have a suitable world to occupy, and the **colonization time**, t_{col}, which describes how long it takes to establish a presence before moving on to another star system.

Their capability for space travel is modeled by describing the **maximum acceleration**, $a_{\max}$, and **maximum speed**, $v_{\max}$, of their spacecraft along with their **endurance**, τ, which describes how long they are able to travel in their spacecraft. Using the spacecraft parameters, $a_{\max}$ and $v_{\max}$ along with their endurance τ, relativistic rocket equations can be used to compute the maximum distance that they can travel, called their **reach**, δ. Using the reach, δ, we can compute the speed, v_{dom}, at which the civilization's domain radius R grows. This allows us to compute the civilization's distance from Earth, which is located at $(x_e, y_e, z_e) = (0, -26000, -45)$ ly.

4. Diffusion and Growth Models

While we can expect that nearby systems will be favored over distant ones, the unpredictable nature of the selection process will make colonization a relatively random process, well-described as a diffusion process. The probability that a given star system is occupied will evolve according to Fick's second law

$$\frac{\partial}{\partial t} p(r,t) = D\nabla^2 p(r,t), \tag{1}$$

where D is the diffusion constant. However, the fact that colonies are growing, maturing, and venturing out on their own adds a source term to the evolution of this probability. Since each colony is a potential colonizer, the rate of growth will be proportional to the occupation probability $p(r,t)$. However, the rate of growth will also be proportional to the probability of the opportunities for occupation, which is given by $\mathscr{R} - p(r,t)$. This results in the addition of a source term on the right

$$\frac{\partial}{\partial t} p(r,t) = D\nabla^2 p(r,t) + \rho\,(\mathscr{R} - p(r,t))\,p(r,t), \tag{2}$$

where ρ is the constant of proportionality amounting to a growth rate. The resulting partial differential equation is known both as Fisher's equation (Fisher 1937) and the Kolmogorov-Petrovsky-Piskunov (KPP) equation (Kolmogorov *et al.* 1937).

To get some idea of what the solutions to this partial differential equation look like, we first consider the case where there is no diffusion, so that $D = 0$ and r is not a function of time t. This gives us the differential equation

$$\frac{d}{dt} p(r,t) = \rho\,(\mathscr{R} - p(r,t))\,p(r,t), \tag{3}$$

which is the logistic differential equation (Verhulst 1838) where ρ is the growth rate. Unlike (2), this differential equation is easily integrated resulting in

$$p(r,t) = \frac{\mathscr{R}}{1 + (\frac{\mathscr{R} - p_o}{p_o})e^{-\rho(t-t_o)}} \tag{4}$$

where $p_o = p(r, t_o)$ is the occupation probability at 'initial' time t_o. The logistic equation is a sigmoid function which initially grows exponentially and then saturates at an occupation probability given by the resilience, $p(r,t) = \mathscr{R}$, due to the fact that colonization options are limited by their resilience.

When diffusion is included, we expect that the population density will expand outward from their homeworld while the population density of the interior of their domain increases logistically saturating at $p(r,t) = \mathscr{R}$. The domain boundary, or frontier, will be an expanding wave with a shape related to both the logistic function and a three-dimensional Gaussian.

However, the fact that the partial differential equation (2) cannot be solved analytically means that we have to make some simplifications. Since we are mainly interested in whether the civilization finds Earth, we are going to ignore the growing population density of the interior, modeling the civilization as a three-dimensional Gaussian distribution, and focus on rate at which the domain boundary grows. The domain velocity is expected to be bounded by

$$v_{\mathrm{dom}} \leq \frac{\delta}{\tau + t_{\mathrm{col}}}. \tag{5}$$

However, one would expect that expansion will not be radial to the boundary, but that spacecraft will zig-zag from system to system resulting in (Bennett and Shostak 2007; Maccone 2012)

$$v_{\mathrm{dom}} = \frac{\langle z \rangle}{\tau + t_{\mathrm{col}}} = \frac{k\delta}{\tau + t_{\mathrm{col}}}. \tag{6}$$

Computing the expected value of z by treating the domain as having a hard boundary results in $\langle z \rangle = 3/8\delta$. However, using a Taylor-series approximation of the Gaussian population profile across the domain reveals that k is not a constant, but that

$$\langle z \rangle = \frac{\delta^2}{5R}(\mathscr{R}^{-1}e^{1/2} - 1)^{-1}. \tag{7}$$

Substituting into (6) we see that the domain expansion velocity depends on the resilience, $\mathscr{R}$, and decreases as the domain radius, R, increases, since the volume to occupy goes as R^3.

Carroll-Nellenback *et al.* (2019) found that to maintain spherical expansion, the normalized density of settleable systems, η, must be greater than a critical value

$$\eta = \mathscr{R}\rho\delta^3 > \eta_c = 0.88, \tag{8}$$

where ρ is the stellar density, which places a constraint on the reach

$$\delta > \left(\frac{0.88}{\mathscr{R}\rho}\right)^{1/3}, \tag{9}$$

so that for lower resilience, one must be able to travel further. A resilience of $\mathscr{R} = 1$ results in a minimum reach of $\delta_{\min} \approx 7.37$ ly.

5. Sampling from the Prior Probabilities

By sampling from the prior probabilities, listed in Table 1, one can evaluate our expectations for the capabilities of interstellar civilizations. For example, by sampling ten million spacefaring civilizations from the priors, we found that 2 219 792 civilizations were capable of interstellar colonization. Therefore the probability that a spacefaring civilization will be interstellar-capable is roughly 22.2%. Of those interstellar-capable civilizations, we found that 9 of those civilizations encountered Earth. Thus the probability that an interstellar-capable civilization will visit Earth is about $\frac{9}{2\,219\,792} \approx 0.000405\%$. That means that out of all of the spacefaring civilizations in the galaxy, only about 0.00009% of them encounter Earth. If we ever learn that someone had visited Earth, these results serve as a sort of Drake equation to estimate that there are about 1.1 million spacefaring civilizations and about 250 000 interstellar spacefaring civilizations in the galaxy.

6. Sampling Civilizations that Found Earth: Nested Sampling

The goal of this paper is to study the statistics of physics-based simulations of interstellar civilizations that encounter Earth. The fact that about only 1 out of 10^6 spacefaring civilizations find Earth, means that to sample 10^6 civilizations that find Earth, we would have to sample 10^{12}, or a trillion, civilizations from the prior. We are going to have to be a bit more clever if we want a million prior-distributed samples of civilizations that find Earth.

To accomplish this, we rely on the nested sampling algorithm (Skilling 2004, 2006), which is a stochastic (Lebesgue) integration algorithm to compute the Bayesian evidence by integrating the likelihood over the prior space. By performing a change of variables so that the prior probabilities are all uniform, the algorithm works by managing a population of N samples drawn uniformly from within a set of nested increasing-likelihood contours.

Instead of a likelihood function, we define a cost function based on the distance between the civilization's domain boundary and Earth. The algorithm then sorts civilizations based on this distance. At the end of the process, we are left with a prior-distributed (uniformly-distributed) population of $N = 10000$ civilizations all of which contain Earth within their domain. By running this algorithm 100 times (with $N = 10000$ samples), we obtain one million prior-distributed civilizations, which have found Earth. The statistics of these one million civilizations describe our best expectations given our current knowledge of physics.

7. Characteristics of Interstellar Civilizations that Encounter Earth

Figure 1A illustrates that maximum speeds, v_{max}, a fraction of the speed of light are sufficient for interstellar exploration and colonization. Over 21 600 civilizations ($\sim 2\%$) were able to find Earth by traveling at speeds less than 10% the speed of light (0.1 c), 75 000 ($\sim 7.5\%$) at speeds less than 0.2 c, and 146 000 ($\sim 15\%$) at speeds less than 0.3 c. Resulting accelerations, a_{max}, remained roughly prior-distributed, so that low accelerations are more probable (Fig. 1B), and reasonable accelerations between 1 to 2 g are in many cases sufficient.

Relatively slow speeds require greater endurance, τ_{max}, which is widely distributed from tens of years to 1000 years (Fig. 1C). Endurance, along with v_{max} and a_{max}, affect the civilizations' reach, δ, which was found to be most probably around 26 and 44 LYs, with a mean of 340 ± 240 LYs (Fig. 1D). While clearly challenging for organisms with human tolerances and lifespans, these capabilities are possible for AI-controlled space probes.

To get an understanding of how large these civilizations are, Figure 2 illustrates the distribution of domain radii (A), R, and the distances between their origin and Earth (B). The most probable domain radius, R, was found to be 260 ly, and the distribution has a mean of 500 ± 330 ly. B. The most probable distance between the civilization's origin and Earth was 248 ly with a mean of 560 ± 410 ly, so that Earth is often well within their domain.

The distribution of lifetimes, T, had a short lifetime spike, due to the Jeffrey's prior, followed by a rather uniform distribution up to 10 million years. The age of the civilization when it found Earth had a similar spike followed by a linear decreasing distribution approaching zero at 10 million years. The time of first contact ranges from 0 (present) to the difference between the lifetime and the age at which they found Earth. Uniformly sampling over this range gives the distribution of first contact times (Fig. 2C), which shows that if we find that extraterrestrials have visited Earth, then it is possible that Earth was found long ago with a mean first contact time of about $1.17^{+2.44}_{-1.02}$ million years ago.

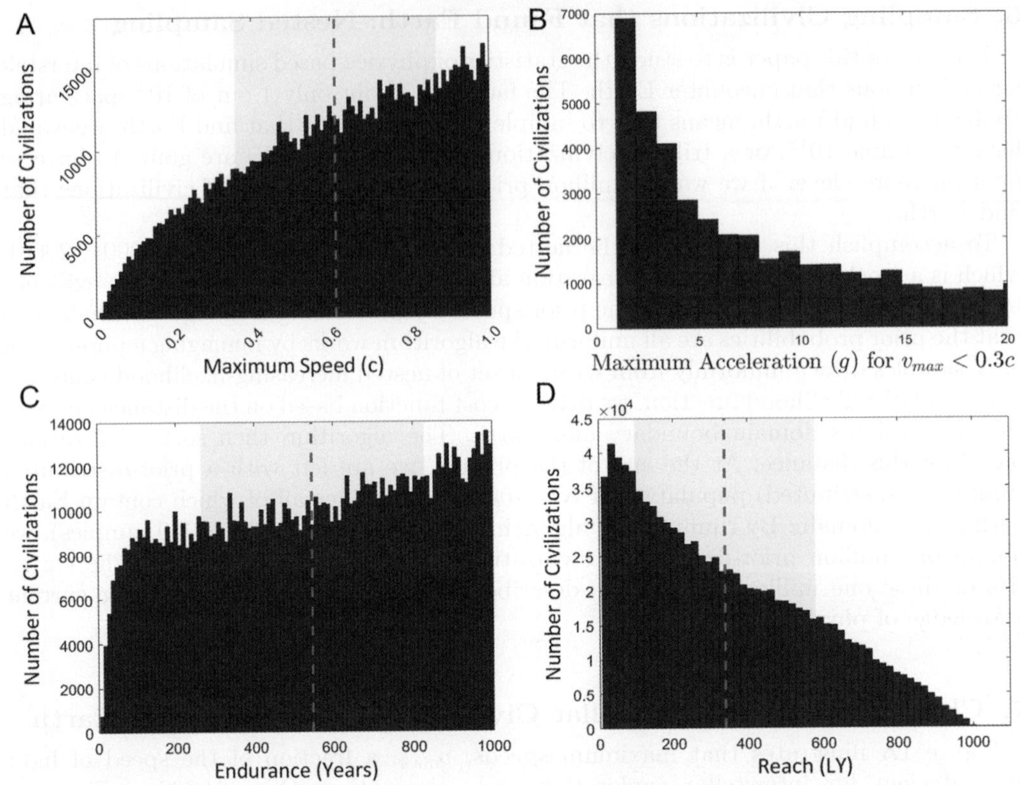

Figure 1. A. The maximum velocities, $v_{\max}$, of the spacecraft reveal that it is better to go faster, but interstellar travel is possible at 10's of percent of the speed of light. Mean $v_{\max} = 0.61 \pm 0.25$ c is displayed with a dashed line and shading. B. Accelerations of spacecraft with $v_{\max} < 0.3$ c reveal that accelerations between 1 to 2 g are in many cases sufficient. C. Endurances, $\tau_{\max}$, are widely distributed. Mean $\tau_{\max} = 545 \pm 282$ y. D. The civilizations' reach, δ, peaks around 26 to 44 LYs, with a mean of 340 ± 240 ly.

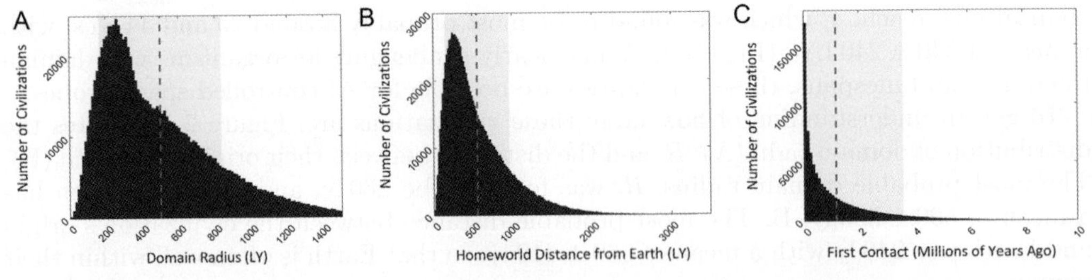

Figure 2. A. The distribution of domain radii, R. Mean $R = 500 \pm 330$ ly. B. The distribution of distance, d_e, between the civilization's origin and Earth. Mean $d_e = 560 \pm 410$ ly. C. The distribution of Earth first contact times, which have a mean of $1.17^{+2.44}_{-1.02}$ million years ago.

8. Discussion

By explicitly stating our expectations of characteristics of spacefaring civilizations as prior probabilities, we have sampled from these priors to determine what our expectations ought to be for interstellar civilizations to spread out throughout the galaxy and encounter Earth. Based on these prior expectations, we have found that only 22.2% of spacefaring civilizations are capable of interstellar travel and sustainable occupation of other star systems. As one would expect, the probability of a civilization finding and visiting, or occupying, Earth is much lower at only about a 0.0004% probability, and that there is an overall probability of 0.00009% that any spacefaring civilization could encounter and visit Earth. If we ever learn that an extraterrestrial civilization had visited Earth, these results serve as a variation of the Drake equation to estimate that there are about one million spacefaring civilizations and about 250 000 interstellar spacefaring civilizations in the galaxy. Practically, what is important is the number of interstellar-capable civilizations within 560 LY of Earth (mean distance between a civilization's origin and Earth), which includes around 1.7 to 3 million star systems ($N = \rho V$).

This analysis is not without faults, and should be treated as a rough (hopefully conservative) estimate. Ignorance about the galactic structure required that we make simplifying assumptions about the galaxy. Unlike Carroll-Nellenback *et al.* (2019), we neglected stellar motions, which they demonstrated could significantly aid domain expansion over extremely long time scales. We employed an average stellar density of $\rho \approx 0.0022\,\text{systems/ly}^3$, which is about half as large as other estimates of $0.004\,\text{systems/ly}^3$ (Gregersen 2010).

We worked to keep the ranges of the civilization parameters broad so that we could observe their importance, and accommodate the potential diversity of civilizations. Since we worked to establish our expectations based on known physics, we considered spacecraft speeds up to, but not exceeding, the speed of light. Since ideas about warp drive have only begun to be explored (Alcubierre 1994; Davis 2009; Bobrick and Martire 2021), it is premature to attempt to model that physics, especially since the speeds of potential warp drives are not yet widely established (Smolyaninov 2011). We used a lower speed limit of $0.001c \approx 1.8 \times 10^6$ km/hr, which is about 3 times faster than our current capability (Parker Solar Probe). The accelerations were set with an upper limit of (an admittedly high) 25 000 g, which is two to five times higher than the estimated accelerations in some well-documented UAP cases (Knuth *et al.* 2019; Coumbe 2022), so that we could explore the consequences of such high accelerations. As one would expect, higher accelerations significantly reduce travel times. For example at a 1000 g acceleration a ship can reach a speed of $0.9c$ in only 17.5 hrs (Knuth *et al.* 2019). However, we found that sustained lower accelerations are also effective. When one considers that a 1 g acceleration is approximately 1 (ly/year)/year, it becomes clear that relativistic speeds are attainable at sustained human-friendly accelerations. Of course, much of this relies on what can actually be engineered, which is expected to be civilization-dependent up to a limit.

Acknowledgements

The author thanks Matthew Szydagis for many helpful discussions about this work.

References

Alcubierre, M. (1994). The warp drive: hyper-fast travel within general relativity. *Classical and Quantum Gravity*, **11**(5), L73.

Batalha, N. M. (2014). Exploring exoplanet populations with NASA's Kepler mission. *PNAS*, **111**(35), 12647–12654.

Bennett, J. and Shostak, S. (2007). Life in the universe.

Bobrick, A. and Martire, G. (2021). Introducing physical warp drives. *Classical and Quantum Gravity*, **38**(10), 105009.

Cami, J., Peeters, E., Bernard-Salas, J., Doppmann, G., and De Buizer, J. (2018). The formation of fullerenes in planetary nebulae. *Galaxies*, **6**(4), 101.

Carroll-Nellenback, J., Frank, A., Wright, J., and Scharf, C. (2019). The Fermi paradox and the aurora effect: exo-civilization settlement, expansion, and steady states. *The Astronomical Journal*, **158**(3), 117.

Clery, D. (2016). U.S. lawmaker orders NASA to plan for trip to Alpha Centauri by 100th anniversary of Moon landing. *Science*. Accessed: 2020-02-13.

Coumbe, D. (2022). *Anomaly. A scientific exploration of the UFO phenomenon*. Rowman & Littlefield Publishers, Lanhman MD.

Daukantas, P. (2017). Breakthrough starshot. *Optics and Photonics News*, **28**(5), 26–33.

Davis, E. W. (2009). Chapter 15: Faster-than-light approaches in general relativity. *Frontiers of Propulsion Science; Progress in Astronautics & Aeronautics Series; American Inst. of Aeronautics & Astronautics Press: Reston, VA, USA*, pages 473–509.

Drake, F. (1980). N is neither very small nor very large. In *Strategies for the Search for Life in the Universe*, pages 27–34. Springer.

Ehrenfreund, P. and Sephton, M. A. (2006). Carbon molecules in space: from astrochemistry to astrobiology. *Faraday Discussions*, **133**, 277–288.

Fisher, R. A. (1937). The wave of advance of advantageous genes. *Annals of eugenics*, **7**(4), 355–369.

Freitas, R. (1983). The case for interstellar probes. *British Interplanetary Society, Journal(Interstellar Studies)*, **36**, 490–495.

Gregersen, E. (2010). *The Milky Way and Beyond: Stars, Nebulae, and Other Galaxies*. The Rosen Publishing Group.

Grosz, I. (2013). Number of galactic civilizations. *The Open Astronomy Journal*, **6**(1).

Guzik, P., Drahus, M., Rusek, K., Waniak, W., Cannizzaro, G., and Pastor-Marazuela, I. (2020). Initial characterization of interstellar comet 2I/Borisov. *Nature Astronomy*, **4**(1), 53–57.

Haghighipour, N. (2007). Extreme habitability: Formation of habitable planets in systems with close-in giant planets and/or stellar companions. *arXiv preprint arXiv:0711.0782*.

Henry, T. J., Jao, W.-C., Winters, J. G., Dieterich, S. B., Finch, C. T., Ianna, P. A., Riedel, A. R., Silverstein, M. L., Subasavage, J. P., and Vrijmoet, E. H. (2018). The Solar neighborhood XLIV: RECONS discoveries within 10 parsecs. *AJ*, **155**(6), 265.

Ivezić, Ž., Sesar, B., Jurić, M., Bond, N., Dalcanton, J., Rockosi, C. M., Yanny, B., Newberg, H. J., Beers, T. C., Prieto, C. A., *et al.* (2008). The milky way tomography with sdss. ii. stellar metallicity. *ApJ*, **684**(1), 287.

Kane, S. R., Hill, M. L., Kasting, J. F., Kopparapu, R. K., Quintana, E. V., Barclay, T., Batalha, N. M., Borucki, W. J., Ciardi, D. R., Haghighipour, N., *et al.* (2016). A catalog of Kepler habitable zone exoplanet candidates. *ApJ*, **830**(1), 1.

Knuth, K. H., Tse, M. K., Choinsky, J., Maunu, H. A., and Carbon, D. F. (2007). Bayesian source separation applied to identifying complex organic molecules in space. In *2007 IEEE/SP 14th Workshop on Statistical Signal Processing*, pages 346–350. IEEE.

Knuth, K. H., Powell, R. M., and Reali, P. A. (2019). Estimating flight characteristics of anomalous unidentified aerial vehicles. *Entropy*, **21**(10), 939.

Kolmogorov, A., Petrovskii, I., and Piskunov, N. (1937). A study of the diffusion equation with increase in the amount of substance, and its application to a biological problem. *Bull. Moscow Univ., Math. Mech.*, pages 1–25. English translation by V. M. Volosov in Selected Works of A. N. Kolmogorov I, pages 248–270. Kluwer 1991, ISBN 90-277-2796-1.

Lineweaver, C. H., Fenner, Y., and Gibson, B. K. (2004). The galactic habitable zone and the age distribution of complex life in the Milky Way. *Science*, **303**(5654), 59–62.

Litchford, R. J. and Sheehy, J. A. (2020). Prospects for interstellar propulsion. *NASA Preprint*. (AAS 20-068).

Maccone, C. (2012). Statistical Fermi paradox and galactic travels. In *Mathematical SETI*, pages 131–144. Springer.

Meech, K. J., Weryk, R., Micheli, M., Kleyna, J. T., Hainaut, O. R., Jedicke, R., Wainscoat, R. J., Chambers, K. C., Keane, J. V., Petric, A., *et al.* (2017). A brief visit from a red and extremely elongated interstellar asteroid. *Nature*, **552**(7685), 378–381.

Mróz, P., Udalski, A., Skowron, D. M., Skowron, J., Soszyński, I., Pietrukowicz, P., Szymański, M. K., Poleski, R., Kozłowski, S., and Ulaczyk, K. (2019). Rotation curve of the Milky Way from classical cepheids. *The Astrophysical Journal Letters*, **870**(1), L10.

Newberg, H. J., Yanny, B., Rockosi, C., Grebel, E. K., Rix, H.-W., Brinkmann, J., Csabai, I., Hennessy, G., Hindsley, R. B., Ibata, R., *et al.* (2002). The ghost of Sagittarius and lumps in the halo of the Milky Way. *ApJ*, **569**(1), 245.

Ohishi, M. (2019). Prebiotic complex organic molecules in space. In *Astrobiology*, pages 11–21. Springer.

Raymond, S. N., Barnes, R., and Kaib, N. A. (2006). Predicting planets in known extrasolar planetary systems. III. Forming terrestrial planets. *ApJ*, **644**(2), 1223.

Sagan, C. (1963). Direct contact among galactic civilizations by relativistic interstellar spaceflight. *Planetary and space science*, **11**(5), 485–498.

Salama, F. (2008). PAHs in astronomy—a review. *Proc. of the Int. Astronomical Union*, **4**(S251), 357–366.

Seager, S. (2013). Exoplanet habitability. *Science*, **340**(6132), 577–581.

Shklovsky, I. S. and Sagan, C. (1966). Intelligent life in the universe. *San Francisco: Holden-Day, 1966*.

Siraj, A. and Loeb, A. (2022). A meteor of apparent interstellar origin in the CNEOS fireball catalog. *ApJ*, **939**(1), 53.

Skilling, J. (2004). Nested sampling. In R. Fischer, V. Dose, R. Preuss, and U. von Toussaint, editors, *Bayesian Inference and Maximum Entropy Methods in Science and Engineering, Garching, Germany 2004*, number 735 in AIP Conf. Proc., pages 395–405. AIP, New York.

Skilling, J. (2006). Nested sampling for general Bayesian computation. *Bayesian Analysis*, **1**(4), 833–859.

Smolyaninov, I. I. (2011). Metamaterial-based model of the Alcubierre warp drive. *Phys Rev B*, **84**(11), 113103.

Tarter, J. C., Agrawal, A., Ackermann, R., Backus, P., Blair, S. K., Bradford, M. T., Harp, G. R., Jordan, J., Kilsdonk, T., Smolek, K. E., *et al.* (2010). SETI turns 50: five decades of progress in the search for extraterrestrial intelligence. In *Instruments, Methods, and Missions for Astrobiology XIII*, volume 7819, page 781902. International Society for Optics and Photonics.

Traub, W. A. (2011). Terrestrial, habitable-zone exoplanet frequency from Kepler. *ApJ*, **745**(1), 20.

Verhulst, P.-F. (1838). Notice sur la loi que la population suit dans son accroissement. *Corresp. Math. Phys.*, **10**, 113–126.

Villard, R. and Sahu, K. (2012). The Milky Way contains at least 100 billion planets according to survey. *Hubblesite, January*, **11**.

Walsh, C., Millar, T. J., Nomura, H., Herbst, E., Weaver, S. W., Aikawa, Y., Laas, J. C., and Vasyunin, A. I. (2014). Complex organic molecules in protoplanetary disks. *Astronomy & Astrophysics*, **563**, A33.

Wang, J. and Fischer, D. A. (2014). Revealing a universal planet–metallicity correlation for planets of different sizes around solar-type stars. *AJ*, **149**(1), 14.

Worden, S., Drew, J., Siemion, A., Werthimer, D., DeBoer, D., Croft, S., MacMahon, D., Lebofsky, M., Isaacson, H., Hickish, J., *et al.* (2017). Breakthrough Listen-a new search for life in the universe. *Acta Astronautica*, **139**, 98–101.

Xu, Y., Newberg, H. J., Carlin, J. L., Liu, C., Deng, L., Li, J., Schönrich, R., and Yanny, B. (2015). Rings and radial waves in the disk of the Milky Way. *ApJ*, **801**(2), 105.

(Toward) Discovery of Life beyond Earth and its Impact
Proceedings IAU Symposium No. 387, 2026
H. Landt, M. Dominik & C. Oliver, eds.
doi:10.1017/S1743921324001340

How probable is Gaia?

Arwen E. Nicholson

Exeter University, Department of Physics and Astronomy, Stocker Rd, Exeter, EX4 4QL
email: arwen.e.nicholson@gmail.com

Abstract. Astronomers are now able to peer into the atmospheres of distant planets and search for signs of life. With the large diversity in exoplanets being found, confidently identifying remote signs of life is a complex process. E.g. while the high level of oxygen in Earth's atmosphere is the product of life, different planetary scenarios can produce oxygen rich atmospheres without any life-processes. Therefore we will need multiple lines of evidence to confidently declare a planet inhabited. Earth has been dramatically shaped by life for over 4 billion years and life plays a key role in regulating Earth's surface chemistry and climate, providing multiple detectable signals a remote observer would be able to detect from afar. The biosphere-Earth co-evolving system is known as 'Gaia'. Understanding the likelihood of Gaian-systems emerging on inhabited planets will inform us on the probability of confidently identifying an inhabited planet.

Keywords. Astrobiology, Gaia, Earth history, biosignatures

1. Introduction

Is life common in the galaxy? This is a question we now have the technology to begin to answer. Using telescope missions such as The James Webb Space Telescope astronomers can now peer into the atmospheres of distant planets and look for signs of life (Snellen et al. 2021; Quanz et al. 2021; Feinstein et al. 2023; Rustamkulov et al. 2023; Alderson et al. 2023; Ahrer et al. 2023). However to confirm any potential detection of alien life we need to be confident that we are looking at an inhabited planet and not some exotic chemistry not known on Earth. The more pieces of observational evidence we have of life, the more confident we can be that we are observing an inhabited planet.

Scientist are compiling lists of chemicals that could represent life-processes if found in the atmosphere of a planet (Seager et al. 2012, 2013, 2016; Sousa-Silva et al. 2020). Suggested biosignature gases include O_2 (Meadows 2017), DMT (Seager et al. 2013), PH_3 (Sousa-Silva et al. 2020) as on Earth these gases are predominantly produced by life. However scenarios exist where a planet can have an oxygen rich atmosphere in the absence of life (Meadows 2017; Krissansen-Totton et al. 2021) and, with the huge diversity of planets being found, any observation of a potential biosignature gas could be due to currently unknown abiotic chemistry not present on Earth. Therefore it's important for potential biosignatures to be considered within their planetary context (Catling et al. 2018) and for additional indicators of life to be identified (e.g. seasonal changes taking place on a planet (Olson et al. 2018)) to confidently identify life. As the only known inhabited planet we can look to Earth to understand how the signatures of life might manifest on other worlds.

Earth has been inhabited for roughly 4 billion years (Dodd et al. 2017) and over this time our planet has been dramatically transformed by life. The Earth-biosphere coupled system is known as 'Gaia', and this system has self-regulatory mechanisms that regulate Earth's atmosphere and ocean chemistry, and Earth's climate Lovelock (1965);

Lovelock & Margulis (1974); Lovelock (1990). This regulation has kept Earth in a habitable state for over 4 billion years despite numerous perturbations, e.g. the increasing luminosity of the sun (Goldblatt & Zahnle 2011), meteor impacts (Becker et al. 2001; Kaiho et al. 2001), and changes in volcanic activity (Liu et al. 2020). An example of Gaian regulation can be found in the carbon cycle which is a vital component of Earth's climate regulation (Lenton 2002). Over geological timescales carbon is moved through Earth's atmosphere and lithosphere; the chemical process of silicate weathering removes carbon from the atmosphere where rocks are exposed to the atmosphere, this carbon is eventually recycled back into the atmosphere via volcanic activity (Berner 1992). The rate of weathering is climate sensitive and increases with temperature and humidity. Therefore if the climate warms, weathering rates can increase, removing more CO_2 from the atmosphere and thus cooling the climate. Life acts to accelerate the rate of weathering (Lenton 2002) and adds additional feedback loops to the carbon cycle as organisms rapidly transfer carbon between the atmosphere and ocean and act as a carbon sink by using it to build their bodies. A lifeless Earth would be a much hotter world – possibly up to 45^oC warmer (Schwartzman & Volk 1989) due to life's impact on atmospheric CO_2.

It is suggested that life might have an even greater impact on Earth's climate by influencing the continental coverage of our planet (Höning et al. 2014) (which impacts the rate of weathering) and playing a role in the establishment / maintenance of plate tectonics (Lenardic et al. 2016) which are a vital part of Earth's climate regulation. If life plays a strong role in shaping its planet to maintain habitable conditions over long time spans, this creates an 'inhabitance paradox' (Goldblatt 2016) – for long term habitability on a planet, the planet must be inhabited. A 'Gaian Bottleneck' (Chopra & Lineweaver 2016; Nicholson et al. 2018) has also been proposed as a window early in a planet's formation where life must emerge and 'catch' an early window of habitable conditions by integrating with its planet to strengthen pre-existing abiotic negative feedback loops and form new ones to regulate the climate and maintain habitable conditions. This hypothesis suggests that if life fails to catch this early habitable window then the planet will become inhospitable due to perturbations experienced by the planet. This hypothesis would predict then that any planet detected with habitable surface conditions will have a very high likelihood of being inhabited. Understanding Gaian systems can not only inform our search for inhabited planets but search for life in the galaxy will provide data to test these hypotheses about Gaia.

As Earth is strongly influenced by life there are numerous biosignatures a distant observer could remotely detect. The disequilibrium in Earth's atmosphere provides evidence of life's metabolisms (Krissansen-Totton et al. 2018), the 'red-edge' is an observable feature in Earth's spectrum due to plant pigmentation (O'Malley-James & Kaltenegger 2018), and temporal variability occurs due to seasonal changes impacting life (Olson et al. 2018). Therefore, Earth has multiple markers of an inhabited planet for a remote observer to detect.

As the ongoing research into the possibility of microbial life on Mars and Venus demonstrates (e.g. Davila & Schulze-Makuch (2016); Limaye et al. (2018)), life acting as a 'passenger' on its planet, e.g. the presence of low levels of microbial life, are hard to identify or rule out with planets within our solar system. When searching for life outside our solar system we will only be able to detect it where it has played a large role in shaping its planet as life on Earth has; we are searching for living planets – other Gaias – not just pockets of life. Our chances of confidently detecting alien life therefore depends on how probable it is for Gaian systems to emerge on planets with life. Whether Gaia is due to 'luck' or is instead 'probable' is an ongoing debate with respects to Earth (Lenton & Wilkinson 2003). A 'lucky Gaia' argues that the long lasting regulatory feedback mechanisms on Earth are mostly due to chance and predicts that Gaias are rare in the galaxy. A 'probable Gaia' instead posits that Gaian regulation is an expected outcome on a planet where life emerges, increasing the number of Gaias we would expect to observe. Here I will argue for a 'probable Gaia' and demonstrate selection mechanisms that increase the probability of planets with life being in Gaian states. Critics of Gaia have questioned why life, if it strongly impacts its environment, would not tend to drive the planet towards inhospitable conditions rather than maintain habitability – after all there are more ways for a planet to be inhospitable than hospitable. However several selection mechanisms have been identified that increase the probability of Gaian regulation on inhabited planets.

Informed by Earth's history I will present three selection mechanisms that increase the probability of Gaia – Selection by Survival (Ford Doolittle 2014), Sequential Selection (Lenton & Wilkinson 2003; Lenton et al. 2018) and Sequential Selection with Memory (Arthur & Nicholson 2022). I will demonstrate these selection mechanisms first using games of dice as this reduces the number of prior assumptions in the model, and then with a model of an evolving ecosystem coupled to its environment to demonstrate these selection mechanisms in more complex systems. These selection mechanisms predict that if a planet is inhabited it is likely to be in a Gaian state, increasing our chances of confidently identifying other living planets in our galaxy.

I will demonstrate these selection mechanisms for Gaia using games of dice as this uses minimal underlying assumptions. In Section 2 I will explain the dice model, then I will show 'games' for each selection mechanism in Sections 3, 4 and 5. In Section 6 I will briefly detail a model of an evolving ecosystem coupled to its environment called the Tangled Nature Model to demonstrate these same selections mechanisms working in a more complex and realistic analogue for planets with life. I will then discuss these results in the context of the search for biosignatures in Section 7.

2. Dice model

Using some simple games we can demonstrate each of the selection mechanisms and their effects on our likelihood of finding other Gaias in the universe. The results for the dice games explained in this section and presented in Sections 3, 4 and 5 are explained in full in Arthur & Nicholson (2023b). In these games we are interested in the likelihood of Gaian regulation emerging on an inhabited planet, rather than the probability of life itself emerging. Therefore we will make the assumption for all simulations that our planets are habitable and that life does emerge on them. We are therefore not considering scenarios where a planet is unsuitable for abiogenesis or where life fails to emerge on a habitable planet.

We will use die to represent planets with life. The number on the top face we take as a representation of the stability of the life-planet coupled system at that point in time. A low value indicates a planet without strong climate regulation indicating that habitable conditions may be transient and the biosphere in an unstable state, and a high value indicates a planet with strong Gaian regulation where habitable conditions and the biosphere can be expected to persist for long timescales. We therefore take the value M rolled by a single dice as the number of timesteps to wait before either ending the game for that dice (indicating a breakdown of habitable conditions and total extinction on the planet) or rolling the dice again. The dice games for each selection mechanism are as follows:

- Selection by survival (SSB) – roll N independent dice M sided dice once each. Each dice i is removed from the game after m_i timesteps where m_i is the face value of dice i.
- Sequential selection (SS) – roll one M sided dice and wait m_i timesteps before rolling again where m_i is the value of the top face of the dice. Repeat this process T times.

• Sequential selection with memory (SSM) – start with a single sided ($M = 1$) dice and roll. Wait m_i timesteps before rolling again where m_i is the value of the top face of the dice and each time $m_i = M$ increase the value of M by 1. Roll the dice a total of T times, waiting m_i timesteps in between each roll.

These games represent a scenario where we survey a large catalogue of inhabited planets over time. Throughout the game we determine the probability of observing inhabited planets in Gaian like states. When observing planets in our galaxy we will be viewing the planet at one single 'snap-shot' of its history. These games can help us understand the probability of any life that is detected on a planet persisting for long time spans.

3. Selection by Survival

Arguments for 'lucky Gaia' (Lenton & Wilkinson 2003) claim that while the Earth does have regulatory feedback loops stabilising its climate over geological timespans, the presence of these feedbacks is due to luck rather than any inherent property of a life-planet coupled system. This suggests that Earth being in a Gaian-state is largely down to the Anthropic principal – Earth must have Gaian regulation in order for us to be here to ask these questions (Watson 1999). 'Lucky Gaia' is also closely related to the 'rare Earth hypothesis' – that conditions in our solar system and galactic neighbourhood have to be just-so for a planet to experience long-term habitable conditions (Brownlee 2003). 'Lucky Gaia' would predict that inhabited planets are rare in the universe. Selection by survival (SBS) (Ford Doolittle 2014), in its simplest representation which we will use here (as outlined in Section 2) is similar to these hypotheses. SBS posits that only those inhabited planets with regulatory feedback loops are likely to survive for long timespans. When life emerges on a planet it can either integrate with and strengthen pre-existing regulatory feedback loops on its planet, and possibly establish new ones, or it could either have a neutral or even a negative impact on the climate regulation of its planet. For our game of SBS we will assume once life emerges on its planet it integrates with its environment and influences the stability of its planet's climate. Each planet is represented by a single dice that is rolled once at the start of the game. The stability of the life-planet coupled system is then indicated by the top face of the dice – a low value indicating an unstable life-planet coupled system at risk of climate breakdown and total extinction, whereas a high value indicates an inhabited planet in a strong Gaian-state.

Figure 1 shows a game of SBS for 25 6-sided die. We remove each dice from the game after the number of timesteps indicated by the top face. This removal represents the biospheres on these planets going extinct after the regulation of habitable conditions breaks down. Figure 1 shows that by time $t = 6$ the only dice left are those that rolled a 6 at the beginning of the game. This represents a situation where at later times the only inhabited planets remaining are those in Gaian states as those lacking regulatory feedback mechanisms have gone extinct.

At the beginning of the game, $t = 1$, all planets are inhabited and as each face had an equal probability of being rolled, the average face value of our planets at time is:

$$\frac{1 + 2 + ... + M}{M}. \tag{1}$$

At $t = 2$ all dice with face 1 have been removed representing the biospheres on these planets going extinct. If we only count the 'inhabited' planets, the average face value is now:

$$\frac{2 + 3 + ... + M}{M - 1}. \tag{2}$$

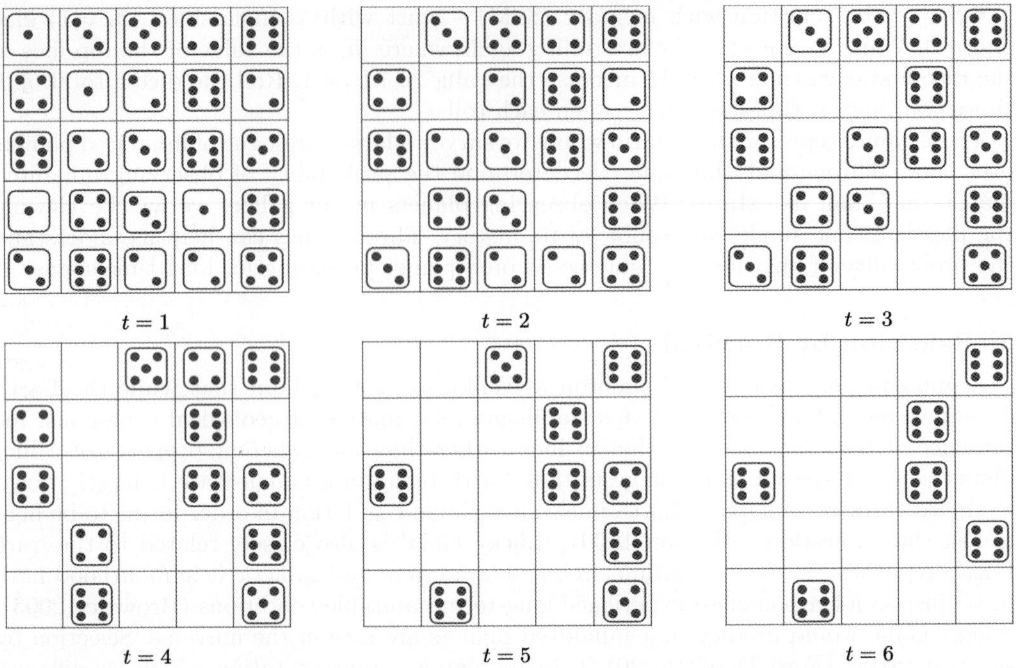

Figure 1. A game of Selection by Survival played for 25 planets, using 6-sided dice to represent inhabited planets. Over time planets in non-Gaian states (represented by low top-face values) go extinct and are removed from the game. Figure from Arthur & Nicholson (2023a).

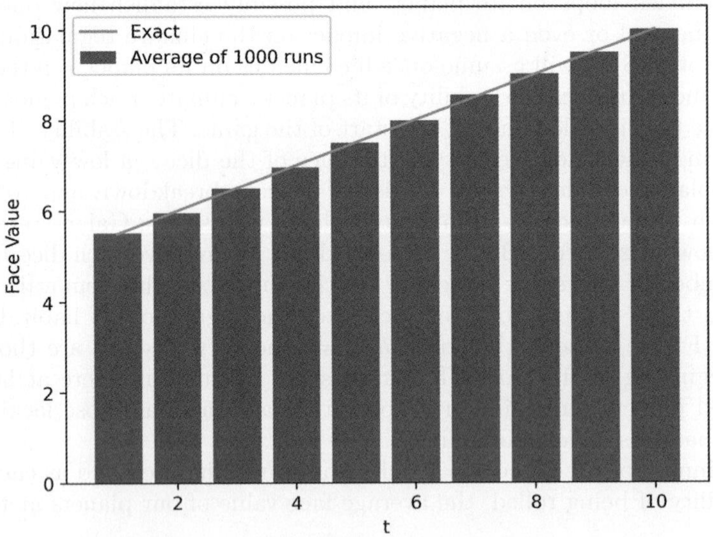

Figure 2. Results of 1000 games of Selection by survival is shown in blue using 10-sided die to represent each inhabited planet. As planets are removed from the game as their biospheres go extinct the average expected face value increases over time. The exact solution as given by Equation 3 is shown in orange. Figure from Arthur & Nicholson (2023a).

At time $t \leq M$ the average face value is:

$$\frac{t + (t+1) + ... + M}{M - t} = \frac{M + t}{2}. \tag{3}$$

Figure 2 shows the average behaviour of our model over 1000 games of SBS now using 10-sided dice to represent our inhabited planets. Figure 2 also shows the exact solution as given by Equation 3. We find that over time the expected face value of the inhabited 'planets' increases as SBS means that those with low face values – indicating low levels of climate stability – experience breakdowns of habitable conditions with life on these planets going extinct. This means at later times only planets in Gaian states are expected to remain inhabited.

While SBS does not make any predictions on the likelihood of inhabited planets acquiring Gaian regulation, it does predict that any inhabited planets persisting for long timescales *will* be in a Gaian state. Therefore in our search for alien life in the galaxy, SBS predicts that if life is out there it will most likely be on a planet with Gaian regulation, especially if we are looking at older planetary systems. This means that any detected alien life is likely to persist and is unlikely to be the product of only a fleeting window of habitability.

In our game of SBS the stability of the life-planet coupled system is determined at the start of the game and remains fixed throughout. However Earth's biosphere, and the regulatory mechanisms determining Earth's climate have dramatically changed over time (Lenton & Watson 2013). The next game allows for a dynamic history for each inhabited planet.

4. Sequential Selection

Now we will consider a selection mechanism called 'Sequential Selection' (SS) (Lenton & Wilkinson 2003; Lenton et al. 2018) that takes place within the history of a single planet. Earth's diverse and abundant biosphere, and the ways it interacts with the non-living parts of our planet, has changed over time. The biosphere on early Earth ~ 4 billion years ago was very different to today's, with a biosphere consisting of simple single celled-microbes, and no atmospheric oxygen (Kharecha et al. 2005; Catling & Zahnle 2020). The rate of various feedback loops on Earth's climate, e.g. the rate of weathering, has also changed over time (Ibarra et al. 2019; Bayon et al. 2022) and the geological record is punctuated with mass extinction events, 5 of which have happened within the last 540 million years where huge percentages of all life went extinct (Raup & Sepkoski Jr 1982). After each mass extinction the biosphere recovered with new lifeforms evolving to dominate the niches previously occupied by now extinct species (Solé et al. 2002). Our game of Selection by Survival (SBS) shown in Section 3 fails to capture the dynamic history seen on Earth, whereas SS is a mechanism that takes place over the trajectory of a single inhabited planet and has the following steps:

(1) Life emerges on a habitable planet
(2) Life evolves
(3) Life impacts its environment
(4) Does this lead to climate stability?
  a. if 'No' – this unstable coupling between life and its planet is short lived, the planet approaches bounds of habitability which causes life to die off. The reduced presence of life removes the destabilising impacts life had on its planet's climate and non-living processes dominate once more returning the climate back to its original condition, which was of an inhabited planet. The remnants of the previous ecosystems remain. Go back to step 2.
  b. if 'Yes' – by definition a stable system will persist. A stable habitable climate will persist on the planet until a perturbation, either external e.g. meteor impact, or internal e.g. evolutionary innovation perturbs the climate, returning us to step 2.

Figure 3. A single game of Sequential Selection for a single planet using a 6-sided dice. 'Gaian-states' (represented by higher top-face values) last longer than non-Gaian states. Figure adapted from Arthur & Nicholson (2023a).

Therefore SS allows for various 're-sets' during a planet's history, with non-Gaian states being transient and Gaian-states being long lived. To represent SS, each inhabited planet is represented by a dice as before, with the top face indicating the stability of the life-planet coupled system. However differing from SBS, once the 'luck' of the biosphere runs out instead of life on the planet going extinct we roll again, representing the upheaval in the biosphere with new life forms emerging after the previous ecosystems collapsed. As with the previous game, the top-face indicates the length of time the current biosphere persists, therefore if we roll a 2, this state persists for 2 timesteps and then we roll again.

For a random time t we want to know what the average expected face value is for our die. The chance of a dice showing k is proportional to the probability of rolling a k, $P(k)$, multiplied by the length of time where the observation could occur. E.g if we observe a 3 we could be observing this value on its first, second or third timestep. We normalise this probability to calculate the expected top-face value as:

$$\sum_{k=1}^{M} k \frac{kp(k)}{\sum_{k=1} mkp(k)}. \tag{4}$$

For a single dice $p(k) = \frac{1}{M}$ and Equation 5 simplifies to:

$$\frac{2M+1}{3} \tag{5}$$

which, when $M > 1$, is higher than the expected face value of a single role of a single dice which is given by $\frac{M+1}{2}$.

Figure 3 shows a single game of SS for a single planet using a 6-sided dice. Initially a '2' is rolled and so the biosphere on the planet persists for 2 timesteps. Then the dice is rolled again, and so on. Higher top face scenarios representing Gaian states therefore last longer than shorter 'non-Gaian' states. Figure 4 shows a plot of the expected face value for 10-sided dice for both the average of 1000 games of SS, and the exact solved equation (Equation 5). We find that the expected face value for planets where SS is taking place is higher than the average for a single roll of the dice.

SS therefore increases the chances of inhabited planets acquiring Gaian-regulation over time and thus increases the timescales for which we would expect any observed alien life on a planet to persist. This increases our chances of detecting other Gaian-systems in our galaxy. This game of SS allows for the ecosystem upheavals seen in Earth-history, however the next roll after each 'reset' is still random. Earth's abundant and diverse biosphere built in complexity over time. Early life had a smaller influence on Earth than it does today, e.g. the high level of atmospheric oxygen in Earth's atmosphere is a recent feature (on geological timescales). Equally, after each great extinction event Earth's biosphere did not revert back to simple non-photosynthetic life, what came next was built upon what came before. In the next game we provide a mechanism for this 'ratcheting' increasing complexity of the biosphere, and its influence on its planet, over time.

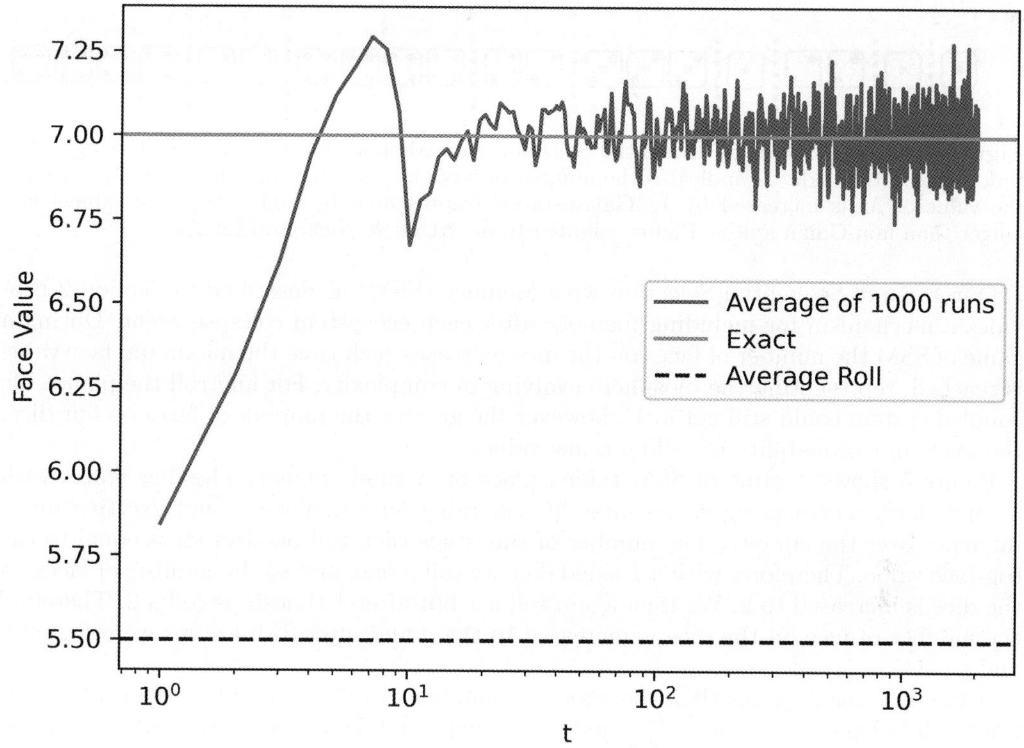

Figure 4. Results from the average of 1000 runs of Sequential Selection (SS) games using 10-sided die in shown in blue. The orange line shows the exact value of the SS game as given by Equation 5. The expected value of a single 10-sided dice is shown by the black-dashed line. Figure adapted from Arthur & Nicholson (2023a).

5. Sequential Selection with Memory

For our game of Sequential Selection (SS) in Section 4, there is an equal probability of rolling a low or a high face-value at every dice re-roll. This would allow for scenarios where complex Gaian-systems, similar to present day Earth, spontaneously emerge on previously lifeless planets. It also means there is a high probability of complex biospheres reverting back to simple prokaryotic ones after any mass extinction. These scenarios do not mirror what we find in Earth's geological record. The diversity of Earth's biosphere has built over time (Benton 1987). Over billions of years Earth's biosphere has increased in complexity and diversity, from the earliest simple life exploiting preexisting chemical gradients present around deep-sea hydrothermal vents (Nisbet 1995), to the prokaryotic methanogenic biosphere living on the oxygen-free early Earth (Kharecha et al. 2005) to present-day Earth's biosphere, with an oxygen-rich atmosphere dominated by life processes and a biosphere with diverse and abundant eukaryotic life. After mass extinctions the new ecosystems on Earth built upon the remnants of what came before (Hull 2015). For example after the Cretaceous–Paleogene extinction event that wiped out non-avian dinosaurs, mammals, who were already present at the time, diversified and took over the vacant niches left behind (Archibald 2011). As the biosphere has increased in complexity its influence on Earth's climate has also increased (Lenton & Watson 2013) – Gaia is built over time and 'remembers' what came before.

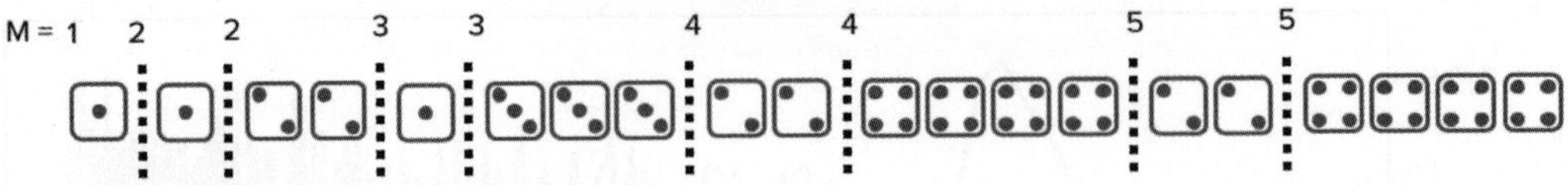

Figure 5. A single game of Sequential Selection with Memory for a single planet using a dice that starts with 1 side. M indicates the number of faces the dice has and each time M is rolled, the value of M is increased by 1. 'Gaian-states' (represented by higher top-face values) last longer than non-Gaian states. Figure adapted from Arthur & Nicholson (2023a).

Our game of Sequential Selection with Memory (SSM) as described in Section 2 provides a mechanism for including memory after each ecosystem collapse event. During a game of SSM the number of faces on the dice increases each time the maximum face value is reached, representing the biosphere evolving in complexity. For any roll the planet-life coupled system could still get a '1', however the greater the number of faces on our dice, the lower our probability of rolling a low value.

Figure 5 shows a game of SSM taking place on a single planet. The dice starts with a single face, representing very simple life emerging on a planet and not exerting much influence over the climate. The number of timesteps each roll persists for is equal to the top-face value. Therefore, with a 1-sided dice we roll a one, and so the number of faces on the dice is increased to 2. We then again roll a 1 but after 1 timesteps roll a 2. Therefore the number of face on the dice is increased to three and after 2 timesteps we roll again, and so on.

If the dice has M sides then the expected number of rolls needed to roll an M, is M. Each roll is expected to last $\frac{M+1}{2}$ and so the expected time needed to wait to increase the number of faces on the dice is:

$$M\frac{M+1}{2}. \tag{6}$$

If we start with a single-sided dice and play a game of SSM then the expected time it would take for a game to produce an M-sided dice is given by:

$$t = \sum_{i=1}^{M} i\frac{i+1}{2} = \frac{1}{2}\left(\frac{M(M+1)(2M+1)}{6} + \frac{M(M+1)}{2}\right). \tag{7}$$

Keeping the highest order terms in M and solving for t gives:

$$M = \sqrt[3]{6t}. \tag{8}$$

Substituting this into Equation 3, the expected face value for a dice in the game of SSM, at time t, is:

$$\frac{2\sqrt[3]{6t}+1}{3}. \tag{9}$$

Figure 6 shows the average top-face value for 1000 games of SSM (shown in blue) and the exact solution as given by Equation 9 (shown in orange). We find that over time the expected face value increases, indicating that inhabited planets are acquiring more Gaian regulation over time.

SSM not only increases the probability of inhabited planets being in Gaian-states but predicts that over time, the stability of the planet-life coupled system will increase over time, and that older planets are likely to be in highly Gaian states with robust climate regulation.

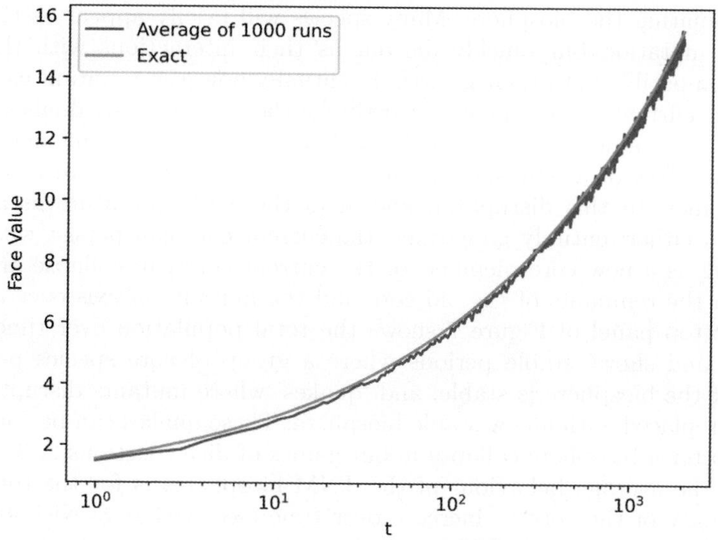

Figure 6. Results from the average of 1000 runs of Sequential Selection with Memory games shown in blue. The orange line shows the exact value the game as given by Equation 9. Figure from Arthur & Nicholson (2023a).

6. Selection in the Tangled Nature Model

These selection mechanisms are imposed upon our games of dice so we can explore how they influence the expected state of inhabited planets over time. However these same selection mechanism emerge spontaneously in a model of entangled ecology known as the Tangled Nature Model (TNM) (Becker & Sibani 2014; Arthur et al. 2017; Jensen 2018). The TNM is an agent based model where individuals are represented by binary strings that encode their interactions with all other species and with their environment. Individuals have a probability of reproduction based on how 'fit' they are to their environment, and each reproduction event has a probability of mutation, which allows for new species to evolve over time. The 'entangled' interactions between individuals determine the fitness of each species and so a species fitness depends not only on the abiotic environment but also on all other extant individuals in the biosphere. This mirrors life on Earth as most life depends on the processes of other organisms for their survival (e.g. mammals are dependant on the high quantities of atmospheric oxygen produced by photosynthesisers). Therefore the TNM is an abstract representation of the high degree of entanglement of life on Earth. The average behaviour of the TNM generalises to an extremely general growth equation (details in Arthur & Nicholson (2022)) increasing our confidence in using the TNM to explore questions of general behaviours of biospheres, both on Earth and other planets. The details of of the TNM are not given here but the work briefly summarised in this section can be found in Arthur & Nicholson (2022).

In Sections 3, 4 and 5 the selection mechanisms are encoded into the dice games via rules imposed on our planets. In the TNM, these same selection mechanisms of Selection by Survival (SBS), Sequential Selection (SS) and Sequential Section with Memory (SSM) emerge from the dynamics of the ecosystem. A typical TNM experiment begins with a single species chosen at random to emerge on the planet. As reproduction and mutation take place, species diversify and an ecosystem quickly establishes itself with a groups of species known as 'core species', who have mutually beneficial interactions with one another, dominating the biosphere. Many species will briefly appear as they come into existence via mutation but quickly die out as their interactions with the established 'core' are incompatible with their growth. Eventually however a mutant can emerge that is suited to the current environment (consisting of the existing individuals and the abiotic factors) and it can rapidly grow. In small numbers the current core members are unlikely to 'feel' much effect of a new species, but as it reproduces the core can be disrupted by the newcomer. In this disruption, known in the TNM literature as a 'quake', the ecosystem can either: entirely go extinct; the current core can persist with the mutant species joining as a new core member; or the current core can collapse and a new core emerges from the remnants of the old core and the mutants in existence at the time of collapse. The top panel of Figure 7 shows the total population over time a single run of the TNM and shows stable periods where a group of core species persist and the population of the biosphere is stable, and 'quakes' where mutants disrupt this stability before being replaced with a new stable biosphere. These quakes can be compared to our dice re-rolls after a biosphere collapse in our games of dice (Sections 2, 3, 4 and 5).

Over time the average behaviour of the TNM biospheres is for the total population and the diversity of the core to increase over time (see Arthur & Nicholson (2022) for full details). The lower panel of Figure 7 shows the average population for 1000 TNM experiments over time, only including non-extinct systems for each time. We see that the average total population increases. Note the logged x-axis. After every 'quake' the new core to emerge is the group with the largest mutually positive interactions giving them the fastest reproduction rate. These new core members out-compete other species alive at the time and dominate the TNM biosphere. This repeated selection gradually increases the average reproduction probability and hence the average population, as observed in Figure 7. The diversity of the core also increases over time (see Arthur & Nicholson (2017, 2022)). As the population increases over time, the number of reproduction events and so the number of mutation events also increases. This increases the number of mutants present in the TNM system at the time of a quake, and thus increases the number of species configurations for the system to 'choose' the new core from. As there are more configurations involving larger number of species, higher diversity in the core is selected for over time. The higher core population has another impact on the TNM dynamics. Larger cores are more stable as they are harder for mutants to destabilise (see Becker & Sibani (2014) for full details). For any mutant to destabilise the current ecosystem it must have overall positive interactions with the core species to be able reproduce to numbers that destabilise can the core. With a larger core this is more difficult and so for a larger core the ecosystem must 'wait' for longer for a destabilising mutant to emerge. Therefore over time a TNM biosphere increases in not only in population and diversity but also in stability. Figure 8 shows the number of quakes (within 1000 timestep windows) over time, averaged over 1000 TNM experiments. We see that over time the number of quakes decreases indicating the increasing stability of TNM ecosystems.

In Arthur & Nicholson (2022) the TNM is adapted to explore SBS, and SS separately. However the default behaviour of the TNM without any adjustments is for SSM to take place. After each core collapse, the new core is built upon the remnants of the previous one. In this way the 'memory' we added to our dice games in Section 5 is already present in the TNM. This increasing stability of the TNM system has been described as an entropic hierarchy due to the TNM system having a larger pool of species to 'choose' from after each quake – e.g. there are more potential configurations and so higher entropy. The entropic barrier in the TNM is the difficulty of generating a mutant with sufficiently large interactions to destabilise the core as it grows in size over time. The behaviour of the TNM with increasing population and diversity over time mirrors patterns found in

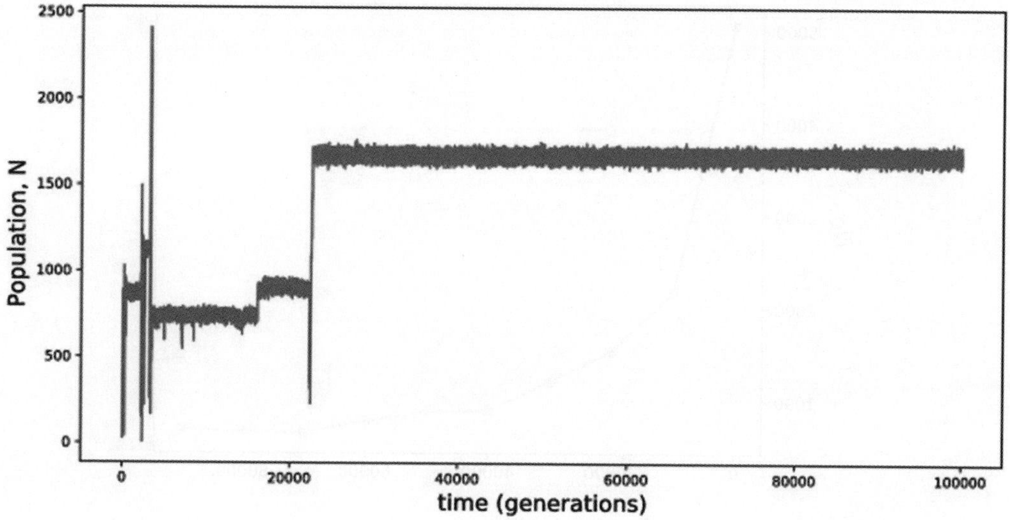

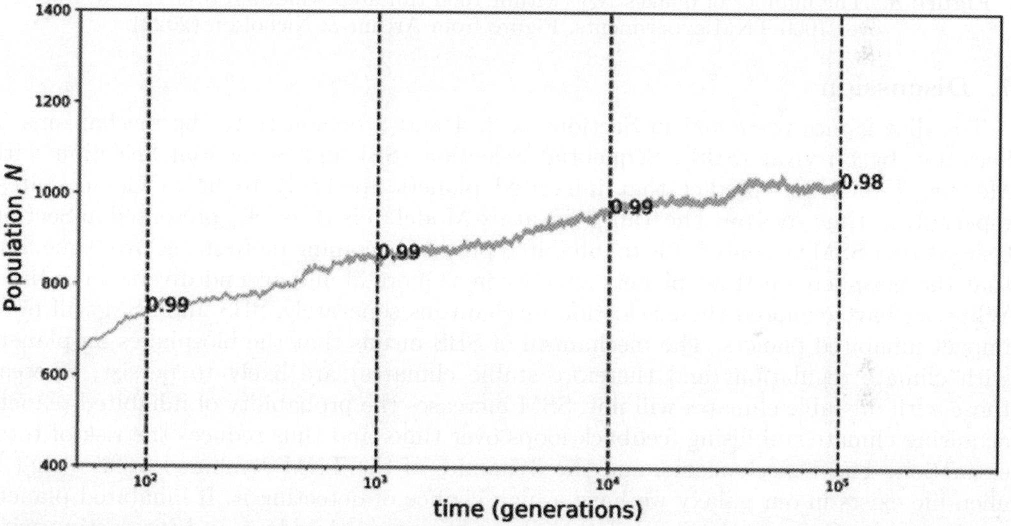

Figure 7. Top panel: total population for a single Tangled Nature Model (TNM) experiment. Lower panel: The average population of 1000 TNM experiments with only surviving systems included. Note the logged x-axis. The numbers at $t = 10^2, 10^3, 10^4, 10^5$ indicate the proportion of non-extinct TNM systems at these times. Figure from Arthur & Nicholson (2022).

Earth's fossil record, which shows increasing diversity over time since the preCambrian (Benton 1995).

SSM emerges from the dynamics of the TNM with the biosphere increasing not only in stability, but also abundance and diversity, over time. Therefore the TNM not only predicts older biospheres to be in highly stable 'Gaian' states, but that those biospheres will be highly diverse and abundant. This increases the chances of these biospheres of being detected; if life on a planet is abundant and diverse, it can be interacting with its planet and shaping it in numerous ways which increases the chances of it leaving remotely observable 'fingerprints' on its planet.

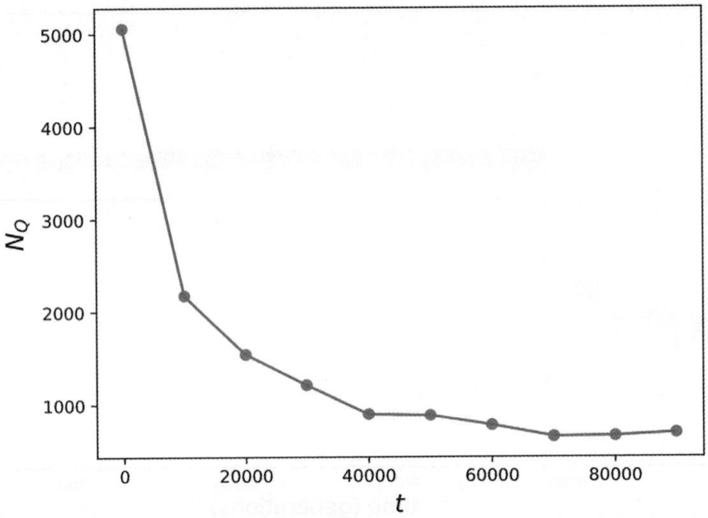

Figure 8. The number of quakes N_Q (within 1000 timestep windows) over time averaged over 1000 TNM experiments. Figure from Arthur & Nicholson (2022).

7. Discussion

The dice games presented in Sections 2, 3, 4 and 5 demonstrate the mechanisms of Selection by Survival (SBS), Sequential Selection (SS) and Sequential Selection with Memory (SSM) and predict that inhabited planets are likely to be in Gaian states, especially as time goes on. The Tangled Nature Model (TNM) briefly presented in Section 6 shows that SSM not only leads to inhabited planets becoming more stable over time, but that the biosphere on these planets also becomes more abundant and diverse over time. While we have explored these selection mechanisms separately, SBS and SSM will both impact inhabited planets. The mechanism of SBS means that the biospheres on planets with climate regulation (and therefore stable climates) are likely to persist, whereas those with unstable climates will not. SSM increases the probability of inhabited planets acquiring climate stabilising feedback loops over time, and thus reduces the risk of total extinction. These mechanisms, and the dynamics of the TNM combined predict that if alien life exists in our galaxy we have a high chance of detecting it. If inhabited planets are in stable, Gaian states, and if their biospheres are abundant and interacting with their planet in diverse ways, this will potentially produce more biosignatures for us to detect. These mechanisms also predict that any alien life we might detect in the future is likely to be part of a stable and long-lived biosphere, rather than representing a fleeting window of habitability on the planet.

As the ongoing questions about the possibility of life on Mars (Petrescu et al. 2018) and Venus (Limaye et al. 2018) demonstrate, when searching for life on distant planets we are highly unlikely to detect low levels of life having only a minimal impact on its environment. When looking for life beyond our solar system we are looking for worlds shaped by life – we are looking for Gaias. We have not explored in this work the probability of life emerging on a habitable planet, and this probability will strongly influence our chances of finding alien life. However this work predicts that on those planets where life does emerge, Gaia is an expected outcome of a planet-life coupled system.

Any single feature of a planet (e.g. oxygen in its atmosphere (Meadows 2017; Krissansen-Totton et al. 2021)) cannot alone confirm a discovery of life, as chemicals that on Earth are produced only by life might have abiotic pathways on planets with different conditions (e.g. Mráziková et al. (2024)). Therefore many lines of evidence will be needed to confidently identify a living planet (Catling et al. 2018). The selection mechanisms presented here predict that if life is common in our galaxy we have a high chance of being able to detect it.

Acknowledgements

This work was funded by the Leverhulme Trust through a research project grant [RPG-2020-82]. Many thanks to my co-author Rudy Arthur who lead this research and to Nathan Mayne for many discussions which helped structure this work.

References

Ahrer E.-M., et al., 2023, Nature, 614, 653
Alderson L., et al., 2023, Nature, 614, 664
Archibald J. D., 2011, Extinction and radiation: how the fall of dinosaurs led to the rise of mammals. JHU Press
Arthur R., Nicholson A., 2017, Journal of theoretical biology, 430, 177
Arthur R., Nicholson A., 2022, Journal of Theoretical Biology, 533, 110940
Arthur R., Nicholson A., 2023a, AstroBiology, 23, 1238–
Arthur R., Nicholson A., 2023b, Astrobiology, 23, 1238
Arthur R., Nicholson A., Sibani P., Christensen M., 2017, Computational and Mathematical Organization Theory, 23, 1
Bayon G., Bindeman I. N., Trinquier A., Retallack G. J., Bekker A., 2022, Earth and Planetary Science Letters, 584, 117490
Becker N., Sibani P., 2014, Europhysics Letters, 105, 18005
Becker L., Poreda R. J., Hunt A. G., Bunch T. E., Rampino M., 2001, Science, 291, 1530
Benton M. J., 1987, Trends in Ecology & Evolution, 2, 153
Benton M. J., 1995, Science, 268, 52
Berner R. A., 1992, Geochimica et Cosmochimica Acta, 56, 3225
Brownlee D., 2003, Rare Earth: Why complex life is uncommon in the Universe. Springer
Catling D. C., Zahnle K. J., 2020, Science advances, 6, eaax1420
Catling D. C., et al., 2018, Astrobiology
Chopra A., Lineweaver C. H., 2016, Astrobiology, 16, 7
Davila A. F., Schulze-Makuch D., 2016, Astrobiology, 16, 159
Dodd M. S., Papineau D., Grenne T., Slack J. F., Rittner M., Pirajno F., O'Neil J., Little C. T., 2017, Nature, 543, 60
Feinstein A. D., et al., 2023, Nature, 614, 670
Ford Doolittle W., 2014, Biology & Philosophy, 29, 415
Goldblatt C., 2016, arXiv e-prints, p.arXiv:1603.00950
Goldblatt C., Zahnle K. J., 2011, Nature, 474, E1
Höning D., Hansen-Goos H., Airo A., Spohn T., 2014, Planetary and Space Science, 98, 5
Hull P., 2015, Current Biology, 25, R941
Ibarra D. E., et al., 2019, American Journal of Science, 319, 1
Jensen H. J., 2018, European Journal of Physics, 40, 014005
Kaiho K., et al., 2001, Geology, 29, 815
Kharecha P., Kasting J., Siefert J., 2005, Geobiology, 3, 53
Krissansen-Totton J., Olson S., Catling D. C., 2018, Science advances, 4, eaao5747
Krissansen-Totton J., Fortney J. J., Nimmo F., Wogan N., 2021, AGU Advances, 2, e2020AV000294
Lenardic A., Crowley J., Jellinek A., Weller M., 2016, Astrobiology, 16, 551
Lenton T. M., 2002, Climatic Change, 52, 409
Lenton T., Watson A., 2013, Revolutions that made the Earth. OUP Oxford
Lenton T. M., Wilkinson D. M., 2003, Climatic Change, 58, 1
Lenton T. M., Daines S. J., Dyke J. G., Nicholson A. E., Wilkinson D. M., Williams H. T., 2018, Trends in Ecology & Evolution, 33, 633

Limaye S. S., Mogul R., Smith D. J., Ansari A. H., Słowik G. P., Vaishampayan P., 2018, Astrobiology, 18, 1181
Liu B., Wang B., Liu J., Chen D., Ning L., Yan M., Sun W., Chen K., 2020, Geophysical Research Letters, 47, e2020GL089416
Lovelock J. E., 1965, Nature, 207, 568
Lovelock J. E., 1990, Nature, 344, 100
Lovelock J. E., Margulis L., 1974, Tellus, 26, 2
Meadows V. S., 2017, Astrobiology, 17, 1022
Mrázikivá K., et al., 2024, Astrobiology, 24, 407
Nicholson A. E., Wilkinson D. M., Williams H. T., Lenton T. M., 2018, Monthly Notices of the Royal Astronomical Society, 477, 727
Nisbet E., 1995, Geological Society, London, Special Publications, 95, 27
O'Malley-James J. T., Kaltenegger L., 2018, Astrobiology, 18, 1123
Olson S. L., Schwieterman E. W., Reinhard C. T., Ridgwell A., Kane S. R., Meadows V. S., Lyons T. W., 2018, The Astrophysical Journal Letters, 858, L14
Petrescu R. V., Aversa R., Apicella A., Kozaitis S., Abu-Lebdeh T., Petrescu F. I., 2018, American Journal of Engineering and Applied Sciences, 11, 78
Quanz S. P., et al., 2021, Experimental Astronomy, pp 1–25
Raup D. M., Sepkoski Jr J. J., 1982, Science, 215, 1501
Rustamkulov Z., et al., 2023, Nature, 614, 659
Schwartzman D. W., Volk T., 1989, Nature, 340, 457
Seager S., Schrenk M., Bains W., 2012, Astrobiology, 12, 61
Seager S., Bains W., Hu R., 2013, The Astrophysical Journal, 777, 95
Seager S., Bains W., Petkowski J., 2016, Astrobiology, 16, 465
Snellen I. A. G., et al., 2021, Experimental Astronomy
Solé R. V., Montoya J. M., Erwin D. H., 2002, Philosophical Transactions of the Royal Society of London. Series B: Biological Sciences, 357, 697
Sousa-Silva C., Seager S., Ranjan S., Petkowski J. J., Zhan Z., Hu R., Bains W., 2020, Astrobiology, 20, 235
Watson A. J., 1999, Geological Society, London, Special Publications, 150, 75

(Toward) Discovery of Life beyond Earth and its Impact
Proceedings IAU Symposium No. 387, 2026
H. Landt, M. Dominik & C. Oliver, eds.
doi:10.1017/S1743921324001455

Escaping the Great Filter: The Future of Civilization and the Search for Technosignatures

J. Haqq-Misra

Blue Marble Space Institute of Science, Seattle, Washington, USA
email: jacob@bmsis.org

Abstract. Human civilization continues to experience rapid growth in energy consumption, while projections of population stabilization remain uncertain. A continued trajectory of exponential energy use would cause direct heating of the planet by ~2300, which would also coincide with a transition to a Kardashev type-I civilization. If such patterns of energy consumption are typical for other technological civilizations, then the lack of evidence for extraterrestrial life suggests that Earth may be among the first. This implies that a "Great Filter" may exist in the near future, which would mark a critical juncture of whether civilization on Earth becomes spacefaring or extinct. Any extant technological civilizations are likely those that have achieved long-term equilibrium with energy consumption and population growth. The search for technosignatures by ongoing ground- and space-based observatories will provide a way to test the Great Filter hypothesis and examine the extent to which energy-intensive civilizations occur in the galaxy.

Keywords. astrobiology, technosignatures, biosignatures, Great Filter, futures studies

1. Introduction

The search for life on other planets is becoming increasingly technologically feasible, with thousands of known exoplanetary systems and a continuous stream of newly-discovered planets providing targeted locations to search. Planets seem to be ubiquitous in star systems (e.g., Cassan et al. 2012; Dressing and Charbonneau 2015) and Earth-sized planets also appear to be commonplace (e.g., Hardegree-Ullman et al. 2019; Bryson et al. 2020). Further efforts at characterizing the atmospheres of exoplanets through spectroscopy have already demonstrated success for giant and terrestrial planets (e.g., Zieba et al. 2023; Madhusudhan et al. 2023), and ongoing searches continue to push the search limits toward finding smaller planets that are orbiting within the liquid water habitable zone (Kasting et al. 1993; Kopparapu et al. 2013) that would be suitable for life.

Earth is the only known example of life, so the search for life must begin with Earth as an example of the kind of inhabited planet to search for, even if such searches then venture into further speculative territory about other possibilities for life. The characterization of exoplanetary atmospheres could reveal potential spectroscopic "biosignatures" that could indicate the presence of extraterrestrial life, such as the simultaneous observation of methane and oxygen or other atmospheric disequilibrium that would be difficult to explain with abiotic processes (see, e.g., Schwieterman et al. 2018). The James Webb Space Telescope is enabling some of the first attempts at searching for potential biosignatures in terrestrial exoplantary systems (e.g., Meadows et al. 2023), and the search for such exoplanetary biosignatures remains one of the primary science drivers of NASA's future Habitable Worlds Observatory (e.g., Tuchow et al. 2024). The actual discovery of any biosignatures beyond Earth may not bear any resemblance to the biosignatures of Earth today or in Earth's past, but these reference points across history provide the only examples of life to calibrate the search and use as a basis for thinking beyond Earth.

The idea of looking for biosignatures in exoplanetary systems includes all observable manifestations that could result from life, which includes the possibility of "technosignatures" as remotely detectable evidence of technology that life has made (e.g., Socas-Navarro et al. 2021; Haqq-Misra et al. 2022c). The search for technosignatures, as with the search for other biosignatures, has Earth as the only known example; however, technology is unique in the sense of being a relatively recent phenomenon on Earth compared to life in general. Earth's technological past can only provide limited guidance on the search for technosignatures, so future projections of plausible trajectories can provide a way to conceptualize a wider possible range of technosignatures.

And yet, the future of Earth's technology remains uncertain as human civilization faces numerous challenges to its long-term sustainability (e.g., Brown 2008). In the context of searching for life, these ongoing challenges facing civilization on Earth suggest the possibility that extraterrestrial civilizations—if they exist—may face similar difficulties to ensuring their long-term survival. The search for technosignatures can provide important information for human civilization by revealing if the phenomenon of technology on a planet is commonplace and long-lived elsewhere, or if a "Great Filter" is likely looming in the future. This chapter examines the extent to which the search for biosignatures and technosignatures can place constraints on such a Great Filter and provide insight into the future trajectories of human civilization.

2. Limits to Growth

This exploration of the Great Filter begins by considering population growth on Earth: what limits our growth, and what are the factors that enable or constrain our expansion into the galaxy. Globally population is nearly 8 billion people today, and United Nations population models continue to project an increase in the total number of people. The rate of population growth is decreasing, with rates differing in each country, but the total population is still approaching almost 10 billion people by 2050, with some spread in the scenario set toward higher or lower values (UN 2023). The range of scenarios in the UN population models indicates uncertainty in the long-term future of population, but all models show near-term growth but at a decreasing rate in most countries.

The rate of growth in regions and globally is determined by the fertility rate, measured as the average number of children per woman. The fertility rate is calculated by conducting statistical surveys to measure the number of births over a large sample of women in a population, which can provide information not only on the total average fertility rate but also the median age for childbearing in different countries. For example, Japan is experiencing population shrinkage with its a fertility rate of 1.3 children per woman, with the most common childbearing years between age 25-34, while the United States is just above replacement value with a fertility rate of 2.1 children per woman; by comparison, a nation with a growing population such as Chad has a fertility rate of 6.3 children per woman and the most common childbearing years between age 20-24 (Warren 2015). Excessive population growth can exacerbate numerous economic and social problems due to overcrowding and the demand for additional resources, while population shrinkage due to lower birth rates or emigration can lead to a country with a larger fraction of elderly people than those being born.

Achieving a stable global population remains a challenging task. China experienced high fertility rates that led to a massive swelling of population up to the 1970's, after which a series of reforms were made that culminated in the "one-child" policy that began in 1979 and lasted until 2015, which brought the fertility rate in China to below replacement rate (Warren 2015). China is unlikely to be a major driver of continued population growth, with nations like India and particularly many countries in Africa as the dominant centers of population growth. Some growth in North American and some European countries is also projected due to immigration. Many nations are tending toward population stabilization by 2050, at least in some United Nations models, with the most significant exceptions in subsaharan and central Africa. Some scenarios provide more optimistic results in which global population approaches a stable upper limit, while others remain open to the possibility that small centers of high fertility rate could drive large global populations even if most nations have achieved replacement value fertility rates.

The uncertainty in future population growth raises an important question for our future: how many people can the planet sustain? This problem of defining Earth's carrying capacity is a complex question and has been estimated in many different ways (Cohen 1995), all of which point toward some upper limit to population growth based on physical limits. The study by Mullan and Haqq-Misra (2019) considered the extent to which food production would limit this carrying capacity, given the uncertainty in UN population models that show a range of scenarios from growth to shrinkage from now to 2100. With the assumption of a subsistence vegetarian diet, Mullan and Haqq-Misra (2019) find that Earth could sustain a population of about 20-30 billion people by using all currently arable land to its maximum potential, which is higher than most upper limits projected by the UN population models and includes consideration of technological innovation to improve agricultural output. If these same considerations are extended to a case of maximizing the agricultural output of all available land on Earth, then the carrying capacity increases up to about 100 billion people, which is far in excess of any of the UN projections.

Population itself could remain below Earth's carrying capacity and even reach stable levels across all nations, but civilization also continues to show trends of increasing energy use. The period from 1850 to the present day show about a 2.6% per year increase in world energy, although this value has slowed down to about 2% per year since about 1980. These historical trends in energy growth suggest the possibility that similar trends might continue into the future. Further projections of these trends from the present day out to the year 2500 are shown in Figure 1. At a 2.6% per year growth rate, human civilization would reach the threshold of a Kardashev Type I civilization—with an energy use equal to the sunlight falling on the planet (Kardashev 1964)—by the year 2350. This transition to a Type I civilization would occur closer to 2500 at a 2% per year growth rate and even later at 1% per year. The rapid growth 7% case is also illustrative, as this is the rate that was assumed in the projections by Von Hoerner (1975) because it was a representative growth rate during the 1950-1970s when these projections were made. This reveals one of the lessons in making linear projections of growth in population or energy: the growth rate that civilization is experiencing at any given moment may not necessarily extend forever. Figure 1 also includes a toy logistic curve as an example of an energy trajectory that reaches an optimal limit and does not continue growing. Such a trajectory could reflect stabilization in global population as well as civilization achieving optimal limits in production and consumption. The future of human civilization remains uncertain, and exercises like this demonstrate the potential for imagining trajectories in which human civilization manages to stabilize all aspects of its growth as well as Type I trajectories in which rapid growth continues.

3. Energy-intensive Civilizations

The idea that technological civilization on Earth could continue on a rapid growth trajectory raises the possibility that such energy-intensive civilizations could develop

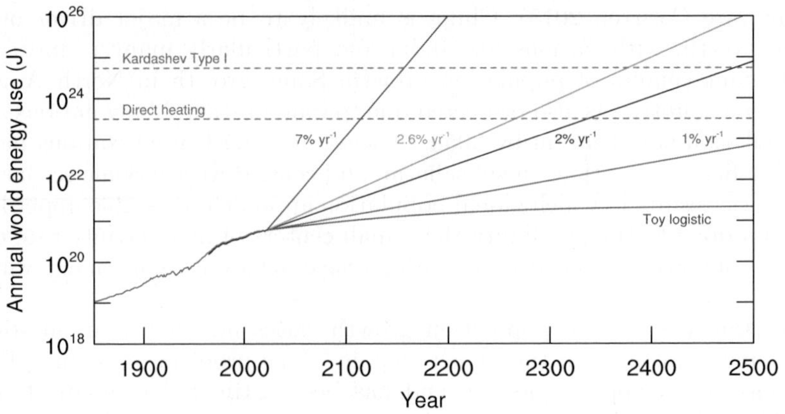

Figure 1. Historic and projected annual world energy use. Data from 1850 to 2015 (grey) are from the Organization for Economic Cooperation and Development, and data from 1965 to 2022 (black) are from the Energy Institute. Projections at different growth rates eventually reach the threshold for direct heating and a Type I civilization, aside from the toy logistic function that reaches net zero growth.

in other planetary systems. Life itself is a process that utilizes energy on Earth, and technology is a development of biological life—humans—that harnesses energy in novel ways. A study by Frank et al. (2017) suggested a framework for viewing Earth as a heat engine that takes low entropy solar radiation as input and outputs high entropy radiation into space as waste; these energy inputs and outputs remain in balance and are used to perform work by various components of the Earth system. Physical free energy is driven by temperature differences on Earth, which provides chemical free energy that can be utilized by life as well as technology. Photochemistry and photosynthesis can also leverage solar radiation directly, and technological civilization can do likewise with photovoltaics. Nonliving, living, and technological systems all energetically interact on Earth, and all contribute toward modification of the planet on a global scale.

This framework views technology as one of many evolutionary developments that can occur on a planet and lead to global-scale modification. The study by Frank et al. (2017) further suggested a classification scheme for describing the various configurations in which geological, biological, and technological processes could occur on a planet. In this scheme, a class I planet is in radiative balance between incoming starlight and outgoing infrared radiation, which would include a planet like Mercury. Class II planets have added components of atmospheric and geologic cycling, which includes planets like Venus and Mars. A class III planet has a think biosphere that includes some biosphere cycling and the presence of biological material—which may or may not be remotely detectable—and which includes a planet like Earth during the Archean eon. A class IV planet has a thick biosphere that includes interactive cycles with geologic and atmospheric processes—which may be more likely to be remotely detectable—and which would includes a planet like Earth during the Proterozoic eon. The final class V planet in this scheme has a thick technosphere that includes interactive cycles with the biosphere, atmosphere, and geosphere—which may be remotely detectable. Earth today is between class IV and class V, with an emerging technosphere that Frank et al. (2017) described as a "hybrid planet."

This scheme for classifying planets can even be defined, at least conceptually, in terms of the free energy for chemical disequilibrium available through geological, biological, and technological processes. A class I planet has no available free energy, while a class II planet has only geologic sources of free energy. A class III planet would have an emerging biological source of free energy, and a class IV planet would have much greater free energy from biological sources compared to geological sources. Earth today as a hybrid planet would likewise have an emerging technological source of free energy that contributes to biological and geological sources, perhaps with a weakly detectable technosphere. Continued trajectories in energy growth likewise suggest the possibility of a future transition to a class V planet in which technological sources of free energy are equal to or greater than those from biological sources, which may have a more strongly detectable technosphere.

The extent to which human activities have already caused global-scale modification led some geologists to propose the "Anthropocene" as a new geological epoch to indicate the onset of this technological transformation (originally conceived by Crutzen (2006), but the term has since been rejected as an official geologic epoch; see, e.g., Ghosh (2024).) The study by Frank et al. (2017) described the Anthropocene as the "hybridization of the planet" in which the energy-intensive activities of human civilization have begun to modify biological and geological cycles. The emergence of a technosphere alongside and intertwined with Earth's biosphere suggests the possibility of a similar trajectory in planetary evolution that could occur elsewhere, which raises an important question: do other energy-intensive technological civilizations exist, or is Earth the only example in the galaxy or universe? This is another way of framing the Fermi paradox (see Webb, in these proceedings), although this framing does not necessarily require expansion and instead focuses on whether or not energy-intensive trajectories occur elsewhere—and if they could be detected.

Numerous attempts at searching for technosignatures—remotely detectable evidence of technology (Tarter 2006)—have been conducted over the past 60 years, primarily focused on searching for narrow-band radio signals or pulsed optical signals that would be indicative of extraterrestrial technology (e.g., Shuch 2011; Gindilis and Gurvits 2019). The lack of any technosignature discoveries during this time does not necessarily imply much about the prevalence or likelihood about extraterrestrial technology, given the limited extent of these prior searches. A study by Tarter et al. (2010) estimated that an appropriate analogy is to consider the experiment of asking the question "are there fish in the ocean?" and then proceeding to examine a single drinking glass of water randomly sampled from the ocean; the lack of a fish in this small sample is insufficient to conclude that the ocean is lifeless, although the experiment was not necessarily ill-conceived. More recent efforts to advance the search for technosignatures have expanded the collection and analysis of data, which Wright et al. (2018) characterized with an updated estimate that the equivalent oceanic volume of the search is now about a hot tub's worth of water. The search for technosignatures continues to advance, but the search space is immense.

Human civilization appears to remain on an energy-intensive trajectory, but no other energy-intensive civilizations have yet been found. One reason may simply be that the search for technosignatures is incomplete, and that technosignature candidates will eventually be found when sufficient resources are available to advance the science in a significant way. For example, a simulation study by Knuth (in these proceedings) showed that even in a galaxy teeming with millions of spacefaring civilizations, the probability of such a civilization encountering Earth during the process of expansion would still be very low. But another reason could be that energy-intensive trajectories are unsustainable, which would imply that any such civilizations that may have come into existence elsewhere have likely already collapsed. This possibility led Hanson (1998) to speculate that "there thus exists a great filter between death and expanding lasting life, and humanity faces the ominous question: how far along this filter are we?" The idea underling the Great Filter is that the evolutionary trajectory between a class I and class V planet includes at least one hard step that is more difficult to overcome than any of the other transitions. The origin of life on a planet is perhaps a difficult transition (from class II to class III), but the transition from an emerging technosphere to a sustainable class

V planet may be even more difficult. (For further discussion on the early Great Filter versus the late Great Filter, see Wilman et al., in these proceedings.) This remains an open question for the trajectory of human civilization: is the hardest step in planetary evolution in the past or in the future?

The lifetime of human civilization could conceivably be limited by a global catastrophic or existential threats, such as cosmic impactors, nuclear war, supervolcanoes, and other events that carry transgenerational risk for the whole of civilization (Bostrom and Cirkovic 2011; Bostrom 2013). Some risk scenarios can be mitigated through technological preparation, but even technological solutions cannot wholly eliminate all scenarios of existential risk. Such scenarios raise the possibility that the Great Filter resides in the future, to the extent that dealing with such challenges is greater than any evolutionary struggle the Earth system has yet faced. If the Great Filter awaits in Earth's future—perhaps due to the challenges in overcoming problems such as climate change and long-term sustainability—and if this is a common evolutionary pattern, then this would imply that energy-intensive civilizations have short lifetimes. On the other hand, if the Great Filter is in Earth's past—such as the origin of life or the development of multicellularity—then targeted searches for technosignatures may find evidence of long-lived extraterrestrial civilizations.

4. Searching for Technosignatures

Earth today, and projections of Earth's future, provide examples of possible technosignatures that current and future missions could search for in exoplanetary systems. Several examples of atmospheric pollutants as technosignatures are discussed in this subsection, which demonstrate that at least some known technosignature candidates could be detectable with known observing capabilities. However, it is also important to note that the library of possible technosignatures is far from complete, and further efforts at developing a comprehensive set of technosignatures will become increasingly important as exoplanet characterization continues. The actual detection of a technosignature or other biosignature may ultimately be an anomalous observation that does not correspond to any specific prior prediction (Cleland 2019), but such theoretical frameworks can still be useful for approaching the search for life by considering a broader range of possibilities than Earth today.

Pollutants that are manufactured by industrial processes represent a broad category of technosignatures that could be discernible in exoplanetary spectral features. One class of such molecules contain both carbon and halogen atoms, known as halocarbons, with chlorofluorocarbons (CFCs) and perfluorocarbons (PFCs) as examples that have led to emissions regulations on Earth. Many CFCs have long atmospheric residence times and are also effective greenhouse gases, so the presence of such atmospheric pollutants can lead to undesirable climatic consequences; however, the detection of CFCs in the atmosphere of an exoplanet would be compelling evidence of a technosignature, as industrially-produced CFCs on Earth have no known abiotic production pathways. A study by Haqq-Misra et al. (2022b) used computational climate models and a synthetic spectral generator to calculate the detectability of the spectral features of CFC-11 (CCl_3F) and CFC-12 (CCl_2F_2) in the atmosphere of TRAPPIST-1e, which found that the James Webb Space Telescope could conceivably resolve such features at Earth-like CFC abundances in about 100 hours of observing time. This is comparable with the time required to observe some biosignature gases. Higher CFC abundances would require less observation time, although human civilization is on a trajectory to lower its CFC emissions rather than increase them.

Some scenarios might involve the intentional release of CFCs into the atmosphere of another planet: a civilization seeking to terraform a planet like Mars might do so by releasing industrial greenhouse gases to intentionally warm the planet. A study by Schwieterman et al. (2024) used a synthetic spectral generator to calculate the detectability of several artificial greenhouse gases including CF_4, C_2F_6, C_3F_8, SF_6, and NF_3, which demonstrated that any of these gases or a mixture thereof that was used for terraforming a planet could be detectable with the James Webb Space Telescope and other future mission concepts with as few as 5 to 20 transit observations. Calculations like these demonstrate the possibility of detecting anomalous spectral features in an exoplanetary atmosphere that could be evidence of industrial processes or waste, although any such observations should also be mindful of possible false positive or unknown abiotic processes that could generate similar features.

Combustion is a technological process that is the dominant source of nitrogen dioxide (NO_2) emissions on Earth, and decreases in NO_2 were even observed in urban areas during periods of inactivity during the COVID-19 pandemic (Bauwens et al. 2020). A study by Kopparapu et al. (2021) used an atmospheric photochemistry model and synthetic spectral generator to demonstrate that present-day levels of NO_2 would require several hundred hours of observation time with a mission like the Habitable Worlds Observatory, which could be challenging but may be comparable to some of the most ambitious observing campaigns that have been conducted. The observation of NO_2 in an exoplanetary system could potentially be linked with combustion, but other processes also produce NO_2 on Earth that could conceivably be operating in exoplanetary environments. It is also worth considering the extent to which a long-lived civilization would continue releasing NO_2 and other pollutants into its atmosphere without finding less harmful alternatives.

Agriculture is among the oldest human technologies that can exert global and potentially detectable changes on a biosphere. Both ancient and modern agriculture is based on applying nitrogen-based fertilizers to soil to enhance crop production, which saturates the soil with excess nitrogen compounds that volatilize into the atmosphere. Agriculture for most of human history was based on the use of manure as fertilizer, but the development of the Haber-Bosch process in the early twentieth century enabled an industrial process for producing ammonia (NH_3) by fixing nitrogen in a high-temperature reaction with hydrogen. Modern agriculture has enabled massive population growth through the enhanced fertilization of soil, by drawing upon the nearly limitless supply of nitrogen and converting it into edible food. A study by Haqq-Misra et al. (2022a) used a synthetic spectral generator to show that nitrogen excesses in the form of ammonia (NH_3) and nitrous oxide (N_2O) could potentially be detectable for large-scale agriculture on an exoplanet if performed at a scale several times larger than on present-day Earth. Neither of these gases would be considered a technosignature by themselves, but the observation of both of these gases together would be compelling evidence for the technological disruption of a global-scale nitrogen cycle.

These examples demonstrate the possibility of examining exoplanetary spectra for potential technosignatures, such as industrial pollutants or other byproducts of technology. This search can be conducted alongside the search for biosignatures, as both are approaches toward characterizing the atmospheres of exoplanets and investigating any anomalous features that arise. Other exoplanetary technosignatures based on Earth today are also possible, such as the observation of nighttime city lights in emission spectra (Beatty 2022) or a belt of orbiting satellite debris much denser than on Earth today (Sallmen et al. 2019). Numerous possibilities exist for potential biosignatures and technosignatures, and any self-consistent scenarios for such exoplanetary spectra can provide guidance for advancing the search.

How many planets must a mission like the Habitable Worlds Observatory observe in order to constrain the prevalence of technosignatures? Observations with current missions

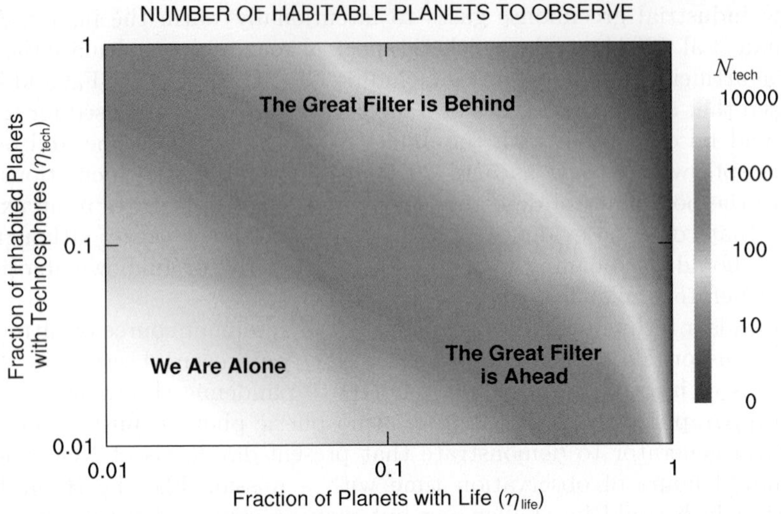

Figure 2. The number of exoplanetary spectra of habitable planets that would need to be observed to constrain the Great Filter (N_{tech}, with 95% confidence) depends on the fraction of planets with life (η_{life}) and the fraction of inhabited planets with technospheres (η_{tech}). Future searches for biospheres and technospheres can provide statistical constraints on whether the Great Filter is in Earth's past ($\eta_{tech} \geq \eta_{life}$) or future ($\eta_{tech} < \eta_{life}$). Figure provided by Haqq-Misra et al. (2020).

like the James Webb Space Telescope demonstrate the capability to search for life in exoplanetary spectra, but future missions will be required to obtain statistically meaningful samples of exoplanetary spectra that could place quantitative limits on the prevalence of biosignatures or technosignatures. A study by Haqq-Misra et al. (2020) considered a statistical approach toward understanding the constraints that the search for biosignatures and technosignatures can place on the Great Filter, which is summarized in Figure 2. Future missions that attempt to search for possible biosignatures and technosignatures, but find nothing, still provide information on the statistical prevalence of life and its implications for the future of life on Earth. If biospheres and technospheres are commonplace on habitable planets (i.e., planets that orbit within the liquid water habitable zone (Kasting et al. 1993; Kopparapu et al. 2013)), then about 10-30 observations would be needed; this would imply that the Great Filter is in Earth's past (Fig. 2, upper right). If biospheres are common but technospheres are rare, then this would require on the order of 100 observations and would imply that the Great Filter is in Earth's future (Fig. 2, lower right). And if a robust search for biosignatures and technosignatures across 1000 or more planets finds that both are absent, then this would increase confidence in the conclusion that life on Earth is rare or unique (Fig. 2, lower left). This approach demonstrates that the search for life can advance by focusing on obtaining statistically meaningful samples of exoplanetary spectra, rather than focusing exclusively on one or two cases, in order to learn about the prevalence of observable features that could be linked with biology or technology.

5. Conclusion

The future remains open, with many possible trajectories for Earth's population and energy that may not necessarily follow prior patterns of exponential growth. The search for technosignatures can provide information about the existence of extraterrestrial technology, which, if discovered, would provide confidence that the Great Filter is in Earth's past and that sustainable trajectories are possible. A comprehensive search for technosignatures might also find nothing, which could imply that extraterrestrial technology is rare, nonexistent, or at least difficult to observe at a distance. Such a result would leave open an even wider range of possible trajectories: the Great Filter could be in Earth's future, or Earth could be one of the first places in the galaxy (or universe) where technology has emerged.

Earth today is characterized by the emergence of energy-intensive civilization as a planetary phenomenon. Even if this is a feature that can evolve on other planetary systems, one possible explanation for the lack of such observations is that any energy-intensive civilizations that currently exist are growing slowly, so as to avoid expanding beyond any carrying capacity limits. The next generation of space telescopes may be able to search for evidence of such civilizations to perhaps find evidence of any that have managed to stabilize growth and technology over long timescales. But if this search finds that technosignatures are rare, and perhaps even unique to Earth, then this only increases the value of life on Earth and heightens the challenge for civilization on Earth to ensure a sustainable future.

Acknowledgments

The author gratefully acknowledges support from the NASA Exobiology program under grant 80NSSC22K1009. Any opinions, findings, and conclusions or recommendations expressed in this material are those of the author and do not necessarily reflect the views of any employer or NASA.

References

Bauwens, M., Compernolle, S., Stavrakou, T., Müller, J.-F., Van Gent, J., Eskes, H., Levelt, P. F., Van Der A, R., Veefkind, J., Vlietinck, J., *et al.* 2020, Impact of coronavirus outbreak on no2 pollution assessed using tropomi and omi observations. *Geophysical Research Letters*, 47(11), e2020GL087978.

Beatty, T. G. 2022, The detectability of nightside city lights on exoplanets. *Monthly Notices of the Royal Astronomical Society*, 513(2), 2652–2662.

Bostrom, N. 2013, Existential risk prevention as global priority. *Global Policy*, 4(1), 15–31.

Bostrom, N. & Cirkovic, M. M. 2011,. *Global catastrophic risks.* Oxford University Press, USA.

Brown, L. R. 2008,. *Plan B 3.0: Mobilizing to save civilization (substantially revised).* WW Norton & Company.

Bryson, S., Kunimoto, M., Kopparapu, R. K., Coughlin, J. L., Borucki, W. J., Koch, D., Aguirre, V. S., Allen, C., Barentsen, G., Batalha, N. M., *et al.* 2020, The occurrence of rocky habitable-zone planets around solar-like stars from kepler data. *The Astronomical Journal*, 161(1), 36.

Cassan, A., Kubas, D., Beaulieu, J.-P., Dominik, M., Horne, K., Greenhill, J., Wambsganss, J., Menzies, J., Williams, A., Jørgensen, U. G., *et al.* 2012, One or more bound planets per milky way star from microlensing observations. *Nature*, 481(7380), 167–169.

Cleland, C. E. 2019, Moving beyond definitions in the search for extraterrestrial life. *Astrobiology*, 19(6), 722–729.

Cohen, J. E. 1995, Population growth and earth's human carrying capacity. *Science*, 269(5222), 341–346.

Crutzen, P. J. The "anthropocene". In *Earth system science in the anthropocene* 2006, pp. 13–18. Springer.

Dressing, C. D. & Charbonneau, D. 2015, The occurrence of potentially habitable planets orbiting m dwarfs estimated from the full kepler dataset and an empirical measurement of the detection sensitivity. *The Astrophysical Journal*, 807(1), 45.

Frank, A., Kleidon, A., & Alberti, M. 2017, Earth as a hybrid planet: the anthropocene in an evolutionary astrobiological context. *Anthropocene*, 19, 13–21.

Ghosh, R. 2024, A fond farewell to the anthropocene. *Issues in Science & Technology*, 40(3).

Gindilis, L. M. & Gurvits, L. I. 2019, SETI in Russia, USSR and the post-Soviet space: a century of research. *Acta Astronautica*, 162, 1–13.

Hanson, R. 1998,. The great filter-are we almost past it. `http://hanson.gmu.edu/greatfilter.html`.

Haqq-Misra, J., Fauchez, T. J., Schwieterman, E. W., & Kopparapu, R. 2022,a Disruption of a planetary nitrogen cycle as evidence of extraterrestrial agriculture. *The Astrophysical Journal Letters*, 929a(2), L28.

Haqq-Misra, J., Kopparapu, R., Fauchez, T. J., Frank, A., Wright, J. T., & Lingam, M. 2022,b Detectability of chlorofluorocarbons in the atmospheres of habitable m-dwarf planets. *The Planetary Science Journal*, 3b(3), 60.

Haqq-Misra, J., Kopparapu, R. K., & Schwieterman, E. 2020, Observational constraints on the great filter. *Astrobiology*, 20(5), 572–579.

Haqq-Misra, J., Schwieterman, E. W., Socas-Navarro, H., Kopparapu, R., Angerhausen, D., Beatty, T. G., Berdyugina, S., Felton, R., Sharma, S., Gabriel, G., *et al.* 2022,c Searching for technosignatures in exoplanetary systems with current and future missions. *Acta Astronautica*, 198c, 194–207.

Hardegree-Ullman, K. K., Cushing, M. C., Muirhead, P. S., & Christiansen, J. L. 2019, Kepler planet occurrence rates for mid-type m dwarfs as a function of spectral type. *The Astronomical Journal*, 158(2), 75.

Kardashev, N. S. 1964, Transmission of Information by Extraterrestrial Civilizations. *Soviet Astronomy*, 8, 217.

Kasting, J. F., Whitmire, D. P., & Reynolds, R. T. 1993, Habitable zones around main sequence stars. *Icarus*, 101(1), 108–128.

Kopparapu, R., Arney, G., Haqq-Misra, J., Lustig-Yaeger, J., & Villanueva, G. 2021, Nitrogen dioxide pollution as a signature of extraterrestrial technology. *The Astrophysical Journal*, 908(2), 164.

Kopparapu, R. K., Ramirez, R., Kasting, J. F., Eymet, V., Robinson, T. D., Mahadevan, S., Terrien, R. C., Domagal-Goldman, S., Meadows, V., & Deshpande, R. 2013, Habitable zones around main-sequence stars: new estimates. *The Astrophysical Journal*, 765(2), 131.

Madhusudhan, N., Sarkar, S., Constantinou, S., Holmberg, M., Piette, A. A., & Moses, J. I. 2023, Carbon-bearing molecules in a possible hycean atmosphere. *The Astrophysical Journal Letters*, 956(1), L13.

Meadows, V. S., Lincowski, A. P., & Lustig-Yaeger, J. 2023, The feasibility of detecting biosignatures in the trappist-1 planetary system with jwst. *The Planetary Science Journal*, 4(10), 192.

Mullan, B. & Haqq-Misra, J. 2019, Population growth, energy use, and the implications for the search for extraterrestrial intelligence. *Futures*, 106, 4–17.

Sallmen, S., Korpela, E. J., & Crawford-Taylor, K. 2019, Improved analysis of clarke exobelt detectability. *The Astronomical Journal*, 158(6), 258.

Schwieterman, E. W., Fauchez, T. J., Haqq-Misra, J., Kopparapu, R. K., Angerhausen, D., Pidhorodetska, D., Leung, M., Sneed, E. L., & Ducrot, E. 2024, Artificial greenhouse gases as exoplanet technosignatures. *The Astrophysical Journal*, 969, 20.

Schwieterman, E. W., Kiang, N. Y., Parenteau, M. N., Harman, C. E., DasSarma, S., Fisher, T. M., Arney, G. N., Hartnett, H. E., Reinhard, C. T., Olson, S. L., *et al.* 2018, Exoplanet biosignatures: a review of remotely detectable signs of life. *Astrobiology*, 18(6), 663–708.

Shuch, H. P. 2011,. *Searching for Extraterrestrial Intelligence: SETI Past, Present, and Future*. Springer.

Socas-Navarro, H., Haqq-Misra, J., Wright, J. T., Kopparapu, R., Benford, J., Davis, R., *et al.* 2021, Concepts for future missions to search for technosignatures. *Acta Astronautica*, 182, 446–453.

Tarter, J., Agrawal, A., Ackermann, R., Backus, P., Blair, S., Bradford, M., Harp, G., Jordan, J., Kilsdonk, T., Smolek, K., *et al.* Seti turns 50: five decades of progress in the search for extraterrestrial intelligence. In *Instruments, Methods, and Missions for Astrobiology XIII 2010*, volume 7819, pp. 13–25. SPIE.

Tarter, J. C. 2006, The evolution of life in the universe: are we alone? *Proceedings of the International Astronomical Union*, 2(14), 14–29.

Tuchow, N. W., Stark, C. C., & Mamajek, E. 2024, HPIC: the habitable worlds observatory preliminary input catalog. *The Astronomical Journal*, 167(3), 139.

UN 2023,. *World Population Prospects 2022: Summary of Results*. World Population Prospects 2022. United Nations Fund for Population Activities.

Von Hoerner, S. J. 1975, Population explosion and interstellar expansion. *Journal of the British Interplanetary Society*, 28(28), 691–712.

Warren, S. G. 2015, Can human populations be stabilized? *Earth's Future*, 3(2), 82–94.

Wright, J. T., Kanodia, S., & Lubar, E. 2018, How much seti has been done? finding needles in the n-dimensional cosmic haystack. *The Astronomical Journal*, 156(6), 260.

Zieba, S., Kreidberg, L., Ducrot, E., Gillon, M., Morley, C., Schaefer, L., Tamburo, P., Koll, D. D., Lyu, X., Acuña, L., *et al.* 2023, No thick carbon dioxide atmosphere on the rocky exoplanet trappist-1 c. *Nature*, 1–4.

(Toward) Discovery of Life beyond Earth and its Impact
Proceedings IAU Symposium No. 387, 2026
H. Landt, M. Dominik & C. Oliver, eds.
doi:10.1017/S1743921324001352

The Great Filter – Is it just a matter of time?

R. J. Wilman[1], A. E. Nicholson[2] and J. Haqq-Misra[3]

[1]Department of Physics, University of Durham, South Rd, Durham DH1 3LE, UK
email: r.j.wilman@durham.ac.uk

[2]Department of Physics & Astronomy, University of Exeter, Exeter EX4 4EN, UK

[3]Blue Marble Space Institute of Science, Seattle, Washington, USA

Abstract. The discussion on the second day of the symposium centred on the Great Filter, a concept proposed by Robin Hanson as a way to reframe the analysis of the Fermi Paradox. It asserts that there must be a least one Great Filter – an evolutionary step that is extremely improbable – somewhere along a chain which starts from a lifeless Earth-like planet, followed by the sequential development of simple, complex, intelligent/technological life, and culminating in an explosive phase of readily-detectable galactic colonization. Some 25 years on from Hanson's proposal, we examine the Great Filter's continuing usefulness as a concept and current thinking on whether any such filter lies in our past (Early Great Filter), or is waiting in our future (Late Great Filter), and what this means for us and our search for life in the Universe.

Keywords. Astrobiology, Gaia, Earth history, biosignatures, technosignatures, Great Filter, futures studies

---

1. Introduction

The second day of the symposium – entitled *Can we detect 'life' beyond Earth? – Uncertainty: Medium* – was perhaps the broadest of the entire week in terms of topical coverage. The session opened with a review of the Fermi Paradox which served to remind us of its role as an implicit foundation for the talks which followed. These covered the nature of exotic life from evolutionary and philosophical standpoints, Gaian regulation in the early history of life, contested evidence for alien technological artifacts and life on Mars, as well the future of our own Earth-based civilization based on projections for population and energy use and its implications for the search for technosignatures. Further presentations covered the technosignatures arising from the construction of alien megastructures beyond the planetary scale, models for the characteristics of extraterrestrial civilizations which encounter Earth, and the possibility of searching for the actions of intelligent alien life in gravitational wave signals.

The discussion at the end of the day sought to bring together these past and future perspectives through the prism of the Great Filter. The latter concept, first proposed by economist and futurist Robin Hanson just over 25 years ago (Hanson (1998a)), is essentially a reframing of the Fermi Paradox, and an economical one at that, as it abstracts away from the details of the specific mechanisms which might underlie it. It asserts that there must be at least one Great Filter – an evolutionary step that is extremely improbable – somewhere along the chain leading from a habitable but lifeless Earth-like planet, through the sequential development of simple life, complex life and intelligent/technological life (approximately our current state) and thereafter to an explosive phase of readily-detectable galactic colonization, as illustrated in Fig. 1.

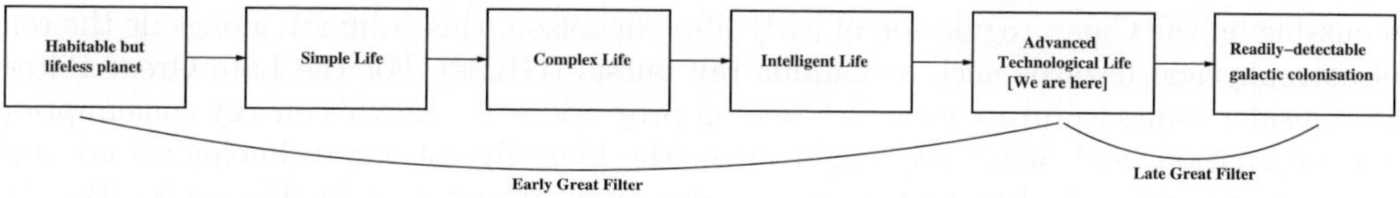

Figure 1. Schematic of the Great Filter, showing evolution from a habitable lifeless planet, through the sequential development of life, leading finally to a colonization explosion. The Early or Late Great Filters would break the development chain prior to, or after, our current position, respectively. Note that Hanson (1998a) identifies a greater number of steps in the early evolution of life.

The long-term fate of humanity and the optimum strategy for the search for life in the Universe revolve around whether any such filter is in our past (Early Great Filter) or still lies ahead of us (Late Great Filter). For the Early Great Filter, the evolution of life to our current stage would have already negotiated the most improbable step, suggesting that our future development as a space-faring civilization will not be as difficult or improbable. This implies that the search for life elsewhere, if it exists (i.e. if any life elsewhere has also passed through the Early Great Filter), should expect to find both planetary biosignatures from less evolved life and technosignatures from more advanced phases of development, as there would be no significant impediments to its long-term evolution, once started.

Conversely, a Late Great Filter would suggest that our further development as a technological civilization will ultimately reach a tight bottleneck, most likely curtailing our future before our descendants can colonize the galaxy. Under this scenario, we would expect planetary biosignatures from all prior stages in the evolution of life to be abundant, along with technosignatures from civilizations at our own stage of development and those more advanced, up to the point – as yet undefined – at which they disappear by whatever process enacts the Late Great Filter. However, Earth's present-day technosignatures are difficult to detect and short-lived over astronomical distances and timescales, respectively. Hence, the closer the Late Great Filter is to our current epoch, the more abundant we would expect biosignatures to be relative to technosignatures, the latter being rare or non-existent as far as observational detectability is concerned.

Such considerations inspired the philosopher Nick Bostrom to pen a provocative article (Bostrom (2008)) subtitled "Why I hope the search for life finds nothing", when referring to the possibility that robotic Martian probes might find evidence for independently-evolved life on the red planet. Such a discovery would suggest that the Great Filter does not lie in the past development of life, but waits ominously in our future. Therefore, according to Bostrom "It would be good news if we find Mars to be completely sterile. Dead rocks and lifeless sands would lift my spirit."

A quarter of a century on from Hanson's proposal, the Fermi Paradox has been placed on firmer scientific and philosophical foundations, with systematic classification of the many possible solutions (Ćirković (2018), Forgan (2019)). The Great Filter nevertheless continues to provide a powerful and concise narrative for many – but not all – of these possibilities, especially in areas of heightened current interest such as artifical intelligence (AI) (e.g. Garrett (2024)) and the large array of other potential threats which could destroy humanity in the coming centuries (Jiang et al. (2023), Ord (2020)).

The structure of this article is as follows. We begin with some background on relevant technical aspects of the original works of Hanson (1998a) and Hanson (1998b) which have been overlooked as the Great Filter concept has diffused to a wider audience in recent years. We also review work by Ćirković (2018) situating the Great Filter in the broader taxonomy of solutions to the Fermi Paradox, along with discussion on its continuing relevance. Thereafter, we consider mechanisms which might cause an Early Great Filter, focussing on the Gaian regulation of early life (Nicholson, this volume), alongside the role of astrophysical hazards such as gamma-ray bursts (GRBs). For the Late Great Filter, we consider limits to future growth based on projections for Earth's energy consumption and population, and how these might affect the longevity of our technological era and hence our prospects for becoming a space-faring civilization colonizing the galaxy. We also review how the observational search for biosignatures and technosignatures may provide constraints on such future developmental trajectories and the nature of the Great Filter. As part of this, we consider the role of AI as a mechanism for the Late Great Filter and its relevance to the 'deadly probes' hypothesis.

2. What's in a name? The Great Filter in context

2.1. *By trial and error: Hanson's hard biological steps*

The original scheme of Hanson (1998a) contains approximately nine 'hard' biological steps, with more granularity than depicted in Fig 1. These steps are assumed to operate independently and sequentially on a 'trial and error' basis, with the probability of completing step i within a time t_i being given by $1-\exp(-f_i t_i) \simeq f_i t_i$ for $f_i t_i \ll 1$, where $1/f_i$ is the typical step completion timescale. If we select on condition of the successful completion of *all* steps within a total time window T, and if all $f_i T$ are small, Hanson (1998a) showed that the joint probability density function over all steps is the product of the f_i terms, and hence that all the hard steps have the same time distribution and take roughly the same amount of time (for a fuller treatment see Hanson (1998b)). In that case (i.e. conditioning on success), the actual time taken for each hard step is not indicative of the difficulty of the step. He argues that the latter is consistent with the roughly equally-spaced periods in the history of life on Earth, with each step taking ~ 0.3 Gyr.

In addition to these 9 hard trial and error biological steps, Hanson (1998a) proposed a further 2 non-biological 'discrete' steps, denoting the probabilities of starting off with a suitable host star-planet configuration (astronomical factors) and of a civilization not destroying itself as the 'end game' of Fig 1 is reached (sociological factors). If each of these steps has a 1% probability, and if each of the 9 biological trial and error steps has a logarithmic average completion time of 30 Gyr (versus an actual completion time in the history of life on Earth of 0.3 Gyr, giving an average $f_i t_i \simeq 0.01$), the overall probability of a given star system giving rise to a galactic colonizing civilization by the current epoch would be roughly the product of 11 factors of 0.01 which is 10^{-22}. Given that there are of order 10^{22} stars in the observable Universe, this would solve the Fermi Paradox at a stroke – we are alone in the Universe, our existence an outrageously improbable fluke. In this case, the term Great Filter would be a misnomer: rather than the filter serving to *stop* the inevitable evolutionary flow (i.e. from left to right in Fig. 1) through a series of steps which otherwise proceed rapidly, the filter would signify the inherent difficulty of the steps themselves, each typically taking much longer than the age of the Universe. The Great Filter would be time.

The challenge would then be to establish why each step in the chain ordinarily takes so long, and hence whether there are other, more likely outcomes apart from those shown in Fig 1. Indeed, Fig 1 may be very incomplete, reflecting just our own evolutionary history; in general, there may be many other ways for life to develop and become complex, many other branches and forks in the evolutionary road which do not culminate in readily-detectable galactic colonization. These other unseen and unimagined pathways may themselves act as the filter, diverting most realisations of life away at one or more steps of Fig 1. The latter possibilities would fit within the category of 'adaptationist' or 'rare mind' solutions to the Fermi Paradox as defined by Ćirković (2018), in which traits such as consciousness and technological toolmaking are temporary 'adaptive' traits which disappear once they cease to offer selective advantage under altered environmental conditions.

2.2. *The Great Filter and the broader landscape of the Fermi Paradox*

Looking to the broader classification landscape of Ćirković (2018), the Early Great Filter shares many features with 'Rare Earth' solutions. The latter are named after the book of the same title by Ward and Brownlee (2000), who proposed that whilst simple microbial life may be widespread in the galaxy, complex biospheres such as our own are rare because they require such an exacting combination of factors (e.g. being in a circumstellar habitable zone, having a large Moon, a Jupiter-mass planet to hoover up comets, to name but three such factors). Individually, each factor may be moderately rare and if the various factors are independent of one another, this leads to an overall very low probability for the development of a complex Earth-like biosphere. The concept of the Gaian Bottleneck (see next section and Nicholson, this volume) is a further development within this framework. As discussed by Ćirković (2018), Rare Earth solutions have been criticised as being anti-Copernican, and also on the grounds of the ceteris paribus fallacy, the erroneous notion that the various factors in the chain could be selected, à la carte, independently of each other (the issue of whether Earth would have been able to form without the presence of Jupiter being one pertinent example- see Ćirković (2017) for a fuller discussion on this point.)

In contrast, the Late Great Filter admits no single monolithic classification, and in the words of Ćirković (2018) it is "whatever can prevent the ascent of an extraterrestrial civilization along Kardashev's scale once the civilization emerges". The options we will consider broadly fall into two silos, namely logistic and neocatastrophic solutions in his terminology. The former reflect inherent and fundamental limitations – be they technological or energetic, biological or sociological – which render readily-detectable galactic colonization unfeasible or vastly more difficult than we might suppose; examples include the energy consumption and induced climate change bottlenecks facing Earth within just a few centuries, to be discussed in section 4 (see also Haqq-Misra, this volume). The neocatastrophic category includes all the familiar 'home grown' threats which could destroy humanity in the coming centuries (nuclear war, climate change, pandemics, unaligned AI to name but a few– see e.g. Ord (2020) who estimates a 1 in 6 probability that humanity will be destroyed within the next century); it also includes natural hazards both terrestrial and cosmic which reset the developmental clock on longer timescales, in terms of biological and/or technological evolution. As discussed by Ćirković (2018), natural hazards may be both random and limited in scope to an individual planet or star, or have a wider reach (e.g. due to gamma-ray bursts) and perform a synchronized reset over a large portion of the galaxy; the latter could ultimately effect what has been termed an 'astrobiological phase transition' and apply equally well as a mechanism for the Early Great Filter, as we discuss below. The final mechanism for the Late Great Filter within the neocatastrophic category, which we will discuss in section 4, is the threat from 'deadly probes'; in some ways, this is an offshoot of the aforementioned threat from unaligned AI, but elevated in scope beyond a geocentric focus to consider the role of superintelligent AI in the postbiological Universe (Schneider (2017).

The above is clearly not an exhaustive review of the Fermi Paradox, nor of mechanisms for the Great Filter, but in keeping with the astronomical focus of the symposium we do not venture into the realm of more exotic possibilities which dispense with aspects of astrophysical naturalism or physical realism. There are critics – including some at this symposium – who object that the whole concept of the Great Filter is flawed, as it is based on the Fermi Paradox which is far from being a paradox as the search has barely begun. But given that the last step in Hanson's chain is a colonization explosion which would have been readily detected by now, the key premise remains intact: one of more of the steps envisaged in Fig. 1, or perhaps the whole scheme, is much more improbable than we might naively suppose. In the words of Hanson (1998a): "Someone's story is wrong" and "It matters who's wrong".

3. The Early Great Filter

An Early Great Filter would place a highly improbable step along the road to complex intelligent life in our past. This hypothesis is related to the aforementioned Rare Earth hypothesis that predicts that complex life will be vanishingly rare in the galaxy due to the very specific conditions needed for it to arise that have come together to allow complex life to evolve on Earth (Ward and Brownlee 2000). These conditions include Earth's location in the galaxy and within our solar-system; the type of star Earth orbits; Earth's size; the presence of Earth's moon; the planetary arrangement of our solar system; and the presence of plate tectonics on Earth. Proponents of the Rare Earth hypothesis argue that each of these requirements alone appear to be rare and so finding them combined in a planetary system will be incredibly rare. Critics of the hypothesis counter that with only one data-point of an inhabited planet, all we know is how complex life evolved *on Earth.* On other planets life might take other pathways (Darling 2001). The role of Jupiter as protecting inner planets from bombardments has been questioned (Horner & Jones 2010) and as astronomers find more and more exoplanets, including rocky planets within the habitable zone of their stars, the conditions for life are looking less rare. Scientists have also identified metabolic pathways life might make use of for planets with very different atmosphere's to Earth (Seager, Bains & Hu 2013) and orbiting stars of different stellar classes to our own (Eager-Nash et al. 2024).

There are several key evolutionary steps that occurred on Earth between life emerging and evolving into humans. These steps include the emergence of life itself, the evolution of simple single-celled life, complex (eukaryotic) single-celled life, sexual reproduction, multicellular life, intelligent tool using animals, and the development of civilisations. If the Great Filter is in our past then this would imply that one of these steps is highly improbable meaning that life on other planets is unlikely to develop past that point. With only a single known inhabited planet – Earth – calculating the probability of any of these steps is impossible. However examining the timings and circumstances of these events in Earth's history can allow us to make educated guesses as to how easy or hard these steps might be.

Geological evidence for conditions on Earth get sparser the further back in time we look as rocks are recycled through Earth's mantle via planet tectonics destroying much of the evidence they contain. The evidence we have suggests that liquid water has been present in large quantities on Earth's surface for over 4 billion years (Lin-gun 2004), and that life emerged on Earth over 3.8 billion years ago (Nisbet & Sleep 2001). Therefore it appears that once conditions for life were established on Earth, life emerged and took hold quite quickly (on geological timescales!). This could make abiogenesis an unlikely candidate for the early great filter. Similarly, an important step in developing complex life on Earth was the evolution of photosynthesis. All photosynthetic life shares a common ancestor, indicating that the machinery key to photosynthesis only evolved once (Olson 1981). This could indicate photosynthesis as a difficult step to navigate on the path to complex life, however there is evidence that photosynthesis is an ancient metabolism, nearly as old as life itself (Blankenship 2010). This rapid evolution could instead suggest that perhaps photosynthesis was not so tricky after all.

Of the evolutionary steps to reach intelligent complex life, Eukaryogenesis – the evolution of eukaryotic life – is most commonly suggested as a candidate for an early great filter.

Eukaryogensis was not a singular event but involved many steps (Dacks et al. 2016). The well accepted theory of simbiogenesis posits that chloroplasts, mitochondria and other cell organelles descended from free living organisms (Margulis 1991) and primary endosymbiosis appears to be a rare event in Earth history (Stephens et al. 2021). All eukaryotic life, which includes all plants, animals and fungi, share a common ancestor (Makarova et al. 2005) and without their evolution it is thought life on Earth would have remained restricted to simple mostly single-celled organisms. Oxygen accumulated in Earth's atmosphere approximately 2.4 billion years ago (Gumsley et al. 2017) (although it would not accumulate to near modern day levels until around 400 million years ago (Lenton et al. 2016)) and this transformed Earth into an environment suitable for the evolution of complex cells. However the earliest fossilised evidence for eukaryotes dates from around 1.78 billion years ago (Knoll et al. 2006). The complexity of eukaryotic cells combined with the time it took for eukaryotes to evolve suggests that this might be a rare process. However the improbability of Eukaryogenesis and its uniqueness as an evolutionary process has been questioned (Booth & Doolittle 2015).

3.1. *Gaian Regulation and the Gaian Bottleneck*

Earth has had life and continuous habitable conditions for most of its existence. While there have been some deviations in Earth's climate where Earth has been frozen, Earth has otherwise maintained conditions at its surface that allow for liquid water. This long uninterrupted habitability has allowed life on Earth to evolve into intelligent life capable of contact with other worlds. The likelihood of such conditions persisting on other planets is therefore of direct relevance to the search for alien life and is an ongoing question in planetary formation and climate research.

Earth has been inhabited for roughly 4 billion years (Dodd et al. 2017) and over this time our planet has been dramatically transformed by life. The Earth-biosphere coupled system is known as 'Gaia', and this system has self-regulatory mechanisms that regulate Earth's atmosphere and ocean chemistry, and Earth's climate Lovelock (1965); Lovelock & Margulis (1974); Lovelock (1990) in the face of perturbations (Becker et al. 2001; Kaiho et al. 2001; Goldblatt & Zahnle 2011; Liu et al. 2020). Earth's carbon cycle acts as Earth's 'thermometer' and is an example of Gaian regulation on Earth (Lenton 2002; Kasting 2019). Over geological timescales carbon is recycled through Earth's atmosphere, oceans and lithosphere in feedback loops that heavily involve life processes. The climate sensitive chemical process 'weathering' removes carbon dioxide from the atmosphere and in warmer wetter conditions the rate of weathering increases, acting to cool the planet over long timescales due to the removal of CO_2 (Berner 1992). The products of weathering then wash into the oceans and become buried on the sea-floor where over long timescales they are recycled back into the atmosphere via plate tectonic movement and volcanic activity. Life accelerates the rate of weathering and acts as a reservoir for atmospheric carbon and thus plays a large role in determining Earth's climate. Models predict a lifeless Earth would be much hotter than today's climate (Schwartzman & Volk 1989).

The carbon cycle that regulates Earth's climate today has changed over time. When the earliest life was on Earth, it is thought that there may have been very little land-surface exposed to the atmosphere (Flament, Coltice & Rey 2008; Dhuime, Wuestefeld & Hawkesworth 2015), which would have made rates of terrestrial weathering, a crucial part of regulating our climate, incredibly low. Similarly Earth's current mode of mobile-lid plate tectonics is not thought to have been active until roughly 3 billion years ago (Condie & Kröner 2008) (although estimates vary hugely (Rollinson 2007)). This would have left early Earth without some of the key processes for climate regulation, during a time where Earth was experiencing numerous impacts, volatile evolution, and thermal instability from Earth cooling after its own accretion. The 'Gaian Bottleneck' hypothesis (Chopra & Lineweaver 2016; Nicholson et al. 2018) posits that this early window of abiotic habitability on a planet is likely to be short lived unless life quickly emerges and 'catches' this window of habitability and establishes climate regulation feedbacks. The Gaian Bottleneck hypothesis would predict that we would find either inhabited habitable planets, or inhospitable planets in our search for life in the galaxy and would place this step as the 'Great Filter' in our past. This prediction differs from an abiogenesis bottleneck as if the emergence of life was rare this would allow for the presence of habitable but uninhabited worlds. The Gaian Bottleneck instead predicts that planets will not remain habitable for long unless life emerges and establishes itself on a planet. If inhabitance is required for the long term maintenance of habitable conditions on a planet this leads to an 'inhabitance paradox' (Goldblatt 2016).

It is suggested that life might have an even greater impact on Earth's climate by influencing the continental coverage of our planet (Höning et al. 2014) (which impacts the rate of weathering) and playing a role in the establishment / maintenance of plate tectonics (Lenardic et al. 2016) which are a vital part of Earth's climate regulation. If life has such a strong influence on its planet's climate evolution then we would predict that any planet detected with habitable surface conditions will have a very high likelihood of being inhabited. Understanding Gaian systems will inform our search for inhabited planets and equally the search for life in the galaxy will provide data to test these hypotheses about Gaia.

3.2. *Natural cosmic hazards & The Great Filter*

Once life has taken hold by successfully negotiating the Gaian Bottleneck, we now consider whether evolutionary 'resets' due to natural cosmic hazards could serve to arrest its development and complexification, and thereby act as a Great Filter. Clearly any such filter could operate either before or after our current location in Fig. 1, and hence be 'Early' or 'Late' in our terminology.

There have been five mass extinctions since the Cambrian Explosion of land-based life circa. 500 million years ago, in which large fractions – but crucially not all – life went extinct (Benton 2023). The most recent of these, the end-Cretaceous mass extinction (66 Mya) which killed off the dinosaurs, has been ascribed to the impact of a ~ 10 km asteroid or comet. Although impacts of this size and severity are thought to strike the Earth roughly every 100 million years – comparable to the typical time interval between these mass extinctions – there is no evidence that any of the others were triggered in this manner. Smaller objects ~ 1 km in size – above which the climatic consequences of the impact become global – strike Earth roughly every 100,000 years (see Bailey (2018) for a review of the asteroid and cometary impact hazard). Whilst such an event would be a major setback for our civilization and could kill perhaps a quarter of the world's population through ecological and food chain collapse, the extent to which it would halt or significantly reverse our technological development is uncertain. In the consideration of such consequences, we should further distinguish between a 'one-off', isolated asteroid impact and an extended episode of multiple large impacts spread over thousands of years. The latter could be caused by the break up of a giant (> 50 km) comet, which enters the inner Solar System every 100,000 years and fragments into a multitude of smaller impactors; the most recent episode of such 'coherent catastrophism' may have occurred as recently as 10-20 thousand years ago (see Bailey (2018); Asher et al. (1994)).

Balancing the hazard posed by impacts are the benefits they offer in terms of creating new niches for the evolution of life and in delivering water (Martin & Livio 2021).

Generalising beyond the case of the Earth to exoplanetary systems, one must consider the extent to which the existence of asteroid belts and asteroid impacts are a universal feature of different systems. Simulations by Martin & Livio (2022) show that this requires that the system contain two suitably located giant planets, but does not require excessive fine tuning of the planetary architecture. Smallwood et al. (2018) assessed the consequences for the rate of asteroid impacts of introducing a Super-Earth (10 $M_{\oplus}$) in the vicinity of the Earth's orbit, finding that the impact rate increases the closer the super-Earth orbit is to the Earth's orbit when interior to it, but decreases significantly when it is moved further out beyond the Earth. Childs et al. (2022) examined the incidence of giant planets beyond the snow-line in M dwarf exoplanet systems – a requirement for asteroid belt formation – and found that, although the giant planet distribution peaks there, none existed in those systems with a planet in the habitable zone. In summary, the picture is mixed concerning whether asteroid impacts could serve as an effective Great Filter mechanism of wide applicability to exoplanetary systems, even if their destructiveness can be shown to outweigh their positive benefits for the development of life.

An arguably more promising Great Filter mechanism lies in the radiative hazard from supernovae and GRBs. The former have a destructive reach of 10 pc whilst the latter can be lethal over kiloparsec scales and hence perform a synchronised reset over a much larger linear extent of the Galaxy. GRBs are, however, highly beamed, with jet opening angles of typically a few degrees, so multiple events would be required to perform a global reset over a large continuous volume of space. Wilman et al. (2018) review this hazard in general terms and estimate that at the present epoch the rate at which 'catastrophic' supernovae and GRBs strike the Earth is approximately 1 event per Gyr, with the current rate dominated by supernovae. Abrevaya & Thomas (2018) review the physics of radiation from such sources in so far as it constrains the development of life in the Universe; for the case of GRBs, the dominant effect on present-day Earth is through the destruction of atmospheric ozone and the consequent increase in surface UVB radiation.

The rates at which such destructive events affect exoplanetary systems clearly depend on the spatio-temporal evolution of star formation over the course of galactic history, with the additional complication that 'long' GRBs tend to occur preferentially in regions of low metallicity star formation. Taking this into account using the Naab & Ostriker (2004) model for the build up of the Milky Way, Wilman et al. (2018) found that the rate at which catastrophic GRBs strike Earth increases from ~ 0.2 events per Gyr at the present epoch to 1 event per Gyr when the Earth formed; the extinction rates are generally higher in the lower metallicity outskirts of the Galaxy (see also the results of Gowanlock (2016) based on largely similar assumptions). It should, however, be noted that these estimates do not take into account that the Earth's atmospheric oxygen composition was substantially below its present level more than 400 Myr in the past, and that in the absence of ozone shielding the direct radiative effects of the GRB may instead dominate. In consequence, any land-based life at such times may already have evolved in a higher-radiation environment, and sea-based life would in any case benefit from higher levels of shielding. Using slightly different assumptions, Piran & Jimenez (2014) estimate that there is a 50% chance that a catastrophic GRB has struck Earth in the past 500 Myr, causing a mass extinction. Melott & Thomas (2009) argue that this could be the late-Ordivian mass extinction some 440 Mya.

To assess whether the decline in the catastrophic GRB rate as the Galaxy evolves could be sufficient to bring about an 'astrobiological phase transition' in which the timescale for the development of sufficiently advanced life drops below the interval between successive GRB extinction events, requires new simulations. These should combine the latest hydrodynamic models for the formation of the galaxies resembling the Milky Way, tracking the evolution of star formation, metallicity and galactic satellite mergers (e.g. using the EAGLE simulations and others of its kind; Schaye et al. (2015), Sawala et al. (2016)), with models for the impact of GRBs on planetary atmospheres at different points in their evolutionary history.

4. The Late Great Filter

4.1. *Technosignatures and Limits to Growth*

The Rare Earth explanation for the Great Filter remains a possibility, and observations of exoplanetary systems to search for biosignatures or technosignatures is one of the only ways to experimentally test the Rare Earth hypothesis. Versions of the Rare Earth hypothesis, including some of the arguments by Ward and Brownlee (2000), may focus too narrowly on conditions that have contributed to the development of life, intelligence, or technology on Earth but may not necessarily be limiting factors if not present. But another important consideration is the "degree of rareness" that is implied by any version of the Rare Earth hypothesis. One possibility is that the universe itself could be teeming with life, but each galaxy may have only one inhabited planet on average. In this case, human civilization could be alone in the Milky Way but in a universe full of life and even technology, which may suggest that the best way to find life is to search for evidence of Kardashev Type II or Type III civilizations that would be detectable at intergalactic distances (Sagan 1973). Another possibility is that life or technology is unique to Earth, even in the entire universe, which would imply that the search for biosignatures or technosignatures will find nothing. In both cases, any null results that are obtained in the the search for life will at least provide statistical constraints on the Rare Earth hypothesis; however, search strategies should consider a wider range of possibilities than Rare Earth alone.

Searching for technosignatures can help to provide constraints on the Late Great Filter. If technosignatures are found to be abundant, then this would suggest that the Great Filter is in the past, such as biogenesis or a Gaian bottleneck. But if no technosignatures are found, this does not necessarily mean that extraterrestrial technology is nonexistent. The framework of the Drake equation (see, e.g., Vakoch & Dowd 2015) considers the number of presently communicative civilizations in the galaxy in terms of several factors, some of which are observable quantities. Factors such as rate of star formation, fraction of stars with planets, and number of habitable planets per system are now known from observations; however, factors such as the fraction of planets that develop life or the fraction of inhabited planets that develop technology are generally unconstrained and may be difficult to resolve through remote observation (Haqq-Misra & Kopparapu 2018). The most advanced fleet of telescopes could observe a potentially habitable planet and find no biosignatures or technosignatures, but this would not logically imply that the planet is devoid of life or technology. The search for biosignatures and technosignatures not only requires observers to recognize signs of life if they are observed but also requires life itself to manifest in ways that are remotely detectable. If the galaxy is teeming with life that is impossible to detect, then the observational consequences would be identical to some versions of the Rare Earth hypothesis.

The statistical sense of the Great Filter is also important to consider when searching for technosignatures and biosignatures. The analogy of the Great Filter describes the hardest steps in evolutionary trajectories in general, with possible bottlenecks or other risks that may be commonplace or frequent for any planet in the galaxy or universe in which the conditions for life are present. But the Great Filter does not necessarily imply the existence of any specific hard steps in the evolutionary trajectory of individual species on Earth; there is no "filter" that could be invoked to explain the concentration of tool use, for example, among a subset of animals on Earth. Instead, the Great Filter analogy applies to evolution at a planetary scale, and describes limitations to hypothetical populations of habitable and inhabited planets. It could also be possible that there are multiple filters of equal "greatness," perhaps with both an Early and Late Great Filter of comparable difficulty. Any limitations on the evolutionary trajectory of a biosphere or technosphere that apply as general principles to life beyond Earth would correspond to limitations on any observations of biosignatures or technosignatures in the search for life.

Conducting a statistically meaningful search for biosignatures and technosignatures to constrain the Great Filter could also provide insight for thinking about the future of human civilization. One possibility is that such a search reveals that technosignatures are about as prevalent as biosignatures—such as all habitable planets tend to have both biosignatures and technosignatures—or perhaps even more prevalent than biosignatures—such as a case involving the migration of technology to other star systems. This would serve as evidence that the Great Filter is in Earth's past, which could serve as a source of confidence for human civilization to continue striving toward survival for geological or astronomical timescales. The knowledge that no Great Filter awaits in the future, and that other civilizations have already demonstrated the potential for long-lived technospheres, could help to inspire continued problem-solving on Earth to ensure a sustainable future. Such a discovery might help to remedy feelings of nihilism or helplessness toward global problems, as it would demonstrate the existence of viable trajectories for managing problems like climate change, sustainable development, and other planetary-scale issues. But another possibility is that a comprehensive search for life finds neither biosignatures nor technosignatures, or only biosignatures. The latter scenario would serve as evidence of a Late Great Filter. This would not necessarily imply that human civilization is is doomed to extinction in the near-term future, but it does mean that managing this future trajectory may be even more challenging given the lack of observable precedent to demonstrate that it can be done. If human civilization is the first in the galaxy or universe to embark on such a trajectory toward a long-lived technosphere, then a search for life will find no technosignatures, and the sustainability of this trajectory will remain unknown.

The assumption of exponential growth is a feature of many projections of Earth's future and contributes to speculation about possible technosignatures that could be detectable elsewhere (Haqq-Misra, this volume). If current exponential energy use were to continue increasing on Earth at 2.6% per year, then this energy use itself would begin to exert direct heating on the planet via the second law of thermodynamics by the year 2250 and would reach the threshold of a Kardashev Type I civilization by 2350 (Fig. 1, Haqq-Misra, this volume). The timing of these transitions would differ if the rate in energy use were to decrease, and a net-zero trajectory even remains plausible in which the rate in energy use continues to decrease to an eventual constant energy consumption per person with a stable global population. A study by Kopparapu et al. (2024) noted that these exponential energy trajectories would require the energy-generation equivalent of covering over 20% of Earth's land with solar panels within about 100 years; this is not to imply that other energy sources should not be used but simply illustrates the scale at which energy infrastucture would need to grow to meet exponentially increasing demands. However, if global population were to stabilize (even at several times greater than today) with a constant per person energy consumption of 75 GJ per year (enough to live a high quality life), then global energy demands could be met with the energy-generation equivalent of less than 10% solar coverage. This suggests the possibility that exponential extrapolations of future growth or energy use may overestimate the extent to which humans will continue along recent growth trajectories. Perhaps human civilization, and even civilizations in general, will tend to optimize energy consumption rather than maximizing it, which may prevent any technosphere from reaching even the threshold of a Kardashev Type I civilization. Stability in both population and energy consumption may be a requirement for any long-term technosphere, and any challenges in achieving such stability could be one reason for a Late Great Filter.

Exponential growth itself cannot continue unbounded for any spacefaring civilization, and if not limited at the planetary scale, then an expanding civilization would most likely be forced to halt its growth rate (or collapse) upon settling across an entire galaxy. This is the "sustainability solution" to the Fermi paradox, which suggests that any extraterrestrial civilization that has attempted to grow exponentially across the galaxy has overextended and collapsed before they could be detected (Haqq-Misra & Baum 2009). This would be an even later Great Filter and would suggest that any existing civilizations that are attempting to expand would be growing at a much slower rate that remains within the limits of their environmental carrying capacity. A civilization could also choose to expand across the galaxy and then cease growing, so as to avoid collapse (Haqq-Misra & Fauchez 2022). In any of these cases, the physical limitations of having no nearby places for expansion would inevitably halt any exponential growth trajectories at the galactic scale.

4.2. *AI and Deadly Probes*

Although not addressed in detail during the Great Filter discussion segment at the conference due to time limitations, an issue which naturally follows on from the considerations of energy and population growth is the role of AI and that of 'post biological' intelligence more generally.

Quite apart from any dangers posed by AI, we must the confront the impact of its voracious energy demands. As reviewed by Ammanath (2024), the computational power used by AI is currently doubling every 100 days, leading to an annualised growth rate in its energy needs of 26–36%. If sustained, this alone will ultimately push energy consumption trends beyond the upper end of those considered in section 4.1, absent major breakthroughs in quantum computing.

Turning to the dangers of AI itself, Garrett (2024) has argued that its rapid rise contrasts starkly with the much slower pace of development in space-faring technology. As a result, he argues that we are likely to witness the emergence of Artificial Superintelligence (ASI) before we have managed to establish a sustained multi-planetary presence as a civilization. The entire future of humanity would therefore be at stake if the ASI destroys its biological parent, an event which he suggests could limit our longevity as a technological civilization to $100 - 200$ years, and therefore serve as a mechanism for the Great Filter. Applying this figure to the Drake Equation as the lifetime of the radio communicative civilization, along with optimistic assumptions for the other parameters, implies that the number, N, of radio-communicating civilizations in the galaxy at the present time is $\sim 1 - 2$. We would thus be essentially alone in the galaxy, resolving the Fermi Paradox. That said, any of the other 'home grown' threats covered in section 2.2, or a combination of them (see also Ord (2020)), could have much the same result.

Whilst Garrett (2024) is correct in highlighting the possibility of an existential threat to biological humanity posed by the rise of ASI within a few centuries, the implications for SETI, the Great Filter and the Fermi Paradox are less clear-cut. As reviewed by Schneider (2017), it has been suggested frequently that any alien life we encounter, via SETI programmes, technosignatures or other manifestations, is likely to be some form of post-biological ASI. This stems from: (i) the short timescales for ASI development; (ii) the fact that there are habitable planets billions of years older than the Earth; (iii) the superiority of silicon and other as-yet-unknown mediums for ASI computation compared with the human brain; (iv) the superior fitness of artificial postbiological entities for long-duration interstellar space travel, relative to biological creatures. Yet the issue of whether such post-biological ASI would pose an existential threat to itself, or other independently-developed ASI on other planets, in addition to the biological parent that created it, is a more speculative matter. The implications of the discovery of AI or ASI technosignatures for the Great Filter hinge on the extent of the technological leap from our present-day capabilities. If this technological leap is not far into the future of human civilization, then the discovery of AI technosignatures would be essentially the same as for other technosignatures and would provide confidence that the Great Filter is in Earth's past, or a long-way into our future.

But if the development of ASI requires a major technological leap from present-day technology, then this transition itself could be the Great Filter; in this case, the discovery of ASI technosignatures, especially those *not* associated with biosignatures, would be evidence of a Late Great Filter along the lines suggested by Garrett (2024) and perhaps indicative of the high risk involved in managing generalized AI. Distinguishing between these two possible AI-based technosignatures may not be immediately obvious, but further research may help to understand possible observable properties that could arise from postbiological technospheres.

This brings us to the discussion of 'deadly probes'. As reviewed extensively by Ćirković (2018), this is a dystopian post-biological scenario in which galaxies come under the hegemony of malignant, probably stealthy, self-replicating von Neumann probes. By accident or design, such probes seek to destroy emergent civilizations at an early stage of their development, explaining the lack of evidence for galactic colonization. It would thus only be a matter of time until Earth is destroyed in this way. Noting that this scenario is consistent with all the evidence but has received remarkably little attention since the seminal work of Brin (1983), Ćirković (2018) refers to the possibility that – absent an enforced universal prohibition on their construction – our galaxy might become forever infected by one or more populations of such probes, which would become subject to their own complex ecosystem dynamics.

5. Conclusions

In summary, for more than a quarter of a century the Great Filter has provided a valuable conceptual framework for the analysis of the Fermi Paradox. Its utility stems from its conciseness and essential simplicity, in its ability to encompass in a single arc both the ancient past and far future development of life in the Universe, as well as the multitude of existential threats facing humanity at the current epoch. Yet, there are limitations in its approach. There may not be a single Great Filter evolutionary step, but rather multiple steps of comparable difficulty. Moreover, the proposed evolutionary 'flow' from simple life to our current state and onwards to a spare-faring civilization capable of galactic colonization may be just one imagined trajectory out of many, perhaps more likely, possibilities.

Only the search for technosignatures and biosignatures can provide observational constraints on the Great Filter. But actually finding any evidence of life depends on some extent of serendipity in space and time between human civilization and any others that might be observed. If biospheres or technospheres are commonplace but are too far away, then they will be almost impossible to discover. The emergence of life and technology could also occur sporadically across time, so that the probability of any two biosphere or technospheres co-existing at the same time is nearly zero. In such scenarios, even the most comprehensive search for life would find nothing and would be unable to place any constraints on the Great Filter. But a data-driven approach by searching for signatures of life and technology provide the only path toward a scientific resolution to this problem.

Acknowledgments

RJW acknowledges financial support via a Fellowship in 2023 at the Centre for Apocalyptic & Post-Apocalyptic Studies (CAPAS), Heidelberg University, which was instrumental in shaping the background to this work. JHM acknowledges support from the NASA Exobiology program under grant 80NSSC22K1009. AEN acknowledges financial support from a Leverhulme Trust research project grant: RPG-2020-82.

References

Abrevaya X.C., Thomas B.C. 2018, *Radiation as a Constraint for Life in the Universe*, in Gordon R. & Sharov A.A. (eds), in Astrobiology Exploring Life on Earth and Beyond (Series), *Habitability of the Universe Before Earth*, Academic Press, p.27–46

Ammanath B. 2024, *How to manage AI's energy demand — today, tomorrow and in the future*, World Economic Forum, April 25, 2024, `https://www.weforum.org/agenda/2024/04/how-to-manage-ais-energy-demand-today-tomorrow-and-in-the-future/` (accessed 10/7/24)

Asher D.J., Clube S.V.M., Napier W.M., Steel D.I. 1994, *Coherent Catastrophism*, Vistas Astron. 38, p1–27

Bailey M.E. 2018, *Asteroid and Cometary Impact Hazards* in R.J. Wilman & C.J. Newman (eds) *Frontiers of Space Risk. Natural Cosmic Hazards & Societal Challenges*, CRC Press, p. 15–35

Becker L., Poreda R.J., Hunt A.G., Bunch T.E. and Rampino M. 2001, *Impact event at the Permian-Triassic boundary: Evidence from extraterrestrial noble gases in fullerenes*, Science 291:5508, p1530–1533

Benton M.J. 2023 *Extinctions: How Life Survives, Adapts and Evolves*, Thames & Hudson

Berner R.A. 1992, *Weathering, plants, and the long-term carbon cycle*, Geochimica et Cosmochimica Acta 56:8, p3225–3231

Blankenship R.E. 2010, *Early evolution of photosynthesis*, Plant physiology 154:2, p434–438

Booth A. and Doolittle W. Ford 2015, *Eukaryogenesis, how special really?*, Proceedings of the National Academy of Sciences 112:33, p10278–10285

Bostrom N. 2008, *MIT Technology Review*, May/June, 72 `https://nickbostrom.com/papers/where-are-they/` (accessed 10/7/24)

Brin G.D., 1983, *Q. J. Roy. Astron. Soc.*, 24, 283

Childs A.C., Martin R.G., Livio M. 2022, *Life on Exoplanets in the Habitable Zone of M-dwarfs*, Astrophysical Journal Letters, 937, L41

Chopra A. and Lineweaver C.H. 2016, *The case for a Gaian bottleneck: the biology of habitability*, Astrobiology 16:1, p7–22

Ćirković M. 2017, *'Some Contemporary – and Persistent – Fallacies and Confusions about Astrobiology'*, in A. Losch (ed.) *What is Life? On Earth and Beyond: On Earth and Beyond.* Cambridge University Press, pp. 156–177.

Ćirković M. 2018, *The Great Silence*, Oxford University Press

Condie K.C. and Kröner A. 2008, *When did plate tectonics begin? Evidence from the geologic record*, Geological Society of America Special Papers Geoscience 440, p281–294

Dacks J.B., Field M.C., Buick R., Eme L., Gribaldo S. and Roger A.J., Brochier-Armanet C. and Devos D.P. 2016, *The changing view of eukaryogenesis–fossils, cells, lineages and how they all come together*, Journal of cell science 129:20, p3695–3703

Darling D. 2001, *Life Everywhere: The Maverick Science of Astrobiology*, New York: Basic Books

Dhuime B., Wuestefeld A. and Hawkesworth C.J. 2015, *Emergence of modern continental crust about 3 billion years ago*, Nature Geoscience 8:7, p552–555

Dodd M.S., Papineau D., Grenne T., Slack J.F., Rittner M., Pirajno F., O'Neil J. and Little C.T.S. 2017, *Evidence for early life in Earth's oldest hydrothermal vent precipitates*, Nature 543:7643, p60–64

Eager-Nash J.K., Daines S.J., McDermott J.W., Andrews P., Grain L.A, Bishop J., Rogers A.A., Smith J.W.G., Khalek C., Boxer T.J., Mak M.T., Ridgway R.J., Hébrard E., Lambert H.F., Lenton T.M and Mayne N.J. 2024, *Simulating biosignatures from pre-oxygen photosynthesizing life on TRAPPIST-1e*, Monthly Notices of the Royal Astronomical Society 531:1, p468–494

Flament N., Coltice N. and Rey P.F. 2008, *A case for late-Archaean continental emergence from thermal evolution models and hypsometry*, Earth and Planetary Science Letters 275:3–4, p326–336

Forgan D.H. 2019, *Solving Fermi's Paradox*, Cambridge University Press

Garrett M. 2024, *Acta Astronautica*, 219, 735

Goldblatt C. and Zahnle K.J. 2011, *Faint young Sun paradox remains*, Nature 474:7349

Goldblatt C. 2016, *The inhabitance paradox: how habitability and inhabitancy are inseparable*, arXiv:1603.00950

Gowanlock M.G. 2016, *Astrobiological effects of gamma-ray bursts in the Milky Way galaxy*, Astrophysical Journal, 11, p.855–873

Gumsley A.P., Chamberlain K.R., Bleeker W., Söderlund U., De Kock M.O., Larsson E.R. and Bekker A. 2017, *Timing and tempo of the Great Oxidation Event*, Proceedings of the National Academy of Sciences 114:8, p1811–1816

Hanson R. 1998a, `https://mason.gmu.edu/~rhanson/greatfilter.html` (accessed 10/7/24)

Hanson R. 1998b, `https://mason.gmu.edu/~rhanson/hardstep.pdf` (accessed 10/7/24)

Haqq-Misra J.D. and Baum S.D. 2009, *The Sustainability Solution to the Fermi Paradox*, Journal of the British Interplanetary Society 62, p47–51

Haqq-Misra J. and Kopparapu R.K. 2018, *The Drake equation as a function of spectral type and time* in Gordon R. & Sharov A.A. (eds), in Astrobiology Exploring Life on Earth and Beyond (Series), *Habitability of the Universe Before Earth*, Academic Press, p307–319

Haqq-Misra J. and Fauchzez T.J. 2022, *Galactic settlement of low-mass stars as a resolution to the Fermi paradox*, The Astronomical Journal 164:6, p247

Höning D., Hansen-Goos H., Airo A. and Spohn T. 2014, *Biotic vs. abiotic Earth: A model for mantle hydration and continental coverage*, Planetary and Space Science 98, p5–13

Horner J. and Jones B.W. 2010, *Jupiter: friend or foe? An answer*, Astronomy & Geophysics 51:6, p6–16

Jiang J.H., Rosen P.E., Lu K., Fahy K.A., Obacz P., 2023, *J Huma Soci Scie*, 6(2), 59

Kaiho K., Kajiwara Y., Nakano T., Miura Y., Kawahata H., Tazaki K., Ueshima M., Chen Z. and Shi G.R. 2001, *End-Permian catastrophe by a bolide impact: evidence of a gigantic release of sulfur from the mantle*, Geology 29:9, p815–818

Kasting J.F. 2019, *The Goldilocks planet? How silicate weathering maintains Earth "just right"*, Elements: An International Magazine of Mineralogy, Geochemistry, and Petrology 15:4, p235–240

Knoll A.H., Javaux E.J., Hewitt D., Cohen P. 2006, *Eukaryotic organisms in Proterozoic oceans*, Philosophical Transactions of the Royal Society B: Biological Sciences 361:1470, p1023–1038

Kopparapu R., Kofman K., Haqq-Misra J., Kopparapu V., Lingam M. 2024, *Detectability of solar panels as a technosignature*, The Astrophysical Journal 967:2, p119

Lenardic A., Crowley J.W., Jellinek A.M. and Weller M. 2016, *The solar system of forking paths: bifurcations in planetary evolution and the search for life-bearing planets in our galaxy*, Astrobiology 16:7, p551–559

Lenton T.M. 2002, *Testing Gaia: the effect of life on Earth's habitability and regulation*, Climatic Change 52:4, p409-422

Lenton T.M., Dahl T.W., Daines S.J., Mills Benjamin J.W., Ozaki K., Saltzman M.R. and Porada P. 2016, *Earliest land plants created modern levels of atmospheric oxygen*, Proceedings of the National Academy of Sciences 113:35, p9704–9709

Lin-gun L. 2004, *The inception of the oceans and CO2-atmosphere in the early history of the Earth*, Earth and Planetary Science Letters 227:3-4, p179–184

Liu B., Wang B., Liu J., Chen D., Ning L., Yan M., Sun W. and Chen K. 2020, *Global and polar region temperature change induced by single mega volcanic eruption based on community earth system model simulation*, Geophysical Research Letters 47:18

Lovelock J.E. 1965, *A physical basis for life detection experiments*, Nature 207:4997, p568–570

Lovelock J.E. and Margulis L. 1974, *Atmospheric homeostasis by and for the biosphere: the Gaia hypothesis*, Tellus 26:1–2, p2–10

Lovelock J.E. 1990, *Hands up for the Gaia hypothesis*, Nature 344:6262, p100–102

Makarova K.S., Wolf Y.I., Mekhedov S.L., Mirkin B.G. and Koonin E.V. 2005, *Ancestral paralogs and pseudoparalogs and their role in the emergence of the eukaryotic cell*, Nucleic acids research 33:14, p4626–4638

Margulis M. 1991, *Symbiosis in Evolution: Origins of Cell Motility: Origins of Cell Motility*, Evolution of Life: Fossils, Molecules, and Culture, Springer, p305–324

Martin R.G., Livio M., 2021, *How much water was delivered from the asteroid belt to the Earth after its formation?*, Monthly Notices of the Royal Astronomical Society, 506, L6-10

Martin R.G., Livio M., 2022, *Asteroid & Life: How special is the Solar System?*, Astrophysical Journal Letters, 926, L20

Melott A.L., Thomas B.C., 2009, *Late Ordovician geographic patterns of extinction compared with simulations of astrophysical ionizing radiation damage*, Paleobiology, 35, p.311–320

Naab T., Ostriker J.P., 2006, *A simple model of the evolution of disk galaxies: The Milky Way*, Monthly Notices of the Royal Astronomical Society 366, p899–917

Nicholson A.E., Wilkinson D.M., Williams H.T.P. and Lenton T.M. 2018, *Gaian bottlenecks and planetary habitability maintained by evolving model biospheres: The ExoGaia model*, Monthly Notices of the Royal Astronomical Society 477:1, p727–740

Nisbet E.G. and Sleep N.H. 2001, *The habitat and nature of early life*, Nature 409:6823, p1083–1091

Olson J.M. 1981, *Evolution of photosynthetic reaction centers*, Biosystems 14:1, p89–94

Ord T. 2020, *The Precipice. Existential Risk and the Future of Humanity*, Bloomsbury.

Piran T., Jimenez R., 2014, *Possible role of gamma-ray bursts on life extinction in the universe*, Physical Review Letters, 113, 231102

Rollinson H. 2007, *When did plate tectonics begin?*, Geology Today 23:5, p186–191

Sagan, C. 1973, *On the detectivity of advanced galactic civilizations*, Icarus, 19, 3, p.350–352

Sawala T. et al. 2016, *The APOSTLE simulations: solutions to the Local Group's cosmic puzzles*, Monthly Notices of the Royal Astronomical Society, 457, 2, p.1931–1943

Schaye J. et al. 2015, *The EAGLE project: simulating the evolution and assembly of galaxies and their environments*, Monthly Notices of the Royal Astronomical Society, 446, 1, p. 521–554

Schneider S., 2017, *Superintelligent AI and the Postbiological Cosmos Approach 1*, in A. Losch (ed.) *What is Life? On Earth and Beyond* Cambridge University Press, p. 178–198.

Schwartzman D.W. and Volk T. 1989, *Biotic enhancement of weathering and the habitability of Earth*, Nature 340:6233, p457–460

Seager S., Bains W. and Hu R. 2013, *Biosignature gases in H2-dominated atmospheres on rocky exoplanets*, The Astrophysical Journal 77:2, p95

Smallwood J.L., Martin R.G., Lepp S., Livio M. 2018, *Asteroid impacts on terrestrial planets: the effects of super-Earths and the role of the ν_6 resonance* Monthly Notices of the Royal Astronomical Society, 473, 295–305

Stephens T.G., Gabr A., Calatrava V., Grossman A.R. and Bhattacharya D. 2021, *Why is primary endosymbiosis so rare?*, New Phytologist 231:5, p1693–1699

Vakoch, D.A. and Dowd, M.F. 2015, *The Drake Equation: Estimating the Prevalence of Extraterrestrial Life through the Ages*, Cambridge University Press

Ward P. and Brownlee D. 2000, *Rare Earth: Why Complex Life is Uncommon in the Universe*, Copernicus

Wilman R.J., Dayal P. and Ward M.J. 2018, *Hazards & Habitability: Galactic Perspectives* in R.J. Wilman & C.J. Newman (eds) *Frontiers of Space Risk. Natural Cosmic Hazards & Societal Challenges*, CRC Press, p. 77–104

(Toward) Discovery of Life beyond Earth and its Impact
Proceedings IAU Symposium No. 387, 2026
H. Landt, M. Dominik & C. Oliver, eds.
doi:10.1017/S1743921324003296

Investigating Cosmic-Ray Induced Radiation in Exoplanetary Environments

Conor Kennedy[1,2] and Robert Brose[2,1]

[1]Dublin City University, School of Physical Sciences, Dublin 9, Ireland.

[2]Dublin Institute for Advanced Studies, Astronomy & Astrophysics Section, DIAS Dunsink Observatory, Dublin, D15 XR2R, Ireland

email: conor.kennedy@stud.uni-heidelberg.de

Abstract. The study of cosmic rays (CRs) and their interactions with exoplanetary atmospheres and stellar environments can provide essential insights into the habitability and atmospheric chemistry of these distant worlds. For instance, on Earth, radiation from cosmic rays constitutes only about 10 percent of the total radiation exposure at ground level, but this percentage can be significantly higher for planets orbiting stars close to cosmic ray sources. Furthermore, the modulation of cosmic rays varies considerably among different stars. This research focuses on modelling the modulation of CRs by various stellar types, particularly F, G, K, and M stars, to understand the impact on exoplanets in their habitable zones. Employing an analytical one-dimensional model of the Cosmic Ray Transport Equation, which allows us to study a wide range of parameters, we analyse how different stellar parameters, such as wind speed and magnetic field strength, influence the cosmic ray environment. The developed model simulates the cosmic ray spectra and mean free path, considering factors like the star's relative motion through the interstellar medium. This work aims to contribute to the understanding of radiation conditions on exoplanets, which is crucial for assessing their potential to support life. It offers a preliminary but meaningful exploration into the complex interactions between cosmic rays and exoplanetary environments, shedding light on factors that might influence the habitability of these alien worlds.

Keywords. Cosmic Rays, Exoplanets, Life

1. Introduction

Cosmic rays (CRs) are high-energy particles originating from a variety of astrophysical sources, including our Sun, Supernovae, massive stars and active galactic nuclei (Blasi 2013; Aharonian et al. 2019). These particles continuously bombard planetary atmospheres, where Earth's magnetic field protects us from many of the low-energetic CRs from the sun. Only CRs with energies above a few 10s of GeV, are able to contribute to ground-level radiation but originate from outside our solar system. Here, the solar wind provides effective protection from CRs with energies below a few GeV, which dominate CRs in terms of flux.

However, the recent discovery of numerous exoplanets, particularly those orbiting M-dwarf stars, raises new questions about the exposure of these planets to cosmic rays. M-dwarf stars, known for their intense magnetic fields and unique stellar environments, could influence the level of cosmic ray shielding in ways that differ substantially from the Sun (Han et al. 2023; Cranmer and Saar 2011).

Further, while our Galaxy is filled with a reasonably well-studied "sea" of CRs, that all stellar systems are exposed to, the CR-density can differ substantially close to the actual sources of CRs. These sources have to have considerable higher fluxes in their vicinity, leading to higher radiation doses on nearby exoplanets, and have harder spectral indexes of their CR-spectra, meaning that planets in their vicinity are exposed to relatively more CRs that can contribute to ground-level radiation.

This research explores the impact of cosmic rays on exoplanetary environments, with a particular focus on how stellar bubbles modulate CRs. Given that most known exoplanets orbit M-dwarf stars, which are cooler and less luminous than our Sun but possess stronger magnetic fields, understanding these modulation effects is crucial. This study aims to develop and refine a model that simulates CR modulation across various stellar classes, starting with validation against solar system data before extending the analysis to different star types and their corresponding habitable zones.

By investigating the interaction of cosmic rays with different stellar environments, this research seeks to provide deeper insights into the habitability of exoplanets, contributing to the broader field of Astrobiology.

2. Methodology

We build a analytical description of the modulation of CRs in the astrospheres of different stars depending on their wind-velocity and surface magnetic field. In the following we describe our model for CR diffusion (sections 2.2), for the astrospheres (section 2.1) and the validation against the measured CR-modulation in the Solar system (section 2.3).

2.1. *Stellar bubble and Habitable Zone Modelling*

Cosmic rays encountering any stellar-wind bubble have to diffuse against the outflow of the stellar wind. Hence, before solving the actual equations for the CR-transport its essential to have a model for the wind-bubble itself. In this work, we assume that stellar astrospheres consist of two zones: An inner zone (Zone A), where the stellar wind is freely expanding - with a constant wind-velocity, a density dropping as $\propto r^{-2}$ and a magnetic field $\propto r^{-1}$ - and a region of a shocked wind (Zone B), where the wind-velocity is $\propto r^{-2}$, the density and magnetic field spatially constant.

Next, it is important to define the extension of both zones. Since the probability of a star being at rest relative to the ambient interstellar medium (ISM) is basically zero, we assume all stars in this work to be moving. The extension of Zone A is then defined by the balance between the winds ram-pressure and the pressure in the ISM plus the dynamical pressure from the movement

$$R_1 = \sqrt{\frac{L_w}{v_w \cdot 2\pi \cdot (nTk_b + \rho_{\mathrm{ISM}} v_*^2)}} \tag{1}$$

where R_1 is the radius of the wind termination shock, L_w the star's mechanical luminosity, v_w the terminal wind velocity, n and ρ being the number and mass density of the ISM and V_* being the stellar velocity relative to the ISM.

There is no analytical solution for the position of the contact-discontinuity of a stellar bow-shock that basically marks the outer boundary for our model. Hence, we assume $R_C = 1.3 \cdot R_1$ based on numerical simulations by Mackey (2022). In our model R_1 and R_C depend only on the velocity of the star v_*, the stellar mass-loss rate $\dot{M}_*$ and the terminal wind-velocity v_w.

The second ingredient needed for the evaluation of habitability, is the position of the habitable zone, which basically depends on the spectral class of the central star. We use the Habitable Zone framework developed by Kopparapu et al. (2013) and use data from Pecaut and Mamajek (2013), which provides effective temperature and luminosity values for a wide range of spectral classifications. We take the position of the HZ as the average

Table 1. Simulation Stellar Parameters.

SpV	$T_{\rm eff}$ (K)	$\dot{M}_*$ ($M_\odot$/yr)	$R/R_\odot$	$\log L/L_\odot$	V_w (km/s)	$B_{\rm min}$ (G)	$B_{\rm mean}$ (G)	$B_{\rm max}$ (G)
F5V	6550	5×10^{-14}	1.473	0.56	500	0.8	2.2	3.3
G5V	5660	2.4×10^{-14}	0.977	−0.05	400	0.2	24.5	47.9
K5V	4440	2.99×10^{-15}	0.701	−0.76	300	1.5	36.5	78
M5V	3060	1.92×10^{-16}	0.196	−2.52	200	16	51	406
Sol (G2V)	5778	6.8×10^{-14}	1.0	0.0	400	–	1	–

of the outer and inner value defined by Kopparapu et al. (2013). We summarised the different parameters of stars we considered in 1.

Additionally, we consider five different values of V_* for each spectral class, namely 10 km/s, 20 km/s, 30 km/s, 60 km/s and 120 km/s.

2.2. *Solving the Cosmic Ray Transport Equation*

The choice of a 1D, steady-state, spherically symmetric model is motivated by the need to simplify the complex three-dimensional nature of CR modulation for initial validation purposes. This approach, while reducing complexity, may introduce limitations in accurately capturing anisotropic effects, which are acknowledged as potential areas for future model refinement. This simplification, commonly used in the literature (Rodgers-Lee et al. 2021), facilitates code development in Python and allows for model validation against existing data.

The cosmic-ray transport equation (CRTE) in steady state and without a source term and in 1-D spherical symmetry is Ginzburg and Syrovatsky (1961):

$$\frac{1}{r^2}\frac{\partial(r^2 V_r N)}{\partial r} = \frac{1}{r^2}\frac{\partial(r^2 D_r \frac{\partial N}{\partial r})}{\partial r} \tag{2}$$

After integrating and assuming negligible constants at large distances, the solution is:

$$N(r,E) = N_0(E)\cdot\exp\left(\frac{V_r}{D_c(r,E)}r\right) \tag{3}$$

The constant N_0 can be derived from the boundary conditions. We assume power-law behaviour of the form

$$V_r = V_w\cdot\left(\frac{r}{r_0}\right)^a \text{ and } D_c(r,E) = D_0\cdot\left(\frac{B}{B_0}\cdot\left(\frac{r}{r_0}\right)^b\right)^c\cdot\left(\frac{E}{E_0}\right)^d \tag{4}$$

for the stellar-wind velocity V_r, the diffusion coefficient of CRs D_c and the magnetic-field B.

We assume that all the astrospheres constitute of two zones, the free-expanding wind (Zone A) and the shocked-wind (Zone B). The values for the scaling-parameters a, b, c and d depend on the spatial behaviour of v_r and B in the different regions of the wind. Applying the values outlined in section 2.1, we obtain

$$N(r,E) = \begin{cases} N_{\rm R1}(E)\cdot\exp\left(\alpha\cdot\frac{3}{2}R^{\frac{2}{3}}\right) & \text{for } R_* < r \le R_1 \text{ (Region A)} \\ N_0(E)\cdot\exp\left(-\beta\cdot\frac{1}{R}\right) & \text{for } R_1 < r \le R_C \text{ (Region B)} \end{cases} \tag{5}$$

where R_* is the stellar radius, R_1 is the wind termination shock, and R_C is the contact discontinuity within the stellar bubble. The constants in equation 5 are defined by the terminal wind-velocity v_w, the surface magnetic field of the star B_*, the local interstellar spectrum of CRs and the normalisation constant D_0 that relates the magnetic-field

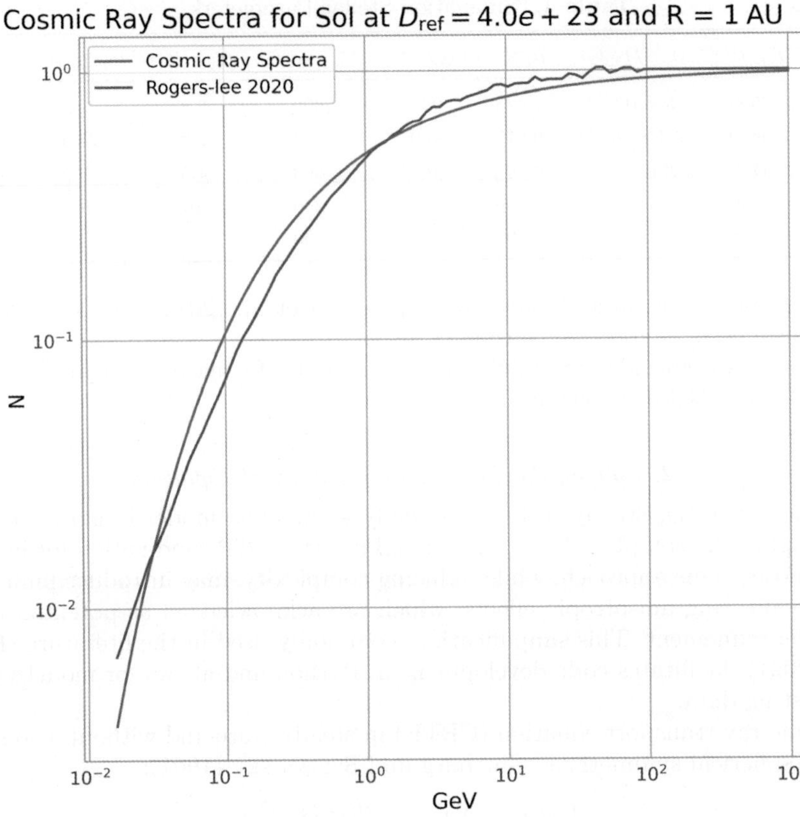

Figure 1. Showing the CR spectra plot for Sol at 1 AU, the D_0 parameter from equation (5) was adjusted to reduce the residual when compared to the reference spectrum from Rodgers-Lee et al. (2021). This adjustment ensures the model accurately reflects the solar system's CR environment. In the plot, the spectrum calculated by the model is represented by the blue line, while the reference spectrum from is indicated by the purple line.

strength in the astrosphere to the diffusion coefficient of the CRs. In this work, we assume that D_0 is uniform across all stellar-systems. However, the turbulence driving in a stellar wind is a complex process that is depending on many parameters and not yet fully understood.

2.3. *Validation and Application*

We tweaked our model that effectively has only D_0 and d as a free parameters to match the observed CR-modulation in the Solar system. We used Sol's average surface $B - Field$ strength at the equator of 1 Gauss, its wind velocity of approximately 400 km/s, and its the mass-loss rate of around $6.8 \times 10^{-14}\ M_\odot\ \mathrm{yr}^{-1}$, with an relative velocity towards the interstellar medium of 22 km/s.

We found that $d = 0.5$ described Solar modulation best, which corresponds to Kraichnan-like diffusion. Figure 1 shows our best-fit model for the Solar system compared to the model of (Rodgers-Lee et al. 2020).

As can be seen from figure 1, the maximum deviation from our reference model is $\leq 10\%$. Hereafter, we assume that the diffusion coefficient D_0 is constant across different stellar environments, although this assumption may require revision due to variations in stellar wind and magnetic field properties among stars. Despite the inherent limitations

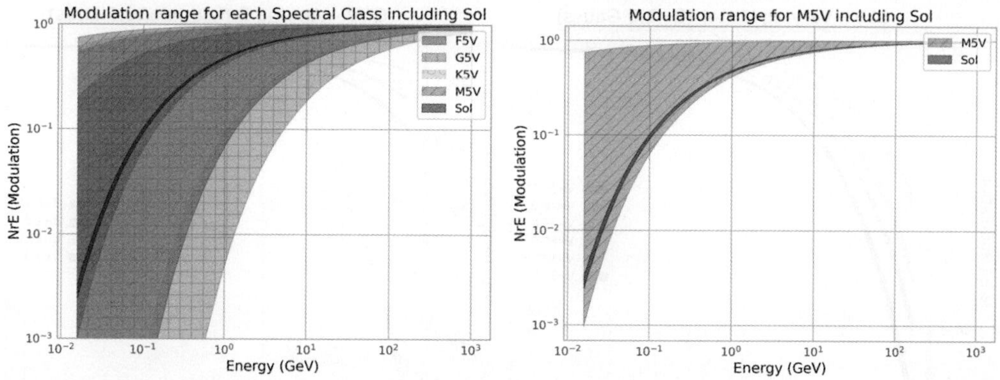

Figure 2. Left panel: Simulations across various **B**-fields V_* and spectral ranges, demonstrating the influence of **B**-field strengths, spectral classes, and stellar velocities on CR modulation. **Right panel:** Modulation spectra for various stellar parameters for M5V red dwarfs. Many M-dwarfs offer less protection against CRs in their habitable zones compared to the Sun.

of using a one-dimensional model for a fundamentally three-dimensional problem, such as heliospheric CR modulation, the maximum residuals observed were within acceptable margins for standard astrophysical modelling practices.

3. Results and Discussion

We first compare the ability of different stars to shield their planets by directly comparing their modulation curves and then by evaluating the surface-level radiation dose that originates from the modulation.

3.1. *Impact of stellar parameters*

We summarized all of our models in figure 2 where we show, sorted by spectral type, what range of modulation we expect for each spectral class given the uncertainties in the stellar parameters. The curves are evaluated in the habitable zone for each spectral class, meaning that for cooler, smaller stars, the spectra are evaluated much closer to the star then compared to the hotter spectral classes. Hotter stars generally exhibit superior CR modulation. However, substantial uncertainties exist due to the broad range of stellar parameters and the complex nature of CR interactions. The case of the best CR-modulation in Figure 2 typically originates from the case with the slowest-moving star and the strongest magnetic field for each stellar type, while the weakest modulation is the fastest-moving star and the weakest field.

As can be seen, our solar system is in the middle of the observed modulation range, but basically better than all M-type stars, which represent the most-abundant type of star in our Galaxy. Despite their stronger magnetic fields, M-dwarfs modulate CRs less effectively than the Sun. It requires extreme conditions (maximum surface $B-$ field and minimum stellar velocity) for M-dwarfs to match the Sun's modulation effect.

Figure 3 illustrates the impact of the magnetic field and the stellar velocity with taking the G5V spectral class as an example.

The right panel of Figure 3 shows how a stronger magnetic field leads to a stronger modulation. This is natural as a stronger field reduces the gyro-radius and hence the mean-free path of the CRs, making it harder for them to deeply penetrate the astrosphere. The study faced challenges due to the vast uncertainty in magnetic field measurements

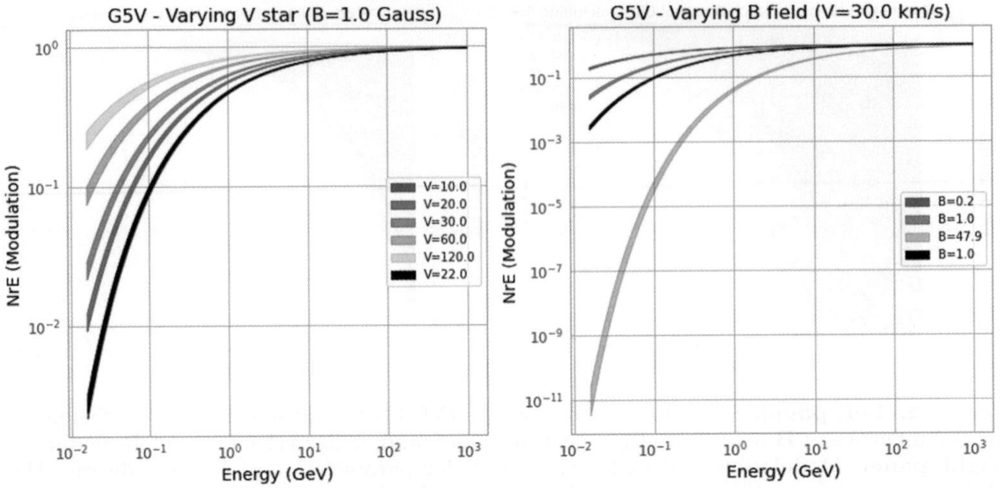

Figure 3. Left panel: Modulation of a G5V star for different stellar velocities that lead to different sizes of the astrospheres.
Right panel: Modulation of a G5V star for different surface magnetic fields of the stars.

(spanning a factor of 100), as shown in Table 1. This uncertainty translates into modulation variations extending over 10 orders of magnitude, however, the total effect on the radiation dose is smaller than this as this is weighted by the energy of the particles.

The range of velocities relative to the ISM that we considered here represents on one hand the spread of velocities observed in or Galaxy for most stars, albeit some stars with way higher velocities exist (Anguiano et al. 7 01; Quispe-Huaynasi et al. 2022). At the same time, the velocity of a star is degenerate with the density of the medium the star is moving through. Here, we assumed $n = 1\,\mathrm{cm}^{-3}$ but the density of the ISM widely varies (Dobbs et al. 2008). So, $V_* = 120\,\mathrm{km/s}$ could correspond to $V_* = 20\,\mathrm{km/s}$ with $n_{ISM} = 36\,\mathrm{cm}^{-3}$, highlighting the role of ISM density in CR modulation. This relationship is particularly significant as the solar system currently traverses a region of below-average ISM density, which might influence CR flux and, consequently, the stellar system.

3.2. *Radiation Doses*

Inferring the exact radiation doses is beyond the scope of this work, however, we can infer the energy deposited in a planets atmosphere from integrating the modulation curves over energy and dividing them by the integral over the Sun's modulation curve.

$$e_{\mathrm{hab}} = \frac{\int_{10\ \mathrm{MeV}}^{3\ \mathrm{PeV}} E \cdot N_{\mathrm{CR}}(E) \cdot NrE_*(E)\,\mathrm{d}E}{\int_{10\ \mathrm{MeV}}^{3\ \mathrm{PeV}} E \cdot N_{\mathrm{CR}}(E) \cdot NrE_{\odot}(E)\,\mathrm{d}E}, \tag{6}$$

where e_{hab} is the energy deposited in an exoplanets atmosphere, relative to the energy deposited in Earths atmosphere due the Solar modulation under the same boundary conditions, N_{CR} the Galactic CR spectrum as given in Rodgers-Lee et al. (2020) and NrE is the stellar modulation.

Figure 4 illustrates the deposited energies for the Galactic CR-spectrum for all spectral classes and stellar velocities. Typically, stars shield shield their planets slightly better than the sun or allow for $\approx 50\%$ more energy deposited by Galactic CRs. However, this is not surprising, given the turnover of the Galactic CR-spectrum at low energies (Rodgers-Lee et al. 2020), where the modulation differs most between different spectral classes.

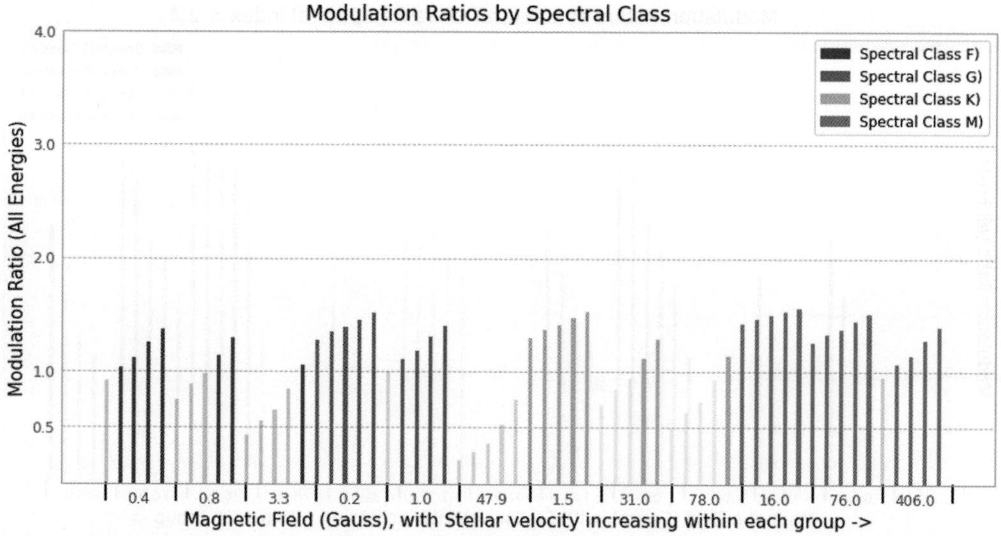

Figure 4. CR modulation ratios for energy deposited on a planet by modulated CR-flux from 10 MeV to 3 PeV, corresponding to the energy deposited in the atmosphere of an exoplanet.The relatively small differences in energy deposition can be explained by the low-energy turnover in the LIS, which results in a softer spectrum that is more uniformly modulated across different spectral classes. Strong colours indicate cases where more energy is deposited as for Earth at the moment, lighter colors indicate better shielding with respect to Earth.

Figure 5 illustrates the energy deposited for all the cases considered in our study close to a source with a source index (s) of 2.4. This source index is characteristic of environments proximal to CR sources, which might include SNRs, star clusters, or colliding wind binaries, as observed in gamma-ray studies. These sources are more prevalent in regions like galactic spiral arms, where the ISM density could also potentially be higher which corresponds to a higher effective v_* in our model.

The values of deposited energy in this case varies by a factor of ≈ 10 between the most effectively and least effectively shielded case and considerably stronger than for Galactic CRs (see Figure 4). Close to a source, as indicated in the case considered in Figure 5 the total amount of particles, is also higher, indicating that the variability of deposited CR-energy throughout a stars lifetime is much higher for a number of stars and especially M-dwarfs.

Our calculations indicate, that none of the stellar classes considered here, has a such insufficient ability to shield their planets, so that life on them is prevented by the direct effects of CRs. However, indirect effects on the climate, like increased cloud coverage and lightning strike rates due to enhanced ionisation of the atmosphere at sea level (Kumar et al. 2023; Svensmark 2006), could effect ecosystems and cause (mass)-extinction events. CRs can produce secondary effects on Earth-like planets by altering the ionization rate of the lower atmosphere. Melott and Thomas showed that an SNR (or a series of SNRs) located at 50 pc from Earth might have altered the atmosphere's conductivity and hence the rate of lightning and consequently wildfires. Their line of reasoning is supported by ^{60}Fe deposits in sediments that coincide with soot and other carbon-related sediments. These sediments are associated with a period of conversion of woodland to savanna across various continents, which in fact might have supported the development of bipedalism in hominids.

For systems with a weaker CR-modulation the indirect effects that were just mentioned, would appear stronger and importantly with a larger difference between the situation

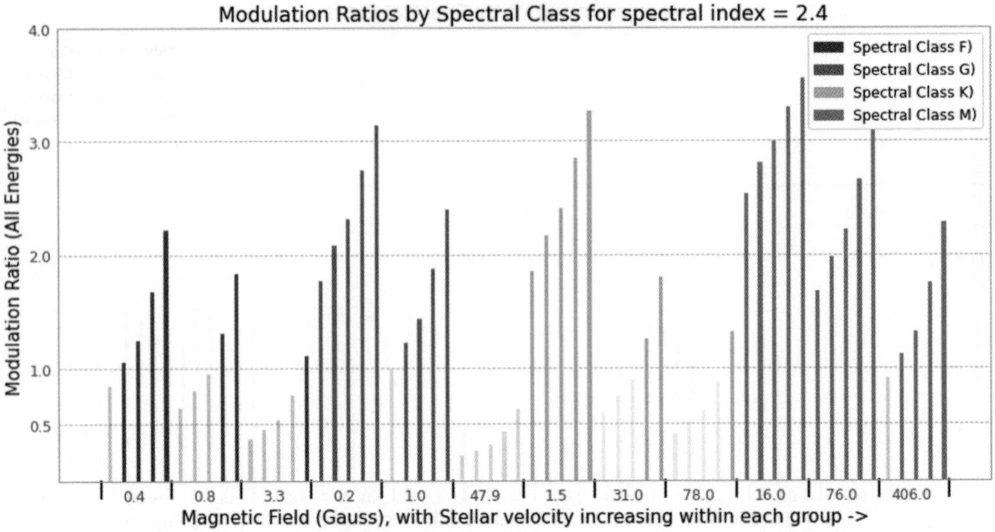

Figure 5. CR modulation ratios for energy deposited on a planet by modulated CR-flux from 10 MeV to 3 PeV, corresponding to the energy deposited in the atmosphere of an exoplanet. Strong colours indicate cases where more energy is deposited as for Earth at the moment, lighter colors indicate better shielding with respect to Earth.

where the background CRs dominate compared to where the stellar system in close to CR-source.

4. Conclusion

This project has systematically explored the modulation of CRs across different stellar systems. The development of a one-dimensional spherically symmetric model and the integration of this model with varying stellar parameters has enabled a nuanced understanding of CR modulation in various stellar environments.

Key findings include the significant variability in CR modulation depending on stellar spectral class, magnetic field strengths, and velocities relative to the ISM. Notably, M-dwarf stars, despite their prevalence as hosts to exoplanets, might offer less protection against CRs compared to stars like our Sun. Further, the expected variation in the energy deposited into a planets atmosphere between different spectral classes is much larger for most spectral classes considered here.

The results highlight the complexity of factors influencing CR penetration in exoplanetary atmospheres and underscore the importance of considering a wide range of stellar and interstellar variables when assessing exoplanetary habitability.

References

Aharonian, F., Yang, R., & de Oña Wilhelmi, E. 2019, Massive stars as major factories of Galactic cosmic rays. *Nature Astronomy*, 3(6), 561–567. Publisher: Nature Publishing Group.

Anguiano, B., Majewski, S. R., Hayes, C. R., Prieto, C. A., Cheng, X., Bidin, C. M., Beaton, R. L., Beers, T. C., & Minniti, D. 2020-07-01, The stellar velocity distribution function in the milky way galaxy. *AJ*, 160(1), 43.

Blasi, P. 2013, The Origin of Galactic Cosmic Rays. *The Astronomy and Astrophysics Review*, 21(1), 70. arXiv:1311.7346 [astro-ph].

Cranmer, S. R. & Saar, S. H. 2011, TESTING a PREDICTIVE THEORETICAL MODEL FOR THE MASS LOSS RATES OF COOL STARS. *The Astrophysical Journal*, 741(1), 54.

Dobbs, C. L., Glover, S. C. O., Clark, P. C., & Klessen, R. S. 2008, The ISM in spiral galaxies: can cooling in spiral shocks produce molecular clouds? *Monthly Notices of the Royal Astronomical Society*, 389(3), 1097–1110.

Ginzburg, V. L. & Syrovatsky, S. I. 1961, Origin of Cosmic Rays. *Progress of Theoretical Physics Supplement*, 20, 1–83. ADS Bibcode: 1961PThPS..20....1G.

Han, E., López-Valdivia, R., Mace, G. N., & Jaffe, D. T. 2023, Magnetic field measurements of low-mass stars from high-resolution near-infrared IGRINS spectra. *The Astronomical Journal*, 166(1), 4.

Kopparapu, R. K., Ramirez, R., Kasting, J. F., Eymet, V., Robinson, T. D., Mahadevan, S., Terrien, R. C., Domagal-Goldman, S., Meadows, V., & Deshpande, R. 2013, Habitable zones around main sequence stars: New estimates. *The Astrophysical Journal*, 765(2), 131.

Kumar, V., Dhaka, S. K., Hitchman, M. H., & Yoden, S. 2023, The influence of solar-modulated regional circulations and galactic cosmic rays on global cloud distribution. *Scientific Reports*, 13(1), 3707.

Mackey, J. 2022,. Interaction between massive star winds and the interstellar medium. arXiv:2211.08808 [astro-ph].

Melott, A. L. & Thomas, B. C. 2019, From Cosmic Explosions to Terrestrial Fires? *Journal of Geology*, 127, 475–481. ADS Bibcode: 2019JG....127..475M.

Pecaut, M. J. & Mamajek, E. E. 2013, Intrinsic Colors, Temperatures, and Bolometric Corrections of Pre-main-sequence Stars. *The Astrophysical Journal Supplement Series*, 208, 9. ADS Bibcode: 2013ApJS..208....9P.

Quispe-Huaynasi, F., Roig, F., McDonald, D. J., Loaiza-Tacuri, V., Majewski, S. R., Wanderley, F. C., Cunha, K., Pereira, C. B., Hasselquist, S., & Daflon, S. 2022,. High Velocity Stars in SDSS/APOGEE DR17. arXiv:2209.03560.

Rodgers-Lee, D., Vidotto, A. A., Taylor, A. M., Rimmer, P. B., & Downes, T. P. 2020, The galactic cosmic ray intensity at the evolving earth and young exoplanets. *Monthly Notices of the Royal Astronomical Society*, 499(2), 2124–2137.

Rodgers-Lee, D., Vidotto, A. A., Taylor, A. M., Rimmer, P. B., & Downes, T. P. 2021,. Stellar versus galactic: The intensity of energetic particles at the evolving earth and young exoplanets.

Svensmark, H. 2006, Cosmic rays and the biosphere over 4 billion years. *Astron. Nachr.*, 327(9), 871–875.

(Toward) Discovery of Life beyond Earth and its Impact
Proceedings IAU Symposium No. 387, 2026
H. Landt, M. Dominik & C. Oliver, eds.
doi:10.1017/S1743921325001504

Cognitive Astrobiology

Pauli E. Laine

University of Jyväskylä
email: pauli.erik.laine@gmail.com

Abstract. Traditionally the search for extraterrestrial intelligence (SETI) has been mostly a kind of engineering task, based on two assumptions: the existence of other technological culture somewhere in our Galaxy and that this culture is using radiotechnology that can be detected with our receivers. However, SETI endeavors could benefit more by expanding its scope into two-folded interdisciplinary research program, one exploring the nature of human, other animals, and artificial cognition, and the another exploring all possible and hypothetical ways and channels that potential ETI could use to manifest itself. This kind of new research program could be named cognitive astrobiology in order to better describe its research areas.

Keywords. SETI, Cognitive Science

1. Introduction

We don't know how unique creatures we are in the Universe. We don't even know how unique phenomenon life is in the Universe. But we know that humans are the only creatures on Earth that have created science and technology. We have languages that are capable of limitless expression and efficient communication. Although many animals behave clearly intelligently, no other animal can human linguistic utterances ((Impey C. Ed. 2010)). Is language the enhancer of our thinking or even enabler of our abstract symbolic cognition? Why such rich languages have evolved on humans only among all possible species and during billions of years of evolution? What does this mean for the possibility of creatures that are functionally equivalent of humans to exist somewhere else in the Universe ((Duner David. 2013)DunerDavid2014DunerDavid2011)?

Human evolution and cultural history are not uniform, linear evolution towards modern (western) civilization. Rather modern civilization is the product of cumulative information and skills recorded and transferred extrasomatically from different cultures and traditions throughout the ages. It is said, that stone age human had already all the cognitive abilities as modern-day humans. Co-evolution of cognition and language made it possible to permanently record and efficiently transfer ideas and information from human to human, from generation to generation. This led to co-evolution of brain and culture. How human cognitive abilities then differ from other animals? Is the difference only quantitative, not qualitative? Is the ability to learn and use language the only thing that distinguishes us from other higher animals in cognitive level? Are we just the first specie to achieve science and technology on Earth, or are we product of series of improbable coincidences?

2. Cognitive Astrobiology

Cognitive astrobiology is an early state research area combining two multidisciplinary research areas: cognitive science and astrobiology. One of the confluences of these two areas is the research of human specific cognitive abilities and the possibility of the emergence of similar abilities in somewhere else in the Universe.

The Search for Extraterrestrial Intelligence (SETI) has been searching for artificial signals for more than 50 years. During these years SETI research has assumed that (or at least is possible that) human- like intelligence and technology could evolve also somewhere else in the Milky Way. However, intelligence is controversial and vague concept.

Using single IQ value for instance does not talk much about logical, productive and creative minds. Cognitive astrobiology can clarify the concept of intelligence by searching the features of the most general, multiple realizable rational cognition, functionally equivalent of humans.

We do not know what possible evolutionary paths there to intelligent society and technological culture are. Does technology always come along when a certain point in social or cultural evolution has been passed? What environmental, physiological, and anatomical features are required for technical culture to emerge? In a way, this is a similar question to the origin of life. We have only one example of life and technological culture, and it is hard to imagine alternatives. Besides complex cognition, technological civilization requires complex manipulative organs too. Otherwise, we would end up with super-intelligent culture of e.g. octopus- or whale-like creatures that could be in some of the nearest exoplanets without us knowing anything about them.

3. Discussion

Fermi's paradox says that if there are other advanced civilizations in the Milky Way, then where are they? This is a serious question to any claim about how life would eventually evolve to something that could make it more observable than just biochemically. How meaningful then is the search for extraterrestrial intelligence? There are countless challenges to classic radio SETI; like that it relies on accidentally picking up signals leaked from other civilizations. It seems more likely, that advanced technological culture would probably limit such wasted energy. After all thinkable searching solutions are tested; we still don't know whether extraterrestrial civilizations exist. Justification for the search comes from the cost-effectiveness: we only must keep our eyes open for any anomalies.

Classical approach of cognitive science has an idea about multiple realizability which states that cognitive processes do not depend on the physical implementation of an intelligent system ((Fodor F. 1975)). This statement allows intelligence and mind to very different entities, e.g. robots, green slime and Boltzmann brains. While this cognitivism has been criticized for forgetting the biological and sociocultural sides of the mind, intelligence *per se* could be multiple realizable emergent property of different cognitive agents in different environments.

Astrobiology has been mainly focused on disciplines like biology and astronomy. However, if we want to add SETI, or just technosignature search into astrobiology research area, then we need also co-operation with cognitive science, neuroscience, comparative zoology, psychology, AI research, etc. to assess the possibility of multiple realizable intelligent creatures, functionally equivalent of humans, somehow manifesting themselves in our Galaxy. Cognitive astrobiology is searching for universal features that make up human cognition. For example, there is proposed universal grammar (UG) by Chomsky (N. 1966). Evolution in very different environments could produce similar cognitions to cope with the survival pressure from it. Brainstorming with researchers from multidisciplinary areas like cognitive science could produce some insight into what could be universal to all possible cognitions, on Earth, and in the Universe.

References

Chomsky, N. 1966 Cartesian linguistics. Harper & Row, Publishers

Dunér, David. 2013. "Introduction: Extraterrestrial Life and the Human Mind." In The History and Philosophy of Astrobiology: Perspectives on Extraterrestrial Life and the Human Mind, edited by David Dunér, Joel Parthemore, Erik Persson, and Gustav Holmberg, 1–25. Newcastle-upon-Tyne: Cambridge Scholars Publishing.

Dunér, David. 2014. "Interstellar Intersubjectvity: The Significance of Shared Cognition for Communication, Empathy, and Altruism in Space." In Extraterrestrial Altruism: Evolution and Ethics in the Cosmos, edited by Douglas A. Vakoch, 141–167. Berlin/Heidelberg: Springer.

Dunér, David. 2011. "Cognitive Foundations of Interstellar Communication." In Communication with Extraterrestrial Intelligence, edited by Douglas A. Vakoch, 449–467. Albany NY: State University of New York Press.

Fodor, F. 1975 The Language of Thought, MIT Press: London.

Impey, C. Ed. 2010 Talking about Life. Conversations on Astrobiology. Cambridge University Press: Cambridge, UK, pp. 143–164.

(Toward) Discovery of Life beyond Earth and its Impact
Proceedings IAU Symposium No. 387, 2026
H. Landt, M. Dominik & C. Oliver, eds.
doi:10.1017/S1743921324002096

New Optical Ground Stations for ETI Communications

Eamonn Ansbro

Space Exploration Ltd., Kingsland, Boyle, Co. Roscommon, F52E954, Ireland
email: kingslandobservatorio61@gmail.com

Abstract. Numerous programmes have been conducted to search for techno-signatures using optical wavelengths. PANOSETI provides the opportunity for a link and follow up to new infrared laser optical ground stations (OGS) in the event of first contact for ETI communications within our local neighborhood of the Solar System. The OGS has a transceiver communications system that can provide the opportunity for post-contact protocols to be initiated with a potential ETI.

Keywords. Optical ground station, ETI, optical laser, ETI communications, Solar System

---

Introduction

Numerous programmes have been conducted to search for techno signatures using optical wavelengths. A Panoramic All-sky All-time Near InfraRed and Optical Techno signature Finder.

(PANOSETI) provides the opportunity for a link and follow up to new infrared laser optical ground stations (OGS), in the event of first contact for ETI communications within our Solar System. PANOSETI is a dedicated SETI observatory that is being used between 350–1650 nm in search for nano-to milli-second transient events for optical wavelengths (Wright 2018). OGS currently uses discrete wavelengths at 800 nm,1064 nm 1550 nm, 1604 nm and have been successfully used with Meter-class ground telescopes that serve as receivers and transmitters (Ansbro 2024).

The opportunity of linking to current OGS are now in place and have the capabilities for communications within our Solar System. However, even if a ETI laser signal is detected by PANOSETI and confirmed by the OGS, there is the opportunity of a follow up with communications protocols within the infrared wavelengths. Post Detection contact protocols that can be operated on a 24 hour cycle, addressing any interrupted flow in communications.

Instrumentation

The site location is at 1500 metres,5000 feet altitude in Monte De Las Animas, Sierra Sur de Jaen, Andalucía. The first OGS will be operational from 2026 The first of new sites in Spain will support optical networks. The site has 300 clear nights. Sky quality SQM is 22.0; seeing resolution (DIMM) is 1 arc second. The optical telescopes range in size from half to one metre.

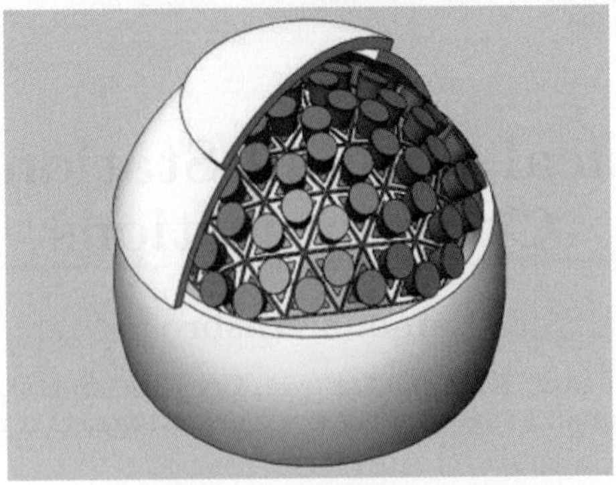

Figure 1. PANOSETI.

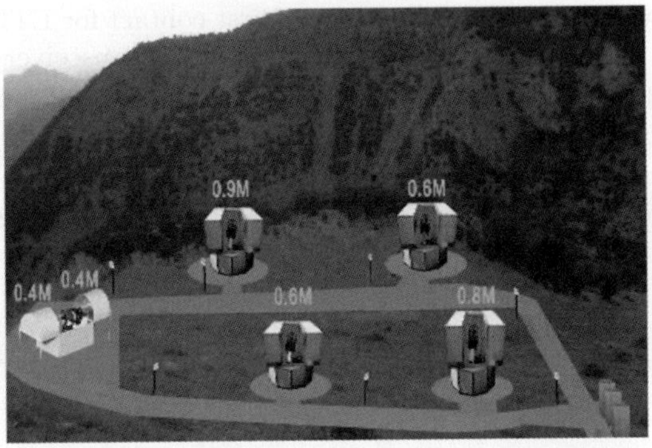

Figure 2. Optical Laser Ground Stations.

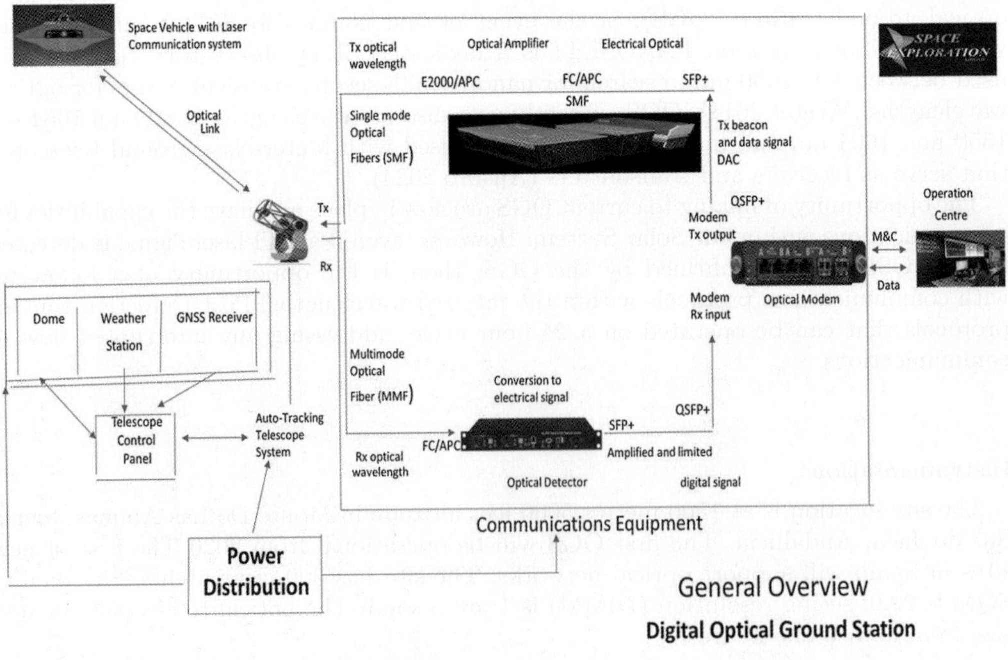

Figure 3. The design of an optical laser ground station.

Communications

The optical laser ground station requires a telescope and optical transceiver. The optical transceiver has a laser for transmission and receiving. One operational one metre class telescope is limited to a certain laser power needed for near distance space. Multiple one metre array or one large 5.5m telescope are a requirement for distances to Mars and beyond. See Figures 1 (PANOSETI) and 2 for telescope array (Ansbro 2024).

See Figure 3 for the design of an optical laser ground station.

Figure 4 shows one telescope for a Telescope Array that can creates one big telescope as pictured in Figure 5.

Figure 4. The individual telescope within the Telescope Array pictured in Figure 5.

Figure 5. Telescope Array that creates one big telescope.

References

Wright, S.A. et al; 2018, 'Panoramic optical and near-infrared SETI instrument: overall specifications and science program', Conference: Ground-based and Airborne Instrumentation for Astronomy VII, DOI: 10.1117/12.2314268.

Ansbro, E.; 2024, 'Optical laser satellite ground stations'. `www.spaceexplorationltd.com`.

(Toward) Discovery of Life beyond Earth and its Impact
Proceedings IAU Symposium No. 387, 2026
H. Landt, M. Dominik & C. Oliver, eds.
doi:10.1017/S174392132400200X

The challenging implications for Christian theology and ethics

Jacques Arnould

Centre national d'études spatiales, 2 place Maurice Quentin, F-75001 Paris
email: jacques.arnould@cnes.fr

Abstract. Until now, there has been no evidence of the existence of life forms other than on Earth. The philosophical, ethical and theological corpus that has been built up on this subject over several millennia is therefore based on a thought experiment or a posture of belief. Between dealing with a heresy and applying the precautionary principle, extraterrestrials represent a special opportunity to question the boundaries of our systems of thought, and sometimes to shake up dogmatism.

Keywords. astrobiology, Christian theology, ethics

Who goes there?

During this symposium devoted to the discovery of life beyond Earth, we asked ourselves several times how to define life. This question is as old as biology but also philosophy or ethics, since it is, in one way or another, at the origin of each of these disciplines.

To the various definitions proposed over these five days, explicitly or implicitly, I would like to add the following proposition, which is not a definition, but could be an essential property of the living being: a living being is capable of asking the question of the other. Who goes there? In my opinion, this property is necessary but not sufficient to declare a reality alive.

All living beings on our own planet, whatever their level of consciousness, ask themselves this question. Far from being an exceptional ability that would turn all living beings into budding or potential philosophers, it is a question of survival. The different ways of answering this question could constitute a method for distinguishing and classifying living beings.

Is it possible or even relevant to apply this proposition to the search for hypothetical existences beyond Earth? Is the ability to answer each other's questions a subject for investigation by astrobiologists? I leave it to them, of course, to judge the scientific value of my proposal.

However, I would like to use this characteristic of the living to address the subject of my talk: the theological and ethical implications of the possibility of discovering extraterrestrial life forms.

I must make it clear that my comments are those of a Christian theologian. I'm not ignoring the many other religious traditions that deal with our subject; out of respect for them, I'll only allude to them, and won't offer any reflections or speculations that might "engage" them†.

Philosophy and theology questioned

Albert the Great, a 13th-century theologian, had no hesitation in declaring that the question of the plurality of worlds is "one of the most astonishing and noble in Nature‡"; a statement that Arthur C. Clarke endorsed as "one of the supreme questions of philosophy§". Tracing the history of this subject over nearly three millennia and, if possible, across several cultures is a challenge that several authors have taken up in recent decades¶, demonstrating that, far from being merely a subject for Hollywood blockbusters, the existence of extraterrestrial life is the motif of a very ancient and serious reflection on the human condition.

Beliefs are no exception. I can testify to having been asked many times: "What will become of religions the day we discover life elsewhere than on Earth?"

Jill Tarter, who is often considered the "papess" of the Search for Extra Terrestrial Intelligence (SETI), is convinced that the assured existence of extraterrestrial intelligence would mean the end of monotheistic religions. According to the American astronomer, these religions would be challenged by extraterrestrial religious and philosophical systems - probably wiser than our own, she adds. "Their longevity is inconsistent with organized monotheistic religions typical of Earth, Tarter explains. Such religions are responsible for the longest lasting warfare and destruction we have witnessed‖." In other words, planet Earth will be saved from terrestrial religious violence by extraterrestrial science.

Ted Peters, a Lutheran theologian with a decades-long interest in extraterrestrial life, unidentified aerospace phenomena and UFOs, undertook an in-depth survey of representatives and members of seven religions or religious currents. The conclusion of his work is different from Tarter's statements: "Responses from persons self-identifying with one of seven major religious traditions report that they do not fear an impending collapse in their own religious belief system**" A very large majority of believers surveyed (over 80% of the 1300) explain that official confirmation of the discovery of an extraterrestrial civilization would not call into question either their own faith, or the religious tradition to which they belong. Even if the ways in which such a discovery is received and interpreted religiously and theologically vary according to denomination.

Peters notes that the most fundamentalist believers (in his survey, these are mainly Protestants and Muslims) appear to be the most vulnerable, both in their faith and

† As a very short bibliography, I propose : Andrew Davison, *Astrobiology and Christian Doctrine. Exploring the Implication of Life in the Universe*, Cambridge, Cambridge University Press, 2023; J örg Matthias Dettermann (dir.), *Islamic Theology and Extraterrestrial Life: New Frontiers in Science and Religion*, Londres, I. B. Tauris, 2024.

‡ Quoted by Steven J. Dick, *Plurality of Worlds. The Origin of the Extraterrestrial Life Debate from Democritus to Kant*,Cambridge/London/New York, Cambridge University Press, 1982, p.xx.

§ Arthur C. Clarke, *Report on Planet Three and Other Speculations*, New York, Signet Books, 1973, p. 79.

¶ Cf. Michael J. Crowe, *The Extraterrestrial Life Debate, 1750–1900*, Mineola (NY), Dover Publications, 1999 ; Steven J. Dick, *Plurality of Worlds. The Origin of the Extraterrestrial Life Debate from Democritus to Kant*, Cambridge/London/New York, Cambridge University Press, 1982 & *Life on Other Worlds. The 20th-Century Extraterrestrial Life Debate*, London/New York, 1998.

‖ Cf. Jill, Tarter, 'Implications of contact with ETI far older than humankind', in A. Tough (ed.), *When SETI succeeds: The impact of high-information contact*, Bellevue (WA), Foundations for the Future, 2000, p. 45.

** Cf. Ted Peters & Julie Froehlig, *The Peters ETI Religious Crisis Survey Report*, 2008, `http://www.counterbalance.org/etsurv/PetersETISurveyRep.pdf`.

in their tradition. One evangelical Protestant asserts that nothing in Christianity rules out the possibility of extraterrestrial intelligent life, while another claims that extraterrestrials do not exist as material creatures, but are the forms taken by Satan and his demonic creatures. The Mormons point out that extraterrestrial life is already part of their religion, while Buddhists explain that, according to their tradition, extraterrestrials are sentient beings like any others and possess the Buddha nature. Finally, the "myth" of extraterrestrial salvation seems to have little influence in religious circles.

Based on his survey, carried out with Julie Froehlig, Peters can therefore conclude that non-believers have a far more negative view of the consequences for religions of the discovery of proven extraterrestrial life than believers.

Indispensable from a sociological point of view, Ted Peters' work is not enough to remove the problematic nature of the question of the plurality of worlds, as Albert the Great and Arthur C. Clarke asserted. Personally, I had even go so far as to say that this question could be considered fundamentally heretical†!

I propose to deal with this subject in three parts:

- The threatened only son
- The three-body problem
- "You are not alone"

The only child syndrom

Common to all living beings on our planet, the question of the Other has taken a dramatic turn with the emergence of human consciousness. It is no longer simply a question of identifying what comes up against the biological boundaries of an individual or a group in order to develop the most appropriate response, but now of taking into account a new perspective, that offered by the imagination. When, standing on their two feet, our ancestors began to stare at the horizon, they wondered whether, beyond the savannah, over the hill, across the river or the sea, in the celestial world, other beings existed, whether similar to them or different from them. Would these strangers become friends, *alter egos* or enemies, servants or masters? And why not gods?

Different and even contradictory experience has been associated with this question of the Other: that of human solitude. Crushed by the spectacle of the celestial vault, surrounded by immense forests or vast deserts, relegated (as we now know) to the outskirts of some galaxy, humans have always been able to feel, to suffer the experience of solitude. They are alone to face up to telluric and cosmic forces, alone to face up to the challenges of life, of daily survival. How do we cope with such an dramatic experience?

In his book *Par-delà nature et culture* (*Beyond Nature and Culture*), Philippe Descola provides a detailed analysis of the responses of the human species, depending on whether the "physicalities" and "interiorities" of humans and non-humans are considered similar or different‡. The West, argues the French anthropologist, has objectified nature by positing a difference in interiorities, in other words, by setting humans apart from nature. By giving them the status of an only child.

This is the theological and anthropological meaning of one of the best-known passages in the *Book of Genesis*:

"Then God said: Let us make man in our image, after our likeness. And let them have dominion over the fish of the sea and over the birds of the heavens and over the livestock

† Cf. Jacques Arnould, Qui va là ? *L'humanité face à l'extraterrestre*, Bruxelles, Éditions de l'Académie Royale de Belgique, 2022 : ≪ La question extraterrestre serait-elle hérétique ? ≫.

‡ Cf. Philippe Descola, *Par-delà nature et culture*, Paris, Gallimard, 2005 ; *Beyond Nature andCulture*, Chicago, University of Chicago Press, 2013.

and over all the earth and over every creeping thing that creeps upon the earth†." Images of God, installed "at the summit" of God's creation, humans are his lieutenants on Earth.

Psychologists who have studied the situation of the only child don't necessarily paint a very flattering picture. In 1970, the Spanish-French psychoanalyst Julian de Ajuriaguerra wrote: "The only child is fragile, capricious, shy, tyrannical with his family, indolent; he has difficulty adapting to his peers and finding it hard to fit into a group; he frequently reacts in the form of disarray (withdrawal into fantasy life) or revolt‡."

Without claiming perfect concordance, we have to admit that the traits given here tothe only child bear some resemblance to those of the human being, as described bymonotheistic traditions, particularly Christian ones, and as they have been reproached§ .To overcome the disarray caused by the experience of solitude, to escape the feeling ofbeing the recluse of a cave (Plato) or the prisoner of a dungeon (Blaise Pascal), there'snothing like the affirmation of being the master of the domain whose power is conferred, validated by the Creator himself, our Divine Father.

The tragedy that threatens the only child, and the remedy he has found for hisdistress, is of course the arrival, the existence, even imagined or supposed, of anotherchild. The one-child syndrome is followed by the "cataclysm" (the word comes fromchild psychiatrist Marcel Rufo) of the arrival of a little brother or sister. Regression andjealousy are the experiences to which (former) only children are subjected. Biblical tradition alone is not lacking in examples of whim, jealousy, fear of losing a form ofprimacy, and concern for genealogical or racial purity¶.

So, what would happen to human beings if an extraterrestrial appeared?

A three-body problem

A similar cataclysm was experienced by believers at the end of the 19th century, at the time of the first scientific discoveries concerning the origin and history of the human species. Marcellin Boule, one of the first French prehistorians, summed up this crisis in a provocative phrase published in a daily newspaper, *Le Matin*, on December 26, 1908: "Isn't it infinitely more honorable to descend from a perfected ape than from a fallen angel‖ ?" He was addressing the Christians of his day, who refused to accept "prehistoric" man as their ancestor. Is this not the whim of an only child who insists on remaining perfect and refuses to recognize the identity of his ancestors as well as that of his cousins? The birth, development and persistence of creationist movements, the multiplication of debates and even trials, are enough to show that this problem is truly serious. They are probably the best illustrations of the troubles and worries caused to certain believers by extraterrestrials, but also of the theological interest of this subject. I'll summarize this interest with reference to celestial mechanics, precisely to the three-body problem.

† Cf. *Genesis* 1: 26 (English Standard Version).

‡ Cf. Julian de Ajuriaguerra, quoted by Jean-Pierre Almodovar, ≪Existe-t-il un syndrome du fils unique ? ≫, *Enfance*, 1973, tome 26, n°3-4, p. 235.

§ Cf. Lynn White, ≪ The Historical Roots of Our Ecologic Crisis ≫, *Science*, 155 (1967) 3767, p. 1203–1207 and Jacques Arnould, ≪ Lettre à Lynn White et à ceux qui s'en réclament ≫, *Pardès*, 2005/2, n°39, p. 211–219.

¶ To name but a few: the birthright exchanged for a plate of lentils between Esau and Jacob (*Genesis* 25:29), the apostle Peter's question to Jesus about the "fate" reserved for the apostle John (*John* 21:20), the Valladolid controversy over the status of the indigenous peoples of America, and so on

‖ The Swiss anatomist Edouard Claparède wrote something similar in an 1861 issue of the *Revue Germanique*: "We would rather be a perfected ape than an Adam degraded by the fall of biblical tradition"

Of course, I won't go into a scientific presentation of this problem, which we know has become a popular theme in fiction†; I'll just mention one of its main "components": the solutions to Newton's three-body equations of motion are essentially impossible to explain, as they are sensitive to initial conditions.

Far be it from me to put God into an equation or formula‡. However, if the believers take seriously the hypothesis of extraterrestrial life and, better still, extraterrestrial intelligence, they find themselves confronted with a three-body problem: is it possible to conceive of relationships that might exist between God, the believer and the extraterrestrial... and that take all three entities into account?

Until now, the relationship between God and earthlings has been the main and sometimes sole focus of theological treatises and liturgical books. So much so, in fact, that the mere idea of a third "body" modifying and perhaps even upsetting the relationship between God and his human creatures has led many Christian thinkers to deny the existence of extraterrestrial beings, and even to affirm their non-existence. The idea of God maintaining singular, original relationships with consciousnesses and intelligences other than those of could overshadow and even jeopardize the special and unique relationship between human creatures and their Creator§! It's not difficult to diagnose this religious malaise: it's a religious form of the syndrome of the only child who believes or claims to be unique... and who insists on remaining so.

We can go further. The mention and role of initial conditions in the scientific presentation of the three-body problem is of particular interest to philosophical and theological reflection. Indeed, long before the birth of science and the questions of the origin of the universe and the origin of life, religions showed an interest in the question of origins; indeed, they largely contributed to the elaboration and dissemination of a very rich mythology.

One of the main lessons to be learned from this long history of reflection and debate is that we must not reduce this question to just the beginning. Origin is not only "chronological" (when?), it also designates what constitutes the singular, perhaps unique, character of a reality. So it's hardly surprising that the discovery of others, and even alter egos, can influence our understanding of our own identity, our own originality, our own origin. How are we to understand the biblical affirmation that God created human beings "in his likeness"? Is it a likeness laid down at the beginning that time can only degrade? Or is it a likeness that has a history, that of the existence of each human being? Is it a likeness to be evolved, rather than preserved at all costs? Is it a resemblance whose originality lies in this history itself?

And how do I think about my own originality in relation to the other originalities also created by God? Should we think of it in terms of mediocrity, and follow the Copernican principle? Or should we think of it in terms of singularity, in the name of the relationship established between the Creator and each of his creatures?

This raises a question that philosophers and theologians have asked very early on in the history of human thought: how does God create? What place does He give to chance, randomness and initial conditions? Should we, as in the three-body problem, consider the world (or worlds?) as a reality sensitive to initial conditions, as deterministic chaos? But is it admissible that God, confessedly the master and organizer of chaos, could also be

† Cf. Liu Cixin, *The Three Body Problem*, New York, Tor Books, 2014.

‡ Cf. Michio Kaku, *The God Equation. The Quest for a Theory of Everything*, New York, Doubleday, 2021 ; Igor & Grichka Bogdanoff,*L'équation Dieu*, Paris, Éditions Grasset, 2019 ; Jose Rodriguesdos Santos, *La Formula de Dios*, Gravida, 2006.

§ The link between geocentrism and this one-child claim was dramatizedby Bertolt Brecht in the sixth scene of his play *Lebens des Galilei* (*Life of Galileo*, 1939), in which a very old cardinal prattles on: "I walk with a sure step on firm Earth, it is fixed, it is at the center of the universe, and I am at the center and the eye of the creator is fixed on me and me alone."

His creator? The mere mention of the possibility of extraterrestrial life calls into question the very foundations of religion.

We know that one of the reasons given for rejecting the plurality of worlds by philosophical and theological schools is precisely the refusal to leave any room, however small, for randomness. As just one example, I quote Thomas Aquinas in his *Summa Theologica*, where he argues against the plurality of worlds in the following terms:

"Only those who recognize no ordering wisdom, but believe in chance, can affirm that there are many worlds, following the example of Democritus, who said that this world, alongside countless other worlds, was the product of a chance meeting of atoms†." In other words, for the Christian theologian of the Middle Ages, since no place can be given to chance, randomness or indecision in divine creation, the hypothesis of a plurality of worlds must be ruled out.

Unfortunately, believers' desires cannot be confused with reality. And we must seriously consider extraterrestrial existences if we are not to deserve Arthur C. Clarke's "judgment": "A faith unable to survive an encounter with truth is not worth regretting‡."

"You are not alone!"

After inviting theologians to think afresh about the origin of the world or worlds, the extraterrestrial question also leads them to revisit a notion central to theologies based on the biblical tradition: the notion of covenant. Isn't this what Oliver Jenkins was inviting us to do when he addressed possible extraterrestrial existences?

In 2017, to mark the fortieth anniversary of the launch of *Voyager 1*, NASA organized a contest inviting people to write a message of less than sixty characters, intended to be transmitted by the interplanetary probe, the first Earth-built probe to leave our solar system. The message written by Oliver Jenkins, "a guy with Asperger's Syndrome" as he describes himself, was selected and read by one of the actors from the famous Star Trek TV series before being transmitted to *Voyager 1*: "We offer friendship across the stars. You are not alone!" The friendship offered by Jenkins echoes, it seems to me, the idea of covenant,another theological notion fundamental to biblical traditions since, even before introducing the covenants with Abraham, Moses and David, the Bible introduces thecovenant with Noah which, after the Flood, binds the Creator to all his creatures and notjust to a particular group of humans, to a chosen people :

"Then God said to Noah and to his sons with him : 'Behold, I establish my covenantwith you and your offspring after you, and with every living creature that is with you –the birds, the livestock and every beast of the earth with you, as many as came; it isevery beast of the earth. I establish my covenant with you, that never again shall allflesh be cut off by the waters of the flood, and never again shall be a flood to destroy theearth§.' "

Understood in this way, the idea of covenant has an ethical dimension: it not onlyprohibits God from destroying his creatures again, as the Flood did; this so-calledNoachic covenant also establishes a responsibility on the part of humans towards theother creatures that inhabit the Earth, in order to maintain order on Earth. Today, thisinterpretation is taken seriously by believers concerned with protecting the Earth'sbiosphere¶. Is it possible to extend this covenant to extraterrestrial beings and creatures‖?

† Thomas d'Aquin, *Somme théologique*, Ia, question 47, art.3.

‡ Arthur C. Clarke, *L'exploration de l'espace*, Paris, Denoël, 1953, p. 261.

§ *Genesis* 9, 8–11 (English Standard Version).

¶ 20 Cf. Pope Francis, *Encyclical Letter Laudate Si*, 2015. `https://www.vatican.va/content/francesco/en/encyclicals/documents/papa-francesco_20150524_enciclica-laudato-si.html`.

‖ Cf. Guy Consolmagno & Paul Mueller, *Would you baptize an Extraterrestrial ? Astronomiers' In-Box at the Vatican Observatory*, Image, 2014.

Christian morality is founded on faith in Christ, in his work of redemption; thus, to establish rules for Christian behavior towards food, the apostle Paul puts forward the following argument: "For if your brother is grieved by what you eat, you are no longer walking in love. By what you eat, do not destroy the one for whom Christ died†." The question then arises: for whom did Christ die? Is Christ's work of salvation limited to our planet? Or, on the contrary, those of divine creation, as astronomical science is discovering them today?

What Christian theology calls "cosmic Christology" has been thought of since theearliest days of Christianity, in a context that rejected the idea of a plurality of worlds,but also more recently, in the 20th century, when extraterrestrial life became ascientifically founded hypothesis. Pierre Teilhard de Chardin dared to put forward thehypothesis of a Christ who would incarnate several times in order to save severalworlds. In a letter dated February 24, 1918, the Jesuit scholar and chaplain to Frenchsoldiers in the First World War wrote: "Given that the cosmos is certainly inseparable, and that Christianity is no smallerthan the Cosmos, we must admit a certain 'polymorphic' manifestation of the cosmic Christ on various worlds, depending on the suitability of these worlds for integrationinto the celestial Universe. The human Christ would then be only one facet of thecosmic Christ. - Otherwise, Christ (if he only supported the Earth) would be smallerthan the World† ."

The confession of a cosmic Christ, a Christ larger than our human scales, is the foundation of a theology of creation that goes beyond the boundaries of our knowledge and especially those of our fears, the foundation of a theology that welcomes the idea of extraterrestrial existences.

Beyond extraterrestrials, Omega!

Teilhard de Chardin's thinking not only leads us to accept, in Christian terms, the hypothesis of extraterrestrial existences; it also invites us to consider a historical dimension to this possibility. It's not enough to ask ourselves about the initial conditions that enabled the emergence of living forms on our planet and elsewhere: as well as questioning the beginnings, the "Alpha" of the universe's history, we need to look at its "Omega", its end. In other words, to introduce the question of the finality of Creation, the existence of life and the universe. And, for the French Jesuit, the answer is the Christ Omega, the Great Attractor.

"Who goes there?": this question, which has followed human beings like a shadow since their emergence on the surface of the Earth, doesn't stop at the question of the plurality of worlds or the existence of extraterrestrial beings. It has never ceased to permeate philosophical, ethical and theological debates, because it leads each and every one of us to wonder about our own future. The question "Who goes there?" is thus closely associated with two other questions: "Why?" and "Who do you believe in?".

† *Romans* 14, 15 (English Standard Version)

† Cf. Pierre Teilhard de Chardin, *Lettres intimes à AugusteValensin, Bruno de Solages, Henri de Lubac et André Ravier*, Paris, Éditions Aubier Montaigne, 1974, note 5, 7, p. 40.

(Toward) Discovery of Life beyond Earth and its Impact
Proceedings IAU Symposium No. 387, 2026
H. Landt, M. Dominik & C. Oliver, eds.
doi:10.1017/S1743921324001960

International Law and the Search for Extraterrestrial Intelligence

Leslie I. Tennen, Esq.

Law Offices of Sterns and Tennen
9524 West Camelback Road, Suite C-130, Number 347
Glendale, Arizona U.S.A.
email: Ltennen@astrolaw.com

Abstract. The search for extraterrestrial life, especially intelligent life, is among the most profound scientific endeavors. Although the international community achieved consensus within a few years of the launch of Sputnik on legal principles fundamental to the peaceful exploration of outer space, including the Moon and other celestial bodies, those principles did not directly address the subject of SETI. Nevertheless, principles relevant to SETI are expressed in the law of outer space, including treaties in force as well as non-binding forms of international instruments. As the search for extraterrestrial intelligent beings has developed and a myriad of issues has arisen, the relevance of many of these legal principles has become apparent.

Keywords. space law; metalaw; SETI; extraterrestrial intelligence; international law; Outer Space Treaty; OST; protocols; COSPAR; planetary protection

1. Introduction

This article will discuss the background and general principles that comprise the framework of the law of outer space. Legal considerations on the use of Moon specifically for SETI will then be identified, followed by an examination of the issues that may arise in the exploration and use of Moon for SETI as well as for other purposes. Next is the question that is central to this Symposium: what happens in the event of a confirmed detection, with emphasis on the guidance provided by the Protocols developed by the International Academy of Astronautics, and the rules of Metalaw as articulated by Andrew Haley and Ernst Fasan.

2. Background and general principles of the legal framework

Space law is considered one of the newest branches of international law, with declarations of guiding principles by the United Nations and the adoption of multilateral treaties within a short time following the launch of Sputnik in 1957. However, lawyers and scholars (and science fiction authors) began considering the legal ramifications of the exploration and use of outer space and celestial bodies decades earlier. One of the founders of space law was Eugene Pepin of France, who also was at the forefront of the development of aviation law. The first stand-alone treatise on space law was written when rocketry was in its infancy ((Mandl 1932).† The American legal pioneer Andrew Haley discussed the subject of metalaw and relations with an alien intelligence in 1956 during the Congress of the International Astronautical Federation (IAF)(Haley 1956). By the early 1960's, Haley had been instrumental in the formation of the IAF, as well as the associated organizations the International Academy of Astronautics and the International

† The legacy of Mandl is tarnished by his overt antisemitism (Plavec 2011).

Institute of Space Law, and he published a landmark book on space governance (Haley 1963).

The United Nations was quick to respond in the wake of Sputnik, and within months of that launch the U.N. General Assembly adopted Resolution 1348 creating the Ad Hoc Committee On the Peaceful Uses of Outer Space (COPUOS), which was made a permanent committee in 1959 by Resolution 1472. Two years later COPUOS drafted, and the General Assembly approved, Resolution 1721, titled "International co-operation in the peaceful uses of outer space" which, *inter alia*, established a procedure for the registration of objects launched into space. This was followed in 1963 by Resolution 1962, the "Declaration of Legal Principles Governing the Activities of States in the Exploration and Use of Outer Space." This Declaration set forth some fundamental principles of space law, including that activities in space shall be conducted in accordance with international law, and that outer space, including the Moon and other celestial bodies, is not subject to national appropriation.

General Assembly Resolutions can be important statements of international sentiment but are not legally binding on states. COPUOS addressed this by developing five treaties which incorporated, refined, and elaborated on the principles previously approved in General Assembly resolutions. The first of these binding instruments was the Treaty on Principles Governing the Activities of States in the Exploration and Use of Outer Space, Including the Moon and Other Celestial Bodies (OST)†, which entered into force in 1967 and continues to stand as the essential charter for the activities of mankind in space. In 1968 the Agreement on the Rescue of Astronauts, the Return of Astronauts, and the Return of Objects Launched Into Outer Space (RRA)‡ was concluded, followed by the Convention on International Liability for Damages Caused by Objects Launched Into Outer Space (LC)§ in 1972, the Convention on Registration of Objects Launched into Outer Space (RC)¶ in 1976, and the Agreement Governing the Activities of States on the Moon and Other Celestial Bodies (MA).‖ Although the MA was opened for signature in 1979, several of its provisions were not met with universal acceptance, and it was not until 1984 that the requisite number of states had ratified the instrument necessary for it to enter into force.

The resolutions adopted by the General Assembly, and the treaties developed through COPUOS, were products of the cold war, and the OST, in particular, reflect an accommodation to the interests of the East and the West. For example, article VI of the OST provides that non-governmental entities are legally able to conduct activities in space, but the appropriate state is obligated to authorize and continuously supervise, and is internationally liable for, their private sector activities.

The treaties adopted subsequently to the OST defined, developed, and provided guidance to the concepts contained in the OST in relation to their specific subject matter. However, the treaties have overlapping but different constituencies, for example, as of January 1, 2024, the OST has been ratified by 114 countries, and signed by an additional 23 states, but the MA has been ratified by only 17 states and signed by an additional four.∗∗ Moreover, none of the treaties have provisions that expressly refer to the search for extraterrestrial intelligence (SETI), or extraterrestrial intelligent life inso many words.

† `https://www.unoosa.org/pdf/gares/ARES_21_2222E.pdf`

‡ `https://www.unoosa.org/pdf/gares/ARES_22_2345E.pdf`

§ `https://www.unoosa.org/pdf/gares/ARES_26_2777E.pdf`

¶ `https://www.unoosa.org/pdf/gares/ARES_29_3235E.pdf`

‖ `https://www.unoosa.org/pdf/gares/ARES_34_68E.pdf`. Article 1, paragraph 1 of the MA provides that the treaty applies to other celestial bodies within the solar system, other than the Earth, unless specific legal norms enter into force for any particular bodies.

∗∗ Saudi Arabia had been a party to the MA but signed the Artemis Accords and withdrew from the MA effective 2024.

The RRA and RC only have tangential applicability to SETI activities, and the LC may have potential but very attenuated applicability.†

While not expressly mentioning the subject, both the OST and MA set forth principles that are relevant to SETI. The treaties generally are considered to apply to rockets and payloads launched into orbit and beyond. As such, they clearly are applicable to instruments such as telescopes, robotic craft and other objects launched into space, as well as to activities conducted on the Moon. They also would apply to active messaging to a possible extraterrestrial civilization by sending a tangible, physical item into space, like the record of messages and images attached to the Voyager spacecraft.† The applicability is less clear, however, for intangible signals and messages sent from Earth, such as the Arecibo experiment.† Ground-based segments of these activities would be subject to national laws and not necessarily international space law.

The search for extraterrestrial intelligence generally is discussed in terms of detecting an active as opposed to an extinct civilization but cannot exclude the possibility of a past ETI. While it is probably unlikely that probes within the solar system will discover evidence of an active extraterrestrial civilization, whether indigenous or from elsewhere (*e.g.*, 2001 A Space Odyssey), astrobiological studies search for evidence of present or past life or the precursors thereof, not restricted to microbes.‡ Several articles of the OST and MA are applicable to SETI activities conducted in space, including on or orbiting a celestial body, in Earth orbit or other orbits such as helio and LaGrange halo orbits, or sent into deep space.

Provisions of the OST and MA which may be relevant to these activities are as follows:

OST Article I:

The exploration and use of outer space, including the Moon and other celestial bodies, shall be carried out for the benefit and in the interests of all countries, irrespective of their degree of economic or scientific development, and shall be the province of all mankind.

Outer space, including the Moon and other celestial bodies, shall be free for exploration and use by all States without discrimination of any kind, on a basis of equality and in accordance with international law, and there shall be free access to all areas of celestial bodies.

There shall be freedom of scientific investigation in outer space, including the Moon and other celestial bodies, and States shall facilitate and encourage international co-operation in such investigation.

The MA echoes these principles, and in article 4 adds that states shall give due regard for the interests of present and future generations, and in article 11 that the Moon and its natural resources are the common heritage of mankind. The MA also provides, in article 1 paragraph 3, that it does not apply to extraterrestrial materials which reach the surface of the Earth by natural means. Would an alien radio signal or artifact found on the Earth or Moon be considered as"extraterrestrial materials"?

OST Article II:

Outer space, including the Moon and other celestial bodies, is not subject to national appropriation by claim of sovereignty, by means of use or occupation, or by any other means.

There is growing pressure on the non-appropriation principle as more states conduct activities in space with increasing competition to occupy specific locations for purposes

† In addition to the treaties, COPUOS has approved statements of principles for particular subjects, including remote sensing, direct satellite broadcasting, nuclear power sources, and international cooperation, which are commonly referred to as "soft law" instruments.

† See `https://voyager.jpl.nasa.gov/golden-record/`.

† See `https://www.seti.org/seti-institute/project/details/arecibo-message`.

‡ See *infra* for a discussion of the COSPAR Planetary Protection Policy.

that may be unrelated to SETI, including commercial ventures. Conversely, there is an expanding recognition of the need to exclude or limit activities in specific areas that could be uniquely beneficial for SETI and other scientific endeavors (which are discussed *infra*). The MA provides in article 9 that states shall use only that area of the Moon which is required for the needs of the station. The term "station" is not defined in the MA, while articles 8, 10, 11, 12, and 15 refer to "facilities," "stations" and "installations." In light of the purposes of the MA, as well as the substantive articles, the limitations on the use of areas should extend to facilities and installations in addition to stations.

OST Article III:

States Parties to the Treaty shall carry on activities in the exploration and use of outer space, including the Moon and other celestial bodies, in accordance with international law, including the Charter of the United Nations, in the interest of maintaining international peace and security and promoting international cooperation and understanding.

OST Article IV paragraph 2:

The Moon and other celestial bodies shall be used by all States Parties to the Treaty exclusively for peaceful purposes. . . .

OST Article V paragraph 3:

States Parties to the Treaty shall immediately inform the other States Parties to the Treaty or the Secretary-General of the United Nations of any phenomena they discover in outer space, including the Moon and other celestial bodies, which could constitute a danger to the life or health of astronauts.

OST Article VI:

States Parties to the Treaty shall bear international responsibility for national activities in outer space, including the Moon and other celestial bodies, whether such activities are carried on by governmental agencies or by non-governmental entities, and for assuring that national activities are carried out in conformity with the provisions set forth in the present Treaty. The activities of non-governmental entities in outer space, including the Moon and other celestial bodies, shall require authorization and continuing supervision by the appropriate State Party to the Treaty. . . .

OST Article VII:

Each State Party to the Treaty that launches or procures the launching of an object into outer space, including the Moon and other celestial bodies, and each State Party from whose territory or facility an object is launched, is internationally liable for damage to another State Party to the Treaty or to its natural or juridical persons by such object or its component parts on the Earth, in air space or in outer space, including the Moon and other celestial bodies.

OST Article VIII:

A State Party to the Treaty on whose registry an object launched into outer space is carried shall retain jurisdiction and control over such object, and over any personnel thereof, while in outer space or on a celestial body. Ownership of objects launched into outer space, including objects landed or constructed on a celestial body, and of their component parts, is not affected by their presence in outer space or on a celestial body . . . .

OST Article IX:

States Parties to the Treaty shall be guided by the principle of cooperation and mutual assistance and shall conduct all their activities in outer space, including the Moon and other celestial bodies, with due regard to the corresponding interests of all other States Parties to the Treaty. States Parties to the Treaty shall pursue studies of outer space, including the Moon and other celestial bodies, and conduct exploration of them so as to avoid their harmful contamination and also adverse changes in the environment of the Earth resulting from the introduction of extraterrestrial matter and, where necessary, shall adopt appropriate measures for this purpose. . . .

The MA, in article 7, clarifies this obligation and provides:

States Parties shall take measures to prevent the disruption of the existing balance of its environment, whether by introducing adverse changes in that environment, by its harmful contamination through the introduction of extra-environmental matter or otherwise.

MA Article 7 also provides that states:

shall report to other States Parties and to the Secretary-General concerning areas of the Moon having special scientific interest in order that, without prejudice to the rights of other States Parties, consideration may be given to the designation of such areas as international scientific preserves for which special protective arrangements are to be agreed upon in consultation with the competent bodies of the United Nations.

OST Article IX (continued):

If a State Party to the Treaty has reason to believe that an activity or experiment planned by it or its nationals in outer space, including the Moon and other celestial bodies, would cause potentially harmful interference with activities of other States Parties in the peaceful exploration and use of outer space, including the Moon and other celestial bodies, it shall undertake appropriate international consultations before proceeding with any such activity or experiment. A State Party to the Treaty which has reason to believe that an activity or experiment planned by another State Party in outer space, including the Moon and other celestial bodies, would cause potentially harmful interference with activities in the peaceful exploration and use of outer space, including the Moon and other celestial bodies, may request consultation concerning the activity or experiment.

The MA, in articles 8 and 15 obligates states to participate in requested consultations.

OST Article XI:

States Parties . . . agree to inform the Secretary-General of the United Nations as well as the public and the international scientific community, to the greatest extent feasible and practicable, of the nature, conduct, locations and results of such activities. On receiving the said information, the Secretary-General of the United Nations should be prepared to disseminate it immediately and effectively.

The MA in article 5 clarifies and defines the required disclosures to include the time, purposes, locations, orbital parameters, and duration of the mission, which are to be made as soon as possible after launching. Scientific results shall be furnished upon completion of the mission, and at thirty-day intervals for missions lasting more than 60 days, with significant additions for missions of 6 months or more.

The MA has one provision that directly references extraterrestrial life. Article 5, paragraph 3 provides: In carrying out activities under this Agreement, States Parties shall promptly inform the Secretary-General, as well as the public and the international scientific community, of any phenomena they discover in outer space, including the Moon, which could endanger human life or health, as well as of any indication of organic life.

OST Article XII:

All stations, installations, equipment and space vehicles on the Moon and other celestial bodies shall be open to representatives of other States Parties to the Treaty on a basis of reciprocity. Such representatives shall give reasonable advance notice of a projected visit, in order that appropriate consultations may be held and that maximum precautions may be taken to assure safety and to avoid interference with normal operations in the facility to be visited.

The MA, in article 15, paragraph 2, eliminates the condition of reciprocity.

Finally, it is worth noting that article 11 of the MA, in addition to declaring the Moon and its natural resources to be the common heritage of mankind, calls on states to "undertake to establish" an international regime, including appropriate procedures, to govern the exploitation of the natural resources. One of the purposes of this international regime is to ensure an equitable sharing by all States Parties in the benefits derived from lunar resources.

3. Legal considerations on the use of the Moon specifically for SETI

The farside of the Moon is a unique location to conduct SETI. This orb made of rock is a natural shield blocking radio waves created by human activities on the Earth and environs. In 1974 the International Telecommunications Union recognized a protected shielded zone on and above the lunar farside to prevent radio-frequency interference. This was reconfirmed as recently as 2023 during the World Radiocommunication Conference in Dubai.

During the 1990's French astronomer Jean Heidman proposed that the crater Saha on the lunar farside be set aside for dedicated SETI and other scientific uses (Heidman 2000). After Heidman's untimely passing, the mantel of this endeavor was accepted by noted Italian mathematician and astronomer Claudio Maccone, who modified the proposal to expand the designated set aside area from a singular crater to a "Protected Antipode Circle." This PAC would consist of an area with a diameter extending from 30 degrees north to 30 degrees south latitudes and comprise approximately one-third of the farside. Near the center point of the PAC is the crater Daedalus, 80 km in diameter, which has been identified as a unique site for an observatory (Maccone 2008).

In 2023 the International Academy of Astronautics created the Lunar Farside Committee, under the chairmanship of Dr. Maccone. The goal of this Committee is to obtain recognition by the international community of defined areas of the Moon to be set aside for scientific use and exploration, and for an agreement of a whitelist of approved frequencies and a blacklist of frequencies that are forbidden for use by spacecraft flying above the lunar farside. His preferred form of this recognition and agreement is a new treaty (Maccone 2024).

There are several structural impediments to drafting and adopting a new treaty. The five existing space treaties were all developed through COPUOS, which operates by consensus. Thus, any individual state member of the Committee can prevent the approval of a new treaty text. COPUOS has not drafted a proposed new treaty since the MA in 1979. After approval by COPUOS, the text of proposed treaties is presented to the General Assembly to be approved and opened for signature. This procedure, even if successful, can take a considerable amount of time.

Moreover, the signature and ratification process for the requisite number of states for the instrument to enter into force will further extend the time frame. In the case of the MA, this signature and ratification process took five years. Extended delay in resolving these issues can negatively impact scientific investigations on the lunar farside as activities, particularly by commercial interests, multiply and impact the pristine environment.

Since the drafting of the MA, COPUOS has been able to achieve consensus on a number of soft law instruments. This is an alternative means of achieving international agreement and is more flexible and adaptable than a treaty and the treaty making process but does not result in binding obligations. Another alternative, with particular relevance to whitelisting or blacklisting frequencies, is to utilize the structure of the ITU, which can adopt radio regulations that are enforceable. The ITU is also in a better position to determine whether the whitelist or blacklist should be temporary or variable, and to modify the lists to respond to changing circumstances.

Implementation of a proposal for a territorial set aside on the lunar farside presents several legal issues. The non-appropriation principle in Article II of the OST prevents unilateral action by a state or a small group of states to deny access by others to areas of celestial bodies. The Artemis Accords take a different path and call for the imposition of so–called safety zones around activities on the Moon in Section II paragraph 7. The concept of safety zones derived from earlier proposals for "keep out" zones, which were an obvious violation of Article II. The defining of areas as "safety zones," in and of itself, does not violate the non-appropriation principle provided that it merely gives notice by a state that activities by other entities in the designated areas could cause interference with their own activities, and is not a predicate for denying other states from free access to the area. Such a notice, on the other hand, would be a predicate for the states to engage in consultations as an implementation of the principles of non-interference, cooperation, mutual assistance, and due regard as provided in Article IX of the OST.

It should be noted that interference can be intentional but also unintentional as an incidental result of activities that create dust, vibrations, electromagnetic interference, and other impediments. It also should be noted that the MA, in article 7, paragraph 3, provides for a form of internationally sanctioned keep out zones by recognizing the possibility of designating: areas of the Moon having special scientific interest . . . as international scientific preserves for which special protective arrangements are to be agreed upon in consultation with the competent bodies of the United Nations.

4. Exploration and use of Moon for SETI and other purposes

The Moon has unique attributes beneficial to scientific investigations and explorations in addition to SETI including cosmology, astrobiology, and planetary defense. The focus of discussions has revolved around the concept of sustainability. For example, the IAU has drawn attention to "low-frequency radio, infrared, and gravitational wave detections, which cannot be matched on the Earth or in Orbit," and stated that "Many commercial and national exploration activities are incompatible with the scientific need to be free of noise and interference." The IAU determined that "Only $\sim$3 sites $>$200 km dia., with smooth terrain on farside - craters Mendeleev, Korolov, and Mare Moscoviensis" are particularly suitable for telescopes for cosmological studies, and that there are "Only a few 50 km-scale cold traps" identified at the poles which are permanently shadowed regions.

States are rushing to conduct missions to the lunar south pole but are exercising their right to conduct exploration and use without any overarching coordination. An expanding roster of public and private entities are conducting these activities, creating the potential for overlapping and conflicting activities and uses of areas of the Moon. In addition to scientific explorations are commercial missions to explore and use the Moon, including resource collection and extraction activities.

Among the most troublesome activities are commercial missions to take and deposit novelty and other otiose payloads on the Moon. These useless and gimmicky items include human and animal remains, assorted miscellaneous trinkets, advertisements, and publicity stunts, which would be called and considered litter if left in most places on Earth. One

- confirmation of detection should be established by "the consensus of the other investigators involved and to a degree of certainty judged by the discoverers to be credible that a signal or other evidence is due to extraterrestrial intelligence"
- the discoverer has the privilege of making the first announcement of the discovery, and has an obligation to make a full and complete report in an open manner to the public, the scientific community, and Secretary General of the UN, and to submit a formal report to the IAU
- data bearing on the evidence of detection and verification should be recorded, stored permanently, and made available to observers and the international scientific community
- appropriate frequencies should be protected through the facilities of the ITU if detection is in the form of electromagnetic signals
- a Post Detection Task Group under the auspices of the IAA SETI Permanent Study Group [now known as the SETI Committee] is to be established to assist and support the scientific and public analysis
- signatories to the Protocols will not respond without first seeking guidance and consent of a broadly representative international body, such as the UN

It was recognized almost immediately after they were adopted that the Protocols would need further updating. A significant omission is any consideration of the Internet including social media (Sterns 2016) . Since the original Declaration and the Draft Reply Declaration were drafted there has been a revolution in media, with virtually instantaneous world-wide distribution of information and disinformation on devices carried by billions of people in their pockets. One consequence of the Internet age is that scientists can become instant celebrities but also instant targets for doxxing, swatting and other abusive conduct. The potential for misleading or deceptive information to reach the global public has mushroomed in recent years with the refinements in artificial intelligence and deep fakes. The IAA SETI Committee has been engaged in updating the Protocols, but the process has been difficult and slow, and the Post Detection Task Group has yet to be formed. Post-detection studies and activities have recently been taken up by other entities, notably the St. Andrews Post-Detection Hub and the IISL SETI Working Group.

6. Confirmed Contact – Metalaw and Relations With Alien Intelligence

Metalaw is that body of legal principles that will apply to relations between humankind and an intelligent extraterrestrial civilization or civilizations. In 1748 Baron Charles de Secondat Montesquieu discussed relations between different intelligent beings in *De l'espirit des lois* (The Spirit of Law). He identified this relationship as a legal relationship that is subject to laws that exist independent of the beings. Andrew G. Haley introduced the term "metalaw" and stated that metalaw is based on a moral foundation. He observed that virtually every religion has a form of the "Golden Rule" – treat others as you want to be treated. But he also observed that to treat an intelligent extraterrestrial being according to how we want to be treated is to impose our anthropocentric will in disregard of the needs, wants and desires of the ET. Haley thus derived the Golden Rule of Metalaw: treat intelligent ETs as they would have you treat them (Haley 1956).

Ernst Fasan expanded on the work of Haley in the seminal book *Relations with Alien Intelligences The Scientific Basis of Metalaw* (Fasan 1970). Fasan examined metalaw relative to the theory of natural law, and applied the construct of Immanuel Kant's Categorical Imperative -a rule of conduct that is absolute and not dependent on the actor, the validity of the claim, or any desired outcome. According to Fasan, this is a standard of rationality that gives rise to moral requirements, and these natural law imperatives should be deduced and detected by any and (eventually) every ET civilization.

Fasan further postulated that any ET race† with whom human contact may be possible must have these characteristics: life; intelligence; detectability (by the other race) and a means of communication; three-dimensionality; and the will to live. These characteristics enable any intelligent race to arrive at the essential rules of metalaw to govern the relationship with another intelligent race. It is implicit in metalaw that both parties intend to engage in communication and enter into a relationship.

Based on the foregoing, Fasan identified and articulated 11 Rules of Metalaw:

(1) No partner of Metalaw may demand an impossibility
(2) No rule of Metalaw must be complied with when compliance would result in the practical suicide of the obligated race
(3) All intelligent races of the universe have in principle equal rights and values
(4) Every partner of Metalaw has the right of self-determination
(5) Any act which causes harm to another race must be avoided
(6) Every race is entitled to its own living space
(7) Every race has the right to defend itself against any harmful act performed by another race
(8) The principle of preserving one race has priority over the development of another race
(9) In case of damage, the damager must restore the integrity of the damaged party
(10) Metalegal agreements and treaties must be kept
(11) To help the other race by one's own activities is not a legal but a basic ethical principle

Haley and Fasan provide an important foundation for the further development of metalegal principles. However, Haley's Golden Rule and Fasan's 11 Rules of Metalaw are subject to all the criticisms of their underpinnings in natural law theory, Kant's Categorical Imperative, and the reliance on morality as a motivator for human action (Korbitz 2010). Application of these rules can result in outcomes that may be immoral, unacceptable, or otherwise contrary to the intent and purpose of Metalaw (Sterns 2004). The rules developed by Haley and Fasan are altruistic and aspirational, but not descriptive of human history and interactions between cultures (Lyall 2009, pp. 556–58).

The subject of Metalaw will continue to be a source of reflection and elaboration by legal scholars and social scientists alike (Bohlander 2023; Bohlmann 2016; Harrington 2019; Vakoch 2014). The advent of artificial intelligence presents a new element to the discussion, as an extraterrestrial intelligence may not be biological in nature, but could be a machine. This raises a plethora of new considerations, including whether the AI is a being with rights to be respected, or whether it is simply a conglomeration of hardware and programming with no inherent rights that can be disassembled, modified, and damaged or destroyed at will. The answers to these questions necessarily will depend on the definitions of life and consciousness - concepts which are discussed in other submissions to this Symposium.

References

Artemis Accords 2020, `https://www.nasa.gov/wp-content/uploads/2022/11\penalty-\@M/Artemis-Accords-signed-13Oct2020.pdf?emrc=653a00`.

Bohlander, M. 2023, Contact with Extraterrestrial Intelligence and Human Law: The Applicability of Rules of War and Human Rights.

Bohlmann, U. 2016, Anthropomorphism in the Search for Extra-terrestrial Intelligence, presented to the 67th International Astronautical Congress.

† The terminology in this section is Fasan's.

Cockell, C.S., Horneck, G. 2004, A Planetary Park System for Mars, Space Policy 20, 291
Committee on Peaceful Uses of Outer Space (COPUOS), Report 2019, U.N. Doc. A/74/20, `https://www.unoosa.org/oosa/en/oosadoc/data/documents/2019/a/a7420_0.html`.

COSPAR 2021, Policy on Planetary Protection, `https://cosparhq.cnes.fr/assets/uploads/2021/07/PPPolicy_2021_3-June.pdf`.

Fasan, E. 1970, Relations With Alien Intelligence The Scientific Basis of Metalaw, *text reproduced in* Sterns, P.M., Tennen, L.I., eds. 2016, Private Law, Public Law, Metalaw and Public Policy in Space, A Liber Amicorum in Honor of Ernst Fasan.

Haley, A.G. 1956, Space Law and Metalaw.

Haley. A.G. 1963, Space Law and Government.

Harrington, A. 2019, SETI and International Space Law, presented to the 70th International Astronautical Congress.

Heidmann, J. 2000, Recent Progress on the Lunar Farside Crater Saha Proposal, Acta Astronautica, 46, 661.

IAU 2024, Statement delivered to COPUOS Scientific and Technical Subcommittee, `https://www.unoosa.org/documents/pdf/copuos/stsc/2024/Statements/11_IAU.pdf`.

Korbitz, A. 2010, The Limits of Metalaw and the Need for Further Elaboration, presented to the 61st International Astronautical Congress.

Lyall, F., Larsen, P.B. 2009, Space Law: A Treatise.

Maccone, C. 2008, Protected Antipode Circle on the Farside of the Moon, Acta Astronautica, 63, 110.

Maccone, C. 2024, "Moon Farside Science" Treaty, presentation to COPUOS Scientific and Technology Subcommittee, `https://www.unoosa.org/documents/pdf/copuos/stsc/2024/ListTechnicalPresentations/2024_STSC_technical_presentations_slides/f4_-\/-_v35_Maccone_-_COPUOS_2024_Presentation_during_January_or_February_2024._5.pdf`.

Mandl, V. 1932, Das Weltraum-Recht: Ein Problem der Raumfahrt (Space Law: A Problem of Space Travel)(in German), translation by NASA, `https://ntrs.nasa.gov/api/citations/19850008388/downloads/19850008388.pdf`.

Nygren, B., President of the Navajo Nation (2024), Statement on Human Remains to be sent to Moon Despite Call for Consultation, https://opvp.navajo-nsn.gov/statement-on-human-remains-to-be-sent-to-moon-despite-call-for-c onsultation/.

Plavec, M. 2011, Beginnings of Rocket Development in the Czech Lands (Czechoslovakia), Acta Astronautica 69, 905.

Sterns, P. M. 2004, Metalaw and Relations with Intelligent Beings Revisited, Space Policy 20(2), 123.

Sterns, P.M., Tennen, L.I. 2016, Social Media and the SETI Post Detection Declaration of Principles, presented to the 67th International Astronautical Congress.

Tennen, L.I. 2024, in: P.J. Blount, T. Masson-Zwaan, R. Moro-Aguilar, K.-U. Schrogl (eds.), The Promotion of Space Sustainability Through National Licensing Regimes, Proceedings of the International Institute of Space Law 2022, p. 283.

Vakoch, D.A. ed. 2014, Extraterrestrial Altruism, Evolution and Ethics in the Cosmos.

(Toward) Discovery of Life beyond Earth and its Impact
Proceedings IAU Symposium No. 387, 2026
H. Landt, M. Dominik & C. Oliver, eds.
doi:10.1017/S1743921325000274

Life as a cosmic phenomenon facing human culture

Martin Dominik[1], István Praet[2], Erik Persson[3], Ricky J. Lee[4], Olli-Pekka Vainio[5], Daniela de Paulis[6], Perig Pitrou[7] and Shoaib Ahmed Malik[8]

[1]University of St Andrews, St Andrews Centre for Exoplanet Science (StA-CES)
School of Physics & Astronomy, North Haugh, St Andrews, KY16 9SS, United Kingdom
email: md35@st-andrews.ac.uk

[2]Durham University, Department of Anthropology
Dawson Building, South Road, Durham, DH1 3LE, United Kingdom
email: istvan.praet@durham.ac.uk

[3]Lund University, Department of Philosophy
LUX, Helgonavågen 3, Building B, 223 62 Lund, Sweden
email: erik.persson@fi.lu.se

[4]Faculty of Law, University of Canterbury, Christchurch 8041, New Zealand
email: ricky@rickyjlee.com

[5]University of Helsinki, Faculty of Theology
Fabianinkatu 33, 00014 Helsingin Yliopisto, Finland
email: olli-pekka-vainio@helsinki.fi

[6]SETI Institute, SETI Artist in Residency Program
Mountain View, CA 94043, United States of America
email: selavyrose@gmail.com

[7]Research Professor, CNRS, Maison Française d'Oxford, United Kingdom
Head of the Team "Anthropology of Life", Collège de France, PSL University, France
email: perig.pitrou@college-de-france.fr

[8]University of Edinburgh, New College
School of Divinity, Mound Place, EH1 2LX, Edinburgh, United Kingdom
email: shoaib.malik@ed.ac.uk

Abstract. We present perspectives from six panellists on "Life as a cosmic phenomenon facing human culture", contrasting our experience and knowledge of life as found on Earth with the vastness of the Universe and the fact that Earth-centric and/or anthropocentric views have repeatedly proven untenable. How does an outwards view of projecting Earth-based experience into the cosmos combine with the inwards view of the potential detection of life beyond Earth telling us who we are?

Keywords. Extraterrestrial Intelligence, Astrobiology, Sociology of astronomy

1. Introduction

While the detection of planets orbiting stars other than the Sun provided new perspectives and context on planet Earth, our home still remains the only place within the vast Universe where life is known to us to exist. This means that as far as evidence goes, life remains an Earth-bound phenomenon. This much contrasts with our understanding that the fundamental laws of Nature should be universal. Our observations and exploration of the Cosmos support physics to be universal and chemistry to be universal, but is biology universal as well? The plurality of worlds has been postulated in human culture and expressed in science fiction, but we are still just at the beginning of scientifically putting that hypothesis to the test. On the other hand, evidence could be found as early as tomorrow.

We invited six panellists to share their views, providing perspectives based on their various expertise and experience, namely Erik Persson, Ricky Lee, Olli-Pekka Vainio, Daniela de Paulis, Perig Pitrou, and Shoaib Ahmed Malik. The panel session was moderated by Martin Dominik and István Praet, and it involved two key lead questions, which refllect an outwards and an inwards view, respectively:

(1) As we shift our view from life being an Earth-bound phenomenon to life being a cosmic phenomenon, are our existing perceptions fit for purpose, do we need to expand these, or do we need to rethink the foundations? Are we ready for the unknown and unexpected?
(2) What would the detection of extraterrestrial life tell us about ourselves? Does it matter, and why would we care?

These two questions are discussed in Sects. 2 and 3, respectively, while Sect. 4 captures further debate prompted by questions from the plenum. We deliberately provide a collection of perspectives rather than some consensus opinion.

2. Looking outwards: Life beyond Earth?

2.1. *Perspective I: Erik Persson*

Are our existing perceptions fit for shifting our view from life being an Earth-bound phenomenon to life being a cosmic phenomenon? In one sense, most of us have already made this shift in our minds. A survey from YouGov (2015) indicates that 52 % of the general public in the UK, 54 % in the USA and 56 % in Germany believe life exists outside Earth. Interestingly, another poll by the Pew Research Center (2021) indicates that 65 % of the adults in the USA believe there is intelligent life beyond Earth. The belief in extraterrestrial life among the younger generation seems to be even higher. In a survey of Swedish high school and university students, it was found that 90 % of the respondents were convinced that there is life outside of Earth (Persson et al. 2019), while a similar study from Peru showed that 92 % of the surveyed university students believe in life outside Earth (Chon-Torres et al. 2020).

Having done this shift in our minds is of course one thing. Living in a world where we have actually discovered extraterrestrial life might be another thing. If it is, then how do we prepare for it? Is it even meaningful to speculate about what it would be like or try to prepare our minds in advance for this event?

The number of possible scenarios is close to infinite, and most of them can probably not even be imagined by the most imaginative science fiction writers. However, it might be useful to infer some kind of classification. I suggest that the most decisive factor for how and to which degree most people's worldview will change when we first discover extraterrestrial life is the impact the discovery will have on our minds in the form of the degree of interaction between us and the aliens. The more interactive the encounter will be, the higher the impact it will have on our thoughts and emotions, which in turn means it will have a higher impact on our perceptions of life, the universe and ourselves.

I also suggest that the degree of interaction with alien lifeforms can be fairly well approximated by the following three parameters: (1) how we find them (or are found by them), (2) the kind of life we encounter, and (3) the distance between us and this life. These three parameters may need to be substituted, complemented or subdivided as a result of more knowledge, but we need to start somewhere and this seems like a good place to start.

Let us begin by looking at the question of how we will discover extraterrestrial life for the first time. in other words, how we become aware that life really is a cosmic phenomenon. The reason why the way of discovery is included among the parameters, is partly that discovery is in itself a form of interaction, and partly because the way of discovery can work as a predictor for our future means of interaction.

The encounter that probably most people think of, and that would also be the most engaging way of finding extraterrestrial life would be in the form of direct contact with another lifeform. This could happen if humans land on another world and step out of their spaceships and directly into a bustling ecosystem of organisms easily recognisable as being alive. It could also happen if an extraterrestrial spaceship lands on Earth and some alien lifeform steps out. Either way, the impact on the perceptions of the humans who experience either of these events would be enormous. The impact of such an event on the rest of us would still be immense though smaller than for those who are there since the interaction for the rest of us would be more indirect via reports and, hopefully, via pictures, preferably moving pictures.

The next most engaging way of discovering extraterrestrial life would be if we send an unmanned spaceship equipped with cameras (sight is after all our most favoured sense) and these cameras capture and stream back to us, pictures (preferably moving pictures) of an ecosystem of organisms easily recognisable as being alive.

Unfortunately, this is probably not how it is going to happen. Most research in astrobiology takes the form of searches for biosignatures. A biosignature is a phenomenon that is not alive but is an indication of life. It can be the discovery of a chemical compound that is associated with life as we know it on Earth and that is difficult to explain by other means on the world in question. If that is how we find extraterrestrial life, it will probably take the form of collecting a sufficient number of sufficiently good biosignatures to eventually reach a point where the scientific community think we can be certain enough that there is life on the celestial body in question. Unfortunately, though, it is probably also the least engaging way of finding extraterrestrial life, at least for anyone not engaged in the search.

Biosignatures can come in different forms and degrees of certainty, however. A special form of biosignature is technosignatures. A technosignature is a phenomenon that is clearly created by an intelligent lifeform. It can, for example, be a radio message, an alien space probe or a construction like a Dyson sphere. Though a technosignature would rank lower on the interaction scale than direct contact, it would clearly rank higher than other biosignatures and maybe even be on the same level as indirect contact in the form of pictures of extraterrestrial living beings.

2.2. *Perspective II: Ricky Lee*

Besides whether we as individual humans on Earth ought to be or are receptive to the possibility of there being life other than that on Earth, what interests me as an international lawyer on a greater level is whether we as a society globally or a civilisation altogether prepare in a legal or political way. It is very hard to prepare structures or frameworks in anticipation of what has not happened. Law tends to be reactive. We do not tend to make laws in anticipation of things that have not happened. But it is important to note in that context that what is more relevant and more controlling in relation to what humanity does about the discovery or contact or relationship with extraterrestrial life or extraterrestrial civilisations is not space law, but general international law. And in terms of general international law, we do not have representative institutions on behalf of the Earth as a civilisation or as a human race. The United Nations is not vested with powers that represent Earth to the Galaxy at large. And for that matter, we do not have an international framework for planetary defence if extra-terrestrial life as in every Hollywood movie turns hostile. I remember that when I was very young, I once attended a function where then UN Secretary General Boutros Boutros-Ghali was present, it was one of the 50th anniversary events of the United Nations, and he was visiting Australia at the time. And on a whim, I asked "Do you think that there will be true unity in the world when we are attacked by aliens?". I was a teenager at the time, so clearly I have been contemplating the question of how we would deal with extra-terrestrial intelligence at a very young age. And he said to me: "No, we will be arguing over who is going to lead our defence force." And sadly, that may very well be the future we will have to confront assuming that we will have the time to even think about how we would react and deal with the prospect or the actual discovery of extra-terrestrial intelligence. We as a planet, we as a civilisation, we as humanity are quite unprepared for that.

Not that I see that changing, not that I see a global summit of international leaders all making speeches at the UN General Assembly in New York on what structure should be in place when aliens arrive. But this does not mean that we as humanity should not contemplate that possibility and how we would like to see our respective governments act.

2.3. *Perspective III: Olli-Pekka Vainio*

The first question posed was whether we are ever truly ready for the unexpected. However, by definition, if we were ready for it, it would not be considered unexpected. In reality, we are never fully prepared for unforeseen events. Yet, just because something is unexpected does not mean it must be shocking.

Let me provide a historical example. As a theologian, I participated in a NASA-funded project a few years ago, which explored the social implications of astrobiology. One of my roles was to examine potential points of connection within the Christian tradition concerning astrobiological discoveries that might cause tension or even a crisis in response to such an event.

Many might be surprised to learn that religious traditions are often quite adaptable when it comes to cosmological matters. During this conference, I heard several people claim that in the pre-Copernican world, humans were considered to be at the centre of the Universe. In a certain sense, this was true. If we reflect on the Ptolemaic model, which was endorsed by the early Church because it was the most advanced scientific understanding of the time, Earth was at the spatial centre. However, we can interpret the concept of being "at the centre" in various ways.

For instance, if we think about the centre in terms of value, the perspective changes. The Ptolemaic system was based on the four Platonic elements, which were arranged in a hierarchy. Earth and water were understood to move downward, while air and fire ascended, symbolising a value system. Therefore, being at the spatial centre also meant being in the place of least value. When this worldview shifted, it was as if medieval people collectively sighed in relief: we are not as insignificant as we once thought.

More intriguingly, the Copernican revolution had minimal impact on the core of the Christian worldview. While it certainly changed the way universities approached mathematics and astronomy, it had almost no theological repercussions. People largely went on with their lives as though nothing of great significance had occurred. One possibility is that societally nothing really changes. This would be, of course, quite nice, and it is not totally unreasonable to think that the change could a go smoothly, like it did it in the past.

2.4. *Perspective IV: Daniela de Paulis*

During the first two and a half days of the symposium, we have talked about life outside of Earth. So we have been looking outside. But it is really important to also reflect back on humanity at this time in history, which is quite crucial. Since 2019, within just 5 years, the world has changed radically. We have seen the World Trade Organization, the World Health Organization, as well as the United Nations, institutions that we thought could help humankind to start thinking globally, failing. They did not come up with a viable approach to coordinate major challenges across all countries and relevant organisations. Therefore, there is a great deal to be done in terms of working towards global thinking, enabling new ways to coordinate between different governments and different cultures. This is a huge challenge. This applies not only when dealing with extraterrestrials, or a potential discovery of extraterrestrial life, but for every single global issue, we need to make that step.

One of the things that attracted me to SETI as an artist is having the privilege to work with many different people from neuroscientists to astronomers, philosophers, and computer scientists. It is one of the major perks of my work. I much enjoy listening to the multitude of arising perspectives, while experiencing how different people look at the world in different ways, and moreover the general public being involved in my work.

Within the field of SETI, great work is done in academic studies. But there are key players who are not in the picture yet, including politicians or world leaders, or the general public. They do not enter the scholarly conversation on SETI yet, but they should be part of it.

We might not be prepared, but we have plenty of opportunities to train ourselves to understand other ways of thinking about living and behaving here on Earth. Recent research on non-human animals raises a lot of questions about the concept of intelligence and the role of language. We still know very little, or as Thomas Nagel phrased it, "What is it like to be a bat?" (Nagel 1974). It is really hard...

2.5. *Perspective V: Perig Pitrou*

I will address this question from my disciplinary field, anthropology, and more precisely, the anthropology of life. What is the anthropology of life, and how does it contribute to answering the question? Unlike biology, which studies life itself, the anthropology of life examines humans' attempts to understand the effects of life. Throughout history, humans have experienced several vital processes, such as reproduction, growth, decay, and death. In all societies, we observe "theories of life," which are attempts to better understand the causes of these phenomena. Anthropology is relevant here because it aims to compare the different ways humans analyse the effects of life on living organisms. At the Collège de France, we created the team Anthropology of Life, consisting of students and colleagues conducting fieldwork worldwide in non-Western societies and scientific labs. In our comparative endeavour, we also study how scientists develop their knowledge of life.

Thanks to these collective activities, we were invited in 2016 by our colleagues at the Observatoire de Paris to participate, with István Praet, in an interdisciplinary project of PSL University led by Stéphane Mazevet on the Origins of Life. Since 2023, I have also been involved in a French national program (PEPR Origins: From Planets to Life), led by astrophysicist Alessandro Morbidelli (Collège de France). Our goal is to develop a collective and interdisciplinary definition of life, incorporating perspectives from history and anthropology. In this context, the epistemological question changes. When studying the anthropology of life on Earth, we explain how humans deal with phenomena that are evident and perceptible at many scales in our bodies and the environment. With astrobiology, the situation is different because we try to imagine what life could be like on other planets, despite having no current evidence of extraterrestrial life. Anthropology, being a science of comparison and shifting perspectives, can provide insights for researchers conceptualising life on other planets by comparing it to how we perceive and conceptualise life on Earth. Anthropologists are accustomed to analysing one reality from another point of view, making them well-suited for interdisciplinary research in astrobiology.

Astrobiology establishes an interdisciplinary dialogue between sciences (chemistry, geology, physics, biology, ecology), each developing different definitions of life based on their empirical data and analytical concepts. Social sciences, particularly anthropology, offer reflexivity on this activity, highlighting how knowledge production depends on historical, technical, and cultural contexts. They also examine the impact of scientific knowledge production on human societies and how it influences our definitions of life and the connection between our planet and the universe. Anthropological investigations are crucial for addressing these new conceptions.

2.6. *Perspective VI: Shoaib Ahmed Malik*

From an Islamic perspective, the concept of life as a cosmic phenomenon is not entirely foreign. The Qur'ān emphasizes the vastness of God's creation, often referencing *al-samāwāt wa-al-arḍ* (the heavens and the earth) as a testament to the divine omnipotence. The idea that life could exist beyond Earth aligns with the Qur'ānic portrayal of a universe teeming with signs of God's creativity and power. Verses such as Qur'ān 16:8, which mentions "*and He creates that which you do not know (wa-yakhluqu mā lā ta'lamūn)*", leave open the hermeneutic possibility of extra-terrestrial life, suggesting that humanity's knowledge is limited and that there could be forms of creation beyond our current understanding.

Moreover, Islamic theology has long recognised the existence of beings beyond the human realm, such as angels (*malā'ika*) and jinn, who interact with the material world in ways that transcend human perception. Although these beings are not considered extra-terrestrial in the contemporary scientific sense, their acknowledgement in the Islamic tradition demonstrates a conceptual openness to the existence of life forms beyond our immediate experience.

The potential discovery of extra-terrestrial life does not necessitate a fundamental rethinking of Islamic theology but rather an expansion of its interpretive frameworks. Islamic scholarship, particularly within the Ash'arī tradition, emphasises God's absolute will and creative power. The possibility of life beyond Earth can be seen as an extension of God's creative will, as all contingent realities, whether known or unknown, ultimately derive from God's necessary existence.

This perspective reinforces the principle of *tawḥīd* (the oneness of God), which underscores the unity of all creation under a single Creator. Whether life exists on Earth or elsewhere in the cosmos, it remains part of the same divine order, reflecting the same divine wisdom and purpose. Thus, from an Islamic standpoint, our existing perceptions are indeed fit for purpose but are also flexible enough to incorporate new discoveries that reveal the breadth and depth of God's creation.

2.7. *Intervention*

Question (Erik Persson):

What to expect from world leaders if we were attacked? If we faced a civilisation that can travel between star systems, we are probably screwed. If we encounter life resembling Earth's microbial life on Enceladus or another Solar System body, is there any discussion going on about the legal implications? What can we do with that, who would have the right to experiment with it?

Reply by Ricky Lee:

Let us remember that space law, as it currently exists, deals with human activities in space as well as human activities directly related to activities in space. In terms of contact with life that does not have the capability to destroy us, and talking about experimenting with life in space, as long as those are activities in space, they are governed by space law. To that extent, the Outer Space Treaty, signed by 111 out of 193 UN member states, says that if you conduct activities that may have an impact on Earth, particularly on life on Earth, then you have an obligation to inform and consult with the international community at large. How enforceable that is, as for any other body of international laws, is another problem. As for bringing life from outside the Earth to the Earth, the same principle applies. You are conducting activity of bringing extra-terrestrial life to Earth, you have an obligation to inform and consult.

If space law did not exist, if the Outer Space Treaty did not exist, and as I pointed out earlier, it is not space law that is controlling but general international law that is controlling, there have always been established principles in international law that if you are aware of or being in control of something that may cause harm to another country, you have a duty to warn and prevent. If you cannot prevent, you have to warn. If you can prevent, you should. Those obligations have always been there as a matter of customary law. And the International Court of Justice in the 1950s spelled that out in very clear terms.

So with regard to any discussions or any legal framework that exist in dealing with what we might do with finding extra-terrestrial microbes or similar forms of life, there are not there. But it is not that hard to find analogous principles in law to apply to those circumstances. Therefore, absence of a specific framework does not mean that you can do whatever you like. There are existing legal principles that address issues specifically of warning and prevention when it comes to potential harm. A week or two ago, there was an attack in a theatre in Russia, and one of the first things that happened afterwards was that the United States announced that they had warned Russia of a potential terrorist attack on civilian population in Russia. In fact, they had done so a couple of weeks earlier. I do not see any kind of pre-emptive discussion taking place at any serious political level, but it doesn't mean to say that the non-politician community of the world cannot have ideas or cannot have those discussions ahead of time.

There are two big obstacles that we need to confront if we want to think about preparing ourselves, both socially and legally to the prospect of discovering extrat-terrestrial life. One is general, and one is very specific. The general one is that we live in a world where we adhere to the ideal that the internet and social media as it exists now and the disruption that they continue to bring, bring forth a community of ideas in which people can put forward their viewpoints without being constrained in access to the media. This is great and it sounds great, but the problem is that it comes with the by-product of an equalisation of knowledge or an equalisation of sources of knowledge. But this is fundamentally not ideal for civilisation development. For example, we treat medical knowledge about vaccination from a doctor, from an immunologist, from the researchers on pharmaceuticals, from Kim Kardashian on exactly the same level, and there are people who give them equal weight. Just as much as we read books by non-lawyers about law, as well as books by non-medical experts about what to eat and how to feed ourselves. In the Netflix movie "Dont look up!", this is both funny and scary because we can all see that coming true, that we end up being inactive in dealing with a global extermination threat because of how we have not yet worked out how to deal with the destructive effects of social media.

The specific problem is that science is finding it increasingly difficult to sell itself in a world where it feels like you can learn everything from Wikipedia. It feels like if you are not sure about something you can just ask. We do not seek knowledge from the sources we should because that is no longer the world we live in, and we are increasingly removing the traditional tools by which we seek that knowledge. Two weeks ago I was in Washington DC, and that week happened to coincide with the last week that the library at NASA headquarters in Washington was going to be open to the public. They are closing that space for budgetary reasons, and the resources are being moved to God. Just in the day that I was there, I saw streams of children going in and out. looking at books and learning about space science and knowledge, and the closure of that library means that that channel would be lost, they will have to google.

So, the best thing, or the most important thing about preparing society, preparing politicians, preparing leaders, preparing our civilisation as a whole, no matter what our viewpoint and our emotional or philosophical response may be to the discovery of extraterrestrial life, is to be sure that we as humanity are prepared socially in terms of how we deal with knowledge, how we expand that knowledge, and how we disseminate that knowledge within our civilisation. Further when we do make contact with other civilisations, we at least know that we got our house in order to know how we are going to invite others to our house, and what we may do to welcome our guests if welcoming our guests is what we want to do.

3. Looking inwards: Who are we?

3.1. *Perspective I: Erik Persson*

That it matters which kind of life we find seems obvious, but how does it matter? Which kind of alien life would have the highest impact on our perception of life, the universe and ourselves? According to the suggested scale, the crucial question would be how much we can interact with it. More complex and more intelligent life, or just life that has a bigger repertoire of behaviour would thus rank higher on the scale, at least to a certain extent. The most important factor is probably its degree of similarity to us. This means that the assumption that the more complex or the more intelligent an alien lifeform is, the higher the impact, is probably only true within certain limits. A lifeform that is so much more intelligent than we are that we have no common conceptions of anything would be impossible to communicate with. This may well mean that the discovery of such a lifeform will have less impact on our perceptions than a lifeform that is on about the same level of intelligence as we. Even organisms on the same level of intelligence as we would be more or less engaging, however, depending on how similar they are in other respects. If their physical appearance is very different, if they spend most of their life under water, or if they communicate in a way that needs a lot of decoding, it will detract from their relatability and thus their impact on our minds.

Finally, the distance to the discovery will make a difference. If we discover microbial life through observations of biosignatures in the form of chemical compounds in the atmosphere of an extrasolar planet, we may never have any other interaction with this lifeform than through observing these biosignatures. This can still be fascinating for an astrobiologist but will not have a big impact on the perceptions of most people on Earth. If we receive a message in the form of radio signals from a civilisation 500 lightyears away, it will probably be more interesting than observing chemical compounds in the atmosphere of a planet even a few lightyears away but also have significantly less impact than the same radio signals would have if they came from only a few lightyears away.

It is still interesting to ponder about how much and which kind of impact a detection of a civilisation 500 lightyears away can have on us. Consider that if we answer it will take 1000 years to get a reply back and by that time those who received the original message and sent our answer will be long gone. I lean towards thinking that the impact this will have on the minds of our descendants will be mostly in the form of myths myths about the second coming, to borrow a concept from religion. That is, myths transferred and transformed between generations about if and when the answer will come, and what it will say. It may even lead to conflicts between different groups that have their own ideas of what the answer will be and whether it will come at all. These myths may well function like religious myths today and take a very strong hold on people on Earth. A worst-case scenario is that it will lead to devastating armed conflicts between groups favouring different myths. In that case, the assumption that the longer the distance, the lower the impact will prove totally wrong.

Finally, there is one more possibility that cannot be ignored. What if we find out that we are alone in the universe after all? There are two possibilities for this. One is that we discover a law of nature that precludes life from existing anywhere else than on Earth. This seems utterly implausible, however. The other possibility is that we keep looking without success and eventually decide that we have searched long enough and thoroughly enough to conclude that we are alone.

Exactly what impact such a conclusion would have, how it would affect our conceptions of life, the universe and ourselves, is of course impossible to know. It seems plausible, however, that if it ever happens, the impact will be at least as powerful as a discovery of extraterrestrial life would be, even one ranking at the top of the interaction scale. Considering that since most of us, as we saw above, have already embraced the idea that life is a cosmic phenomenon, we are probably even less prepared, both practically and mentally for such an event.

One result that is often suggested, is that we would feel more motivated to take good care of the life we have on our own Earth. This might be the case, but if that really changes our behaviour is a different question. Whether or not the universe is teeming with life, we Earthlings still depend for our survival on Earth's life-supporting ability. Considering that even this insight does not seem to be sufficient motivation to substantially change our ways, it might be questioned whether an insight that Earth is the only planet with life in the universe, would be enough to make us change.

Another alternative is that we will put terraforming near the top of our agenda as a species. That is, we will put a lot of effort into trying to make other worlds habitable and then seed them with life. Whether that is possible will of course depend on why we are alone in the universe in the first place.

The, in my opinion, most plausible chain of events, however, is that we will never accept the conclusion that life only exists on planet Earth. Accepting such a conclusion depends, as pointed out above, on us deciding that we have reached a sufficient number of negative results from high-quality research to be convinced that we are alone. Though the research itself is a matter of empirical science, the decision as to when we are certain enough is basically a subjective operation. It is ultimately a matter of what it takes to convince us (see Persson 2014). Considering, as noticed before, that most of us are already convinced that extraterrestrial life exists, and considering the unknown but probably immense impact such a decision would have on our conceptions, I doubt that we will ever be convinced by any amount of non-findings that we are alone in the universe.

3.2. *Perspective II: Ricky Lee*

When I was talking about getting our house in order, I actually meant to refer to our societal attitudes towards how the way in which we deal with knowledge, both existing knowledge and how we evaluate different ideas, reacts to discoveries and scientific advances. I put that in the context of a world currently being disrupted by internet and social media. We have not yet found our feet as to how we re-establish the way in which we pursue and expand and disseminate knowledge, in the same way as we took a bit of time to react to the invention of moveable type and the printing press. Each time we make an advancement towards our technological ability to disseminate knowledge in a better way, there is disruption in the way in which we react to that as a society. Getting our house in order must not be equated to a totalitarian fascist state by which we sway our loyalty to one supreme leader who is fully entrusted by the entire humanity to speak for us and negotiate for us.

There are plenty of fictional stories we can find about the way in which we react to extra-terrestrials and how they might become boring over time, for example. The equivalent for young people now might be the that Jesus has been living in South Park, and after 3 weeks Jesus is just any other resident there. I think there are a significant number of people in the world who will react positively to the discovery of extra-terrestrial life, but there is actually one worse outcome than people reacting negatively. And that is those reacting indifferently, which might become a growing segment. That might be the scarier prospect for us as humanity and as a society that what would have been the most significant historical development to human civilisation to find out we are not alone, to find out that we are not cosmically unique, is treated with indifference like any other news story, jammed into a half an hour nugget for us to digest on the evening news. That might be the scarier thought.

If my memory serves me right, in the 1990s, there was a new story about a meteorite that was discovered, where scientists believed that this meteorite carried signs or evidence, or whatever word the media chose, that there might be life on Mars. Now, the reaction or the later scientific analysis is not the reason why I raise this. What I found significant is that we did not see an international reaction that said "Life can exist outside of Earth, so that might be an existential threat for us; we should do something about that or that", nor one that said "Oh wow! We are not alone; how does that transform the way in which we understand the Universe? How does that transform the way in which we understand our place in it?". None of that happened. So to me, when I refer to getting our house in order, it is not so much that we all must unify in one voice and answer to the great leader, it is that we need to help elaborate, explain, disseminate science. We need to help understand the way in which we teach younger generations about pluralism and science and knowledge, and how they all coexist together as a human value. Then we may find ourselves reacting to the news of discovery of extra-terrestrial life not just innately positively, but from a learned place, from a place of understanding, from a place of appreciation.

3.3. *Perspective III: Olli-Pekka Vainio*

The implications of discovering extra-terrestrial life would depend on the nature of the life we encounter. Such a discovery could influence various philosophical pursuits, particularly how we approach questions regarding human uniqueness or the meaning of life. Many potential avenues for academic discussion have already been explored, and any future discovery might further refine or limit the directions these discussions could take. However, it is possible that our reaction to such a discovery would reveal more about ourselves than about the discovery itself. Yet, I suspect it would not expose anything radically new about human nature.

Earlier, I presented a more optimistic scenario. But there are alternative perspectives. For instance, the movie "Dont Look Up" (2021) portrays humanity's failure to address a global crisis, reflecting our current understanding of how we might respond to potential dangers or other life-altering events. We have discussed the importance of controlling the narrative around contact and the need for preparedness to avoid societal chaos. While order may seem preferable to chaos, we should consider this: even if we are unified and well-prepared, we could collectively misinterpret the situation – for example, assuming that extraterrestrials come in peace when they do not. Collective errors are always possible, and of errors they are often the most dramatic kinds of errors.

In this context, some degree of chaos might actually be beneficial. Biologists understand that chaos can drive evolution by forcing adaptation and the search for alternatives. The desire to control everything is not inherently good. During the COVID-19 pandemic, not all government responses were rational; many were driven by emotion, opportunism, or haste. There is still debate about whether lockdowns worsened the situation and whether more liberal approaches, such as Sweden's, were more effective than the stricter measures implemented by the U.S. or the UK. While it is indeed useful and rational to have plans, striving for a rigid framework that assumes total control may not be the best strategy and could lead to unintended negative consequences down the line.

3.4. *Perspective IV: Daniela de Paulis*

If we consider the scenario of discovering microbial life on another planet, I think that most people would be enthusiastic about this discovery. Just because it is such a great and different news, for sure it would make a lot of people thrilled for a while. There is a short story by an Italian novelist called Ennio Flaiano called "A Martian in Rome" (Flaiano 1960). It is quite a humorous story, and it is also quite provocative. It basically revolves around a Martian landing in Rome. At first, everyone is extremely curious, wants to talk to the Martian, but after just a few weeks it becomes boring; nobody is interested in this Martian anymore. So, the news wears off and the new reality is absorbed quite quickly.

Finding a signal from extraterrestrial civilisations, which implies that there really is some intelligent life form out there, which we know little about, might be more unsettling. But still, I think that most people will react quite positively, however such a scenario might never emerge.

Our reaction would likely be shaped by the many science-fiction novels and films that we have been reading or watching over the past decade. We would be living in one or several of these films.

We are all working towards the event of a post-detection scenario. But how is it going to look like, and what will its timeframe be? Perhaps it is going to constitute a longer process rather than a punctual event. It will not be something that we would be dealing with for a few weeks and months, and then it will be settled, but rather a process that would be absorbed, discussed, metabolised over a long period of time, potentially involving cognitive revolutions. Such processes are not linear, but rather involve zig-zag lines with steps forwards, backwards, and sideways. Whatever the exact post-detection scenario will be, it is probably something that will carry on over many decades before we can fully assess how that will change our understanding, what it can tell us about ourselves as humans, or how we might be reacting to it.

3.5. *Perspective V: Perig Pitrou*

Usually, I do not answer that question because I consider it not entirely scientifically available, although it is very interesting to wonder or ask what would happen if. For me, it is very speculative. As an anthropologist, I prefer to analyse what is actually happening when researchers try to discover life in other parts of the universe. Describing the real world and humanity's activity in searching for life on other planets is a challenge that holds significant interest. Instead of attempting to provide a speculative answer, I believe it is more valuable to explain how researchers and different human groups communicate and establish international programmes around the topic of life. Despite our cognitive and biological limitations, we, as humans, with our cumulative knowledge built year after year, century after century, can make significant discoveries, such as observing remote parts of the Universe and detecting exoplanets.

Telling this story, not just as a narrative but delving deeper into the complexity of this human experience, also involves political aspects and impacts. Even if we cannot provide definitive answers, the potential discoveries from this collective endeavour are fantastic. Astrobiology, for example, offers new insights for humanity. It is widely accepted that the heliocentric theory and Darwin's theory significantly changed our understanding of our place in the Universe. Similarly, astrobiology represents a revolutionary shift in how we perceive our place in the Universe and among living beings. Recognising this revolution requires us to share this knowledge beyond the academic audience. The main issue for me is how we can integrate diverse scientific approaches and communicate this scientific knowledge across different cultures. This is a crucial challenge for the modern world.

We can no longer take political and geopolitical decision based on the only distinction between those who possess scientific knowledge and those who lack it. While it is essential to defend science, we must also understand how science can contribute to a broader framework, a global elaboration of not only the knowledge of life but a collective understanding of what it means to live on our planet and possibly on other planets in the future. Here, anthropology is crucial to explain how conception of life and the universe are involving many cultural, religious, artistic representations.

There is one last point I would like to stress about the origins of life. It is not only about discovering how life emerged on Earth or other planets. Scientists are engaged in activities of modelling life and developing experiments where humans could demonstrate their power to study life, detect signs of life, and potentially create it from inanimate material. While this is not yet the case, projects in synthetic biology or protocell fabrication in microfluidics indicate serious progress. In these modelling projects, understanding vital processes improves by confronting different evolutionary scenarios, both chemical and biological, corresponding to conditions on various planets where life could have emerged.

In this context, life is not an alien object but a product of human activity. Its origin is not dependent on an external causality but on human power and intentionality. Even though we are still far from this achievement, this type of realisation raises numerous technical, legal, and political questions, further justifying the insights of anthropology. This discipline also helps study how contemporary societies develop imaginaries about the origins of life and the possibility of its existence elsewhere in the universe, as seen in science fiction. While imagination is a sphere of activity generated by free thought, it remains influenced by cultural determinants that anthropology has learned to decipher.

3.6. *Perspective VI: Shoaib Ahmed Malik*

The potential detection of extra-terrestrial life presents both challenges and opportunities for exploration within Islamic thought, compelling scholars and believers to reflect deeply on the implications of such a discovery for our understanding of existence, theology, and the divine order.

One of the primary challenges would be reconciling the existence of extra-terrestrial intelligent life (ETI) with the traditional Islamic narrative that emphasises the unique status of human beings in the cosmos. The Qur'ān describes humans as *khalīfa* (vicegerents) on Earth, a role that suggests a special responsibility and connection to God. The discovery of other intelligent beings with similar or superior qualities could challenge this notion, raising questions about the universality of the Islamic message and whether it applies exclusively to humans or also to other sentient beings.

Another challenge lies in the theological implications regarding the eschatological destiny of ETI. Islamic eschatology is human-centric, with detailed descriptions of the afterlife tailored to human deeds and beliefs. The existence of ETI would necessitate a rethinking of whether these beings would also be subject to divine judgment and, if so, how Islamic teachings might apply to them. This raises complex questions about the scope of God's guidance and mercy, and how these might manifest in a broader cosmic context.

Moreover, the integration of ETI into the existing theological framework could lead to debates over scriptural interpretations. Some might argue that the Qur'ān's silence on the explicit existence of ETI implies their irrelevance to the Islamic narrative, while others could view this as an opportunity to explore non-literal or broader interpretations of Qur'ānic verses that suggest the possibility of other forms of life. This could lead to differing opinions on the extent to which Islamic theology should adapt to new scientific realities.

Despite these challenges, the discovery of ETI also presents significant opportunities for expanding and enriching Islamic thought. One of the key opportunities is the chance to explore the universality of Islamic principles in a new light. The concept of *tawḥīd* (the oneness of God), which underpins the unity of all creation, could be expanded to encompass a cosmic scale, reinforcing the idea that all beings, regardless of their location in the universe, are part of the same divine plan.

Furthermore, the detection of ETI could stimulate new discussions on the nature of prophethood and revelation. If other intelligent beings exist, the question arises as to whether they have received divine guidance specific to their circumstances, similar to the way humans have received the Qur'ān. This could lead to a broader understanding of revelation, one that considers the possibility of multiple forms of divine communication tailored to the unique needs and contexts of different creatures. Additionally, the encounter with ETI could inspire ethical reflections within the Islamic tradition. Questions about the treatment of other intelligent beings, the extension of moral and legal principles beyond Earth, and the responsibilities humans might have towards other forms of life could lead to a richer and more nuanced Islamic ethic that is responsive to the realities of a more interconnected universe.

In the context of extra-terrestrial intelligence (ETI), decolonial research offers critical tools for addressing the otherisation that often accompanies encounters with the unfamiliar or unknown. Historically, colonial encounters were marked by the otherisation of indigenous peoples, who were often seen as inferior or as objects to be dominated. This mindset, which has historically influenced Western thought, has the potential to be replicated in humanity's approach to ETI. However, by integrating decolonial perspectives — within which Islamic thought plays a vital role — we can foster a more inclusive and ethical approach to the possibility of encountering other forms of intelligent life.

Decoloniality encourages us to critically examine the processes of otherisation that have historically framed interactions with those deemed other or outside the normative boundaries of the familiar. This is particularly relevant in the discourse on ETI, where the tendency might be to view extra-terrestrial beings through the same lens that colonial powers once used to view indigenous populations: as curiosities, threats, or beings to be dominated and studied rather than as equals.

Islam, with its emphasis on the inherent dignity of all of God's creation, offers a counter-narrative to this tendency. The Qur'ān speaks of the unity of creation and the shared purpose of all beings, which can be extended to include potential extra-terrestrial life. By applying Islamic principles through a decolonial lens, scholars can challenge the notion that ETI would necessarily be "other" in a negative or inferior sense. Instead, these beings could be seen as fellow participants in the divine plan, with their own unique perspectives and contributions to the cosmic order.

Islamic teachings, when combined with decolonial thought, can also promote the recognition and inclusion of multiple viewpoints at a cosmic scale. The colonial legacy often imposed a singular, Eurocentric perspective on the world, dismissing or marginalising other ways of knowing and being. In the study of ETI, this same dynamic could lead to a narrow interpretation of what constitutes life, intelligence, or civilization, based solely on Earth-centric models.

Decolonial research, grounded in Islam's rich intellectual tradition, advocates for a plurality of perspectives that respect the diversity of creation. This approach can be instrumental in shaping how humanity might engage with ETI. Rather than imposing a singular worldview or set of values, an Islamic decolonial perspective encourages us to consider that ETI may possess entirely different forms of knowledge, social structures, and spiritual understandings, all of which deserve respect and careful consideration.

By promoting a multiplicity of viewpoints, decolonial research can help prevent the replication of colonial dynamics on a cosmic scale. It advocates for an approach to ETI that is not only scientifically informed but also ethically and theologically nuanced, recognising the potential for multiple truths and realities within the vast expanse of the universe.

In conclusion, while the detection of extra-terrestrial life presents significant theological challenges, it also offers profound opportunities for exploration within Islamic thought. It encourages Muslims to expand their understanding of God's creation, to rethink traditional interpretations in light of new discoveries, and to explore the ethical implications of a universe filled with diverse forms of life. This process of reflection and adaptation can ultimately deepen the Islamic worldview, making it more inclusive and responsive to the evolving human understanding of the cosmos.

4. Further discussion

Question:

What do you think about a scenario where we would all receive a transmission that we would be fairly sure is artificial, but would not be able to track it for years and years on ends. Do you think that we would actually think about sending a reply to a known source of incoming transmission? How do you think you would react? Would we be extremely disappointed?

Reply (Erik Persson):

I would say, it depends on who "we" are. I think there would be lots of very different reactions around the world. I think some would want to do everything to interpret it, while others would say that even just reading it can be dangerous. Or answer back and say "we do not understand, please send it in another format. There would be room for loads of different reactions, and these will interconnect and need to be dealt with.

Question:

In the 6 oclock BBC TV news headlines for the day of the announcement of the microfossil in a Mars meteorite (in August 1996), this big discovery was only the third item. Apparently, something appears to be wrong with the order. Ronald Reagan notably broke the ice with Mikhail Gorbachev during the 1985 Geneva summit by asking "If we were attacked by aliens, would you come to our assistance?, to which Gorbachev replied "Yes, sure. Times have changed, how would this unfold nowadays?

Reply (Daniela de Paulis):

The idea of being attacked by an alien civilisation appears to be very anthropocentric. Where do we get the idea from? Mostly from concepts of colonialism. Receiving an extra-terrestrial signal, rather than facing an invasion, would in my opinion be the most interesting scenario. That would really shift our understanding. We would most likely face a situation where we would not know how to deal with these data. From my experience with people who are trying to decode the message I created as part of one of my projects, it emerges that they do not deal very well with uncertainty, they want to know the solution. And this is just an artistic experiment. So what if it was a real message? I suspect that some people would really not be able to cope with the idea that there is something out there that we just cannot understand. So this great uncertainty would really be quite unsettling, but in my opinion I would personally not be interested in the idea that we would be attacked by aliens. First of all, if they are much more advanced than us, that is it, the game is over. But otherwise this is just a scenario we imagine because that is what we create here on Earth. But where do we get the idea from? That might be the most interesting question.

Reply (Perig Pitrou):

We should also take into account the fact that we keep discovering unknown forms of life on Earth. And it has already been the case with the discovery of a certain organism in the deep ocean; every time we discover a new organism somewhere on the planet where we did not expect to find it, this has also changed the definition of life. So I think we should also add this idea to the reflection we can have on our own planet. And there is another point which is very important: the fact that at least in France the national programme of astrobiology is a cluster of researchers, where one work package is working on the observation of exoplanets, but another work package is on how can we make synthetic life in a laboratory. And they actually work together, so the idea is observing the Universe will help us to design experiments on synthetic life. So another question we should ask in relation with astrobiology is what would happen the day human scientists will be able to make synthetic life, because here in my field work, I remember the statement of my colleague of chemistry that there is no theoretical obstacle to the synthetisation of life. The proof is they are living organism on earth, so Nature was able to make synthetic life, and therefore there is no obstacle to do it on Earth. There are a lot of political, technical, and social effects that should maybe be anticipated. Researchers are working on the synthetisation of life, and we should have some extra care and extra thought on what is happening on these kind of projects.

Question:

Do you think there is a danger that a first contact, perhaps by radio signal or whatever, could be exploited politically for purposes?

Reply (Ricky Lee):

Yes, I am sure it will. In fact, I would be very surprised if it is not. We talked about contact with alien civilisations and about how our impression that they may have hostile intent may be more a projection of how we behave, and how we think, and that for all intents and purposes, most alien civilisations are not going to travel hundreds of light years to Earth or to our Solar system and start fighting. Will Smith said that, yes, I am quite aware. It is not the aliens that I do not trust, it is us. We as human civilisations tendered in history not to react well to outsiders, meeting us for the first time, or coming to join our living space for the first time. I was born in a colony, and I live and grew up in a country that evolved from colonialism, be it in Hong Kong or Australia. Of course it is a very different colonial experience to that of, say, India or most of Africa. But both the colonisers and the colonised behave in ways that in history, no matter where we are given the great variety of different ways in which they went about colonising and the different ways they went about being colonised, almost none of which are encouraging if that is how we react to contact or extra-terrestrial intelligence. So I think your question as to whether we see political exploitation or possible political exploitation of receiving an alien message for the first time, I would be very surprised if it is not exploited, because if you think aliens are coming, you would want to position yourself in the greatest position of strength possible via the others. That is unfortunately historically how we have behaved across cultures, across geography and across time. If anything, that is the scariest part to me. Not that the aliens might come with amazing weaponry, they probably can and would. But how we react and how we deal with each other in the face of that knowledge, that to me is the greater fear.

Question:

Many discoveries in science are quite gradual, long discoveries. We could certainly get a signal tomorrow, and it is unambiguous, and within a few months everybody has agreed. An astrobiology site that I looked at considers that as one of the less plausible scenarios. In one of the more plausible scenarios, we find some biosignatures on observing exoplanets, and a very long period of time passes requiring new technological innovations, and it will take around 50 years until a scientific consensus forms that it is almost certainly life. In that kind of scenario, we do not need to be prepared because we are going to have 50 years.

Reply (Olli-Pekka Vainio):

This brings us back to Cixin Liu's "Three-Body Problem" series, where it takes hundreds of years from the moment of first contact to the actual encounter with alien beings. During this time, human cultures rise and fall, and by the time we finally meet the aliens, the culture that makes contact bears little resemblance to our present-day societies. Along the way, things get very messy. Returning to our initial questions, I believe we will witness all possible reactions happening simultaneously. This is something we should be prepared for, as there is no way to control every aspect of these events. It is crucial to be ready for the complexity and chaos, as well as for the long process of thinking through all the related issues.

Question:

Considering the human history and how we approached exploration before, which was not always considerate but pretty destructive, if a contact happens, can we actually have kind of good neighbourly relationships unless we deal with our own imperialism? There is a lot of space law, international institutions, and protocols. Do we think that this would all go poof? Once contact happens would different countries go their own way, or would there still be consensus? And who would enforce that consensus and the UN rules and so on?

Reply (Olli-Pekka Vainio):

I may have already answered this question, but let us consider the history of colonialism, which is an incredibly complex topic. There were both positive and negative aspects throughout the colonial centuries, and in fact colonialism is not over. We are now entering a new era of colonialism, which will shape our history for hundreds of years to come. We might also reflect on our tendency to idolise potential alien cultures, imagining them as vastly different from us – perhaps unified under a galactic government, living in harmony without internal conflicts. In contrast, they might be more like us: disorganised, quarrelling, and maybe in the need of a saviour.

As I have mentioned before, we are likely to see a wide range of reactions all at once. A plausible scenario is that while governments are still deliberating on how to respond, someone like Elon Musk has already sent a series of messages. It is also possible that the message we have received from extraterrestrials comes from their own version of

Elon – an independent actor who has the means to send signals across galaxies. In that case, the glorious first contact might end up being our Elon and their Elon exchanging funny memes.

Comment:

One of the big questions was how will people react, and I would like to encourage looking into risk analysis and risk perception research. There are a lot of findings out there on how people perceive certain risks and how tolerant they are towards these risks. I am a risk and crisis communication scholar. We could use that research applied more systematically to those different scenarios and actually learn something that we can use for preparation scenarios. You do not have to read all the sci-fi literature to anticipate that there are some systematic data out there. What I can tell from the crisis research is that panic is very unlikely. Most people do not panic in face of a crisis unless the big saucer is above the White House and melts it down. Probably for now the best we can do in terms of preparation and becoming resilient is to fight the erosion of loss of institutional trust that is happening at the moment, where malevolent actors in different countries are trying to destroy the credibility of traditional and quality journalism and science for political and other reasons. Working on that is one important factor, and then there are many other suggestions that I would have.

Reply (Daniela de Paulis):

The scenario we are going to face will be completely different from what we prepared for.

References

Chon-Torres, O. A., Ramos Ramírez, J. C., Hoyos Rengifo, F., Choy Vessoni, R. A., Sergo Laura, I., Ríos-Ruiz, F. G. A., Murga-Moreno, C. A., Alvarado Pino, J. P. L., & Yance-Morales, X. 2020, Attitudes and perceptions towards the scientific search for extraterrestrial life among students of public and private universities in Peru. *International Journal of Astrobiology*, 19(5), 360–368.

Flaiano, E. 1960,. *Un marziano a Roma.* Einaudi, Torino.

Nagel, T. 1974, What Is It Like to Be a Bat? *The Philosophical Review*, 83(4), 435–450.

Persson, E. 2014, What does it take to establish that a world is uninhabited prior to exploitation? – A question of ethics as well as science. *Challenges*, 5, 224–238.

Persson, E., Capova, K. A., & Li, Y. 2019, Attitudes towards the scientific search for extraterrestrial life among Swedish high school and university students. *International Journal of Astrobiology*, 5(3), 280–288.

Pew Research Center 2021, Most Americans believe in intelligent life beyond Earth; few see UFOs as a major national security threat. `https://www.pewresearch.org/short-reads/2021/06/30/most-americans-believe-in-intelligent-life-beyond-earth-few-see-ufos-as-a-major-national-security-threat`. Accessed: 2024-7-16.

YouGov 2015,. You are not alone: most people believe that aliens exist. `https://yougov.co.uk/society/articles/13379-you-are-not-alone-most-people-believe-aliens-exist`. Accessed: 2024-7-16.

(Toward) Discovery of Life beyond Earth and its Impact
Proceedings IAU Symposium No. 387, 2026
H. Landt, M. Dominik & C. Oliver, eds.
doi:10.1017/S1743921324002126

Four Central Topics in Contemporary Christian Astrotheology

Olli-Pekka Vainio

University of Helsinki, Faculty of Theology, Fabianinkatu 33, 00014, University of Helsinki, Finland
email: olli-pekka.vainio@helsinki.fi

Abstract. This article examines four different points where an astrological or astrobiological discovery might propose a challenge for Christian religion. These points of contact are the size of the cosmos, human uniqueness, Christological considerations and the ultimate fate of the cosmos. I argue that the most popular interpretations of these challenges are not very serious, while there are certain issues that may genuinely push Christian religion and theology to reconsider some topics in their belief system.

Keywords. Christianity, SETI, ETI, Fermi paradox

If extraterrestrial life is discovered, what implications does this hold for religions? Such a question often appears to assume a simple, perhaps negative, answer. One of the most renowned works expressing such sentiments is Thomas Paine's (1737–1809) "Age of Reason," wherein he unequivocally argues that Christian theism is absurd in light of 17th-century cosmology.† However, the matter is far from straightforward. I will attempt to elucidate why this is so shortly. In this concise overview, I will delve into four themes pertinent to Christian theology and the discoveries in astrobiology. These themes encompass the vastness of the cosmos, the uniqueness of humanity, soteriology, and eschatology. Some of the issues are more threatening to the rationality of Christian religion than the others.

First, it is essential to define key concepts. Astrobiology is a discipline dedicated to studying the origin, evolution, and distribution of life in the universe. It integrates various fields, including astronomy, biology, geology, chemistry, and physics, aiming to comprehend the conditions conducive to life beyond Earth's immediate sphere of influence.‡ Astronomy, on the other hand, is the scientific exploration of celestial bodies and phenomena beyond Earth. It encompasses the study of stars, planets, moons, asteroids, comets, and other cosmic entities, along with their movements, composition, origin, and evolution. Within astronomy lies cosmology, a branch that investigates the structure, genesis, and expansion of the universe on a grand scale.§

The term UFO (unidentified flying object) has recently been supplanted by UAP (unidentified aerial/anomalous phenomena). UAP typically denotes atmospheric phenomena lacking a clear explanation. These phenomena encompass observations and recordings of peculiar objects by, e.g., the US armed forces, which have also become

† Paine 1995, 710. Against this, Tyler McNabb and Chad McIntosh, for example, have argued that theism is more likely to predict the existence of other intelligent species than naturalism. See. McNabb 2024

‡ Domagal-Goldman 2016. Catling 2013.

§ Jayawardhana 2011.

accessible to the general public in recent years. Nevertheless, UFO can still be employed more broadly to denote individuals' mysterious encounters, such as luminous phenomena and, in extreme cases, abduction experiences.

SETI (Search for Extra-Terrestrial Intelligence) is a project specifically employing radio telescopes to detect potential signals or other indicators of extraterrestrial civilizations. ETI (Extra-Terrestrial Intelligence) is a general term referring to any intelligent life beyond Earth. It need not necessarily be organic life, as many researchers deem it more probable that the initial ETI encountered would be artificial intelligence or a non-organic life form.†

The term 'astrotheology' was initially introduced in 1715 by William Derham in his book *Astrotheology: Or a Demonstration of the Being and Attributes of God from a Survey of the Heavens*. In his work, Derham argues for God's existence, drawing from natural theology and citing a new comprehension of astronomical facts. Today, astrotheology pertains to the examination of astronomical and astrobiological discoveries vis-à-vis theological beliefs and doctrines. Astrotheological inquiry in the English-speaking realm has been pursued by various scholars, including Ted Peters, renowned for his popularized writings on the subject.‡ Andrew Davison has also penned a comprehensive overview of the field.§ Additionally, NASA's chief historian Stephen J. Dick, has broached theological matters in numerous writings.¶ Various introductory works and articles on the topic are readily available.‖

I will now address four essential questions that frequently emerge in astrotheological literature. Additionally, I will present the fundamental options and solutions proposed to potential problems or tensions. Finally, I will also briefly evaluate UFO experiences from the perspective of the religious worldview.

The Size of the Cosmos

The Old Testament already contains descriptions of the relationship between humanity and the cosmos (Psalm 148; Isaiah 40:22; Jeremiah 33:22; Job 38), reflecting the smallness and insignificance of humanity in relation to the broader reality. In the texts of the early Church Fathers, cosmology is depicted in various ways, with no singular cosmological model universally accepted. Noteworthy statements from a historical theological standpoint include Origen's (184–253) advocacy for an allegorical interpretation of Genesis, Basil the Great's (330–379) assertion that Moses has not seen it worthwhile to address these matters, and Augustine's (354–430) admonition to Christians not to conflate theology with natural science.** Both Boethius (d. 524) and Moses Maimonides (d. 1204) emphasize humanity's incapacity to grasp the scale of the cosmos, viewing this incapacity as a source of humility.††

David Hume (1711–1776) and Friedrich Schleiermacher (1768–1834) offer tentative assessments regarding the significance of the cosmos's size for theism. One of the characters in Hume's dialogues speculates that the vastness of the cosmos could serve as evidence for theism. Schleiermacher contends that the order of the cosmos holds greater existential importance than its size.‡‡

† Schneider 2016.

‡ Peters 2016. Peters 2014. Peters 2018.

§ Davison 2023.

¶ Dick 2000. Dick 1998.

‖ See e.g. Wilkinson 2013. O'Meara 1999. Playford 2024. Thigpen 2024.

** Origen, De Prin. 4.16. Basil the Great, Six Days of Creation (Hexaemeron), 1.9. Augustine, De. Gen. ad. Lit. 1.19.38 to 20.40; 2.9.20.

†† Boethius, Consolations of Philosophy, II, VII. Moses Maimonides, Guide for the Perplexed, III, 14.

‡‡ Hume, Discourses on Natural Religion, 5.2. Schleiermacher 1893, 67.

Few modern philosophers have endeavored to develop an antitheistic argument based on the size of cosmos. Nonetheless, Nicholas Everitt has argued that theism presupposes an anthropocentric cosmology, wherein the world is only 6,000 years old and the Earth occupies the center of the cosmos.† Thus, modern cosmology would contradict the presumptions based on theism. If indeed the case, the argument from scale of the cosmos would hold weight. However, the challenge lies in demonstrating that theism actually necessitates the described cosmology, which is far from self-evident. Therefore, it can be asserted that the size of the cosmos and awareness thereof do not pose a theologically significant challenge. On the contrary, they have been historically regarded as indications and manifestations of God's magnificence.‡

The Location of Humanity in the Cosmos and Human Uniqueness

Frequently, particularly in popular sources, the cosmological implications of the Copernican revolution for theology are overstated. This misinterpretation stems from the association of the cases of Copernicus and Galileo with well-known ecclesiastical scandals, which have been somewhat exaggerated in later historiography.‡

If factors unrelated to cosmology are disregarded, the question of humanity's position in the cosmos emerges as central. It is widely recognized and indisputable that the Church predominantly embraced Ptolemaic cosmology because, for over a millennium, it represented the most sophisticated scientific model of planetary cosmology. However, it gradually became evident that this model could not account for all celestial phenomena, prompting the slow pursuit of a superior model. However, the Church's opposition was never directly aimed at the position ascribed to humanity within the new model. Rather, it was a philosophical commitment to the principle of "saving the phenomena." This principle dictated that sensory perceptions should not be dismissed without truly compelling reasons. If we observe the sun rising and perceive the Earth as stationary, substantial evidence is required to challenge this perception. Additionally, biblical verses describing the world in similar terms supported the notion of the phenomenal world. Yet, in Philip Melanchthon's (1497–1560) early refutations of the Copernican model, theological considerations played only a minor role. Instead, the primary arguments revolved around general philosophical principles of science. Consequently, Melanchthon swiftly altered his stance and integrated calculations based on the Copernican model into science education at Lutheran universities.§

C. S. Lewis (1898–1963), deeply immersed in Renaissance literature and worldview, asserts that the influence of the Copernican revolution on people's worldview was practically negligible. Instead, one might perceive the departure from the Ptolemaic model as a relief. In this model, the notion of Earth's specialness and consequent value due to its position was not entertained. Rather, material existence was deemed less significant than spiritual reality. Even theological perspectives constructed upon this premise emphasized that Earth's "central" position merely signified humanity's peripheral and insignificant existence in the cosmos. Implicit in this view was the Platonic concept of the elemental hierarchy: earth and water gravitate downward, while air and fire ascend. In Dante's *Divine Comedy*, Satan is located at the center of the cosmos, encased in ice, reinforcing the notion of immobility as indicator of value, whereas ascent toward heaven takes place through increasing movement and temperature signifying higher value.¶

† Everitt (2004).
‡ For more information on the size argument, see Vainio 2018a.
‡ McMullin 2005. Finocchiaro 2010.
§ See Vainio 2018b, 45–46. Melanchthon 1550.
¶ Lewis 1964.

Earth-centered cosmology, nonetheless, aimed to present an aesthetically pleasing depiction of reality conducive to intellectual contemplation. It drew its inspiration from Platonist philosophy, which cannot be regarded as particularly anthropocentric. Even within this framework, human value was contingent upon comprehending the deeper nature of the universe, with location holding little relevance.

The concept of humanity as the image of God (*imago Dei*) has often been juxtaposed with astrobiological discoveries, yet this juxtaposition stems from a misunderstanding of the doctrine of *imago Dei*. Although this doctrine, lacking formal conciliar formulation, may underscore human uniqueness, it does not inherently imply exclusivity in being made in God's image. In the New Testament, the term exclusively refers to Christ, not humanity. Instead, the concept serves as a teleological description of humanity's destiny: to grow perpetually towards the likeness of God. This does not preclude the possibility of other beings being made in God's image or the manifestation of God's likeness evolving gradually or differently in various beings.†

Christology and salvation

Questions about Christology can be divided into two parts. The first concerns the exclusivity of divine revelation, while the second pertains to the applicability of Christ's atonement to non-human beings. Though closely related, it is worthwhile to consider them separately.

The issue of exclusivity poses a challenge that already confronts virtually all philosophical and religious worldviews. Why should I believe that my perspective on X is true if there are intelligent individuals who hold differing views? This dilemma is known as the problem of disagreement in epistemology.‡

Presently, various religions and worldviews grapple with this issue, necessitating some form of justification for maintaining their beliefs if they are unable to persuade dissenters of the validity of their views. The discovery of extraterrestrial intelligence (ETI) does not fundamentally alter this predicament; it merely potentially expands the range of possibilities.

Regarding Christ's atonement and the message of salvation, the options can be summarized as follows:

1. Humans are the only intelligent species in the universe.
2. ETI exists, but humanity will never meet it.
3. ETI exists, and it can communicate with humanity, but a physical meeting is not possible due to cosmic distances.
4. A physical encounter with ETI is possible.
5. ETI has not fallen and does not need salvation in the same sense as humanity.
6. ETI has fallen, but it is not part of the economy of salvation.
7. ETI has fallen and belongs to the same economy of salvation in such a way that the incarnation of Christ on Earth is applicable also to ETI.
8. ETI has fallen and belongs to the same economy of salvation in such a way that Christ incarnates in an appropriate way in different ETI worlds.

Chad McIntosh and Tyler McNabb approach this question through the concept of narrative incoherence, which means the following: While we may consider the two arguments to be theoretically compatible, there may still be a tension between them that renders

† Lovin 2015. McIntosh 2023.
‡ Pittard 2019.

their compatibility implausible in some relative sense. Although narrative incoherence does arise in certain scenarios, according to them, it is not insurmountable.§

The first two options do not yield any incoherence. Points three and four merely describe varying degrees of encounter, which practically result in the following alternatives. An analogy for points five and six can be found in angelology. Angels do not belong to the same salvation economy as humans; they are saved by aligning with God or turning away from Him. Christ's involvement is not essential in their case. Additionally, the choice solidifies their condition, preventing a good angel from becoming evil, or vice versa. Hence, the similarity of ETI to angels in either sense does not engender significant narrative incoherence. C. S. Lewis speculated on such possibilities in his sci-fi trilogy.†

Point seven would necessitate "cosmic evangelism," wherein the message of Christ should be preached to ETI. On this topic, Mary Doria Russell has written a science fiction novel, *The Sparrow.* This scenario also aligns with scenario three. In these cases, there is somewhat more narrative incoherence, but its significance varies based on individual intuitions.

Regarding point eight, however, Thomas Aquinas already contemplated the possibility of multiple incarnations (*Summa Theologiae* 3.3). According to him, it is fundamentally possible for the Logos to assume multiple natures, akin to how a person wears many coats. Thomas posits that this does not result in any Christological heresy. He argues that the Incarnation, or the salvation economy chosen by God for humanity, is not inherently necessary; it was merely a "fitting" (*conveniens*) means of achieving salvation as it corresponded to humanity's fallen state. Thus, Thomas' theology of the Incarnation opens up the possibility of broader speculation about alternative salvation economies.‡

Eschatology and the end of the cosmos

According to current knowledge, our universe will eventually meet its end billions of years from now, with the cosmos either collapsing (a big crunch) or continuing to expand until all existing matter is evenly distributed, resulting in particles no longer influencing each other and the cosmos reaching a state of maximum entropy, also known as heat death.

Of course, the eschatological descriptions found in the New Testament do not anticipate such a reality, at least not directly (cf. 2 Peter 3:10). Describing the different stages of Christian eschatology and their sequence is notoriously challenging, and various denominations lack a shared description of end-times events. Major denominations have traditionally steered clear of speculating on future events, with interpretations of events in the Book of Revelation, for example, often framed as describing ongoing or perennial occurrences rather than future events. In Reformed theology, however, end-time speculation has been more prevalent.§

Regardless of whether eschatology is viewed through a speculative lens or not, the central Christian teaching is that Christ will return to judge the living and the dead. This event not only determines the course of human history but also impacts the entire history of the cosmos. Thus, the final event in history is Christ's return, not some natural cosmological occurrence.

How plausible is it to consider that events on Earth could affect the entire cosmos in this manner? Andrew Davison suggests that *Parousia* (Christ's return) might be interpreted as a local event focused solely on humanity, with the rest of the cosmos potentially

§ McIntosh 2021. See also Davison. 2023, 188.

† Especially in the first two volumes, Out of the Silent Planet and Perelandra.

‡ Pawl 2016. Brazier 2014.

§ See, for example, articles in Walls 2010.

continuing to exist relatively unaffected.¶ On the other hand, why should the notion of one objects influence over the entire cosmos be considered implausible? Within the framework of chaos theory, broad models can be constructed in which a small change, often referred to as the "butterfly effect," can disrupt the entire system.

Alternatively, it's possible that organic life in this cosmos is inherently short-lived, with developed cultures destined to destroy themselves. This perspective offers a response to the so-called Fermi paradox: if the cosmos possesses nearly unlimited resources to foster life, why does it appear devoid of other civilizations? One answer is that as cultures advance, they become increasingly vulnerable, with technological advancement outpacing moral development. Furthermore, advanced cultures become more reliant on technology, making them susceptible to rapid collapse in the face of technological disruptions. Our own world could be obliterated in an instant by a virus outbreak, a nuclear weapon, a comet impact, or a solar storm. Therefore, it's unsurprising that we haven't encountered other civilizations; they may flourish briefly before fading away. This idea aligns well with Christian eschatology.

Another approach is to interpret eschatological teachings purely symbolically, avoiding any direct reference to future events. While this may circumvent problematic literal interpretations of passages from the Book of Revelation, it raises questions about the Christian salvation narrative, which seems to necessitate some form of concrete historical conclusion. Can this be reconciled with the idea that humanity's story may continue for hundreds of thousands of years or longer? The answer to this question depends partly on the interpretation of Gospel's teaching on Christ's return. Initially, the disciples anticipated Christ's imminent return, but over time, the belief evolved into the notion that history could end at any moment, albeit with no precise timeframe. This aligns well with the lengthy temporal journey of humanity, as depicted in Gene Wolfe's *Book of the New Sun* series.

Evaluating the Religiosity of UFO Experiences

The scholars of religious studies have analyzed numerous UFO movements and cults. For instance, some groups, like Heaven's Gate or Raëlism clearly exhibit characteristics of a cult, but not all religiosity surrounding UFO's is this extreme. However, this is not a place to pass judgment on the religiosity of UFO movements. Instead, I am merely highlighting the tendency of individuals who have had UFO experiences to interpret them as spiritual or transcendent events, while there seems to an absence of a shared rituals or any other overtly religious practices among UFO believers.

For instance, Diana Pasulka has conducted research and interviews with several individuals she calls as "experiencers."† In these encounters, UFOs are not always necessarily perceived as conventional metallic ships with nuts and bolts, but as something entirely different. What stands out is the frequent comparison of these experiences in literature to encounters with angels or demons. In classical angelology, angels are not material beings in the same sense as humans or other objects in the physical world. Nevertheless, they can assume forms that appear material and exert influence on the material realm.

Experiencers do not form a homogeneous group, yet despite this diversity, many of them display a notable inclination to interpret their experiences within a deeply religious framework.‡ Thus, UFO experiences do not necessarily diminish the religiosity of experiencers; in many instances, they appear to intensify it. However, it remains unclear what kind of religiosity UFO experiences engender. Pasulka recounts numerous cases where the outcome is, at the very least, an increase in spiritual awareness, perhaps with an

¶ Davison 2023, 351–366.
† Pasulka 2023. Pasulka 2019.
‡ See, for example, the Netflix documentary Encounters (2023).

occult twist, while for others, it has led to a deeper commitment to Roman Catholic Christianity. Studying this topic is complicated by the reticence of experiencers to discuss their encounters openly, often leaving many details undisclosed when recounting their experiences.

Reflections

In this article, I have not addressed issues such as ethics, politics, and jurisprudence within the context of astrobiology.† While these questions may possess a religious component, they cannot be considered fundamentally religious in the same manner as actual inquiries into Christian doctrines. For instance, inquiries into whether the natural states of exoplanets can be altered through terraforming processes, or considerations regarding the procedures for collecting or preventing space debris, are significant. However, they can, at least initially, be approached independently of a religious framework.

A separate yet parallel question pertains to the attitudes of various other religions and worldviews toward extraterrestrial intelligence. This constitutes a broad inquiry requiring an in-depth analysis of each tradition and its internal variations, a topic that has been explored in several studies.‡

The four tenets of Christian theology examined in this article exhibit notable differences. While the first two often feature in popular criticism, in reality, there is no real challenge there for Christian cosmology. The genuine challenges arise in the realms of soteriology and eschatology. However, the eschatological question is, in a sense, trivial. When the world ends, it ends, and only then will we ascertain who was correct and who was mistaken (assuming such discernment is even possible). In other words, this issue is not actualized in the same manner as potential soteriological challenges that concern the specificity and centrality of Christ.

If we think about ETI and us humans along the lines where we are more or less culturally in the same space, the societal and theological challenges for Christianity and possible narrative incoherence could be assessed using the platform offered in this article. But if we consider some other cosmic scenarios, the incoherence will grow. These scenarios include the darker solutions to the Fermi paradox, including variations of Zoo and Dark Forest hypotheses and possibilities that we live in some kind of simulated or artificial reality. In these cases, the evidence that Christian religion is based on could still support the truthlikeness of Christian doctrinal claims in intra-world sense yet fail to be true in the ultimate sense. This means that given the state of the world as we perceive it now Christian worldview has at least some support so that it is a rational worldview to hold, but if the facts about the world change, then it may lose this support. However, even this scenario would not necessarily disprove all the claims included in the Christian worldview, or lead to the end of religion. Religions are to a certain extent akin to philosophies or philosophical statements so that they are not necessarily strongly affected by empirical considerations. For example, the existence of moral facts or the existence of God cannot be argued based on the empirical phenomena even if some phenomena can be taken as confirming or disconfirming these claims in cumulative sense. All this is to say that we may expect Christian religion, and other religions too, to adapt to possible astrobiological or cosmological findings, and how much they can adapt is affected by the nature of these discoveries. However, at this point of history, we can only speculate, and it is also only after some confirmed facts, things might get truly interesting.

† Of these, see e.g. Wilson 2015.

‡ Weintraub 2014. See also Peters 2018, 208-270. Playford, Bulllivant, &; Siefert 2024.

Literature

Brazier, Paul (2014). C. S. Lewis: The Question of Multiple Incarnations, Heythrop Journal, LV, 391–408. https://doi.org/10.1111/heyj.12049

Catling, David C. (2013). *Astrobiology: A Very Short Introduction.* Oxford: Oxford University Press. https://doi.org/10.1093/actrade/9780199586455.001.0001

Davison, Andrew (2023). *Astrobiology and Christian Doctrine.* Cambridge: Cambridge University Press. https://doi.org/10.1017/9781009303187

Derham, William (1715). *Astrotheology: Or a Demostration of the Being and Attributes of God from a Survey of the Heavens.* London: Willian Innys.

Dick, Steven J. (1998). *Life on Other Worlds: The 20th-Century Extraterrestrial Life Debate.* Cambridge: Cambridge University Press.

—— (2000). Cosmotheology: Theological Implications of the New Universe. Stephen J. Dick (toim.), *Many Worlds: The New Universe, Extraterrestrial Life, and the Theological Implications.* West Conshohocken, PA: Templeton Foundation Press, 191–210.

Domagal-Goldman, Shawn D., and Katherine E. Wright (2016). The Astrobiology Primer 2.0, *Astrobiology*, 16, 561–653. https://doi.org/10.1089/ast.2015.1460

Everitt, Nicholas (2004). *The Non-Existence of God.* London: Routledge. https://doi.org/10.4324/9780203643785

Finocchiaro, Maurice A. (2010). *Defending Copernicus and Galileo: Critical Reasoning in the Two Affairs.* Dordrecht: Springer.

Jayawardhana, Ray (2011). *Strange New Worlds: The Search for Alien Planets and Life Beyond Our Solar System.* Princeton: Princeton University Press.

Lewis, Clive Staples (1964). *The Discarded Image: An Introduction to Medieval and Renaissance Literature.* Cambridge: Cambridge University Press.

Lovin, Robin (2015). Astrobiology and Theology. Stephen J. Dick (toim.), *The Impact of Discovering Life Beyond Earth.* Cambridge: Cambridge University Press, 222–32. https://doi.org/10.1017/CBO9781316272480.018

McIntosh, Chad (2023). A Spectrum View of the Imago Dei, *Religions*, 14. https://doi.org/10.3390/rel14020267

McIntosh, Chad, and Tyler Dalton McNabb (2021). Houston, Do We Have a Problem?, *Philosophia Christi*, 23, 101–24. https://doi.org/10.5840/pc202123110

McMullin, Ernan (2005). *The Church and Galileo.* South Bend, IN: Notre Dame University Press.

McNabb, Tyler Dalton, and Chad McIntosh (2024). Houston, Naturalists Have a Problem: ETI and Evidence for Theism, *Space, Philosophy and Ethics* (tulossa).

Melanchthon, Philip (1550). *Initia Doctrinae Physicae.* Wittebergae.

O'Meara, Thomas F. (1999). Christian Theology and Extraterrestial Intelligent Life, *Theological Studies*, 60, 3–30. https://doi.org/10.1177/004056399906000101

Paine, Thomas (1995). *Age of Reason.* New York: Library of America.

Pasulka, D. W. (2019). *American Cosmic: UFOs, Religion, Technology.* Oxford.

—- (2023). *Encounters: Experiences with Nonhuman Intelligences.* St. Martin's Essentials.

Pawl, Timothy (2016). Brian Hebblethwaite's Arguments Against Multiple Incarnations, *Religious Studies*, 52, 117–30. https://doi.org/10.1017/S0034412514000626

Peters, Ted (2014). Astrotheology: A Constructive Proposal, *Zygon*, 49, 443–57. https://doi.org/10.1111/zygo.12094

Peters (2016). Astrobiology and Astrochristology, *Zygon,* 51, 480–96. https://doi.org/10.1111/zygo.12249

Peters (2018). *Astrotheology. Science and Theology meets Extreterrestial Life.* Eugene, Or: Wipf and Stock.

Pittard, John (2019). *Disagreement, Deference, and Religious Commitment.* Oxford: Oxford University Press. https://doi.org/10.1093/oso/9780190051815.001.0001

Playford, Richard, Stephen Bulllivant, and Janet Siefert (2024). *God and Astrobiology.* Cambridge: Cambridge University Press. https://doi.org/10.1017/9781009296175

Schleiermacher, Friedrich (1893). *On Religion. Speeches to its Cultured Despisers.* Tr. John Oman. London: Paternoster house.

Schneider, Susan (2016). Alien Minds. Steven J. Dick (toim.), *The Impact of Discovering Life Beyond Earth.* Cambridge: Cambridge University Press, 189–206. https://doi.org/10.1002/9781118922590.ch17

Thigpen, Paul (2024). A Very Short Introduction to the History of Catholic Debates About the Multiverse and Extraterrestrial Intelligence, University of Notre Dame, Accessed 14.3.2024. https://churchlifejournal.nd.edu/articles/a-very-short-introduction-to-the-history-of-catholic-debates-about-the-multiverse-and-extraterrestrial-intelligence/?utm_content=285750796&utm_medium=social&utm_source=facebook&hss_channel=fbp-128985777176957.

Vainio, Olli-Pekka (2018a). The Argument from Scale Revisited, *Theology and Science*, 16, 439–46. https://doi.org/10.1080/14746700.2018.1525223

Vainio, Olli-Pekka (2018b). *Cosmology in Theological Perspective. Understanding Our Place in the Universe.* Grand Rapids, MI: Baker Academic.

Walls, Jerry L. (2010). *The Oxford Handbook of Eschatology.* Oxford: Oxford University Press.

Webb, Stephen (2002). *If the Universe Is Teeming with Aliens ... Where is everybody?: Fifty Solutions to the Fermi Paradox and the Problem of Extraterrestrial Life.* New York: Praxis.

Weintraub, David A. (2014). *Religions and Extraterrestrial Life: How Will We Deal With It?* Heidelberg: Springer. https://doi.org/10.1007/978-3-319-05056-0

Wilkinson, David (2013). *Science, Religion, and the Search for Extraterrestrial Intelligence.* Oxford: Oxford University Press. https://doi.org/10.1093/acprof:oso/9780199680207.001.0001

Wilson, Elspeth M., and Carol E. Cleland (2015). The Moral Subject of Astrobiology. Stephen J. Dick (toim.), *The Impact of Discovering Life Beyond Earth.* Cambridge: Cambridge University Press, 207–21. https://doi.org/10.1017/CBO9781316272480.017

(Toward) Discovery of Life beyond Earth and its Impact
Proceedings IAU Symposium No. 387, 2026
H. Landt, M. Dominik & C. Oliver, eds.
doi:10.1017/S1743921325103050

A Sign in Space: Analogies between the Project and a Real First-Contact Scenario

Daniela de Paulis
SETI Institute, SETI Artist-in-Residence Program
email: selavyrose@gmail.com

Abstract. On 24 May 2023, a simulated extraterrestrial message was transmitted towards Earth by the Trace Gas Orbiter, a spacecraft of the European Space Agency. The radio signal containing the message was received by several terrestrial radio antennas, as part of the interdisciplinary project "A Sign in Space". I conceived the project in 2019, and developed it in collaboration with the SETI Institute, the Green Bank Observatory, the Italian Institute for Astrophysics and the European Space Agency. The message designed for the project required one year to be decoded by citizen scientists. The public reaction to the project provides preliminary insights into the unfolding of a concrete First-Contact scenario.

Keywords. SETI, art project, space

Acronyms

SETI: Search for Extraterrestrial Intelligence
IAA: International Academy of Astronautics
SETI: Search for Extraterrestrial Intelligence
GBO: Green Bank Observatory
GBT: Green Bank Telescope
ATA: Allen Telescope Array
INAF: Italian National Institute for Astrophysics
ESA: European Space Agency
ET: Extraterrestrial
BL: Breakthrough Listen

Early Stages and Objectives of the Project

"A Sign in Space" was conceived as a realistic representation of a First-Contact scenario, engaging SETI researchers and the general public in the detection, decoding and interpretation of a simulated message from an extraterrestrial civilisation. From its early stages, the project aimed at actively engaging the world-wide community of SETI researchers, creating a global network of scientific institutions and independent specialists collaborating towards a concrete scheme, as if in the event of a confirmed First-Contact scenario. In its initial planning phase, the project also aimed at actively involving the scientific community by implementing the various stages of the IAA Post-Detection Protocol [1]. "A Sign in Space" was thus conceived both as a concrete application of the Post-Detection Protocol developed by the IAA SETI Committee, and as a theatrical event, staging a fictional scenario with the cooperation of prominent scientists and engineering resources. The project aimed at highlighting and exploring the societal, cultural and scientific impact of a First-Contact scenario, as well as testing the intellectual framework devised by specialists to regulate such event. Since 2015, I have been actively participating in SETI research as a member of the IAA Committee. The work conducted by various teams in the Committee, searching for technosignatures, and assessing the possible societal reaction to the detection of a signal from an extraterrestrial civilisation, have allowed me to establish a well-informed structure to build the project upon. During the early stages of the work, I formed several specialist teams: the scientific team, comprised of ESA space engineers, radio astronomers at GBO, INAF and ATA; the outreach team, comprised of communication specialists from the SETI Institute, INAF and GBO; the message team, comprised of specialists from various fields, including anthropology, philosophy, computer science, astrobiology. All advisors helped critically assess and develop the message composition process and the project in general [2].

Live event

"A Sign in Space" required four years development before the launch event on 24 May 2023. Coordinated and hosted by the SETI Institute, the launch event was streamed live on the SETI Institute's YouTube channel, with a world-wide following [3]. The livestream featured some of the project's direct collaborators, as well as astronomy outreach specialists and entrepreneurs connected to the SETI Institute. The event, structured as a panel conversation, reflected upon the collaborative structure of the project, emphasizing its scientific and artistic objectives. During the live programme, the message designed for the project was transmitted by the Trace Gas Orbiter towards Earth, reaching the collaborating radio telescopes – Allen Telescope Array, Medicina Radio Astronomical Antenna, Green Bank Telescope – after sixteen minutes. The reception of a signal from an extraterrestrial intelligence by various facilities is the ideal First-Contact scenario, as declared in the IAA Post-Detection Protocol. After being detected at one facility, the signal should be monitored by other telescopes, in order to ascertain that the signal is transmitted by an extraterrestrial intelligence, and thus rule out other possible sources. During the live event, the reception of the signal was visualised in real time on the computer screens of the Medicina Radio Astronomical Antenna. Possibly influenced by the movie "Contact" (R. Zemeckis, 1997) - in which the scientist who plays the main character listens to the sound of a signal from an extraterrestrial civilization, while it is being received by a radio telescope - people in the YouTube comment section asked why they could not hear the sound of the signal. This was one of the several instances during the project's presentation in which the public's perception of a First-Contact scenario seemed profoundly primed by a renowned science fiction work. The conversations among people active in the decoding process on Discord were also informed by popular science fiction works, showing how influential films and novels about ET might be in guiding the public reaction to a possible First-Contact scenario.

The verisimilitude and scientific accuracy of the project's presentation aimed at creating a vivid narrative around the discovery of an extraterrestrial civilisation in the mind of the international public. The global media greatly contributed in shaping the narrative, by covering the 24 May 2023 launch event and by covering various aspects of the project's scientific and collaborative approach. People from all over the world could experience the presentation of a First-Contact scenario comparable to that envisioned by the scientific community. Several media articles drew analogies between "A Sign in Space" and a scenario with an actual reception of an ET intelligent signal [4] [5]. Before the public launch of the project, it was agreed among all collaborating institutions that the project would be announced as a theatrical simulation, in order to avoid any possible risk associated with the report of a real First-Contact event. As the societal repercussions of the global COVID-19 pandemic were still echoing by the time the project was to be inaugurated, it became ethically significant that "A Sign in Space" would not contribute in fueling the conspiratorial sentiment that had emerged during that time. The detection of a real ET signal could prove as destabilising or even more alarming to societies around the world than a global pandemic. The narrative reported by the global media around "A Sign in Space" remained consistent and accurate concerning the simulation. As to be expected in a global media coverage, detailed information about the project were however disregarded or distorted. In the case of a real ET detection, it is possible that, despite all efforts by the global media to accurately report the facts, parallel fictions might infiltrate or even dominate the main narrative, as in the COVID-19 pandemic.

Data Management

As soon as the radio signal was received at the GBT, ATA and Medicina Radio Antenna, radio engineers and astronomers collaborating in the project downsized the data received before sharing them in the public domain. The data were initially released on the Breakthrough Listen Open Data archive [6] and later on Zenobio. The data of the raw signal from the three facilities were downloaded by people from around the world. Given the considerable size of the raw data, one of the scientists collaborating in the project advocated for complementing the open data repository options with storage space provided by a blockchain company, in order to facilitate the downloading of the signal from locations with a reduced Internet service [7]. This option was revoked shortly after, in order to prevent commercial exploitation of the data of the signal and of the project.

I find that the strong interest by a private company to host the data of the message on their website could be used as an important lesson towards an actual First-Contact scenario. Should a real ET signal be detected, data would become extremely precious, and could be targeted for commercial or other forms of exploitation. Without a well-established and properly executed procedure for the storage and handling of the data by members of the scientific community, the content of a confirmed ET signal could be intentionally corrupted or manipulated, to serve political, ideological, financial gains. The lack of detailed measures on how to store and secure the data could permanently damage decades of scientific investment in ET search.

In "A Sign in Space", the raw data of the signal were tackled by members of the general public active on a dedicated Discord channel. After a few days, a small group of citizen scientists managed to successfully extract the message data, and a confirmation of the correct process and outcome was announced on the channel. While dealing with the considerable challenges of decoding the message designed for the project, members of the public continued analysing the raw data of the signal to review the early stages of the process, and ensure that the data of the message were properly extracted. The confirmation from the project's team however assured the public that the process was accomplished correctly. In the case of a real detection, it might be extremely challenging distinguishing the intentional content of a potential alien message from the raw signal data. The obstacles posed by the crucial steps of extracting the message from the signal data might hinder the entire decoding process: without any feedback on the certainty of the correct extraction of the content from the raw signal, the translation of such content might lead to flawed interpretations. Inevitable uncertainties posed by the decoding process of an extraterrestrial intelligent signal, might pave the way to opportunistic exploitation of the data, to interpretations supporting controversial ideologies, to polarising ideas or dogmas, with long lasting societal, political, cultural consequences, as in the case of the COVID-19 pandemic.

In "A Sign in Space", the public played a crucial role in decoding and interpreting the simulated extraterrestrial message. The IAA Post-Detection Protocol does not explicitly address the question of open data sharing beyond the scientific community. In the case of a real ET signal detection, the scientific community might prefer keeping the data confidential within restricted specialist circles, analysing the content at facilities removed from public scrutiny. This approach however raises security concerns for the specialists involved in the decoding process. The option of keeping the data confidential within a limited intellectual elite could lead to distrust and even emotional instability among members of the general public. Conspiracy theories and anti-science sentiments might thrive as a result, as during the recent pandemic. The option of sharing the data in the public domain as in "A Sign in Space", might help mitigate these risks, however it could lead to an even more problematic series of social responses that are arduous to anticipate. In "A Sign in Space" the decoding process took place in a relatively controlled environment, with moderators and procedures in place. Also, the public was aware that the project was an artistic endeavour and a simulation, therefore no clear analogies between the project and a real First-Contact scenario can be drawn at this point, in relation to consequences associated with the possible misinterpretation and misuse of the data of a confirmed extraterrestrial intelligent signal by members of the general public.

The general public offered a great variety of decoding approaches to the simulated extraterrestrial message. The cooperation between specialists and the general public could be beneficial in an actual First-Contact scenario. Complementing the skills and perspectives of professionals with those provided by citizen scientists could help broaden the limits of a specialised approach that ultimately might prove unsatisfactory when decoding a signal designed by an extraterrestrial intelligence: the greater the number of approaches, the wider the spectrum of interpretations, the more likely the possibility of identifying the meaning of a signal conceived by a civilization with skills and cognitive abilities very different from those of humans.

Public Reactions

Throughout the entire duration of the project, I experienced overall positive reactions by the general public, even delight to the possibility of encountering a more advanced civilization that might help humans overcome their endless struggles. The conversations on Discord have been generally respectful and science/knowledge driven, with some highly engaging and interesting examples. In most cases, the dialogues tackled existential, scientific and philosophical questions related to the possible discovery of an extraterrestrial intelligence. In a few instances, some members of the decoding community attempted to stir the conversation towards various declinations of pseudo-science and alien encounter narratives, however these instances were short lived and dismissed by the majority of the decoding community. It required more than one year for the message to be decoded since the launch of the project [8]. During this time, the decoding process evolved through different stages among the decoding community: determination and optimisms fluctuated with frustration and discouragement. Due to the absence of reference points, decoders moved through the entire process in darkness, with any decoding direction being equally misguided or correct. Only a few citizen scientists undertook a specific decoding direction for a sustained period of time, while struggling to see progress or resolution in their quest. Some people active in the inquiry reported lacking sleep because of the challenges posed by the decoding process, having dreams informed by the simulated extraterrestrial message, or generally having become obsessed with the task. The sense of frustration and gloom became very palpable at times, and this could have escalated further if the message had not been decoded at last. In the case of a real First-Contact scenario, the dissatisfaction and resentment resulting from an enigmatic and frustrating decoding process, with many decoding possibilities open yet equally uncertain, combined with the lack of clear reference points, could escalate both within the scientific teams and the citizen scientist community engaged in the quest.

Following the successful decoding of the simulated extraterrestrial message designed for the project, "A Sign in Space" remains open-ended: the community of people who gathered on Discord to decode the signal is now attempting to interpret the message through various theories, and to ideally converge towards a specific meaning. People from different cultural backgrounds and fields of expertise are attempting to agree on the cultural significance of the message, advancing the investigation beyond the mathematical language. Should a real intelligent message from space be received here on Earth, different and contradictory interpretative and decoding efforts might emerge. Humankind might not be able to agree on a specific meaning to an ET message, across all cultures. The project hints at a prospect in which the study of a confirmed intelligent signal from space might lead to inconclusive results, questioning whether humankind might be able to cope with profound uncertainties when dealing with a First-Contact scenario in the long term.

Following the launch of "A Sign in Space", some members of the general public believed that a real intelligent signal from space was detected. Their reaction, expressed through posts in the message sections of online articles and YouTube videos reporting about the project, shows that they welcomed the news with curiosity. These members of the public were however a few exceptions, whereas the majority of people who engaged with the project seemed properly informed about its fictional nature. Some people posted public messages displaying apathy, sarcasm or anger at the use of public funding for a type of endeavour they deemed pointless. It is possible that a First-Contact scenario might be welcomed with a similar spectrum of reactions, from positive and enthusiastic, to apathetic or negative.

Since the launch event, I received many emails, drawings, interpretations, whose content was mostly enthusiastic and constructive. Members of the general public wrote to share their interest in the project and to express their opinion on the possible discovery of extraterrestrial intelligent life. Some of these people shared a long-term fascination with the topic of extraterrestrial intelligence and life beyond Earth. Although not necessarily specialists, most members of the public who actively engaged with the project seemed already versed on the topic of SETI. It is possible that the public announcement of an actual First-Contact scenario will mostly impact people with an intellectual, economic, political or ideological interest in the topic, whilst leaving the majority of people mildly or completely unengaged. Possible societal threats to the announcement of a First-Contact scenario might emerge instead from the sustained manipulation and biased communication of data and information, stirring public opinion towards a potentially controversial assimilation of the event.

Conclusions

In this paper I presented some initial and informal insights into the possible analogies between "A Sign in Space" and a concrete First-Contact scenario. Greater research is required, also in collaboration with specialists, such as data analysts, anthropologists, philosophers, sociologists, to properly assess the compatibility of the two events.

This paper is a complementary text to the previous publication about the project, presented as part of the Billingham Cutting Edge Lecture at the International Astronautical Congress in Milan in October 2024 [9].

For more than a year, the project engaged a broad public, prompting the creative and intellectual participation of citizen scientists from various cultural backgrounds. A large amount of written and visual material has been produced as part of the project, providing useful tools for future research on the cultural impact of a First-Contact scenario.

Bibliography & Notes

[1] Protocol for an ETI Signal Detection https://iaaseti.org/en/declaration-principles-concerning-activities-following-detection/ (last retrieved on 11 February 2025). This is the version of the Protocol that informed the project.

[2] Daniela de Paulis et al. "A sign in space: An interdisciplinary exploration of the potential reception of an extraterrestrial signal", Volume 2012, Acta Astronautica.

[3] SETI Live: A Sign in Space - Simulating First Contact https://www.youtube.com/watch?v=zgQq36IR-Lk&t=1050s, 24 May 2023 (last retrieved on 17 February 2025).

[4] Sigal Samuel, A message just arrived from outer space: can we decode it? VOX https://www.vox.com/future-perfect/2023/5/24/23736170/aliens-message-seti-extraterrestrial-intelligence, 24 May 2023 (last retrieved on 17 February 2025).

[5] Taylor Nicioli, Alien-like signal from 2023 has been decoded. The next step is to figure out what it means, CNN https://edition.cnn.com/2024/11/11/science/seti-alien-signal-decoded-sign-in-space/index.html#openweb-convo 11 November 2024 (last retrieved 17 February 2025).

[6] Astronomers at the Green Bank Observatory used the Breakthrough Listen backend to receive and record the signal transmitted by TGO. BL Open Data Archive https://breakthroughinitiatives.org/opendatasearch (last retrieved 17 February 2025).

[7] https://blockworks.co/news/blockchain-storing-messages-from-mars (last accessed on 13 February 2025).

[8] ESA 'Alien' signal decoded, 22 October 2024, https://www.esa.int/ESA_Multimedia/Images/2024/10/Alien_signal_decoded (last accessed on 17 February 2025).

[9] Daniela de Paulis et al. "Global outreach and cultural impact of A Sign in Space, an interdisciplinary simulation of a First Contact scenario" International Astronautical Congress 2024, Milan (IT).

(Toward) Discovery of Life beyond Earth and its Impact
Proceedings IAU Symposium No. 387, 2026
H. Landt, M. Dominik & C. Oliver, eds.
doi:10.1017/S1743921324001996

Life in the clouds of Venus: The phosphine debate and its scientific, social, and psychological impact

Clément Vidal and Daliah Bibas

Center Leo Apostel, Vrije Universiteit Brussel, Belgium
emails: contact@clemvidal.com, daliah.raquel.bibas@vub.be

Abstract. We examine the phosphine on Venus debate using a multidimensional impact model covering scientific, social and psychological dimensions. We illustrate the results of the analysis with two spider diagrams.

Keywords. Phosphine, Venus, Social Impact, Extraterrestrial Life

---

1. Introduction

The potential for life on Venus has been envisioned (e.g. Grinspoon 2003; Schulze-Makuch et al. 2004; Limaye et al. 2018; Morowitz and Sagan 1967), however, no indication has ever been discovered until recently. In September of 2020, Dr. Jane Greaves and her international team published a paper in *Nature Astronomy* about their discovery of phosphine in the clouds of Venus (Greaves et al. 2020). Phosphine, a molecule associated with anaerobic environments on Earth, suggested potential biological or unknown chemical processes on Venus. However, the discovery faced immediate scrutiny and scepticism. In this article, we discuss the phosphine debate and its scientific, social, and psychological impact on humanity, using the multidimensional impact model for the discovery of extraterrestrial life proposed by Vidal (2015). We illustrate the model and its application to the phosphine debate with spider diagrams in Fig. 1 and 2.

2. Empirical Dimensions

To determine the impact of the phosphine debate on Venus, we must first look at the ten dimensions characterising the potential lifeform discovered. In terms of (1) *Distance*, Venus is in the inner solar system, neighbouring Earth, which presents a higher impact than a distant discovery. The candidate microbial life form is simple on a (2) *Complexity* scale and expected to be small in (4) *Size*, similar to microorganisms that we know exist on Earth. Microorganisms are typically found in great (5) *Numbers*, and the detection of phosphine suggests that they are (6) *Alive*, as the relatively short atmospheric lifetime of the reactive gas phosphine would imply that it is being replenished continuously or was produced very recently.

The (7) *Clarity* of the discovery was low and rather hypothetical. Although it was recently suggested that the presence of phosphine in the atmosphere of a rocky planet is a promising indicator of life (Sousa-Silva et al. 2020), phosphine is found in the reducing atmospheres of giant planets (Bregman et al. 1975; Tarrago et al. 1992) and there may be some unknown chemical processes that can produce phosphine abiotically on Venus (Greaves et al. 2020; Sousa-Silva et al. 2020). Since we're dealing with a potential biosignature, the dimensions of (3) *Intent*, (8) *Communicative Intent*, and (9) *Knowledge*

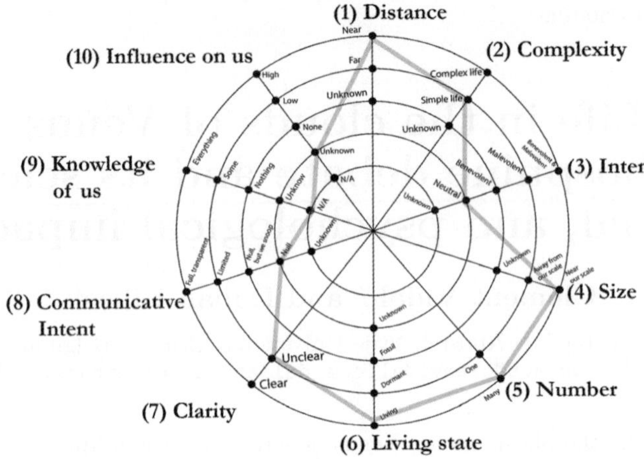

Figure 1. Spider diagram of the dimensions of impact of the phosphine debate on Venus. Dimensions (1-10) describe the kind of extraterrestrial life. Details about the generic impact model, its dimensions and options are in (Vidal 2015).

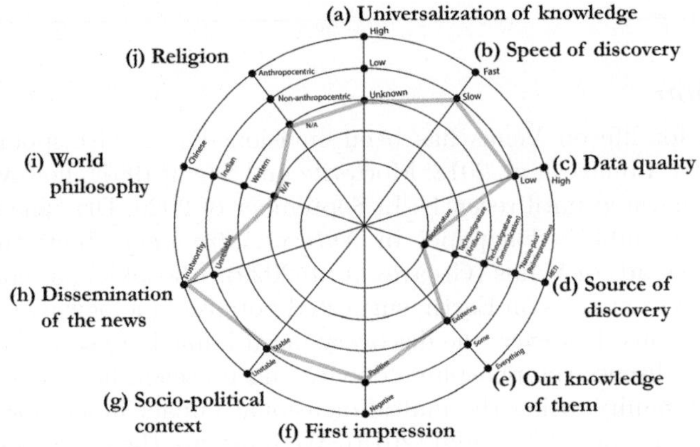

Figure 2. Spider diagram of the dimensions of impact of the phosphine debate on Venus. Dimensions (a-e) describe the scientific impact and dimensions (f-j) describe the social dimensions of impact. Details about the generic impact model, its dimensions and options are in (Vidal 2015)

of Us are inapplicable. The inapplicability of these dimensions suggests a lower overall impact than a technosignature where such dimensions might be applicable. However, it is important to consider the potential (10) *Influence on us*. Since Venus and Earth are neighbouring planets and have both existed for long time scales, and Venus might have been habitable (e.g. Way et al. 2016), potential exchanges of material through lithopanspermia could have taken place. It is therefore possible that potential life on Venus and life on Earth could share a common ancestor.

3. Scientific Impact

The scientific impact of the phosphine discovery on Venus can be characterised first, by the (a) *Universalization of Knowledge* dimension, which would be low if the lifeform turns out to be very similar to known microorganisms on Earth, or high if it turns out to have a different make up and mechanics (e.g. no RNA, DNA, or ribosomes). The (b) *Speed of Discovery* has been relatively slow as it took time for independent teams to re-analyze the data, and of course the "discovery" remains in debate (Bains et al. 2021; Encrenaz et al. 2020; Lincowski et al. 2021; Snellen et al. 2020; Thompson 2020; Trompet et al. 2021; Villanueva et al. 2021). Furthermore, it would take even more time to send a sampling mission to Venus to analyse and test such potential Venusian life. Despite checks from two independent telescopes, the (c) *Data Quality* and its processing was low according to many experts, and this was aggravated by the fact that the ALMA telescope had some mistakes in their data release that they had to correct. The (d) *Source of Discovery* is a biosignature, which is a kind of signature that is typically expected in the astrobiology community. If confirmed without return-sample missions, (e) *Our Knowledge* of Venusian life forms would likely be limited to existence.

4. Social Impact

The (f) *First Impression* of the discovery was positive as it was unanimously welcomed. Although many were sceptical and scrutinised the initial findings, it was overall a positive first impression and it did not cause chaos. At first, major media outlets like *CNN*, *BBC*, and *The New York Times* reported the discovery enthusiastically (Amos 2020; Stirone et al. 2020; Strickland 2020), however, as independent teams re-analyzed the data, new outlets, such as *Forbes* and *Wired*, covered the controversy and highlighted that the initial discovery has been disproven (Beall 2020; Siegel 2020). Despite the COVID19 pandemic, the (g) *Socio-Political Context* can be argued to have been stable with minimal impact on space sciences. The (h) *Dissemination of the News*, as often, was rather sensationalist. The authors of the initial paper were careful when putting forth the possibility of life on Venus, but scientists complained that it was premature to even suggest this possibility (Siegel 2020). The dimensions of (i) *World Philosophy* and (j) *Religion* are not easy to study, as they would imply data gathering on how various human groups with different world philosophy or religion did react to this news.

5. Psychological Impact

Psychological impact is also very difficult to study, as it would imply data gathering on an individual level across various human groups, which is why we do not attempt to draw a spider diagram for these dimensions. However, the subjective impact of finding life on Venus would be high for many scientists as Venus has not been seen as a promising target, and most of the attention and hope was geared towards Mars and icy moons. For such scientists, their (k) *Preconceptions of Extraterrestrial Life* would have been challenged. Discovering life on Venus (or Mars) may lead to major (l) *Worldview* changes, especially because the possibility of cross-contamination from or towards Earth could change our evolutionary theories of the origin of life on Earth. There might be a remote danger of a Venusian microbe impacting Earth's ecosystem, for example, should we decide to make a return-sample mission and accidentally contaminate and dramatically affect Earth's ecosystem. In such a scenario, the human (n) *Hierarchy of Needs* would be impacted at a basic level if it directly affects human health or the food chain supply. To further study a particular individual reaction to the discovery, the (m) *Developmental Stage*, (o) *Age*, and (p) *Personality* of that individual matters. The phosphine debate involves at least two major uncertainties: the actual detection of phosphine itself and the link between phosphine and life. Someone emotionally stable, flexible in their worldview, positively minded, and with high self-esteem would cope with the uncertainties much more easily than someone who is emotionally unstable, with a dogmatic worldview, negatively minded, and with low self-esteem.

6. Conclusion

The potential discovery of phosphine on Venus has had far-reaching impacts across scientific, social, and psychological dimensions. If life on Venus can be confirmed, this could mean that life is common in the solar system, and it would not be far-fetched to argue that it is extremely common in the galaxy. This paradigm shift would challenge our fundamental understanding of life's distribution in the cosmos. Although not yet a confirmed discovery, the impact of the potential discovery is still seen as high because (1) the discovery is nearby and thus follow-ups are not only possible, but already planned (see e.g. the Venus Life Finder mission, French et al. 2022), (2) it has stimulated research, debate, and international collaboration, and (3) it has reignited the scientific community's curiosity, which had largely discounted the possibility of life on Venus. This phosphine debate is highlighting the need to study more systematically the potential presence of life in the solar system. With a renewed focus and the latest push in missions towards Venus, we are on the brink of potentially groundbreaking discoveries with higher levels of scientific, social, and psychological impact.

Acknowledgements

We thank David L. Clements for insightful discussions.

References

Amos, J. 2020, Is there life floating in the clouds of Venus? `https://www.bbc.com/news/science-environment-54133538`.

Bains, W., Petkowski, J. J., Seager, S., Ranjan, S., Sousa-Silva, C., Rimmer, P. B., Zhan, Z., Greaves, J. S., & Richards, A. M. S. 2021, Phosphine on Venus Cannot Be Explained by Conventional Processes. *Astrobiology*, 21(10), 1277–1304.

Beall, A. 2020,. Scientists found signs of life on Venus. Now they're not so sure. `https://www.wired.com/story/venus-phosphine-life-questions/`.

Bregman, J. D., Lester, D. F., & Rank, D. M. 1975, Observation of the nu-squared band of PH3 in the atmosphere of Saturn. *The Astrophysical Journal*, 202, L55.

Encrenaz, T., Greathouse, T. K., Marcq, E., Widemann, T., Bézard, B., Fouchet, T., Giles, R., Sagawa, H., Greaves, J., & Sousa-Silva, C. 2020, A stringent upper limit of the PH3 abundance at the cloud top of Venus. *Astronomy & Astrophysics*, 643, L5. arXiv:2010.07817 [astro-ph].

French, R., Mandy, C., Hunter, R., Mosleh, E., Sinclair, D., Beck, P., Seager, S., Petkowski, J. J., Carr, C. E., Grinspoon, D. H., Baumgardner, D., & on behalf of the Rocket Lab Venus Team 2022, Rocket Lab Mission to Venus. *Aerospace*, 9(8), 445. Number: 8 Publisher: Multidisciplinary Digital Publishing Institute.

Greaves, J. S., Richards, A. M. S., Bains, W., Rimmer, P. B., Sagawa, H., Clements, D. L., Seager, S., Petkowski, J. J., Sousa-Silva, C., Ranjan, S., Drabek-Maunder, E., Fraser, H. J., Cartwright, A., Mueller-Wodarg, I., Zhan, Z., Friberg, P., Coulson, I., Lee, E., & Hoge, J. 2020, Phosphine gas in the cloud decks of Venus. *Nature Astronomy*,, 1–10. Publisher: Nature Publishing Group.

Grinspoon, D. Venus and Mars. In *Lonely Planets: The Natural Philosophy of Alien Life* 2003,, pp. 167–190. HarperCollins.

Limaye, S. S., Mogul, R., Smith, D. J., Ansari, A. H., Słowik, G. P., & Vaishampayan, P. 2018, Venus' Spectral Signatures and the Potential for Life in the Clouds. *Astrobiology*, 18(9), 1181–1198. Publisher: Mary Ann Liebert, Inc., publishers.

Lincowski, A. P., Meadows, V. S., Crisp, D., Akins, A. B., Schwieterman, E. W., Arney, G. N., Wong, M. L., Steffes, P. G., Parenteau, M. N., & Domagal-Goldman, S. 2021, Claimed detection of PH3 in the clouds of Venus is consistent with mesospheric SO2. *The Astrophysical Journal Letters*, 908(2), L44. arXiv:2101.09837.

Morowitz, H. & Sagan, C. 1967, Life in the Clouds of Venus? *Nature*, 215(5107), 1259–1260. Publisher: Nature Publishing Group.

Schulze-Makuch, D., Grinspoon, D. H., Abbas, O., Irwin, L. N., & Bullock, M. A. 2004, A Sulfur-Based Survival Strategy for Putative Phototrophic Life in the Venusian Atmosphere. *Astrobiology*, 4(1), 11–18. Publisher: Mary Ann Liebert, Inc., publishers.

Siegel, E. 2020,. Venus Is Dead! New Analysis Shows Phosphine, A Possible Biosignature, Is Absent. `https://www.forbes.com/sites/startswithabang/2020/10/22/venus-is-dead-new-analysis-shows-phosphine-a-possible-biosignature-is-absent/`.

Snellen, I. A. G., Guzman-Ramirez, L., Hogerheijde, M. R., Hygate, A. P. S., & van der Tak, F. F. S. 2020, Re-analysis of the 267-GHz ALMA observations of Venus: No statistically significant detection of phosphine. *Astronomy & Astrophysics*, 644, L2. arXiv:2010.09761 [astro-ph].

Sousa-Silva, C., Seager, S., Ranjan, S., Petkowski, J. J., Zhan, Z., Hu, R., & Bains, W. 2020, Phosphine as a Biosignature Gas in Exoplanet Atmospheres. *Astrobiology*, 20(2), 235–268.

Stirone, S., Chang, K., & Overbye, D. 2020, Life on Venus? Astronomers See a Signal in Its Clouds. *The New York Times*,. `https://www.nytimes.com/2020/09/14/science/venus-life-clouds.html`.

Strickland, A. 2020,. A gas found on Earth that signifies life has been detected in the clouds on Venus. `https://www.cnn.com/2020/09/14/world/venus-phosphine-gas-clouds-scn/index.html`.

Tarrago, G., Lacome, N., Lévy, A., Guelachvili, G., Bézard, B., & Drossart, P. 1992, Phosphine spectrum at 4-5 μm: Analysis and line-by-line simulation of $2\nu2$, $\nu2 + \nu4$, $2\nu4$, $\nu1$, and $\nu3$ bands. *Journal of Molecular Spectroscopy*, 154(1), 30–42.

Thompson, M. A. 2020, The statistical reliability of 267 GHz JCMT observations of Venus: No significant evidence for phosphine absorption. *Monthly Notices of the Royal Astronomical Society: Letters*, 501(1), L18–L22. arXiv:2010.15188 [astro-ph].

Trompet, L., Robert, S., Mahieux, A., Schmidt, F., Erwin, J., & Vandaele, A. C. 2021, Phosphine in Venus' atmosphere: Detection attempts and upper limits above the cloud top assessed from the SOIR/VEx spectra. *Astronomy & Astrophysics*, 645, L4. Publisher: EDP Sciences.

Vidal, C. A Multidimensional Impact Model for the Discovery of Extraterrestrial Life 2015,. `https://zenodo.org/records/1286707`.

Villanueva, G. L., Cordiner, M., Irwin, P. G. J., de Pater, I., Butler, B., Gurwell, M., Milam, S. N., Nixon, C. A., Luszcz-Cook, S. H., Wilson, C. F., Kofman, V., Liuzzi, G., Faggi, S., Fauchez, T. J., Lippi, M., Cosentino, R., Thelen, A. E., Moullet, A., Hartogh, P., Molter, E. M., Charnley, S., Arney, G. N., Mandell, A. M., Biver, N., Vandaele, A. C., de Kleer, K. R., & Kopparapu, R. 2021, No evidence of phosphine in the atmosphere of Venus from independent analyses. *Nature Astronomy*, 5(7), 631–635. Publisher: Nature Publishing Group.

Way, M. J., Del Genio, A. D., Kiang, N. Y., Sohl, L. E., Grinspoon, D. H., Aleinov, I., Kelley, M., & Clune, T. 2016, Was Venus the first habitable world of our solar system? *Geophysical Research Letters*, 43(16), 8376–8383.

(Toward) Discovery of Life beyond Earth and its Impact
Proceedings IAU Symposium No. 387, 2026
H. Landt, M. Dominik & C. Oliver, eds.
doi:10.1017/S1743921325103049

Would the discovery of extraterrestrial intelligent life have an impact on Catholic theology?

Enrique Solano

Society of Catholic Scientists (https://catholicscientists.org)

Abstract. This study examines the potential impact of the discovery of extraterrestrial intelligent life (ETI) on Catholic theology. The investigation addresses whether the existence of ETIs would challenge core theological principles, particularly the doctrine of Redemption. The study concludes that the discovery of ETIs would not disrupt the foundational tenets of Catholic faith, which is centered on Jesus Christ. The Church has not made any official statements regarding the theological implications of ETIs, reflecting the current speculative nature of their existence. However, the Vatican Observatory is actively involved in research related to potential life-hosting stars, indicating a scientific interest in the topic.

Keywords. Extraterrestrial intelligence, Catholic theology

1. Is the existence of extraterrestrial intelligent life (ETI) a taboo for Catholic church?

No. No pronouncement has been made by the Church on the possible implications of the discovery of ETIs. This is quite logical given that science still has a long way to go to prove the existence of such life forms (if such a proof ever occurs). This, however, does not prevent some members or institutions from expressing their opinion or even from working on this topic. An example of the above is that the Vatican Observatory† dedicates its own resources (the VATT telescope) to carry out characterisation studies of stars potentially hosting planets (Strassmeier, 2023). Also, prominent members of the observatory, such as its current director (G. Consolmagno‡) or the previous director (Lares, 2020) have published reports and papers on this topic.

2. Would the discovery of ETIs affect my relationship with God?

Not at all. If ETIs were to play a role in our salvation, they would be cited in the Bible, which, together with Tradition and Magisterium, is our safe conduct to salvation. By way of comparison, other extraterrestrial creatures whose existence for Christians is beyond doubt as a dogma of faith and who do play a role in our salvation (angels) are mentioned a total of 767 times. Moreover, the dogmatic constitution Dei Verbum tells us that Revelation culminates in Jesus Christ. Therefore, any hypothetical contact with an extraterrestrial civilisation would not affect our salvation history in any way.

God's love is infinite. The Church tells us that Christ died for each of us and would do so again even if I were the only one on Earth. If other creatures exist and are also loved by God, this would not change my relationship with Him at all.

† https://www.vaticanobservatory.org/
‡ https://www.vaticanobservatory.org/wp-content/uploads/2023/07/Intelligent-Life.pdf

3. The problem of Redemption

Making the existence of ETIs compatible with Redemption may be more complex. What could be the mechanism of salvation used by God with these hypothetical creatures? We could think of two hypothetical scenarios:

- Do not need salvation. It might be that ETIs have not committed what we know as original sin and are still in full harmony with God. Perhaps this is the rule in other civilizations and we are the exception as described in the parable of the lost sheep (Mt 18:10-14).
- Do need salvation. Does the Cross have an infinite salvific capacity or is reincarnation necessary for ETIs that moved away from God? We do not know. What we do know for sure is that God's mercy is infinite and that He will have a plan of redemption for them if necessary.

4. Conclusions

Would the discovery of ETIs have an impact on Catholic theology? The answer is **NO**. The positions of the catholic who looks with apprehension at scientific advances for fear that they will break his faith, and the atheist who thinks that astrophysics and astrobiology are on the way to making a discovery that will put the Catholic faith to shame, make no sense at all.

- Do ETIs exist? Not known.
- If they exist, are they aware of the existence of a creator? Not known.
- Assuming they exist, could we communicate with them and, in particular, could we transmit them the experiences of our relationship with God? Not known.
- Do they need Redemption? Not known.

For the time being, and perhaps for a long time to come, we can only speculate.

References

Lares, M., Funes, J. G., & Gramajo, L. 2020, Monte Carlo estimation of the probability of causal contacts between communicating civilizations. *International Journal of Astrobiology*, 19(5), 393–405.

Strassmeier, K. G., Weber, M., Gruner, D., Ilyin, I., Steffen, M., Baratella, M., Järvinen, S., Granzer, T., Barnes, S. A., Carroll, T. A., Mallonn, M., Sablowski, D., Gabor, P., Brown, D., Corbally, C., & Franz, M. 2023, VPNEP: Detailed characterization of TESS targets around the Northern Ecliptic Pole. I. Survey design, pilot analysis, and initial data release., 671, A7.

(Toward) Discovery of Life beyond Earth and its Impact
Proceedings IAU Symposium No. 387, 2026
H. Landt, M. Dominik & C. Oliver, eds.
doi:10.1017/S174392132400245X

Preparing Humanity for the Discovery of Extra-Terrestrial Life: The SETI Post Detection Hub

John Elliott

University of St Andrews, UK
email: john.elliott@st-andrews.ac.uk

Abstract. This paper introduces the SETI Post-Detection Hub, its origins, motivations, and initial undertakings, as well as the rationale for such an endeavour - to explore and develop thorough and integrated approaches in preparation for a potential discovery event in the Search for Extraterrestrial Intelligence (SETI). Officially launched at the University of St Andrews in 2022, it brings together a highly multidisciplinary international team of experts.

From the outset, our rationale has been to weave together insights from diverse fields, including governance, impact strategies, analytics, and the humanities, with particular attention to the inclusion of all humanity's cultural voices – a vital requirement to fully understand our challenge and its solutions. So, in preparation for the discovery of extraterrestrial life, the Hub will endeavour to grow readiness for complex futures by fostering a synergistic environment that encourages imaginative, methodical preparation, to stir transformative interdisciplinary and diverse engagement.

Keywords. Post-Detection, SETI, Extraterrestrial, Life, Hub

Acronyms/Abbreviations

Search for Extraterrestrial Intelligence (SETI)
SETI Post-Detection Hub (Hub)
St Andrews Centre for Exoplanet Science (StA-CES)
Centre for Global Law and Governance (CGLC)
International Academy of Astronautics (IAA)
UK SETI Research Network (UKSRN)

1. Introduction

Well before NASA or SETI embarked on their 'missions', the existence of life beyond our planet is something we [humans] have often contemplated over the generations – probably even from our prehistory. More recent questions posed, regarding life beyond our planet and our preparedness for its discovery, are explored in Winston Churchill's "remarkably scientific contemplation of the potential for alien life" Are There Men on the Moon? (Churchill 1942 – original draft 1939), Oppenheimer and Einstein's (top secret) joint document entitled "Relationships with Inhabitants of Celestial Bodies" Oppenheimer & Einstein (1947), where they include post detection recommendations such as "it would be necessary to establish communication with them through some language or other .... and international law should make place for a new law on a different basis, and it might be called "Law Among Planetary Peoples", as well as in many fictional publications such as War of the Worlds (H.G. Wells 1898; Radio adaption 1938).

So, barring historic cultural pressures, it has only been a lack of the required technology that has prevented earlier generations from 'reaching out' into the Cosmos, to explore these possibilities. A journey that after the past 60 years of active SETI observations is still only now just beginning to explore the vastness of the search space involved - a mere glass or bath's worth (at best) examined from the cosmic ocean.

So, ever since Cocconi & Morrison (1959) brought attention to the potential for us to detect radio signals at interstellar distances – new capabilities driven on by wartime advances in radar and radio technology - and when Frank Drake commenced Project Ozma - the first modern SETI experiment in 1960, actively searching for evidence of intelligent (technologically capable) extraterrestrial life (Drake 1961) - a post-detection initiative and capability has logically been required, so we have strategies in place ready for a discovery when it occurs. And arguably, as such a post-detection capability includes many non-radio based and generic societal impact considerations, this (post detection) provision should exist outside of any specific search strategy or mission, as discovery may lay outside any dedicated programs. So, whether due to project inactivity, technological limitations, or discovery through exploration, discovery may occur outside of an Earthbound SETI program.

More recent advances in technology have now enabled us to realise that whenever we see a pinpoint of light in the night sky [star], regardless of which one we choose, there is a good chance that there is a planet (or planets) nearby (Cassan et al. 2012) – evidence that has had a profound impact on the scientific community and public alike. Where once we based probabilities for extraterrestrial life on the likelihood of other worlds being out there (beyond our solar system), we are now uncovering the reality of the vast number of exoplanets in our galaxy, their makeup, and proximity to parent stars. And, of course, that is not to overlook the possibility of (simple) life being discovered on another planet or moon in our own solar system.

To facilitate such a multidisciplinary provision, the SETI Post-Detection Hub at St Andrews was launched in November 2022, providing (for the first time) a long overdue permanent 'home' for coordinating the research and development required to create a fully comprehensive operational framework. By doing this, the Hub brings together interested members of the SETI and wider academic communities, as well as policy experts, to work on topics ranging from message decipherment and data analytics to the development of regulatory protocols, space law, and societal impact strategies. Science that also has much that can contribute to answering terrestrial questions, as well as reflecting on what makes us Human. The Hub's current research team has already expanded its international profile to include members from across much of the globe: both established SETI researchers and new expert voices.

This paper introduces the 'Hub's' operational rationales, goals, and objectives, as well as how the idea was developed from its original conception (Elliott, 2008) from a personal perspective (as its founder), and the subsequent endeavours of a small UKSRN team to develop its operational parameters, to where it is now and where we go from here – a more general narrative on the wider SETI history is detailed in Elliott et al. (2023)

2. The Road to the Hub

From 1998 to the late 2010's my work focused on developing algorithms, mathematical tools, and resources for detecting and decoding language-like structures that may feasibly be contained within a radio signal – a 'language' detector / decoder for a message from afar. This work involved analysing over 60 human languages that represent all known morphologies and script types, together with animal and artificial [AI robot] communication systems, to identify whether any common identifiable indicators [structures] lie beneath the veneer of the arbitrary sounds and symbols used to convey meaning that could detect the presence of communication - a suite of tools [software, algorithms, theoretical models, etc.] to detect and build from a common underlying detectable construct (Elliott & Atwell 1999, 2000; Elliott 2003, 2004, 2007, 2009, 2010, 2011, 2012, 2015; Herzing, D., Denning, K. & Elliott, J. 2018).

Outline phases that comprise these analytical stages, which demonstrate the potential complexity and protracted timeframe involved include (Elliott, 2008, 2010):

• Post-reception of a confirmed ET signal will result in public knowledge of the event (SETI principle No 3 - *Confirmed detections: If the verification process confirms that a signal or other evidence is due to extraterrestrial intelligence, the discoverer shall report this conclusion in a full and complete open manner to the public, the scientific community, and the Secretary General of the United Nations.*). In this event, a first pass filter to address the most fundamental question of "is there a message or information carrying content" - analysis to determine whether any discernible intelligent content is present - will be required.

• If evidence of intelligent-like [communication] structure is confirmed, the next question soon to be asked will be "what type of phenomenon is it?" To answer this, stages involving a series of entropic-based, structural, and relational measures (including internal characteristics) will need to be completed. Logical steps that are designed to provide as near a continuing flow of new information as possible, to avoid prolonged media 'silences' during this complex process.

• The next stages address more complex internal structural assessments, designed to provide robust methods for confirming the signal's comprising components (including if a crib / primer is present), all layers of 'syntax' hierarchy, as a precursor to bootstrapping information for further decipherment. This culminates in levels of component interaction and categorisation that, if language-like, will possibly provide a measure of the author's cognition. Such stages incorporate unsupervised AI algorithms, specifically developed / adapted for dealing with unknown phenomena.

• Finally, analytical stages move through the increasing complex tasks towards semantic assignment, if we have evidence that provides mapping [crib / primer] or at least probable affinity to known phenomena. If not, these last stages may present us with a semantic 'wall' that cannot be surmounted. Nevertheless, we will have learned much from previous stages [1 to 3] and confirmed that another intelligent communicator exists in the Cosmos.

Notes:

• Signal data gathered, analysed, and communicated, using the D.I.S.C Quotient (Elliott & Baxter, 2010) - a tool for dynamic, rather than static, analysis of information as it is discovered - will dynamically adjust its calculations for the signal's potential significance and impact, if new additional data is received beyond an initial detection capture.

• It is from the point of announcing a confirmed discovery that the SETI Post Detection Hub takes a different perspective and extends its remit to the extensive multidisciplinary requirements needed for discovery analytics (such as above), communication [dissemination] strategies, protocols, planning and scenario modelling, amongst many other aspects involved for post detection.

During this signal [message] modelling and analysis work, it became clear that any information disseminated to the world's public will not just be a single announcement but a series of updates over time (possibly protracted over many weeks and months), as we analyse and understand what has been discovered. For example, in a signal detection scenario, we will find ourselves facing the complexities involved in understanding the comprising constructs and maybe even glimpsing the intellect of the author, while the world's curiosity and expectations are likely to demand immediacy of information. Of course, semantic decipherment may be beyond us, but much can be gleaned from the 'syntax'. Dynamics that require careful planning, so the narrative is timely, clear, transparent, and trusted, across all of humanities diverse cultures and faiths, whilst being mindful of the evolving forces of social media. It is worth noting that work done at this point on post detection strategies, such as the D.I.S.C Quotient (Elliott, 2008; Elliott & Baxter, 2010), were prior to social media influences as they have now developed - dynamics that are now fully integrated in current work.

It was also noted in Elliott & Baxter's DISC Quotient paper (2010) that in announcing a confirmed detection event, an array of human factors will likely result in fundamental change in accepted knowledge (wisdom), behaviour, and attitude; individual and socio-cultural interaction with new information (truths); awareness and implications of the source.

Key Factors influencing information dissemination are therefore likely to include:

- Reinforcement of and / or new adjustment to information and assumptions.
- Perception of relevance and impact on human civilization.
- Conjecture beyond physical, geographical, and decipherable facts.
- Religious/societal impact and reaction.
- Recognition of need for new knowledge.
- Willingness to accept and change as a result of contact.
- Need/willingness to "participate" (be involved) in process.

In the initial phase (e.g., announcement of signal being detected), the location and therefore perceived immediacy of any impact for a potential encounter will facilitate the opportunity to convey the likely vast distance and therefore time separation involved – proximity being a highly weighted impact factor for such an event. However, in the subsequent decipherment phases, if effective communication [media] conduits are available, dynamic metrics such as the DISC Quotient will be able to convey detailed and scientifically robust updates of the decipherment process to the public, and thus (hopefully) reinforce trust. Such factors may be critical in mitigating negative societal and political impact, due to unfounded speculation and disinformation.

3. Some Background Considerations

3.1. *Overview*

If we are to have a framework in place that can be efficient, effective, and comprehensive for the challenges we will face, upon discovering extraterrestrial life, we must revisit and reflect upon all our assumptions, approaches, and current models – a challenge that will undoubtedly evolve with both our technology and (hopefully) improving societal awareness. Much of these are reflected in the established Hub working groups (see section 4.4) that comprise our ongoing work, and will develop and expand, as we progress. The following subsections discuss some of these considerations.

3.2. *The Announcement*

Much pertaining to good practice, regarding the detection for evidence of an extraterrestrial technology (e.g., signal, beacon), is enshrined in the SETI (2010), with transparency at its core. However, this idealistic model for discovery and subsequent announcement, neither considers or plans for non-compliance or subsequent societal impact factors.

If we leave aside the many issues, we are likely to encounter across the myriad of potential scenarios possible and focus on the dynamics of a discovery's announcement, we do have some examples to learn from: Oumuamua (Loeb 2021) and the discovery of an "arsenic-loving bacteria" in Mono Lake, California (Wolfe-Simon et al. 2011).

Although these two examples may seem unrelated at first – one a distant interstellar object that was claimed to be an alien spacecraft; the other a microbial life-form on our own planet - they both present the possibility that the discovery provides evidence for a profound shift in our knowledge – life beyond our known parameters. And with this, the likelihood of significant societal impact, if the continuing dominant scientific narrative supports the confirmation of its discovery. Of course, other factors such as proximity and perceived threat will then weight the effect of this impact, either way. With these two examples, early feedback from independent expert voices were quick to question the headline claims, providing objective balance to media coverage. However, this situation can leave individuals [experts] responding in good faith to media enquiries, with incomplete information, which has the potential for unintended inconsistencies and contradictions that could then add fuel to speculation.

The main issues here being the dynamics for communicating the announcement of a new discovery, without prior scientific rigour, clarity of information, and the implications for the many facets that comprise our [human] global community. Without this, we are failing in our responsibility. Of course, there may be many motives why such a process is ignored or circumvented, so we must account for such behaviour and plan accordingly.

3.3. *Intelligence*

And what of the all too often Human conceit that ET intelligence will look and behave like us? Of course, we must start somewhere, so our own evidence of intelligence on this planet is a good place to start. Nevertheless, to mitigate such a conceit, we also need to remain vigilant to evolve and expand our understanding of what intelligence may 'look' like. To address this, we will need to go beyond the anthropomorphic template – Elliott 2015 initial work on non-human (AI) communication investigates aspects of this.

Alternative forms of intelligence can be evidenced on this planet. The Octopus – a cephalopod with an evolutionary last common ancestor (bilateral flatworm) to us 600 million years ago – has developed an intelligence that is distributed, rather than just one central 'brain'. Hive intelligence shows us that the whole (individuals that comprise a hive or collective animal community) is much more than the sum of its 'parts' – an ant colony or bee hive being prime examples.

How rigid can our definition of life be, when non-organic life - for example independent autonomous 'evolved' artificial intelligence (AI) - could be the remnant remains of an extinct parent biological civilisation, or how it captured its essence to survive in some form. Such AI probes may be the most likely physical encounter for us, as we explore our local galactic region – an AI meeting another AI encounter being a distinct possibility for such distant journeys, given our projected short to medium term technological future.

Life may have also evolved to be naturally independent of a fixed environment (planet, moon, etc) for survival, and roam space without aid or technological support. Their senses may have capabilities we have not considered possible – could an advanced form of synaesthesia, with a far wider sensing spectrum than we possess have evolved to also include inter-dimensional capabilities – even something that awaits our natural human evolutionary future (Homo Deus)? Our imagination may hold the key to realities that seem beyond logic, as we observe the universe through the limited prism of our senses and current technology. It is therefore our scientific responsibility to keep an open mind on such possibilities, as history has taught us that once held 'truths' are often overturned, when we have the tools to detect and understand new evidence. A quote that aptly sums this up, which has resonated through multiple reiterations since the 1920s but has no definitive source (inc. attributions to Franz Kafka, and Frank Zappa), is that "a mind is like a parachute, if it isn't open it doesn't work.

3.4. *Cultural Perspectives*

It is also important that we do not fall into the usual scientific default of forming a contextualised approach and logical constraints based on western and industrialised nations perspectives alone. And intergenerational ignorance is not a reason or excuse for blinkered thinking.

A key factor to indicate cultural erosion occurring is language loss. Bromham et al. (2022) inform us that "as with global biodiversity, the world's language diversity is under threat. Of the approximately 7,000 documented languages, nearly half are considered endangered. In comparison, around 40% of amphibian species, 25% of mammals and 14% of birds are currently threatened with extinction. The processes of endangerment are ongoing, with rates of loss estimated as equivalent to a language lost every one to three months, and the most pessimistic predictions suggesting that 90% of the world's languages will be lost within a century." Ancient and indigenous wisdom can often be uniquely encoded within its language's semantics. So, if lost, may diminish our collective memory and 'forget' an important truth that would otherwise aid our understanding, which must not happen if we can avoid it.

So, a culturally holistic approach is particularly important for our post detection societal and risk strategies. Otherwise, we may diminish our understanding and effectiveness to a world only comprising the dominant political landscape and ignore the opinions, insights, and likely responses of many [indigenous] peoples. We therefore need to listen to and include our collective voices, when mapping and understanding such global dynamics.

Aspects of this approach may be thought of as tangential to the purpose of this work but consider AI. If the sum knowledge of our AI ambassador [probe] is only the product of a cultural [industrial world] homogeneity, will the resulting communication inadvertently be lost in translation. Or could the same diminished knowledge held in this future homogeneity lose our chance to understand a detected [message] from afar, when it comes. The Hub therefore looks to include as many voices as possible.

3.5. *Protocols and Metapolicy*

It is worth noting that at present, after over half a century of SETI activity, we still only have one agreed policy document for handling the detection of an extraterrestrial intelligence (ETI): the SETI Protocol (Declaration of Principles, 1989 – revised 2010). "This situation is in contrast to policy making in other matters of global concern, such as ethics (the International Declaration on Human Rights), the exploitation of outer space (the Outer Space Treaty), and climate change (the Kyoto agreement and others)" (Baxter & Elliott 2012).

This (voluntary, non-enforceable) protocol guides actions for the immediate period following a detection, focusing on the verification of source, data handling, response policy, and transparency. Currently, the discoverer is expected to "report the conclusions in a full and complete open manner to the public, the scientific community, and the Secretary General of the United Nations", but it does not provide any guidance of what they should do with that information. So, in practice, this currently leaves us with a disconnect between the ambitions of the [IAA SETI Declaration of Principles] document and how decision-making bodies such as the UN can liaise with SETI experts to establish an effective operational conduit, whilst the world's curiosity and expectations would demand immediacy of information and (if required) action. This area is a particular focus for the Post Detection Hub.

Baxter & Elliott's paper on a SETI Metapolicy (2012), set out areas in which policies could be developed, including handling ETI artefacts, and approaches to direct contact

- tentative recommendations for developing a wider-ranging 'metapolicy' to cover wider conceptual areas, which have fed into our initial narrative when considering rationales for the post detection hub.

These include:

- Security arrangements, including the issue of active SETI, should be considered and perhaps war-gamed.
- In the case of the receipt of an electromagnetic signal we need to consider how to manage the issues of replying, and the release of translated ETI data into human culture.
- If an ETI artefact is found, questions of ownership and security arise.
- In the case of direct contact with an ETI or a smart probe, contact policies based on diplomatic principles can usefully be developed.

As an update to this and since the Hub's inception, a small team within the IAA SETI Permanent Committee are now currently working on high level update proposals for the SETI protocol, regarding the post detection element; these initial revisions are estimated for late 2024 or 2025. These will not, of course, be addressing the detailed operational aspects of post detection, which are the objectives for the Hub.

4. Hub Beginnings

4.1. *Overview*

Following the termination of the relatively short-lived and rarely convened (only 3 meetings officially recorded, over its lifetime) IAA (International Academy of Astronautics) SETI Post Detection Task Force, (circa 2000 - 2010 – a subcommittee of the IAA SETI Permanent Committee) it was realised that no active post detection initiatives were now in existence. A situation that was inconsistent with principle 7 [Post Detection] of the SETI protocol. To try and rectify this situation, Elliott and Denning presented a proposal for creating a successor Post Detection sub-committee, to the IAA SETI Permanent Committee executive in April 2018. Although this proposal was received with interest, it was subsequently deemed too broad and open-ended a remit.

Despite initially being very disappointing, this feedback then became an important catalyst for taking the decision to draft a proposal for the formation of the first ever dedicated centre for post detection, and presenting these at the 2019 **UK SETI R**esearch **N**etwork (UKSRN est. 2012) conference: outlining the rationale, operational requirements and need for a Post Detection Hub that comprises a very broad 'church' of expertise to investigate the extensive multidisciplinary questions involved. An initiative that must have an open-ended remit, which enables continually evolving research and development, to reflect future scientific advancements and societal changes. In response to this, a small UKSRN team was formed to develop this idea: John Elliott, Stephen Baxter, William Edmondson and Michael Bohlander.

During the following year, this small team drafted the general parameters under which a future post detection hub could form and operate. A final (internal) draft proposal, completed in March 2022, included the following subset of broad principles, which remain among the core guiding themes:

1. The Hub's primary role shall be to serve as a dedicated and cutting-edge forum with the aim of coordinating research for a coherent post-detection framework and providing advice to relevant stakeholders, including the analysis and development of regulatory protocols, operational logistics, data analytics and societal impact strategies.
2. Post-detection presupposes the fact that the discovery is verifiable and assured.

a) The question of whether and in what forms a discovery is at all possible or to be expected shall thus not influence the PD-Hub's mission and activity.
d) The discussions about post-detection regulation shall thus focus solely on processes following from detection in the broader sense.

3. The PD-Hub shall operate on terms of academic and scientific rigour and be open to bona fide researchers from all disciplines and countries.
4. The PD-Hub shall engage with all categories of stakeholders who may be expected to become involved in the actual real-world decision-making process following a discovery, with the aim of informing their decision-making in practice.

Of course, subsequent discussions and work have since contributed to developing and refining these roles.

The Hub's mission therefore also encompasses developing an effective regulatory framework for the entire SETI endeavour. As previously stated, there is no agreed, let alone legally binding, procedure for either the search for extraterrestrial intelligence or the required action in the case of detection. Baxter & Elliott (2010) states that "the early involvement of space lawyers and UN agencies could help improve the chances of having any proposed policy widely accepted, and workable in practice". Subsequently, in-depth work is now under way across these and other areas, such as scenario modelling, which will form the core of a detailed operational framework for the Hub.

4.2. *Organisational 'Nuts and Bolts'*

Complementary to the ongoing regular research meetings (plenary, working group, etc), our day-to-day operations currently include, outreach, presentations, publications, funding applications and drafting proposals for, as well as meetings with, external, and especially with domestic and international political and governmental stakeholders. This will also soon include organising symposia, seminars and conferences, as permanent fixtures on our calendar, designed to synthesise new innovative approaches with more familiar formats, to enable new thinking and solutions.

4.3. *The Team*

From Athens to San Francisco and from Toronto to Sydney, the Hub already has a wide geographical spread of international expertise across many disciplines, which will grow, as we evolve. Members include researchers based at many universities internationally, including St Andrews, Cambridge, Harvard, Stanford, Athens, Freiburg, Durham, York (Toronto, Canada), Birkbeck (London), Birmingham, Manchester (Jodrell Bank), Liverpool, and New South Wales.

Our team is already a good mix of "established and new voices", which both applies to long established expertise with SETI as well as interested newcomers from fields that are historically not well represented in the SETI community. This includes people from across the entire academic age and career stage spectrum: from PhD students to retired, and essentially everything in between (early-career researchers, research fellows, well-recognised senior academics). With new connections already established, and the evolving dynamics arising from our collective energy, new and novel work is already emerging. In addition to this, we are also now making connections to expert practitioners, as well as establishing student networks at the University of St Andrews and internationally.

Examples of our current expertise include: computational linguistics, signal categorisation, decoding & analytics; science communication; risk communication; philosophy (language, mind, ethics, metaphysics, religion); astronomy; astrobiology; science fiction; anthropology, sociology, psychology, and social science; linguistics; space policy and law; humanitarian and human rights law; metalaw; planetary protection policy; biology (zoology, evolutionary biology, marine biology, molecular and cell biology, genetics, neuroscience, bioinformatics); mathematics; data sciences; cognitive science; physical chemistry; nuclear/chemical/biological risk analysis; international relations; strategic foresight and futures studies.

4.4. *Working Groups*

This interdisciplinary endeavour has required careful orientation to effectively combine disparate expertise. Careful familiarisation with the actual practice of SETI science has also formed an essential part of our approach, to produce optimally informed policy recommendations or other forward planning.

Regular meetings, (one or two a month) have forged valuable inter-Hub connections, notably spawning initiatives to co-author research papers and engage in group scenario building and creative works. In addition to this, Hub members are encouraged to contribute to a series of multi-disciplinary working groups.

Currently groups, focussing on specific aspects of post-detection discovery, include:

- Discovery Regulations [Law].
- Beyond Human - Other Biologies and Minds.
- Public Outreach and Science Communication.
- Signal Analytics, AI & Data.
- Human Factors (e.g., Philosophy, Psychology, Sociology).
- Impact & Risk Strategies.
- Dissemination Strategies.
- Emerging Practices (inc. Indigenous Cultures).
- Imagination and Storytelling.

By grouping related themes and disciplines, the aim is to address crucial questions, and to tend the many vital cross-disciplinary intersections between categorical domains, thereby generating viable solutions and paths that inform a flexible, scenario-responsive operational framework.

4.5. *Affiliations*

Current affiliations to the Hub are:

Internally:

St Andrews - the Centre for Exoplanet Science (StA-CES) and the Centre for Global Law and Governance (CGLC).

Externally:

Durham Law School - joined the Hub as a co-hosting organisation in January 2023. It has research strength in the field of international law, which is crucial in devising proposals to be submitted by the Hub to political decision-makers for the purpose of establishing domestic and international regulatory frameworks, such as national or international official policies, legislation or international agreements, in order to put SETI in all its forms on a proper and legitimate footing. This has now focused on contributing efforts through organising a working group attached to the IISL (International Institute of Space Law), where Hub members and IISL members are collaborating to develop legal devices for post detection.

University of Manchester - Jodrell Bank Centre for Astrophysics, University of Manchester joined the Hub in June 2023, where their many years of experience and expertise, contributing to the SETI search endeavour, includes specific elements that will help complete our operational modelling, such as search techniques, data capture and handling, for post detection information 'handshake'.

4.6. *Impact*

The media coverage that followed the official launch of the Hub was beyond expectations, 'flooding' the St Andrews press office with enquiries and information on the hundreds of newspapers covering the event internationally - many of which were non-English speaking and across most continents. Since then, its impact has been significant with ongoing media, public, academic, and SETI community interest – this has included countless interviews and articles, as well as invitations to the British and Irish Science festivals, and Royal Society. Already this often-neglected aspect of SETI, since finding its permanent voice and home, has prompted new initiatives and conversations, including the UK government. Not least, its growing concentration of expertise, which historically [SETI] has only previously seen solo or sporadic short-term and small-scale collaborative projects. It is anticipated that the Hub community, its global reach, and progression, will see significant growth in 2025.

5. The Road Ahead

Our core strategic emphasis for research and development at the SETI Post Detection Hub will, of course, remain on planning for extraterrestrial discoveries and their potential societal impact.

These core areas fall into three main areas:

• supporting and facilitating active research, collaboration, and development activities between Hub members.

• nurturing external alliances.

• Growing the Hub community, including its next generation of researchers.

To achieve many of the objectives that comprise these core aims, communication lies at its centre: both intra-Hub and inter-institutes, organisations, and governmental bodies. It is also important that these connections include both established and new voices, including the launch of our forthcoming student networks, planned for October 2024. Within the Hub, communication amongst members and between working groups is embedded in our philosophy: encouraging collaborations and ensuring that the working groups cross-pollinate. Up until now, our primary focus has been to foster this intra-web of connections within the Hub, with some very important external connections, as well as significant media outreach.

This immediate road ahead will also look to establish outreach platforms and conduits with the World's public, which will include establishing the Hub as a reliable and expert point of media contact and public [online] access for news items around the discovery of life. Our emphasis on societal implications necessitates this outreach to extend beyond organisations within the immediate SETI sphere – proactively seeking input to measure and form real relationships across society: including public talks, across all generations, and imaginative events that welcome a broader audience. It will also endeavour to grow fresh relationships outside of the typical SETI sphere, such as organisations supporting interdisciplinary and Indigenous Studies work.

On our calendar for summer 2025 is slated the Hub's first hybrid conference – an event that intends to look to both member researchers and a general audience by devoting one day to sharing findings from the Working Groups and another to public engagement. Not only this, but also how contributing papers that form the research input for the conference will see new approaches.

Funding is a priority, if we are to sustain this initiative for the long-term, to further develop our research programs and contributions in the field. To this end, we are involved in applications for funding from bodies such as UK Research and Innovation, Leverhulme, and Wellcome. However, to increase our likelihood of securing adequate finding, we will also look to other funding sources, to ensure that those contributing their valuable time to this research are provided with the necessary support – support that will almost certainly accelerate our efforts.

6. Conclusion

Social and academic interest in SETI and the possibility of extraterrestrial life has been increasing considerably, in the past couple of decades. A major catalyst for this almost certainly being the detection of thousands of planets [exoplanets] around distant stars in our own galaxy, providing tangible evidence for millions of potential life-bearing worlds. And, closer to home, new initiatives looking for signs of life, present or past, through the exploration of nearby worlds such as Mars, Venus, and moons around our nearby gas giants. The potential discovery of microbial life in our own 'back yard' [solar system] will likely raise different types of concern than would following the discovery of intelligent life from a distant source – yet, as a species, we are unprepared for either. The time is thus right for consideration of humanity's response – and responsibility – following the detection of both life and intelligence in the Cosmos. So, it became obvious that we should plan now for this eventuality, by setting out and coordinating impact assessments, protocols, procedures and treaties designed to allow humanity to respond responsibly, as upon discovering evidence of an extraterrestrial origin, the scientific and social landscape will change profoundly.

Therefore, in preparation for a detection or contact with ETI, the Hub will endeavour to grow readiness for complex futures by fostering a synergistic environment that encourages imaginative, methodical preparation. Intending to stir transformative interdisciplinary and diverse engagement, the Hub will actively involve policymakers towards and convening work on operational frameworks across society, and towards becoming a trusted point of contact for media in the event of detection in the news.

"It will focus on sharing approaches to pivotal questions within the field, crafting public programs, developing extensive bibliographies, and supporting the work that will be needed towards the refreshing of the IAA SETI protocols through cultivating a deeply engaged culture of diverse expertise. Growing on the aims to engage a wider and diverse public, the Hub will reach out to experts and the public during the development of scenarios and protocols under discussion" Elliott et al (2023)

Will we ever discover life on these other Worlds? Will we ever receive a message from E.T? We do not know. We also don't know when this could happen - ancient microbial life may be beneath the Mars rover, as you read this. But we do know that we cannot afford to be ill-prepared – scientifically, socially, and politically rudderless – for an event that will have such a profound effect on us and one that we cannot afford to mismanage. The science of discovery – our known, unknowns and unknown, unknowns – is one that challenges us to the limit of our current knowledge, perceptions, and technology, whilst endeavouring to avoid any limiting Human conceit: a process that is ever evolving.

References

Anton,A., Elliott, J. & Schetsche, M.(2024) Meeting Extraterrestrials: scenarios of first contact from the perspective of exosociology In: Acta Astronautica. 215, p. 308–314 7 p.

Baxter, S. and Elliott, J. (2012) A SETI metapolicy: New directions towards comprehensive policies concerning the detection of extraterrestrial intelligence, Acta Astronautica, Volume 78, 31–36.

Bromham, L., Dinnage, R., Skirgård, H. *et al.* (2022) Global predictors of language endangerment and the future of linguistic diversity. *Nat Ecol Evol* **6**, 163–173. `https://doi.org/10.1038/s41559-021-01604-y`

Cassan, A., Kubas, D., Beaulieu, JP. et al. One or more bound planets per Milky Way star from microlensing observations. Nature 481, 167–169 (2012). `https://doi.org/10.1038/nature10684`

Churchill, W.S. (1942) Are There Men on the Moon? Sunday Dispatch [article], UK.

Cocconi, G. & Morrison, P. (1959) Searching for Interstellar Communications. *Nature* 184, 844–846, `https://doi.org/10.1038/184844a0`

Drake, F. D. Project Ozma, (1961) Physics Today, vol. 14, issue 4, p. 40–46, 10.1063/1.3057500

Elliott, J., Genevieve, K., Dominik, M., Denning, K., Baxter, S., Profitiliotis, G. & Finer, E. (2023) The SETI post-detection hub: preparing for discovery, IAC-23,A4,2,2,x79324

Elliott, J. Beyond an Anthropomorphic Template, (2015) Volume 116, November–December 2015, Pages 403–407 Acta Astronautica.

Elliott, J. (2012) Constructing the Matrix Acta (2012) Volume 78, 26–30 Acta Astronautica

Elliott, J. & Baxter. S The DISC Quotient, (2012) Volume 78, September–October 2012, 20–25 Acta Astronautica.

Elliott, J. (2011) The Filtration of Inter-Galactic Objets Trouvés and the identification of the Lingua ex Machina Hierarchy, Volume 68, Issues 3–4, February–March, 399–405 Acta Astronautica.

Elliott, J. (2011) First stage identification of syntactic elements in an Extraterrestrial Signal, (2011) Volume 68, Issues 3–4, February, Pages 389-398 Acta Astronautica.

Elliott, J. (2010) Detecting the Signature of Intelligent Life, (2010) Volume 67, Issues 11–12, December 2010, 1419–1426 Acta Astronautica.

Elliott, J. (2009) A Semantic 'Engine' for Universal Translation, `https://doi.org/10.1016/j.actaastro.2009.06.006` [volume 68, February–March 2011 435–440] Acta Astronautica.

Elliott, J. A Post-Detection Decipherment Matrix, (2007) Volume 61, Issues 7–8, October 2007, 712–715 Acta Astronautica.

Elliott, J. (2004) 'Unsupervised Discovery of Language Structure in Audio Signals, *Proceedings of IASTED International Conference on Circuits, Signals and Systems*, (CSS 2004), Clearwater Beach, Florida, USA

Elliott, J. (2003) 'A Human Language Corpus for Interstellar Message Construction', *Proceedings of the 54th International Astronautical Congress*, Bremen, Germany, pp. IAA-03-IAA.9.1.04

Elliott, J & Atwell, E. (2000) 'Is anybody out there: the detection of intelligent and generic language-like features', 53 Journal of the British Interplanetary Society, 13–22

Elliott, J & Atwell, E. (1999) Language in signals: the detection of generic species-independent intelligent language features in symbolic and oral communications: in proceedings of the 50th International Astronautical Congress, pp. IAA-99-IAA.9.1.08: International Astronautical Federation.

Herzing, D., Denning, K. & Elliott, J. (2018) Beyond Initial Signal Categorization and Syntactic Assignment: The Role of Metadata in Decoding and Interpreting Nonhuman Signals, COSPAR conference, Pasadena, USA.

Loeb, A. (2021) On the Possibility of an Artificial Origin for 'Oumuamua, Astrobiology, `https://doi.org/10.1089/ast.2021.0193`

Oppenheimer, J.R., & Einstein, A. (1947) Relationships with Inhabitants of Celestial Bodies, (top secret) USA government document.

SETI Protocol revised (2010) `http://resources.iaaseti.org/protocols_rev2010.pdf`

Wells, H. G. (1898). The War of the Worlds. London, England: Chapman and Hall.

Wolfe-Simon, F., Blum, J.S, Kulp, T.R., Gordon, G.W., Hoeft, S.E., Pett-Ridge, J., Stolz, J.F., Webb, S.M., Weber, P.K., Davies, P.C.W., Anbar, A.D., and Oremland, R.S. (2011) A Bacterium That Can Grow by Using Arsenic Instead of Phosphorus.Science332, 1163–1166(2011). DOI:10.1126/science.1197258

(Toward) Discovery of Life beyond Earth and its Impact
Proceedings IAU Symposium No. 387, 2026
H. Landt, M. Dominik & C. Oliver, eds.
doi:10.1017/S1743921325103037

Finding Intelligent Life and First Contact Scenarios: Towards Public Awareness and Engagement and the Impact of Discovery

Shirin Shaterzadeh Yazdi[1], Saeed Jafari[2,3], Raha Basafa[4] and Hossein Khalili[5]

[1]Department of Physics, Shahid Beheshti University;
email: Shirin.Shaterzadeh@worldspaceweek.ir

[2]Department of Linguistics, University of Kurdistan, Iran

[3]UNESCO's International Decade of Indigenous Languages 2022 - 2032;
email: saeed.jafari@spacegeneration.org

[4]Department of Religion, University of Manitoba;
email: basafar@myumanitoba.ca

[5]IAU OAO/OAD Funded Under Iran Sky;
email: hossein.khalili@worldspaceweek.ir

Abstract. The prospect of human contact with extraterrestrial intelligence (ETI) carries profound societal implications, far surpassing the impact of discovering non-sentient microbes on distant planets. Such an encounter would reverberate throughout human society, challenging established beliefs, including theological doctrines, and reshaping our cosmic perspective. During the last decade, the search for extraterrestrial intelligence (SETI) and non-intelligent extraterrestrial life has received notable attention in digital media and newspapers and influenced public thoughts and perceptions about astrobiology, SETI and planetary science fields. This abstract takes a science communication and society perspective, evaluating our current decision-making processes and policies in anticipation of potential discoveries and their consequential interactions with varying forms of extraterrestrial life, including public attitudes, news dissemination and rumour control, and decoding and messaging ETIs.

Keywords. SETI and science communication; public awareness and engagement; impact of first contact; media and misinformation analysis; societal aspect of discovery

1. Introduction

Human imagination has long been captivated by aliens. The possibility of coming across extraterrestrials has sparked a lot of speculation in recent decades on their possible reasons for visiting, degree of animosity, appearance, communication methods, and even culture. Despite a great deal of research and several scenarios (some of which lie in the realm of science fiction), there is only one sincere and conclusive response to each of these queries: Up to the point of encounter, at least, we know nothing about the extraterrestrial life. Nevertheless, as emphasized by Race (2015), preparation for such an encounter is crucial. While some may argue that these kinds of preparations are unnecessary, the Covid-19 outbreak we recently experienced has shown us how important it is to be prepared for even the most unlikely of catastrophes. Although everyone on the planet was aware of the global pandemics prior to the Covid-19 pandemic, none was ready for it when it struck. Millions of people died tragically as a result of this lack of preparation, and it also had significant effects on global society, theology, and the economy. We have to acknowledge that the finding of extraterrestrial life, in whatever form, is inevitable and take proactive steps to prepare for it in order to avoid a repeat of the encounter. there are different kinds of ETL/ETI that we might encounter. Thus, we must prepare different scenarios for each type of encounter Race (2015).

2. A Brief History of the Extraterrestrial Intelligence

Contrary to common assumption, there are a number of indications throughout human history that suggest the existence of extraterrestrial life and intelligence.Several religious and legendary writings from around the globe make multiple references to the inhabitants of the heavens. The homes of the gods in classical cultures like Greece and Iran serve as excellent illustrations of this. As a result, the idea emerged in some of the most significant science-fiction books in history as well as oral literature and folklore. Actually, Johannes Kepler's 1608 science-fiction book Somnium explores the possibility of new life forms on the moon in great depth. It is thought to be the first science-fiction book in history. The nineteenth century saw the publication of many noteworthy science-fiction books centred on the idea. One of the first books to depict a fictitious spacecraft journey to the moon was Jules Verne's 1865 novel From the Earth to the Moon. In contrast to Jules Verne's innovative and upbeat perspective, H.G. Wells' 1898 novel War of the Worlds, which depicts the Martians' invasion of Earth with their advanced technology and lethal weaponry, offers a more conservative viewpoint. These are only a few instances that illustrate the hopes and fears that people in the industrial age have around the idea of ETI.

When the space age began and more sophisticated instruments were employed in astronomy in the 20th century, the idea underwent a significant metamorphosis. An major conundrum was raised by discoveries like the more precise calculation of the Milky Way's star count and the realisation that certain nebulas, like the Andromeda nebula, are actually distinct galaxies from the Milky Way: If there are multiple galaxies in the universe, each with billions of stars, some of which are older than our sun and forming planetary systems, then it is highly probable that extraterrestrial life has arisen on a few of these planets, evolving to a point of intelligence and civilisation that would have allowed for interstellar travel and communication. "Where is everybody?" would be the question to ask if this assertion were accurate. Enrico Fermi posed precisely this query to his coworkers during a lunch talk on a summer day in 1950. Jones (1985).

This question is now known as the Fermi Paradox, and while there isn't a satisfactory explanation, it marked the start of a new age in which the possibility of extraterrestrial life (ETI) existed and was no longer just a theoretical concept or a fiction. Frank Drake devised his well-known formula, the Drake equation, as a mathematical representation of the chance of encountering extraterrestrial intelligence (ETI) ten years after the Fermi Paradox in 1961. In addition, Drake was a pioneer in the building of radio telescopes and the beginning of the ETI search, which became known as SETI. Radio telescopes have been used in addition to optical telescopes in the past fifty years in the ongoing hunt for extraterrestrial intelligence (SETI).

3. The Impacts of Discovery

Serious repercussions await humanity the instant alien life (ETL) is discovered, which will likely resemble the early stages of the COVID-19 pandemic regardless of the extent of the discovery and the techniques of discovery. There will probably be significant effects on many facets of peoples' life, which we have divided into the following three categories:

3.1. *Theological Implications*

The belief in the presence of invisible others is deeply rooted in human psychology, particularly in the need to alleviate loneliness. As Freud (1984) suggests, people fear loneliness and, in response, project a parental figure to provide comfort and guidance. This psychological tendency is mirrored in religious beliefs, where the notion of an unseen presence offers reassurance. Luhrmann (2020) states that "religion is a practice in which people go to effort to make contact with an invisible other." Within the framework of faith, cognitive science explains that "the great realization of the cognitive science of religion is that people quickly, easily, and automatically generate ideas about agents they can't see when they are scared or startled." Luhrmann (2020) In essence, both religion and psychology suggest that humans are inclined to imagine unseen beings in response to existential uncertainty.

This mechanism is not limited to religious beliefs; it extends to myths and cultural narratives. As Johnston (2015) explains, myths are stories told repeatedly, with their meanings shaped by what listeners already know about the characters. These narratives help people feel as if their gods are real. The repetition of myths and religious stories has historically solidified the existence of saints and champions in the collective imagination, transforming them from fictional or semi-historical figures into perceived realities. The same psychological process applies to extraterrestrial intelligence-repeated storytelling fosters a sense of familiarity, making the idea feel real. Science fiction, conspiracy theories, and speculative discussions reinforce this belief, much like how religious traditions reinforce faith. Just as myths shape religious perceptions, cultural narratives surrounding extraterrestrial life create a framework in which the existence of intelligent aliens feels not just possible but inevitable.

At its core, the search for connection-whether through religion, myths, or extraterrestrial life-addresses a fundamental human need. Being part of a greater whole can help individuals escape the feeling of isolation. This need for connection may explain why some turn to faith, while others look to the stars for answers.

Döbler et al. (2023) discuss studies showing a negative relationship between frequent religious activity and belief in extraterrestrial life (ETL). This suggests that people who are more religious tend to be less likely to believe in ETL. This aligns with previous research indicating that certain religious views are tied to a geocentric worldview, in which Earth is seen as the only inhabited planet. However, religious belief does not universally reject the idea of extraterrestrial life-some theological perspectives incorporate the possibility of life beyond Earth.

Islamic theology presents an interesting contrast to Christian geocentrism. In contemporary Islamic scholarship, some scholars turn to the Quran to argue that references to life beyond Earth existed even before modern space exploration. Unlike Christianity, which traditionally views humans as the singular, unique creation of God, Islam considers humans as the *Ashraf-e-Makhluqat* (the noblest of creations) but acknowledges the existence of other beings. The Quran mentions both visible and invisible beings, such as *Jinn*, who exist within our earthly realm, and angels, who inhabit another world. Additionally, certain Quranic verses suggest that this world and its creatures are not the first or the only creations of Allah, indicating a theological openness to the existence of other life forms.

This theological perspective aligns with findings from another study, Routledge et al. (2017), which shows that individuals who do not identify with a religion are more likely to believe in extraterrestrial intelligence (ETI) than those who do. This suggests that in the absence of religious frameworks providing existential meaning, some people turn to the possibility of extraterrestrial life as an alternative explanation for human existence.

When it comes to belief in ETI, religion can serve as both a facilitator and a barrier. In some traditions, particularly those with a geocentric worldview, religious doctrine may discourage belief in extraterrestrial life by reinforcing the idea of Earth's uniqueness. However, Islam presents a more flexible theological framework. By acknowledging the existence of both visible and invisible beings beyond human perception-such as Jinn and angels-Islam creates a space where the possibility of other intelligent life does not necessarily contradict faith demonstrating that religious belief can, in some cases, accommodate scientific inquiry rather than oppose it.

3.2. *Scientific Implication*

the confirmation of ETL will drastically alter our knowledge of the beginnings and evolution of life and it would demonstrate that habitats other than Earth are capable of supporting life. Some of the circumstances which have the potential for life to adapt are mentioned in Schulze-Makuch (2015). Also, the discovery might result in advancements across several fields. For instance, new studies concentrating on circumstances and mechanisms that might support life in the cosmos would greatly advance the science of Astrobiology, which would benefit greatly from the discovery. Since there is no assurance that life on other worlds has precisely formed and evolved like that which we know on Earth, another result would be a larger definition of life and even intelligence Schneider (2015).

3.3. *Societal Implication*

This impact is highly dependent on the kind and stage of life we end up discovering. It is reasonable to predict that the impact on society will increase with the discovery of more sophisticated forms of life. However, direct contact with a sophisticated civilisation would have an impact on all human societies. For example, the discovery of microbes on a distant planet by spectroscopy might result in a persistent rise of misinformation and deception in the social media.

When ETI is discovered, a wide range of emotions are likely to be experienced in the immediate aftermath. More optimistic people express joy and interest regarding ETI, whereas more conservative people express fear and awe. The effects might be so profound and wide-ranging that they influence not just music but also literature, art, and culture. Regarding humanity's obligation to ETI, new moral and ethical issues would also surface, and these issues have the power to alter social norms Wilson and Cleland (2015).

4. The Phases of Discovery and Decision Making

In order to conduct a more straightforward examination of the entire scenario, we can evaluate it into three distinct categories:

4.1. *Pre-Discovery Phase*

The first phase, which is the one we are currently experiencing, is thought to be the easiest to act upon and make decisions during. The search for evidence of extraterrestrial life is an ongoing endeavor for many scientific projects and initiatives. Space missions seek evidence of past and present life on other planets and moons in our solar system, while experiments conducted in the deepest regions of oceans contribute to our understanding of the origins of life under extreme conditions. Additionally, searches are being conducted by projects such as SETI to look for evidence of advanced life on other worlds.

One thing to bear in mind at this stage is the importance of precise and open communication regarding ongoing scientific endeavors throughout the pre-discovery stage.

Maintaining public interest and confidence through regular updates on the goals, procedures, and discoveries of astrobiology and SETI initiatives could increase financing for additional study in this period.

In this stage, the media's role and the cooperation of scientists and scientific journalists are vital, particularly in terms of publishing the results of ETI-finding projects.

Pre-discovery is the optimum time to anticipate different eventualities and promote community readiness because it is the planning phase for an unavoidable future. Politicians and scientists might collaborate to evaluate possible discovery situations, ranging from discovering microscopic life to coming across a well developed civilization. Through the expansion of global cooperation and the gathering of specialists from many regions, we may establish a cohesive strategy for managing the discovery moment. Creating interdisciplinary communities to manage the problem in the future, akin to a United Nations for extraterrestrial affairs, may be the best case scenario for such cooperation.

Establishing protocols for detecting ETI signals (or any other forms of encounter with an ETI which will be discussed in Post-Discovery phase) has been carried out by *SETI protocols for an ETI signal detection* in 1989 and a revised version was published in 2010 SETI Institute (nd). Also, in the early days of internet, Shostak et el. proposed an Immediate Reaction Plan (IRP) Shostak and Oliver (1999) based on SETI's experience with hoaxes such as the case of EQ Peg Oliver et al. (1998).

Nevertheless, some find SETI's current protocols out of date and abstract as it provides a few unofficial instructions that only applies to members of scientific society who are a part of SETI or are related to one of SETI's programs. Baxter and Elliott (2012) The other issue that the scientific community faces in pre-discovery phase, as Oliver et al. (1998) clearly emphasizes, is what shapes the public perspective is not what scientists believe should happen, but rather what journalists bring into reality.

4.2. *The Phase of Discovery*

As soon as any ETL stage is identified, a precise and quick verification procedure will be essential. First and foremost, the discovery must be able to be verified by rigorous scientific procedures, which will involve independent research in the academic community, peer review, and multiple testing.

The announcement of the discovery would be the next necessary step, requiring open communication. To gain public support and avoid any unintended societal effects, scientists and academic bodies must communicate clearly about the nature of discovery-whether the ETL is a microorganism or an advanced intelligent civilization-the methods employed-whether it is indirect evidence or direct-and preliminary findings.

This phase would need assessing and analysing the opportunities and threats brought forth by the finding. It is possible that the finding has biological or technical concerns, or it could present opportunities for scientific, technological, and cultural breakthroughs.

Policies pertaining to the treatment of ETL/ETI must be developed since any interaction with ETI/ETL needs to be handled carefully and with respect. These regulations need to cover things like technical safety, biological containment, and advanced civilization communication protocols.

4.3. *Post-Discovery Phase*

In the case of post-discovery era, one must keep in mind that we are discussing the undetermined future. Therefore, specifying the exact consequences of this phase would be impossible. On the other hand, although IAA SETI has protocols for pre-discovery and discovery phases, there are no confirmed official protocols for the time of post-discovery

SETI Institute (nd). fortunately, different scenarios have been proposed mostly based on the nature of discovery and we are going to discuss three of these scenarios.

The first scenario is known as the signal scenario, which contains the detection of an artificial signal which is produced and sent by an extra terrestrial intelligence or civilization. Compared to other scenarios, this is the one scenario we have been prepared for the most during the recent decades since finding an ETI signal is the basis of SETI programs.

Before assessing the implications of this scenario, one should keep in mind that according to what we know from information theory and linguistics, even if we receive and confirm an ETI signal, decoding the signal and having access to its information might be impossible Hanser et al. (2004). Thus, we can only assess the signal based on its level of complexity and non-randomness (for example a complicated signal with entropy rank of three and four would indicate the existence of a civilization at least as advanced as human civilizations that use at least one language with a similar structure as human languages.)

Regardless of information theory considerations, two main factors would be effective on the impacts of signal scenario. the first factor would be the distance of the signal's source. The farther the the source is, the less impact it would have on different aspects of human life. This fact becomes more obvious when one think of a signal coming from thousands of light years away: even if we respond to this message, it might reach the ETI once its civilization has been destroyed or the specie has been extinct. But a signal from ten to a few hundred light years away (in other words a signal from our *cosmic neighborhood*) would give rise to public attention in short term, but it would not affect people's everyday life and political assemblies Anton and Schetsche (2023).

Another likely outcome could be huge cultural and theological influence which would vary in each country based on people's cultural and religious backgrounds. Additionally, the psychological effects of the discovery would be a matter of significance. Closer distances is likely to be linked to spread of negative feelings such as fear and anxiety. Therefore, We suggest that protocols be developed for science communicators and journalists to guide them in raising public awareness.

Finally as the scientific implication, We would like to propose another possible consequence related to information theory. Since the public interest to different branches of science and technology rises and falls due to major incidents in the world, it would be not far from mind to consider the possibility of a new wave of interest in information theory for deciphering the ET message.

Technosignature scenario is the second case in post-discovery era. Finding a technosignature includes finding evidence or direct observation of ET technology (e.g., a Dyson sphere). Although this scenario overlaps with signal scenario in cultural and psychological implications, its political effects would be more significant as countries and governments would want to respond to this discovery independently and with different (aggressive and non-aggressive) approaches. Thus, another suggestion for future would be instructions for forming international and multi-governmental political councils that must be responsible for administrating the political issues related to technosignature scenario Anton et al. (2024).

The third scenario with maximum impact on different aspects of human life and civilizations is named encounter scenario. This case is described as the discovery of a non-terrestrial spacecraft, controlled by extra terrestrials or artificial intelligent and maneuvering around or close to the earth. One boldly could say that this is the most interactive scenario in the case of contacting ETI. Moreover, a floating Et craft near the earth is the proof of existence of an extremely advanced civilization which is by far more developed than the residents of the earth. Predicting the consequences of this scenario is stepping into the realm of uncertainties and speculations. All one could anticipate of direct interaction with an ETI would mostly be based on science fiction. The possible political, economical and societal aftermaths of such encounter would vary from the most positive forms (inspired by films like E.T and Arrival) to the most devastating forms (inspired by war of the worlds). Regardless, it is likely to assume the development of space military strategies and new defense protocols for the planet earth Anton and Schetsche (2023).

5. Public Trust

Trust in astrobiology and extraterrestrial civilizations involves intuition, discourse, and interdisciplinary significance. The possibility of extraterrestrial life challenges cultural values and is both awe-inspiring and disquieting. Building trust requires transparent communication about the scientific process, despite differing interpretations. SETI and METI face scientific uncertainty that may challenge public trust, as astrobiological research evolves with new data and theories. The multidisciplinary approach to life beyond Earth connects to public perceptions of life on Earth but can lead to misunderstandings. Public trust in science diminishes when findings clash with deeply held beliefs, and political and cultural influences shape attitudes toward biosignature and technosignature discoveries.

6. Misinformation hit SETI in a Big Way

Sensationalized headlines and inaccuracies fuel rumors, spreading misinformation and shaping public perceptions. Without reliable information, conspiracy theories thrive, undermining trust in astronomy. Misinformation about SETI, astrobiology, and UFOs circulates widely, affecting observatories and SETI institutions through hype, biases, and misleading reports.

Science journalists play a key role in accurate reporting, but UFOs and Ancient Astronauts often get confused with CETI topics. Media prioritizes engagement over accuracy, leading to misunderstandings even among educated audiences. A gap between journalists, scientists, and visual artists worsens public misinterpretations, especially on social media.

Scientists detect unusual signals but lack evidence of extraterrestrial messages. Public reactions to SETI findings remain uncertain. The UK SETI Research Network surveyed opinions on extraterrestrial searches to inform an international protocol for first contact.

7. Importance of Scientific Literacy

Science fiction and sensationalized news shape public perceptions of extraterrestrial life, influenced by culture, religion, and media. These ingrained beliefs sustain outdated ideas like ancient aliens and unexplained phenomena. Media coverage has fueled rumors and misinformation about intelligent life.

Scientific illiteracy remains a challenge, despite efforts by science communicators. Misunderstanding of the scientific process hinders public engagement. Education, trust in science, and social context are key factors in pre- and post-detection scenarios. Preparing for first contact requires a systematic post-detection policy, addressing message decoding, data analysis, and societal impact. Global collaboration is essential for these efforts.

Projects like *A Sign in Space* reveal how cultural narratives shape our readiness for first contact. By engaging the public in decoding, they test collective intelligence while exposing the challenge of separating fiction from scientific reasoning-an essential step in preparing for an actual extraterrestrial message De Paulis et al. (2024).

8. Importance of Science Fiction to Influence Public Perception

Science fiction narratives serve as powerful cultural repositories for speculations about encounters with extraterrestrial intelligence, significantly shaping public imagination and perceptions of SETI Oliver (2003). These stories function as cognitive laboratories where societies explore potential contact scenarios, influencing both scientific inquiry and policy formation. Beyond mere entertainment, these stories function as cognitive laboratories where societies rehearse potential contact scenarios, thereby shaping anticipatory frameworks that influence both scientific inquiry and policy formation.

The genre's persistent exploration of first contact protocols-from optimistic portrayals in works like "Arrival" to cautionary tales such as "Close Encounters of the Third Kind"-establishes predetermined response patterns in collective consciousness that inevitably color institutional approaches when contemplating actual detection protocols. This phenomenon manifests particularly in the disproportionate allocation of resources toward detection methodologies that align with popular science fiction tropes, potentially neglecting more promising but less dramatically compelling search strategies.

Furthermore, science fiction's treatment of alien intelligence-often anthropomorphized despite scholarly recognition that non-terrestrial cognition would likely manifest in profoundly unfamiliar ways-reinforces cognitive biases that may hamper nuanced scientific dialogue. The recursive relationship between fictional narratives and scientific discourse creates feedback loops where speculative elements gradually permeate scholarly literature, subtly influencing research priorities within SETI communities and establishing experiential templates through which the public will inevitably interpret any eventual detection announcement Puranen et al. (2024)

9. Role of Science Journalism in SETI

Science journalism operates as a crucial intermediary in the SETI domain, translating specialized astronomical data into narratives accessible to non-specialists while simultaneously navigating tensions between scientific accuracy and media engagement imperatives. The field faces distinctive challenges when reporting SETI developments, as journalists must contextualize highly technical signal analysis within public discourse already saturated with extraterrestrial mythology and conspiracy frameworks.

During periods of intense media attention, the communication of uncertainty becomes a key issue for both scientists and science journalists, particularly when reporting SETI developments where claims can be easily sensationalized or misinterpreted. The tone of media reporting significantly influences public perceptions, with premature reporting of potential extraterrestrial civilizations potentially seen as questionable Korbitz (2014).

These journalistic practices significantly influence public funding priorities and scientific legitimacy perceptions, while simultaneously shaping institutional communication strategies within the SETI community. Furthermore, the increasing fragmentation of media landscapes has created parallel information ecosystems where foundational SETI concepts undergo distinctive interpretations, potentially complicating unified public responses to any eventual detection announcement through divergent framing mechanisms that reflect ideological rather than scientific considerations.

10. Conclusion

Our world would be drastically altered by the discovery of extraterrestrial intelligence, changing society norms, scientific knowledge, and our beliefs. Science fiction and mythology both reflect our curiosity with extraterrestrials (ETI), underscoring the profound cultural relevance of this possible meeting.

It's critical to be ready for anything that can come up when we carry out our search with programs like SETI. The responses and ramifications of such a finding can be managed with the aid of strong policies, open communication, and public involvement.

In the end, discovering extraterrestrial life forces us to reconsider our role in the cosmos. We may embrace the fresh insights that such a finding would offer and manage the revolutionary changes that lay ahead by cooperating and educating the public.

References

Anton, A., Elliott, J., & Schetsche, M. 2024, Meeting extraterrestrials: Scenarios of first contact from the perspective of exosociology. *Acta Astronautica*, 215, 308–314.

Anton, A. & Schetsche, M. 2023,. *Consequences of First Contact-A Scenario Analysis*, pp. 127–172.

Baxter, S. & Elliott, J. 2012, A seti metapolicy: New directions towards comprehensive policies concerning the detection of extraterrestrial intelligence. *Acta Astronautica - ACTA ASTRONAUT*, 78.

Döbler, N. A., Amri, Y. E., & Carbon, C.-C. 2023, Is there anybody out there? can individual loneliness, need for closure, and religiosity predict the belief in extraterrestrial life and intelligence? *Springer International Publishing*, 3(1), 21–33.

De Paulis, D. *et al.* Global outreach and cultural impact of a sign in space, an interdisciplinary simulation of a first contact scenario. In *International Astronautical Congress 2024* 2024,, Milan, Italy.

Freud, S. 1984,. *On Metapsychology.* Penguin Freud Library. Penguin, London. (Originally published in 1914).

Hanser, S., Doyle, L., McCowan, B., & Jenkins, J. 2004, Information theory applied to animal communication systems and its possible application to seti. *Proceedings of the International Astronomical Union*, 213, 514.

Johnston, S. I. 2015, The greek mythic story world. *Arethusa*, 48(3), 283–311. From Luhrmann's book, Page 27.

Jones, E. M. 1985, "where is everybody. " an account of fermi's question.

Korbitz, A. 2014, Toward understanding the active seti debate: Insights from risk communication and perception. *Acta Astronautica*, 105, 517–520.

Luhrmann, T. M. 2020,. *How God Becomes Real: Kindling the Presence of Invisible Others.* Princeton University Press, Princeton, New Jersey. Page ix.

Oliver, C. 2003,. *SETI AND THE MEDIA: IMPROVING SCIENCE COMMUNICATION.* PhD thesis.

Oliver, C., Sim, H., & Shostak, S. 1998, The case of eq peg: Challenge and response.

Puranen, E. J., Finer, E., Helling, C., & Smith, V. A. 2024, Science fiction media representations of exoplanets: portrayals of changing astronomical discoveries. *Journal of Science Communication*, 23(01).

Race, M. 2015,. *Preparing for the discovery of extraterrestrial life: are we ready?: Considering potential risks, impacts, and plans*, pp. 263–285.

Routledge, C., Abeyta, A. A., & Roylance, C. 2017, We are not alone: The meaning motive, religiosity, and belief in extraterrestrial intelligence. *Springer US*, 41(2), 135–146.

Schneider, S. 2015,. *Alien minds*, 189-206. Cambridge University Press.

Schulze-Makuch, D. 2015,. *The landscape of life*, pp. 81–94.

SETI Institute n.d.,. Protocols for an eti signal detection. Accessed: [Insert Date of Access].

Shostak, S. & Oliver, C. A. 1999, Immediate reaction plan: A strategy for dealing with a seti detection.

Wilson, E. & Cleland, C. 2015,. *The moral subject of astrobiology: Guideposts for exploring our ethical and political responsibilities towards extraterrestrial life*, pp. 207–221.

(Toward) Discovery of Life beyond Earth and its Impact
Proceedings IAU Symposium No. 387, 2026
H. Landt, M. Dominik & C. Oliver, eds.
doi:10.1017/S1743921324001595

Rehearsing Post-Detection Futures: Building Futures Literacy vis-à-vis Post-Detection Considerations for a Responsible Search for Extraterrestrial Life

George Profitiliotis
Blue Marble Space Institute of Science
email: george.profitiliotis@bmsis.org

Abstract. The search for extraterrestrial life, sapient or not, is a multi-phased process comprising pre-discovery, discovery, and post-discovery phases. Post-detection considerations can be conceptualized as contemplations of scientific and non-scientific issues pertinent partially to the discovery phase and to the final post-discovery phase that will be set in motion by a confirmed discovery of extraterrestrial life. To systematically explore the corresponding complex future landscape, scholars have proposed using alternative scenarios. However, this historical approach has actually focused more narrowly on generating specific detection situations, while neglecting the broader contextual environment scenarios that will necessarily encompass the detection. By drawing on Futures Studies, this work argues that a more comprehensive anticipatory approach is needed, involving the parallel delineation of both possible detection situations and possible future contextual scenarios, followed by their integration. Additionally, this work introduces a "Rehearsing Post-Detection Futures" workshop workflow inspired by the "Futures Literacy Laboratory" approach. This ready-to-deploy, interactive, participatory workshop is intended for educators and aims to help students and scholars in relevant disciplines broaden and diversify not only what but also how and why they anticipate when they consider the effects of a detection of extraterrestrial life in the future, particularly during the most urgent and precarious post-detection stage, i.e., the short-term stage right after the detection and its communication, thereby facilitating the cultivation of the participants' futures literacy. Such interventions can support the mindful deployment of a critical-hermeneutic anticipatory perspective towards building a more responsible search for extraterrestrial life, sapient or not.

Keywords. post-detection, futures literacy, futures studies, responsible search for extraterrestrial life

---

1. Prologue: Detection Situations and Scenarios as the Backbone of Post-Detection Considerations

The search for extraterrestrial life, sapient or not, is arguably an extended process comprising three consecutive phases: pre-discovery, discovery, and post-discovery. In the pre-discovery phase, which is the one the search is currently undergoing, it is only theoretical predictions that point towards the possible existence of the phenomenon of extraterrestrial life, while any already existing observations that might be indicative of this phenomenon have yet to be recognized or signified as such. The logically subsequent discovery phase itself consists of three stages, namely, detection, interpretation, and basic understanding of the phenomenon, and possesses not only technical aspects but also conceptual and social ones. In turn, this discovery phase is hypothetically succeeded by the post-discovery phase, in which the discovery's broader impact, alongside any potential reactions to the discovery, would be most prominent (Dick 2018). Hence, extraterrestrial life post-detection considerations can be conceptualized as contemplations of scientific and non-scientific issues, that is, technological, social, cultural, ethical, economic, environmental, legal, and political ones, pertinent to the interpretation and basic understanding stages of the discovery phase, as well as to the final post-discovery phase that will be set in motion by a confirmed discovery of extraterrestrial life. It's worth noting that the issues expected to arise in the first minutes, hours, and days after the discovery, and possibly after its communication, that is, the short-term impact, may be very different from the ones expected later on, that is, the long-term impact (Connors 1976): the former involves "initial reactions to the news, first impressions of the extraterrestrials, attitude perseverance and change, rumor, and collective behavior, including possible panic", while the latter involves more delayed effects on people, societies, and institutions, which can potentially range from a reconsideration of the place of the human in the universe, triggered by the knowledge that there is life elsewhere, to profound changes in philosophy, religion, politics, scientific worldviews, and cultures (Harrison et al. 2000). Beyond doubt, the above points paint a complex picture of the post-detection considerations panorama.

What is more, the swirling brushstrokes of this picture become even more entangled when we realize that any post-detection scene must necessarily take place not in the present but at an undefined time in the unknowable future. In other words, the need for thorough anticipation is inextricably linked with any kind of reflection on post-detection considerations. In this direction, several times in the literature (e.g., (Harrison and Connell 2001), (Voros 2018)) scholars have turned to the use of alternative scenarios as a proposed tool to systematically explore ex ante the possible reactions to a detection of extraterrestrial life and to help proactively manage its impacts. However, as a matter of fact, this historical operationalization of scenarios has essentially attempted to capture the possibility space of potential detection situations by means of different generative combinations of the alternative values of particular distinguishing factors. Some examples of distinguishing factors that have been historically used to generate such scenarios include: solely the level of interaction that could occur between humans and the discovered extraterrestrial life (Gartrelle 2015); the combination of the proximity of the location of discovery, taking the value of proximal or distant, with the complexity of the discovered life, taking the value of "simple" or "complex" (Harrison and Connell 2001); and the combination of three factors, i.e., proximity of the location, complexity of the discovered life, and directness of the encounter, the latter taking the value of direct contact with humans or indirect contact mediated exclusively by technology (Dick 2018). One key underlying assumption shared by the various instantiations of this historical operationalization of scenarios about how a detection of extraterrestrial life may take place is that the alternative values of the relevant distinguishing factors can significantly influence human responses to it (Harrison and Connell 2001), as well as the consequences (Gartrelle 2015), the implications, the impact (Dick 2018), and even the social change (Gartrelle 2015) that can result from it. That said, the undoubted usefulness of this line of thought can surely be amplified by turning to the field of Futures Studies for a reconceptualization of the term "scenario" and its much-warranted distinction from "situation".

The field of Futures Studies is concerned with the systematic and explicit contemplation of multiple alternative futures, especially with the goal of enhancing preparability and increasing human influence over the unpredictable and unknowable future (Bell 1997). "Futures" are used in plural to highlight the strong contrast between dealing with many possible futures and predicting a singular future: the former is a democratic activity that tries to emancipate people's imagination from domineering narratives of an "official" future, thereby opening up the future, while the latter contributes towards controlling and colonizing the future (Gidley 2017). Historically, Futures Studies started with the predictive and mainly quantitative subfield of forecasting that assumes the continuity of the structure of the system under study and of the laws that govern it, which consequently makes possible the extrapolations and projections of the past into short- or long-term time horizons. A second subfield that emerged later is strategic foresight, which is qualitative, non-predictive, and puts the discontinuities related to the system under study to use in the form of a variety of possible futures (Poli 2017). Strategic foresight is a systematic approach to learning and understanding possible futures to challenge the mindset of decision makers, as well as to building shared visions to guide and enable present-day decisions (Iden et al. 2017). This is oftentimes accomplished by means of scenarios, which in Futures Studies are conceptualized, following Porter, as internally consistent views of what the future might turn out to be (Porter 1985), describing possible future states of the external world (van der Heijden 2005) with time horizons ranging from a couple of years to a couple of decades (Bradfield et al. 2005). To phrase it differently, a set of scenarios describes alternative future external environments of the system of interest that are plausibly possible, which is in contrast to the confusing wide and commonplace use of the word "scenario" with various other meanings, including that of a situation. Scenarios, then, provide the "story frame", the broader future context, the surrounding stage for rehearsing stories and situations before they occur (Ramirez and Wilkinson 2016), to which they owe their prevalence as the primary method for navigating deep uncertainty (Bohensky et al. 2006) in a normative or exploratory manner (van der Heijden 2005). Scenarios as broader future states can contain multiple specific situations alongside particular personas and things related to those situations (Smith and Ashby 2020).

Based on the above, the aforementioned historical application of the term "scenarios" to capture the possibility space of potential detection cases can now be seen in a new light as an attempt to generate exhaustive sets of specific possible detection situations. Nevertheless, any putative future detection will not unfold in a vacuum; rather, it will necessarily be encompassed by the future states of the broader economic, political and legal, societal and cultural, environmental, and scientific and technological contextual environment, which must not be handled as just a singular extrapolation of the present-day state of affairs. Thus, discussions of post-detection considerations can be significantly enriched by the parallel delineation of sets of possible detection situations and of sets of possible future scenarios of the contextual environment, followed by their intertwinement into an integrated supporting backbone. Given the need for exhaustiveness, the delineation of sets of possible detection situations can be effectively accomplished by leveraging general morphological analysis, a method originally proposed by the astronomer Fritz Zwicky who applied it, among other cases, to problems related to astronomical phenomena (Zwicky 1957). This modelling method allows for the systematic structuring and analysis of multi-dimensional and non-quantified problems by means of the formulation of a matrix containing all the possible parameters that are relevant to the problem, represented by all their potential discrete values, the various logically coherent combinations of which can generate complete "solutions" Álvarez and Ritchey (2015). Furthermore, the delineation of sets of possible future scenarios of the contextual environment can be effectively accomplished through the "Intuitive Logics" scenarios school of thought that originated in Pierre Wack's initial work at the Royal Dutch Shell and is nowadays considered the "mainstream scenario approach" (Postma and Liebl 2005). This participatory approach focuses on the production and use of qualitative scenarios of equally plausible alternative future worlds by means of intuition, brainstorming techniques, research, and expert opinion to define the logics that underpin the organizing principles pertaining to the matter of interest (Amer et al.(2013). Nevertheless, the appropriate investigation of this highly demanding, two-pronged, strategic foresight task is not further pursued in this chapter but is the focus of an upcoming study.

2. Rehearsing Post-Detection Futures: A Deployable Futures Literacy Workshop Workflow

While a systematic and thorough application of the aforementioned strategic foresight methods to the delineation of sets of possible detection situations and of sets of possible future contextual scenarios, followed by their appropriate intertwinement, can prove highly beneficial for advancing the required post-detection-relevant preparations and plans by decision-makers ahead of time, completing the picture of our presently available options for proactive mobilization necessitates the consideration of one more angle towards the future. The third and currently growing subfield of Futures Studies that emerged more recently is the study of conscious human anticipation, i.e., a forward-looking attitude that informs action in the present; this subfield engages with reality in a way that considers not only what actually exists, but also dispositions, habits, tendencies, and forces that act like latent factors which may become actualized under the proper triggering conditions (Poli 2017).

As conceptualized by Miller, conscious human anticipation comprises two modes: Anticipation-for-the-Future, namely, using the future as a goal in preparation and planning, and Anticipation-for-Emergence, namely, using the future as a disposable construct to see the present from novel viewpoints. What is more, this approach holds that conscious human anticipation can be enhanced by means of cultivating one's "futures literacy", namely, the capability to bring the future into the present through imagination, i.e., to "use-the-future", for different reasons and in a variety of ways (Miller 2018a). There are three main benefits in cultivating one's futures literacy: it can facilitate successful preparation and planning, it makes it easier to sense and make sense of the world, and it enhances the capacity to appreciate complexity, to take advantage of change, to deploy everyday forms of contextual creativity, and to embrace a diversification strategy towards resilience (Ehresmann 2018). The "Futures Literacy Laboratory-Novelty" (FLL-N) is a tailorable interactive tool that has been specifically developed to promote the cultivation of futures literacy in different contexts (Miller 2018a).

FLL-N's are essentially structured, interactive, participatory workshops conducted in three phases: firstly, the "Reveal" phase helps participants make explicit their currently-held implicit ideas about the probable and desirable futures and then interrogate the fundamental assumptions that sustain them; then, the "Reframe" phase exposes participants to an unfamiliar future that is completely freed from those expectations and hopes and helps them rehearse and grasp a bewildering world; finally, the "Rethink" phase prompts participants to compare and contrast their experiences in the previous two phases, to reflect on new facets of the future that were illuminated by this process, and to realize that specific expectations and hopes tend to tacitly constrain the way the present is being seen (Miller 2018b). "Rehearsing Post-Detection Futures" is an accordingly-structured, ready-to-deploy FLL-N-inspired workflow for an interactive, participatory workshop intended particularly for educators working with either students or scholars in astronomy, astrobiology, planetary science, space exploration and allied disciplines, as well as in the social sciences and humanities sub-disciplines devoted to the search for extraterrestrial life, sapient or not. Its goal is to help its participants broaden and diversify not only what but also how and why they anticipate when they consider the short-term effects of a detection of extraterrestrial life in the future with a decade-long time horizon. The longer-term period of the post-discovery phase's more delayed effects that do not refer to the first minutes, hours, and days after the detection and its communication is not explored in this particular implementation and is left to future work.

This work follows the recent recommendation to leverage "Futures Literacy Laboratory"-inspired workshops to help various stakeholder groups anticipate broader and more diverse implications of a future detection of extraterrestrial life to build resilience (Profitiliotis and Theologou 2023), and is closely associated with the astrobiology-related "Moral Vistas to Xenic Beyonds" workshop's workflow (Profitiliotis 2024), whose novel operationalization of Miller's anticipatory assumptions clusters (Miller 2018a) as a "List of Foundations" it adopts. The workshop presented hereinafter has been designed to be scalable in terms of the number of participants it can accommodate and to be interactively implemented in virtual, physical, and hybrid formats through simple audience engagement tools that allow for polling, free-form text data collection, and presentation of outcomes. It's worth noting that this workshop has been tested in two different pilot sessions during international astrobiology-related conferences, namely, the *2022 "First Penn State SETI Symposium"* and the *2024 "Kavli-IAU Symposium (IAUS 387): (Toward) Discovery of Life Beyond Earth and its Impact"*, with around 25 participants per session. During the course of the workshop, participants are not divided into smaller breakout groups, so the facilitation can be undertaken by a single educator. The overall duration of this workshop is 2 hours, and adaptations are encouraged to accommodate particular pedagogical needs and contexts.

The centerpiece of this workshop is the detection situation embedded in the contextual scenario of the unfamiliar future of the "Reframe" phase. In terms of content, the contextual scenario's key characteristics have been formulated to challenge participants' currently-held implicit hopes and expectations and to promote this future world's unfamiliarity. To prevent the introduction of extremely outlandish assertions that might be entirely rejected by the participants, initial inspiration for formulating those key characteristics was drawn from two scenario studies on the futures of research (Elsevier 2022) and the futures of the space sector (The Aerospace Corporation 2021). Likewise, the detection situation's content reverses several presumptions that appear to be prevalent across the academic astrobiology community, namely, that the detection will most likely be about a biosignature (what) observed by a group of academic researchers (who), because of a targeted, astrobiology-related research program (why) focusing on a remote exoplanet (where), and that it will unfold as an ambiguous and prolonged process (how) in a future world that is essentially an extrapolation of the present-day context (when). This particular detection situation is about a technosignature observed by an independent, freelance researcher, because of serendipity during a resources prospecting operation on the lunar far side's surface, and it unfolds as an unambiguous and fast process in an unfamiliar future world reflected on the contextual scenario's key characteristics.

The following points provided inspiration for the detailed construction of the detection situation. Firstly, commercial lunar missions are actively being supported via NASA's Commercial Lunar Payload Services (CLPS) program, among other initiatives. In fact, private actors becoming involved in a potential technosignature detection has been imagined more than once by the technosignatures community: in a speculative fiction story, Benford presented a multi-billionaire investing resources into SETI research for profit (Benford 2008); in a popular magazine article, Loeb discussed the excellent commercial potential of deciphering an alien technology and marketing the same idea to Earth (Loeb 2020); and, in an academic journal article, Gertz contemplated the potential accidental discovery of an extraterrestrial probe on an asteroid by a privately-owned mining mission (Gertz 2016a). Secondly, Earth's Moon has been proposed as an attractive possible location for signs of extraterrestrial artifacts that may have ended up there, as the little evolution of the lunar surface might preserve possible evidence for millions or even billions of years (Socas-Navarro et al. 2021) (Benford 2021) (Davies and Wagner 2013). Thirdly, it is known that scientific discoveries can also occur via serendipity, that is, unintentionally, accidentally, and luckily (Arfini et al. 2020). Fourthly, ground-penetrating radar has been suggested as a potentially useful tool for collecting technosignature-relevant data on the Moon (Davies and Wagner 2013). Finally, a communicative probe residing in our Solar System has been described as an effective way for extraterrestrial agents to deliberately and directly contact Earth (Gertz 2016b), which would leave little room for ambiguity.

In terms of form, participants' involvement in the unfamiliar future is supported by three complementary narrative methods for creativity training which engage their narrative cognition (Fletcher and Benveniste 2022), namely, world-building, perspective-shifting, and action generation; these can respectively "help the mind imagine new environments", "help the mind imagine from different perspectives", and "help the mind imagine possible future actions", thereby making that bewildering future world more accessible. The world-building element is implemented by means of a collaborative exercise wherein participants are asked to create news headlines from the unfamiliar future. The perspective-shifting and action generation elements are jointly implemented by means of an interactive storytelling process that introduces participants to the backgrounds of two characters and then asks them to step into their shoes and make decisions that shape the development of a branching narrative taking place in the unfamiliar future world.

The content of the **"Reveal"** phase of the "Rehearsing Post-Detection Futures" workshop is structured as follows. Workshop participants are given the following instructions to work through a simple audience engagement tool that allows for free-form text data collection and presentation of collected data: *"Let's travel to the future! Please close your eyes. It is now the year 2035. Imagine the search for alien life in its most likely state. In this most likely future of the search for alien life, imagine that you encounter in the news an event that has something to do with the search for alien life, particularly regarding the science and society interface. Write a news headline that describes this event of this probable future!"*.

After the participants have created their news headlines, the collected text data is presented to them, and they are given the following instructions to continue working: *"Reflect on the deeper attributes of the future world to which your event belongs. Which major assumption underlies it in terms of what you believe the future will be like? Write down your major assumption in the following format: 'I think this event is part of the likely future, because I believe that in 2035, compared to today, [something] will be [increased / decreased / the same / fundamentally different]"'*.

After the participants' reflection, the newly-collected text data is presented to them, and they are given the following instructions to vote through a simple audience engagement tool that allows for polling and presentation of outcomes: *"Compare and contrast the assumptions above and refrain from judging them. Overall, what do you think are the deeper foundations of your collective assumptions? Choose any that apply from the following 'List of Foundations'—you may select more than one: 'We built on forecasts, predictions, or projections that extrapolate from the past to the future'; 'We built on ideological or religious narratives that predetermine a fixed future destiny'; 'We speculated on future creative interventions to be undertaken by an individual or organization to solve some currently known issues in their external environment'; 'We speculated on future self-improvements to be undertaken by an individual or organization to overcome some currently known shortcomings within themselves'; 'We relied on a systematic consideration of the possible interplays of various factors to notice novel but recurring patterns as they were emerging'; 'We relied on active, alert awareness and mindfulness to notice something rare or unique as it was emerging"'*.

After the participants' voting and reflection on the outcomes, they are given the following instructions to again work through a simple audience engagement tool that allows for free-form text data collection and presentation of collected data: *"Let's travel to the future once more! Please close your eyes. It is now the year 2035. Imagine the search for alien life but this time in its most desirable state. In this most desirable future of the search for alien life, imagine that you encounter in the news an event that has something to do with the search for alien life, particularly regarding the science and society interface. Write a news headline that describes this event of this desirable future!"*.

After the participants have created their news headlines, the collected text data is presented to them, and they are given the following instructions to continue working: *"Reflect on the deeper attributes of the future world to which your event belongs. Which major assumption underlies it in terms of what you want the future to be like? Write down your major assumption in the following format: 'I think this event is part of my desirable future, because in 2035 I want [something] to be [increased / decreased / the same / fundamentally different], in comparison with today"'*.

After the participants' reflection, the newly-collected text data is presented to them, and they are given the following instructions to vote through a simple audience engagement tool that allows for polling and presentation of outcomes: *"Compare and contrast the assumptions above and refrain from judging them. Overall, what do you think are the deeper foundations of your collective assumptions? Choose any that apply from the 'List of Foundations'—you may select more than one"*. The complete "List of Foundations" used previously is again presented to them for the new voting.

The content of the **"Reframe"** phase of this workshop is structured as follows. After the participants' voting and reflection on the previous activity's outcomes, they are given the following instructions to again work through a simple audience engagement tool that allows for free-form text data collection and presentation of collected data: *"Now, aware of the deeper foundations that sustain our engagements with our likely and desirable futures of the search for alien life, we leave them behind temporarily, as we continue our journey into the futures! Let's travel to the future for the third time. It is now the year 2035. However, the future is neither what we thought likely nor what we wanted. This future of the search for alien life is a strange one. The following are some key characteristics of this strange future world! As you read them, imagine you are experiencing the people, the systems, the places, the things, and the events of this future world. What is going on there, particularly regarding the search for alien life? How does this world work? Think silently!*

- *Funding for exploratory, blue-sky research has been reduced; the emphasis is on rapid development of practical solutions; industry is replacing government as the main source of R&D funding for universities.*
- *Researchers can, at very low cost, source the materials they need to work independently from institutions using micropayments, leading to the emergence of "science-as-a-service".*
- *Online repositories built on preprint servers thrive.*
- *Artificial intelligence-enabled technologies are used in multiple aspects of data-driven research, including peer-review.*
- *The lunar economy is taking off: private "exploration-as-a-service" companies are lowering barriers to entry, enabling the development of a research value chain across the Earth-Moon system.*
- *Multilateral diplomatic successes have enabled global lunar governance via the establishment of lunar economic zones connected to a permit system for time-, place-, and purpose- delimited exclusive rights.*

– *An integrated global network of 3D virtual worlds, i.e., the metaverse, is now a highly popular manifestation of the internet's social sphere.*

In this unfamiliar future of the search for alien life, imagine that you encounter in the news an event that has something to do with the search for alien life, particularly regarding the science and society interface. Write a news headline that describes this event of this unfamiliar future! Also, read the events your fellow participants are observing and build on them!".

After the participants have created their news headlines, the collected text data is presented to them, and they are given the following instructions to continue working: *"Reflect on the deeper attributes of the future world to which your event belongs. Which major assumption underlies it in terms of what you realized about this future? Write down your major assumption in the following format: 'I think this event is part of this strange future, because, while experiencing this world, I realized that [something] was [frequent / rare / non-existent] in 2035"'.*

After the participants' reflection, the newly-collected text data is presented to them, and they are given the following instructions to vote through a simple audience engagement tool that allows for polling and presentation of outcomes: *"Compare and contrast the assumptions above and refrain from judging them. Overall, what do you think are the deeper foundations of your collective assumptions? Choose any that apply from the 'List of Foundations'—you may select more than one".* The complete "List of Foundations" used previously is again presented to them for the new voting.

Next, participants are introduced to the characters whose perspectives they will need to adopt during the branching narrative of the interactive storytelling: *"This is the story of two people living in this unfamiliar future world. Get to know them by reading their backstories below! As their story unfolds, you will need to step into their shoes and make difficult decisions in some particularly problematic situations, as if you were them!*

Main character and backstory: Hannah, 31 years old, female, lives in Berlin, Germany.

Hannah is a PhD in Advanced Remote Sensing for Planetary Applications, currently working independently in lunar resources prospecting under a "science-as-a-service" scheme. The work she conducts is either used for her own postdoctoral research or sold to other parties on Sci-Legion, the world's prime "science-as-a-service" digital marketplace that brings together buyers and sellers from every part of the research value chain from all over the world. Hannah herself buys diverse raw and refined data from "Sci-Legion" to independently conduct her research work, sometimes assisted by AI-enabled processes that augment her workflow.

In 2035, during a routine prospecting operation in one of the economic zones on the far side of the Moon, Hannah initially identifies an anomaly under the lunar regolith that appears interesting for further exploration as a candidate water ice deposit. After buying more detailed and diverse data from Sci-Legion and analyzing it thoroughly, Hannah hypothesizes that the anomaly might in fact be a buried fragment of a metallic asteroid, which might be worth more than water ice. With the first-mover's advantage at hand, Hannah proceeds with the routine process of applying for the necessary permits for short-term exclusive rights for the on-site exploration of the candidate metallic fragment. After her permits are granted, Hannah buys exploration time for a pair of semi-autonomous lunar rovers from a private "exploration-as-a-service" company and remotely orders them into extensive data collection on site. While analyzing the ground-penetrating radar data from the rovers, Hannah identifies extensive anomalies in the data, which appear to be caused by an active modulation of the radar signal upon reflecting on the candidate metallic fragment. Perplexed by the recurrence of this phenomenon in follow-up investigations that corroborate her hypothesis of active modulation of the radar signal by the target, Hannah proceeds with a routine characterization of the morphology of the site's regolith structure, only to conclude that the site has been unperturbed for millions of years.

Reminiscent of a highly-publicized anomalous observation in 2032 that was framed as the potential detection of an extrasolar megastructure, which is yet to be confirmed, Hannah's own observed phenomenon baffles her to a great extent. Having ruled out other explanations, Hannah hypothesizes that the phenomenon must be a technosignature, and the metallic fragment must be an extraterrestrial artifact.

Secondary character and backstory: Sara, 31 years old, female, lives in Berlin, Germany.

Sara is a social media science communicator and an open-science activist, vigorously opposed to the established trend of a research landscape dominated by private companies. Sara is part of a movement supporting the public's right to know about scientific outcomes, the subversion of the domineering role of private companies in research that has been promoting secrecy, and the limitation of the use of AI-enabled automated processes that threaten specific research roles. She is Hannah's cousin and best friend. Sara and Hannah have essentially grown up together for almost three decades and are currently flat-mates in Berlin, Germany.

Then, in the course of the branching narrative, participants are gradually presented with five consecutive decision points at which they can select one of the two options presented to them by voting through a simple audience engagement tool that allows for polling and presentation of outcomes. The majority vote determines which story segment will next be presented to the participants. The first decision point is embedded in the narrative's starting segment. The starting segment and all the other possible segments are presented hereinafter.

[1] Detection

The heavy rainfall was infiltrating Hannah's dreams as she lay in her bed in the twilight zone between awake and asleep. The white noise fueled her restless mind that was still trying to make peace with the last thought she slept on the previous night. Was it true? Had she really found a relic of primordial makers by accident? Should she feel happy about it? She opened her eyes and stared at her smartphone that was charging by her bedside. She hesitated for half a minute and then decided to just have a look: once again, three hours earlier than the alarm clock. A rough mental calculation warned her that, within the past week, she had been spending more time trying to sleep than actually sleeping. Then, like that crow that favored greeting her every morning she would jog through the nearby park—at least when she was still into leaving the house now and then—the thought returned to her. Had she truly discovered an alien artifact on the Moon all by herself?

She carried that thought with her as she sneaked into the shower, tiptoeing past her flat-mate's open door; thankfully, Sara appeared to be asleep. Hannah had been consistently avoiding her for days, while trying to make up her mind on whether she would share the news with her or not. Sara cared for her; Hannah knew it. She was her cousin and her best friend with whom she had grown up for almost three decades. If there was one person in the world Hannah would trust, that was Sara. Yet, the decision was not straightforward. Sara was a strong proponent of open science and truly believed that knowledge should always be shared with the public and never remain locked up in the servers of individuals or powerful technology corporations. If Hannah told her what she had found the other day, would Sara be even capable of keeping it a secret?

A sudden knock on the door collapsed her intricately branching thoughts of how such a potential discussion could take place into an imminent eventuality. The bathroom door was slightly opened, and Sara's voice mingled with the splashing of the shower: "Hannah, you are up already? I think we need to talk. What's going on? Please talk to me. You know I am here for you"

Option A: Hannah tells Sara the truth about her hypothesis of a discovery → **[2]**

Option B: Hannah lies to Sara that her frustration is because of too much work → **[3]**

[2] Hannah tells Sara the truth

Hannah sighed in the shower, the water mingling with tears that were heralding a liberating confession. "Yes. Wait for me in the living room. I need to tell you something. It's important" said Hannah, trying to get her thoughts in order. Sara nodded in relief and headed for the kitchen to prepare some breakfast, as she had been doing for the past week, hoping to lure Hannah out of her sudden solitude. This time, she succeeded.

Hannah joined her in the kitchen, balancing her laptop on her right palm. Dozens of words were piling up behind her lips as she was bringing up the evidence she needed to support her claims. Sara grabbed two plates full of pancakes and followed her silently to the living room, comforted by the thought that progress had at last been made and her dear friend would be talking to her any moment now.

Half an hour later, Sara was staring with astonishment at her cold pancakes that resembled the cratered lunar far side, as if she was metaphorically replicating her cousin's actions that led her to the hypothesized artifact. "This is huge! I mean, it's amazing! I almost can't believe it! It's monumental! We must do something! You must tell the public! This is a game-changer!" babbled Sara, tapping her fingers nervously on the couch. Hannah bit her lip saying nothing, her mind producing faint images of how things might play out in a situation where she publicly revealed her accidental findings and her corroborated hypothesis. She couldn't help but wonder if the outcome would be similar to the viral public reactions which followed the news that a research group was claiming to have detected an extrasolar megastructure some years ago; as that anomalous observation was yet to be confirmed, new podcasts, videos, and articles would still pop up now and then to stir up the waters.

"Hannah? Are you still with me? Come on, say something! You should be proud of yourself! I know I am! You know I will help you! You can count on me!" whispered Sara, taking Hannah's hand into her own. Hannah took a few deep breaths to manage the almost unbearable weight of not knowing which course of action would effectively align her desire to let everyone know that there were other kids on the cosmic block with her own interest of self-preservation.

Option A: Hannah agrees with Sara that she must reveal the evidence and findings to the public → **[4]**

Option B: Hannah thanks Sara for her support but asks her to keep it a secret between friends → **[5]**

[3] Hannah lies to Sara

Hannah sighed in the shower, the water mingling with tears that were flowing from the wellspring of a suppressed confession. "It's nothing, really. I might be working too much. It's not important" said Hannah, trying to conceal her true thoughts as well as possible. Sara exhaled in distress and headed for the kitchen to prepare some breakfast, as she had been doing for the past week, hoping to lure Hannah out of her sudden solitude. Once again, she did not succeed.

Hannah retreated to her room and locked herself up under the pale light of her laptop's screen. As she was trying to justify to herself that blatant lie she had thrown at her best friend, Hannah regressed to an impulsive behavior to evade her emotions: she started going through all the documented steps of her scientific reasoning on the AI-assisted mind-mapping software tool, from the collected evidence to her "extraterrestrial artifact" last remaining explanation. In parallel, Sara, sitting silently in the living room, finally made up her mind. "Do it" she typed in her secure messaging app, ordering the "hacking-as-a-service" group she had commissioned in the dark web to hack her dear friend's computer. In a moment, Sara was able to view Hannah's screen using a remote assistance software exploit.

Half an hour later, Sara was scratching impatiently at the wood grain of Hannah's tightly shut door. "Hannah, please come out! You cannot just hide this! This is huge! You must tell the public! This is a game-changer!" babbled Sara, tapping her fingers nervously on the locked door. Hannah had been silently biting her lip in anger and shame from the moment she heard of the hack, her mind producing faint images of how things might play out in a situation where the public gained access to her accidental findings and her corroborated hypothesis. She couldn't help but wonder if the outcome would be similar to the viral public reactions which followed the news that a research group was claiming to have detected an extrasolar megastructure some years ago; as that anomalous observation was yet to be confirmed, new podcasts, videos, and articles would still pop up now and then to stir up the waters.

"Hannah? Are you mad? Come on, say something! There is nothing to hide! I am sorry for the brutal shortcut, but you know I will help you, right?" whispered Sara, touching her forehead on Hannah's door. Hannah took a few deep breaths to weigh her lie against her friend's deception.

Option A: Hannah begrudgingly agrees with Sara that she must reveal the evidence and findings to the public → [**6**]

Option B: Hannah argues with Sara and demands that she keeps it a secret → [**7**]

[4] **Hannah agrees to reveal detection to the public**

"I'm so glad we talked. Yes, let's do it! I'm truly happy we are together on that!" said Hannah in a low but cheerful tone, extending her arms towards her cousin. Sara smiled heartily and embraced her tightly. "So what's the plan?" asked Hannah, taking a bite out of the cold pancake she held on her fork like a trophy. The sun had still not risen in the rainy Berlin sky. "In case you forgot, rule number one in this home office: let the music flow" replied Sara cheerfully and started their favorite playlist on their home theater system. Hannah had indeed forgotten; all those days she had been repeatedly going over her research workflow in total isolation, the only sounds that would occasionally breach her cocoon of silence were the dripping rain and Sara's pleading invitations for breakfast, lunch, and dinner that went unanswered.

"I'm very sorry for making you worry. I was trying to figure out if I had made a mistake somewhere and I lost track of time" apologized Hannah, while her cousin was feverishly adding ideas for disclosure to her mind-mapping drawing board software. Sara was an experienced social media science communicator, proficient in even the most obscure communication tools and platforms, but she actively avoided using AI-powered processes out of a moral choice to subvert job automation. "This is an AI-free workplace: 5 years without incident" read the sticker on her laptop.

"It's fine! Feast on your pancakes of victory! We have some serious work to do!" replied Sara and ate her pancake in one bite. "If we are really going public on this, you must be prepared to release everything. I mean, probably not the proprietary stuff you bought on Sci-Legion, but their metadata, all your analyses and the data you produced, your methods, your reasoning, and the clear hypothesis you have formulated after ruling out all the other options. This has to be transparent. Sci-Legion will see increased marketplace traffic by others who will want to replicate it, and will promote you further. But the release must happen during a worldwide public event to assert your status as the discoverer. Remember the '3 Minutes with Prometheus' metaverse show? It is broadcasted to all decentralized platforms and recasted on all social media. I can get you in as a breaking announcement. You know, if it's not blockchained, it never happened" jabbered Sara.

Option A: Hannah agrees with Sara to go public on metaverse and social media → [**8**]
Option B: Hannah thanks Sara for her support but goes public on the academic media → [**9**]

[5] Hannah asks Sara to keep it a secret

"I'm so glad we talked. You are always so kind! I need a favor. I want this to be a secret between friends. I want us to be together on that, but let's not involve others, for now!" said Hannah in a low imploring tone, extending her arms towards her cousin. Sara smiled in confusion but embraced her tightly. "So, what's our secret agenda?" asked Hannah, taking a bite out of the cold pancake she held on her fork like a trophy. The sun had still not risen in the rainy Berlin sky. "In case you forgot, rule number one in this home office: let the music flow" replied Sara, masking her disappointment, and started their favorite playlist on their home theater system. Hannah had indeed forgotten; all those days she had been repeatedly going over her research workflow in total isolation, the only sounds that would occasionally breach her cocoon of silence were the dripping rain and Sara's pleading invitations for breakfast, lunch, and dinner that went unanswered.

"I'm very sorry for making you worry. I was trying to figure out if I had made a mistake somewhere and I lost track of time" apologized Hannah, while her cousin was feverishly adding ideas for a potential future disclosure to her mind-mapping drawing board software; she wouldn't give up so easily on convincing Hannah to go public. Sara was an experienced social media science communicator, proficient in even the most obscure communication tools and platforms, but she actively avoided using AI-powered processes out of a moral choice to subvert job automation. "This is an AI-free workplace: 5 years without incident" read the sticker on her laptop.

"It's fine! Feast on your pancakes of covert victory!" replied Sara and ate her pancake in one bite. "First things first! We must protect your rights as the discoverer! You know I am against those paid, Big Tech, end-to-end research management products, but... Here it goes: Sci-Legion offers a blockchain notarization service, doesn't it? We must notarize your whole workflow. Including pointers to the proprietary stuff you bought on Sci-Legion, all your analyses and pointers to the data you produced, your methods, your reasoning, and the clear hypothesis you have formulated after ruling out all the other options. This way, Sci-Legion will protect your interests—good for business—and we will be the only ones to know. You know, if it's not blockchained, it never happened" jabbered Sara. Hannah found it reasonable and halfheartedly agreed.

Option A: Sara deceives Hannah and goes public herself on metaverse and social media → [**10**]
Option B: Sara respects Hannah's request and keeps it a secret → [**11**]

[6] Hannah begrudgingly agrees to reveal detection to the public

"How could you do this to me? Can't you see the risks? You trust your vigilante hackers more than you trust me?" exploded Hannah and flung her door open, her eyes sparkling with tears. Sara's lips immediately tightened, as she faced Hannah for the first time in days. "It's safe! I've worked with them many times! They only hack for good! It was just a simple screen mirroring hack! I needed to know you were not in trouble! I love you!" exclaimed Sara, her voice cracking from emotional strain. Hannah's gaze softened, and she begrudgingly extended her arms towards her cousin—she realized she would probably have done the same. Sara nodded heartily and embraced her tightly. "What is done is done. So, what's the plan now? I guess we disclose?" asked Hannah, striking a pragmatic tone. The sun had still not risen in the rainy Berlin sky. "Some music maybe?" replied Sara hesitantly and went on to start their favorite playlist on their home theater system. Hannah appreciated it; all those days she had been repeatedly going over her research workflow in total isolation, the only sounds that would occasionally breach her cocoon of silence were the dripping rain and Sara's pleading invitations for breakfast, lunch, and dinner that went unanswered.

"I'm very sorry for making you worry. I was trying to figure out if I had made a mistake somewhere and I lost track of time! I know you are sorry too!" apologized Hannah, while her cousin was reluctantly adding ideas for disclosure to her mind-mapping drawing board software. Sara was an experienced social media science communicator, proficient in even the most obscure communication tools and platforms, but she actively avoided using AI-powered processes out of a moral choice to subvert job automation. "This is an AI-free workplace: 5 years without incident" read the sticker on her laptop.

"I'm truly sorry! I only want what's best for you! I do have some ideas, if you are in the mood..." replied Sara, and Hannah nodded understandingly. "If we are really going public on this, you must be prepared to release everything. I mean, probably not the proprietary stuff you bought on Sci-Legion, but their metadata, all your analyses and the data you produced, your methods, your reasoning, and the clear hypothesis you have formulated after ruling out all the other options. This has to be transparent. Sci-Legion will see increased marketplace traffic by others who will want to replicate it, and will promote you further. But the release must happen during a worldwide public event to assert your status as the discoverer. Remember the '3 Minutes with Prometheus' metaverse show? It is broadcasted to all decentralized platforms and recasted on all social media. I can get you in as a breaking announcement. You know, if it's not blockchained, it never happened" jabbered Sara.

Option A: Hannah agrees with Sara to go public on metaverse and social media → **[8]**
Option B: Hannah thanks Sara for her support but goes public on the academic media → **[9]**

[7] Hannah demands that Sara keeps it a secret

"How could you do this to me? Can't you see the risks? You trust your vigilante hackers more than you trust me?" exploded Hannah and flung her door open, her eyes sparkling with tears. Sara's lips immediately tightened, as she faced Hannah for the first time in days. "It's safe! I've worked with them many times! They only hack for good! It was just a simple screen mirroring hack! I needed to know you were not in trouble! I love you!" exclaimed Sara, her voice cracking from emotional strain. Hannah's gaze softened, and she begrudgingly extended her arms towards her cousin—she realized she would probably have done the same. Sara nodded heartily and embraced her tightly. "What is done is done. But this stays between us! You will keep it a secret!" demanded Hannah, striking a pragmatic tone. The sun had still not risen in the rainy Berlin sky. "Your call. I won't tell... I need some music..." replied Sara heartbroken and went on to start their favorite playlist on their home theater system. Hannah appreciated it; all those days she had been repeatedly going over her research workflow in total isolation, the only sounds that would occasionally breach her cocoon of silence were the dripping rain and Sara's pleading invitations for breakfast, lunch, and dinner that went unanswered.

"I'm very sorry for making you worry. I was trying to figure out if I had made a mistake somewhere and I lost track of time! I know you are sorry too!" apologized Hannah, while her cousin was reluctantly adding ideas for a potential future disclosure to her mind-mapping drawing board software; she wouldn't give up so easily on convincing Hannah to go public. Sara was an experienced social media science communicator, proficient in even the most obscure communication tools and platforms, but she actively avoided using AI-powered processes out of a moral choice to subvert job automation. "This is an AI-free workplace: 5 years without incident" read the sticker on her laptop.

"I'm truly sorry! I only want what's best for you! Though, I do have some ideas, if you are in the mood..." replied Sara, and Hannah nodded curiously. "We must first protect your rights as the discoverer! You know I am against those paid, Big Tech, end-to-end research management products, but... Here it goes: Sci-Legion offers a blockchain notarization service, doesn't it? We must notarize your whole workflow. Including pointers to the proprietary stuff you bought on Sci-Legion, all your analyses and pointers to the data you produced, your methods, your reasoning, and the clear hypothesis you have formulated after ruling out all the other options. This way, Sci-Legion will protect your interests—good for business—and we will be the only ones to know. You know, if it's not blockchained, it never happened" jabbered Sara. Hannah found it reasonable and halfheartedly agreed.

Option A: Sara deceives Hannah and goes public herself on metaverse and social media → **[10]**

Option B: Sara respects Hannah's request and keeps it a secret → **[11]**

[8] Hannah goes public on metaverse and social media

"Huh, I guess we're going to need more than three minutes for this, right?" joked the host of the "3 Minutes with Prometheus" popular metaverse show. Hannah's photorealistic avatar made a grimace that resembled an uncomfortable smile. Back at home, the real Hannah was nervously fiddling with her sweaty palms inside Sara's professional metaverse booth. She almost couldn't believe that Sara actually managed to pull off the breaking announcement trick; it took her less than a day to arrange it—the perks of the job.

Hannah did not really suffer from stage fright: she had successfully defended a PhD dissertation, and had survived multiple conferences, both virtual and on-site. The metaverse, though, was an entirely different beast. As Hannah was explaining her research workflow step by step from the virtual podium, hundreds of avatars were watching her carefully, their real forms masked behind animal bodies, superhero costumes, and fantastic combinations of inanimate objects; this was the first time Hannah could not recognize a familiar face in the audience to get some sense of comfort. She felt exposed, using her true form and name among people hidden behind anonymity, some of them potentially malevolent; however, she had to appear credible. Back at home, Sara starteyd preparing pancakes in the kitchen to boost her morale, and the sweet smell lifted Hannah's spirit to continue.

"Which led me to the last remaining explanation: I hypothesize that the phenomenon must be a technosignature, and the metallic fragment must be an extraterrestrial artifact" concluded Hannah, and the host of the metaverse show gulped audibly. "Of course, my whole research workflow, excluding third-party proprietary material, is available for everyone to evaluate and replicate. Just touch this icon. Every file is blockchain-tokenized and available for secure, decentralized transfer. I have nothing to hide" asserted Hannah courageously and then observed silently as waves of avatars proceeded with accessing her packaged files—first by the dozens, then by the hundreds.

Reaction content followed fast: experts and non-experts alike started uploading pieces in almost every imaginable form, from plain text to virtual reality video, to express their views on her work; some of them even live-streamed step-by-step replications of her research workflow. Recordings of her metaverse broadcast started appearing in every conceivable social media platform in a matter of minutes, and related news penetrated the remaining media channels in a matter of hours. Memes, NFTs, deepfakes, and shallowfakes emerged and disappeared like tidal waves, some of them weaponized by obscure cults and pseudoscientific groups. Some were accusing her of saying way less than what she knew; others were convinced that she was just another crisis actor that was perpetuating a hoax to divert public attention away from geopolitical tensions.

Option A: Witnessing her work going viral and public reactions escalating, Hannah limits her online presence → **[12]**

Option B: To publicly defend herself and her work, Hannah expands her online presence → **[13]**

[9] Hannah goes public on the academic media

"An announcement at '3 Minutes with Prometheus'? Are you insane? This is the most popular metaverse show! It's broadcasted and recasted everywhere—I mean, literally everywhere! No, thanks! No offence, love! It's just not my cup of tea!" exploded Hannah half-jokingly and started walking around the room in circles. Hannah did not really suffer from stage fright: she had successfully defended a PhD dissertation, and had survived multiple conferences, both virtual and on-site. The metaverse, though, was an entirely different beast. She pictured herself as a metaverse avatar explaining her research workflow step by step from a podium, hundreds of other avatars watching her carefully, their real forms masked behind animal bodies, superhero costumes, and fantastic combinations of inanimate objects—no familiar face in the audience to get some sense of comfort. She imagined feeling exposed, using her true form and name to appear credible among people hidden behind anonymity—some of them potentially malevolent. No, that could never work. As much as she would have liked to avoid it, she decided to go the "academic" way.

As Sara started preparing pancakes in the kitchen to boost her morale, Hannah summoned her AI-powered virtual research assistant, the sweet smell lifting her spirit to continue. "I am going for a preprint article. I can't withhold all of this, but I want to keep a low profile. Any hints?" she asked Sara, as the AI assistant was writing up several short drafts of her article for her approval. "Just be honest and transparent. Don't use strong language. Make it very clear that your work has not yet been reviewed either by a human or an AI, but you are confident of its soundness and replicability at this early stage. Express your wish to see more research on that, and you are done! Do you want me to help with some modest outreach? Like, do my job?" responded Sara, serving two huge piles of pancakes for her cousin and herself. "No, no, no—I'm good! I'll just take the hints and the pancakes! No outreach for me, thank you very much! I push the paper into the preprint server and then I'm done. That's the safest route!" said Hannah and grabbed a pancake, walking around the kitchen in circles, reading the AI's drafts. Sara laughed out loudly and gently patted her friend on her back in understanding.

Yet, a couple of hours later, Sara would be holding her in consolation. All it took was "DrMcDr", an artificial influencer—just one account using a hive of bots that regularly crawled preprint servers to make 30-second science-y content for multilingual audiences. The paper's reach was phenomenal. Reaction content followed fast: experts and non-experts alike started uploading pieces in almost every imaginable form, from plain text to virtual reality video, to express their views on Hannah's article; some of them even live-streamed step-by-step replications of her research workflow. Remixes of the original 30-second content started appearing in every conceivable social media platform in a matter of minutes, and related news penetrated the remaining media channels in a matter of hours. Memes, NFTs, deepfakes, and shallowfakes emerged and disappeared like tidal waves, some of them weaponized by obscure cults and pseudoscientific groups. Some were accusing Hannah of having published way less than what she knew; others were convinced

that she was just perpetuating a hoax to divert public attention away from geopolitical tensions.
Option A: Witnessing her work going viral and public reactions escalating, Hannah limits her online presence → **[12]**
Option B: To publicly defend herself and her work, Hannah expands her online presence → **[13]**

[10] Hannah is deceived by Sara who goes public on metaverse and social media

"Huh, I guess we're going to need more than three minutes for this, right?" joked the host of the "3 Minutes with Prometheus" metaverse show. Sara's unicorn-headed avatar made a grimace that resembled an uncomfortable smile. Back at home, the real Sara was nervously fiddling with her sweaty palms inside her professional metaverse booth. That popular metaverse show was being broadcasted to all decentralized platforms and recasted on all social media. She almost couldn't believe it herself that she actually managed to pull off the breaking announcement trick in less than a day—the perks of the job.

Sara did not really suffer from stage fright: she had successfully communicated—and leaked—scientific discoveries many times in the past, and had survived multiple conferences, both virtual and on-site. That specific leak, though, was an entirely different beast. As Sara was explaining Hannah's research workflow step by step from the virtual podium, hundreds of avatars were watching her carefully, their real forms masked behind animal bodies, superhero costumes, and fantastic combinations of inanimate objects; this was the first time Sara was worrying about the real people hidden behind their anonymity in the audience, some of them potentially malevolent; however, she had to press on—the public needed to know. Back at home, Hannah started preparing pancakes in the kitchen for the first time in days, and the sweet smell scarred Sara's heart in anticipation of what would come out of her betrayal.

"Which led the discoverer to the last remaining explanation: the hypothesis that the phenomenon must be a technosignature, and the metallic fragment must be an extraterrestrial artifact" concluded Sara, and the host of the metaverse show gulped audibly. "Of course, these are not just my words. The whole research workflow is notarized on Sci-Legion by the discoverer and is available for every Sci-Legion user to evaluate and replicate. Just touch this icon. The blockchain notary will verify that the content is genuine. The rationale is explained, and pointers to all necessary Sci-Legion data products are there. The discoverer has nothing to hide" asserted Sara courageously and then observed silently as waves of avatars proceeded with accessing the leaked content—first by the dozens, then by the hundreds.

Reaction content followed fast: experts and non-experts alike started uploading pieces in almost every imaginable form, from plain text to virtual reality video, to express their views on the leak; some Sci-Legion users even live-streamed step-by-step replications of Hannah's research workflow. Recordings of Sara's metaverse broadcast started appearing in every conceivable social media platform in a matter of minutes, and related news penetrated the remaining media channels in a matter of hours. Memes, NFTs, deepfakes, and shallowfakes emerged and disappeared like tidal waves, some of them weaponized by obscure cults and pseudoscientific groups. Some were accusing Sara and Hannah of saying way less than what they knew; others were convinced that they were just another case of crisis actors that were perpetuating a hoax to divert public attention away from geopolitical tensions.
Option A: Witnessing her work going viral and public reactions escalating, Hannah limits her online presence → **[12]**

Option B: To publicly defend herself and her work, Hannah expands her online presence → **[13]**

[11] Hannah is respected by Sara but exposed by hackers who go public on metaverse and social media

"Huh, security? Can someone help? We are live here!" cried in despair the host of the "3 Minutes with Prometheus" metaverse show, while his flamboyant avatar was being carried away from the virtual podium in the talons of a giant flying goose. A group of five goose-headed avatars silently took over the podium and the show, as virtual security guards were being neutralized by rampant giant geese. That popular metaverse show was being broadcasted to all decentralized platforms and recasted on all social media. Sara's unicorn-headed avatar stood among the virtual audience, making a grimace that resembled an uncomfortable smile. Back at home, though, the real Sara was nervously fiddling with her sweaty palms inside her professional metaverse booth. She had seen the geese taking over virtual shows in recordings, but never before from up close—the perks of the job.

"Citizens of the world, once again you have been deceived. Your scientists have failed you!" announced with a calm and reassuring voice the spokesperson of the goose-headed hackers. "But fear not! The AnonyGoose Collective is always watching. Those with the power to withhold the truth and keep you in the dark shall tremble, for we will seize it from them. Learn this: there are other fish in the cosmic ocean; humanity is rivaled by another instance of intelligence—an instance of intelligence that possessed spacefaring technology long before our first windmill!" continued the hacker on the virtual stage, as Sara stared shivering, drenched in cold sweat, guessing what would follow.

"We are the watchers. We see everything. We hear everything. Our bots are everywhere—undercover, unspoofable, unremovable—guarding your right to know what those in power want to keep hidden. We recently intercepted private oral communication related to the most extraordinary discovery the world has ever seen. Yet, the people involved pledged to keep it 'between them', to keep it 'a secret'! This is not acceptable. At this very moment, we leak everything!" exclaimed the hacker on the podium, as an access icon appeared on the virtual stage.

Waves of avatars, their real forms masked behind animal bodies, superhero costumes, and fantastic combinations of inanimate objects, proceeded with accessing the leaked content—first by the dozens, then by the hundreds. Sara was panicking; she and Hannah had been compromised. Back at home, Hannah started preparing pancakes in the kitchen for the first time in days, and the sweet smell scarred Sara's heart in anticipation of what would come out of this catastrophe.

Reaction content followed fast: experts and non-experts alike started uploading pieces in almost every imaginable form, from plain text to virtual reality video, to express their views on the leak; some Sci-Legion users even live-streamed step-by-step replications of Hannah's research workflow. Recordings of the hackers' metaverse broadcast started appearing in every conceivable social media platform in a matter of minutes, and related news penetrated the remaining media channels in a matter of hours. Memes, NFTs, deepfakes, and shallowfakes emerged and disappeared like tidal waves, some of them weaponized by obscure cults and pseudoscientific groups. Some were accusing Hannah of conspiracy; others were convinced that this was just another hoax to divert public attention away from geopolitical tensions.

Option A: Witnessing her work going viral and public reactions escalating, Hannah limits her online presence → **[12]**

Option B: To publicly defend herself and her work, Hannah expands her online presence → **[13]**

[12] Hannah limits her online presence

"This is a total disaster—a catastrophe! How could... Unbelievable! They have no idea!" mumbled Hannah, erratically packing a suitcase with anything that caught her eye. "Call Dad!" she shouted at her smartphone furiously, as Sara appeared at her door, apparently disturbed by the turn things had taken. Hannah's work had gone viral in half a day, igniting heated debates in the scientific, public, political, and religious spheres all around the globe. Her social media traffic had escalated accordingly, up to the point where she decided to fully deactivate her online presence—she just couldn't handle it. Celebrities, influencers, community religious figures, local politicians, even primary school kids, were constantly bombing her with messages or tagging her to their own content, just to increase their reach. The hashtag "#hannahilation" and its corresponding metaverse icon were breaking record after record across every social media platform and metaverse world, decentralized or not. Hannah could sense the extremes this wildfire could bring about: character assassination, or character veneration. If that proved to be the case, she would only be in the position to hope that these words would stick to their metaphorical meanings.

"Hannah, I'm so sorry..." Sara cried out in a devastated tone. "I really don't have time for that. I should have been more careful. I should have known better. I am moving out, like, right now. You should be packing up too, for your own good!" responded Hannah abruptly, without giving her a single look. The rainy Berlin sky roared and thundered, dimming even further the scarce natural light in Hannah's room. As if rushing in to save the day, Hannah's smartphone screen lit up with an incoming call, brightening up its frugal surroundings—and her prospects. "Dad? It's me! I was trying to reach you too! Yes, don't worry about the news! I'm taking care of it! I will explain later!" babbled Hannah, zipping her suitcase.

She mentally went through her emergency checklist: a handful of cash, a backup phone with an unused number, and copies of her digital files off the cloud. Everything was fine; she would manage; she would be safe. Sara was just looking at her, frozen, heartbroken, and unable to do anything at all to help. As she was arranging her words to say, perhaps, "goodbye", her secure messaging app sent a distinct audible notification assigned only to her most trusted contacts. "There is another... Hannah! My people tell me that a research group has access to another artifact! They are looking around for you... They want to offer you a job in their group, including institutional protection... I am forwarding you the secure contact..." stuttered Sara, as Hannah was putting on her backpack, ready to hit the road. "I'll... I'll think about it. Let's talk when all this stuff goes away... Take care..." responded Hannah hastily, her voice cracking from the rushed farewell.

Option A: Hannah stays in the independent "science-as-a-service" business → **[14]**
Option B: Hannah gets in touch with the research group and accepts their job offer → **[15]**

[13] Hannah expands her online presence

"This is a total disaster—a catastrophe! How could... Unbelievable! They have no idea!" mumbled Hannah, erratically swiping through the new messages in her social media inboxes and her email accounts. "Assistant, draft responses—today's messages only!" she shouted at her smartphone furiously, as Sara appeared at her door, apparently disturbed by the turn things had taken. Hannah's work had gone viral in half a day, igniting heated debates in the scientific, public, political, and religious spheres all around the globe. Her social media traffic had escalated accordingly, up to the point where she needed the AI assistants of both her smartphone and her old laptop to monitor her online presence—luckily, she would always keep old digital devices around as emergency backups. Celebrities, influencers, community religious figures, local politicians, even primary school kids, were constantly bombing her with messages or tagging her to their own content, just to increase their reach. The hashtag"#hannahilation" and its corresponding metaverse icon were breaking record after record across every social media platform and metaverse world, decentralized or not. Hannah could sense the extremes this wildfire could bring about: character assassination, or character veneration. If that proved to be the case, she would need to actively defend herself before those words transgressed their metaphorical meanings.

"Hannah, I'm so sorry..." Sara cried out in a devastated tone. "I really don't have time for that. The window is closing to avoid the worst! I will rebut with one video to invalidate their claims and I will have the AIs monitor reactions and figure it out as it goes. Then, I'm packing up—I'm moving out. You should be too, for your own good!" responded Hannah abruptly, without giving her a single look. The synthetic voices of the AIs were reading aloud their draft responses for Hannah to approve, filling the tiny room with an almost natural patter. As if rushing in to save the day, Hannah's smartphone screen lit up with an incoming call, presently silencing the polyphonic madness. "Dad? It's me! I was trying to reach you too! Don't worry about the news! I'm taking care of it! I will explain later!" babbled Hannah, approving some of the AI's draft responses.

She mentally went through her strategy for the rebuttal video: flag their "ad hominem" attacks, recap her work using transparent reasoning, and refrain from showing any emotion that could be weaponized against her. Everything was fine; she would manage; she would be safe. Sara was just looking at her, frozen, heartbroken, and unable to do anything at all to help. Then, as the AI assistants were going about their reading-out-loud business, something caught her ear. "Wait, wait, wait! Hannah, what was that 'unable to join your lunar expedition' thing? What is the AI talking about?" asked Sara curiously. Hannah, baffled, reviewed the original message. "Oh, it's... Can't believe it... It's Eron Nusk... They are putting together a mission to the artifact... He offers me a job..." stuttered Hannah, as Sara was smiling wholeheartedly. "I'll... I'll think about it... Sara, let's talk when all this stuff goes away..." mumbled Hannah, and Sara exhaled understandingly as she headed for her room to start packing.

Option A: Hannah stays in the independent "science-as-a-service" business → **[14]**
Option B: Hannah gets in touch with the private company and accepts their job offer → **[16]**

[14] Hannah stays independent

The heavy rainfall forcibly postponed the regular visit of Hannah's new crow friend to her attic window. The neighbors were already outside; she saw them getting into their cars from behind her window-pane as they were leaving for work in haste. Luckily, she was working from her home office. "Assistant, read today's appointments and messages" she asked her smartphone's AI in an almost kind tone, as she was preparing pancakes for breakfast—some habits truly died hard. "Usual meeting with Counselor in seven hours. One secure message from Sara" reported the AI's synthetic voice. Hannah frowned. She would really need to talk to Sara at some point—it had been almost a month since her moving out. New house; new town; things would hopefully settle down.

Although she would still receive invitations for interviews or keynote speeches from time to time, she was now old news. The metaverse and the media were buzzing with amateur and professional explorers who were normalizing the transition from a "search" to a "hunt" for extraterrestrial artifacts and were at the same time competing for their influencer niche—competing both against each other and against the artificial influencers that were producing highly effective multimedia material at unimaginable speeds. To Hannah's eyes, these bizarre times of constantly accumulating nuggets of extraterrestrial-related content were more alien than the artifact itself.

Thankfully, that was no longer her problem. She may have opened up the vista to that uncharted shore, but, beyond that, she was not interested in claiming anything more. It was part of her strategy: nature abhors a vacuum; others would follow promptly; the flaming eye of the world would turn away from her. It seemed to work. And although her monumental discovery had made a pretty large dent in history books, she still had a job to do, and she loved it.

"Alright! Research never stops" she whispered to herself setting the pancakes next to her laptop as she was logging in to Sci-Legion to get some tasks done; that client would really use some mined oxygen. Perhaps, if the rain stopped, her new crow friend would visit her attic window, and they could play for a few minutes; or, perhaps, she would skip work altogether and get to know that crow a little better—after all, wasn't "looking for other instances of intelligence" her legacy?

[15] Hannah joins the research group

The heavy rainfall forcibly postponed the regular visit of Hannah's new crow friend to her attic window. The neighbors were already outside; she saw them getting into their cars from behind her window-pane as they were leaving for work in haste. Luckily, she was working from her home office. "Assistant, read today's appointments and messages" she asked her smartphone's AI in an almost kind tone, as she was preparing pancakes for breakfast—some habits truly died hard. "Weekly meeting with the Communications Officer in two hours. Urgent meeting with the Principal Investigator in four hours. Usual meeting with Counselor in seven hours. One secure message from Sara" reported the AI's synthetic voice. Hannah frowned. She would really need to talk to Sara at some point—it had been almost a month since her moving out. New house; new town; things would hopefully settle down.

Although she would still receive invitations for interviews or keynote speeches from time to time, she was now old news. The metaverse and the media were buzzing with amateur and professional explorers who were normalizing the transition from a "search" to a "hunt" for extraterrestrial artifacts and were at the same time competing for their influencer niche—competing both against each other and against the artificial influencers that were producing highly effective multimedia material at unimaginable speeds. To Hannah's eyes, these bizarre times of constantly accumulating nuggets of extraterrestrial-related content were more alien than the artifact itself.

Thankfully, that was no longer her problem. She may have opened up the vista to that uncharted shore, but, beyond that, she was not interested in claiming anything more. It was part of her strategy: nature abhors a vacuum; others would follow promptly; the flaming eye of the world would turn away from her. It seemed to work. And although her monumental discovery had made a pretty large dent in history books, she still had a job to do, and she loved it.

"Alright! Research never stops" she whispered to herself setting the pancakes next to her laptop as she was logging in to her research group's decentralized social server for the usual morning overview of yesterday's worst science press with the other postdocs. Perhaps, if the rain stopped, her new crow friend would visit her attic window, and they could play for a few minutes; or, perhaps, she could take a long lunch break by the window and get to know that crow a little better—after all, wasn't "looking for other instances of intelligence" in her postdoctoral research proposal?

[16] Hannah joins the private company

The heavy rainfall forcibly postponed the regular visit of Hannah's new crow friend to her attic window. The neighbors were already outside; she saw them getting into their cars from behind her window-pane as they were leaving for work in haste. Luckily, she was working from her home office. "Assistant, read today's appointments and messages" she asked her smartphone's AI in an almost kind tone, as she was preparing pancakes for breakfast—some habits truly died hard. "Weekly meeting with the Communications Department in two hours. Urgent meeting with the Chief Science Officer in four hours. Usual meeting with Counselor in seven hours. One secure message from Sara" reported the AI's synthetic voice. Hannah frowned. She would really need to talk to Sara at some point—it had been almost a month since her moving out. New house; new town; things would hopefully settle down.

Although she would still receive invitations for interviews or keynote speeches from time to time, she was now old news. The metaverse and the media were buzzing with amateur and professional explorers who were normalizing the transition from a "search" to a "hunt" for extraterrestrial artifacts and were at the same time competing for their influencer niche—competing both against each other and against the artificial influencers that were producing highly effective multimedia material at unimaginable speeds. To Hannah's eyes, these bizarre times of constantly accumulating nuggets of extraterrestrial-related content were more alien than the artifact itself.

Thankfully, that was no longer her problem. She may have opened up the vista to that uncharted shore, but, beyond that, she was not interested in claiming anything more. It was part of her strategy: nature abhors a vacuum; others would follow promptly; the flaming eye of the world would turn away from her. It seemed to work. And although her monumental discovery had made a pretty large dent in history books, she still had a job to do, and she loved it.

"Alright! Research never stops" she whispered to herself setting the pancakes next to her laptop as she was logging in to the company's secure employee platform to get some tasks done; the expedition rovers needed their shepherd. Perhaps, if the rain stopped, her new crow friend would visit her attic window, and they could play for a few minutes; or, perhaps, she could take a long lunch break by the window and get to know that crow a little better—after all, wasn't "looking for other instances of intelligence" in her job description?

After finishing the branching narrative, participants are given the following instructions to again work through a simple audience engagement tool that allows for free-form text data collection and presentation of collected data: *"How did you relate to this unfamiliar future world, while accessing those events and situations occurring in it? Write down a summary of your first-person, subjective perspective in this strange future, i.e., your thoughts, emotions, experiences, and actions"*.

The content of the **"Rethink"** phase of this workshop is structured as follows. Participants are given the following instructions to again work through a simple audience engagement tool that allows for free-form text data collection and presentation of collected data: *"Your journey into the future(s) has almost reached its end. Reflect on the previous activities by comparing and contrasting the likely, the desirable, and the unfamiliar futures. What new insights have you discovered while accessing that unfamiliar future of 2035, as compared to what you initially considered probable and/or desirable for 2035? These insights can be related both to how and to what we anticipate when we engage with the future(s)!"*.

After the participants have written down their insights, the collected text data is presented to them, and they are given the following instructions to continue working: *"You are now back to the present day! Reflect on the collective insights that you scavenged from your trips to the futures. Do any of these insights help see any present matters in a new light? Do you see any ideas that could be acted upon in the present to help proactively shape the future of the search for alien life, particularly regarding the science and society interface? What actions must be taken to this end? Write down your ideas, and let's discuss!"*. This activity concludes the workshop.

Following the FLL-N framework, this workshop has been designed to produce the following outcomes, all of which are expected to appear in the collected data of the "Reflect" phase's activities or in the ensuing discussion. The first outcome refers to the kind of assumptions that fuel the imaginings of the probable, the desirable, and the unfamiliar futures. Participants should be able to notice that different items from the "List of Foundations" underlie more prominently the assumptions of each one of these distinct types of future; namely, the first two are more prominent in the probable future, the next two in the desirable future, and the last two in the unfamiliar future. This observation might be complemented by the realization that the first two items which are more closely related to the probable future can also intervene in the imaginings of the desirable future, and that the first two and the next two items, which are more closely related to the probable future and the desirable future respectively, can also intervene in the imaginings of the unfamiliar future, despite its bewildering character. Thus, the first outcome can be enriched by this new understanding that conscious human anticipation builds on specific foundations of assumptions, therefore consciously discerning these foundations can help participants isolate and deploy the ones which are most appropriate to the specific type of future that needs to be imagined, to allow for less interference and decreased constraints on such imaginings. The second outcome refers to gaining the new understanding that implicit imaginings of the probable and desirable futures tend to constrain how participants see the present. This means that, by consciously challenging and changing their own imaginings of the future, participants can create new lenses for engaging with the present in novel ways. Taken together, these outcomes are expected to enhance the participants' futures literacy, thereby empowering them to consciously rehearse different types of post-detection futures to discover new insights about futures they need to prepare for and futures they need to plan for, as well as to notice, understand, and utilize new patterns and unique instances of change as they emerge in the present to invent and apply more agile outlooks towards the short-term effects of a detection of extraterrestrial life in the future. The latter is of crucial importance, since the future's fundamental unknowability means that the actual manifestation of a future detection can always be something completely unthought-of.

3. Epilogue: Towards a Responsible Search for Extraterrestrial Life

The "Rehearsing Post-Detection Futures" workshop's workflow that was described in detail in the previous section is a particular example of a readily deployable and scalable, futures literacy-promoting tool vis-à-vis post-detection considerations. Depending on the pedagogical context, other futures literacy interventions may be comparably and complementarily beneficial. One such possibility is a future-oriented role-playing simulation built around a particular detection situation embedded in a particular contextual scenario of the future, but opting for a divergent thinking method for creativity training (Fletcher and Benveniste 2022), such as Edward de Bono's Six Thinking Hats technique (de Bono 1985), as part of its mechanics—the other alternative being the engagement of participants' narrative cognition, which was preferred in the "Rehearsing Post-Detection Futures" workshop. Irrespective of the specifics, though, properly-designed futures literacy interventions can be a very promising avenue for enhancing conscious anticipation in astrobiology-relevant target groups. In turn, the attained increased futures literacy can support the mindful deployment of the most appropriate approach to anticipatory practices, i.e., the critical-hermeneutic one, to inform the responsible advancement of the search for extraterrestrial life as a new and emerging science, given its current relatively early stage of development. The reason for this is that the critical-hermeneutic anticipatory perspective which is interlaced with futures literacy appears to be a promoter of (self-)reflexive capacity and a key substrate for the proper use of different representations of the future in the various ways that can support more responsible research in new and emerging sciences (Urueña 2021).

Nanoscience (Charitidis 2020), quantum science (Roberson 2023), and synthetic biology (Taylor and Woods 2020) are a few examples of disciplines whose natural sciences facet has already started to become deeply intertwined with their social and ethical responsibility facet. In this vein, given the ostensibly vast and diverse implications of a potential future detection, the search for extraterrestrial life in particular, and perhaps astrobiology in general, can draw inspiration from the rich published discourse on the "responsibilization" of other new and emerging sciences to leapfrog to an effective fusion of its natural sciences and its social and ethical responsibility strands. Post-detection considerations can't afford to ignore either the ongoing developments that expand the value range of the various factors whose combinations comprise the multiplicity of possible detection situations or the constantly changing conditions in the broader environment whose arrangements comprise the alternative contextual scenarios of said situations. The "Rehearsing Post-Detection Futures" workshop focused exclusively on the most urgent and precarious stage, namely the short-term stage right after the detection and its communication. This rapid post-detection stage leaves no room for last-minute, a posteriori remedies: it is vital to swiftly prioritize it in any responsible search for extraterrestrial life, or even responsible astrobiology, frameworks, both because of its own high stakes and as crucial groundwork for scaffolding other preparatory actions towards the more delayed, long-term stage. Appropriate and timely futures literacy interventions targeting students, scholars, and other pertinent communities, especially vulnerable ones, can support the mindful deployment of a critical-hermeneutic anticipatory perspective towards building a more responsible search for extraterrestrial life, sapient or not.

Acknowledgements

The author would like to gratefully acknowledge Prof. Kathryn Denning for the enlightening, insightful, and inspiring discussions during this workshop's development stage in preparation of its first pilot test session, as well as the Chair and the Members of the 2024 Kavli-IAU Symposium (IAUS 387) Organising Committee for their pivotal support in evaluating and implementing this workshop in the "Futures I - Resilience and flexibility" symposium session. The author would also like to sincerely thank Prof. Carol Oliver and Dr. Andreas Schwarz for their ideas and contributions as co-designers and co-facilitators of the "Futures II - What are we missing?" symposium session which used a future-oriented role-playing simulation to complement and augment this workshop's outcomes.

References

Álvarez, A., Ritchey, T. 2015, Applications of general morphological analysis, *Acta Morphologica Generalis*, 4(1), 1–40

Amer, M., Daim, T.U., Jetter, A. 2013, A review of scenario planning, *Futures*, 46, 23–40

Arfini, S., Bertolotti, T., Magnani, L. 2020, The antinomies of serendipity how to cognitively frame serendipity for scientific discoveries, *Topoi*, 39(4), 939–948

Bell, W. 1997, Foundations of futures studies: Human science for a new era. vol.1: History, Purposes, and Knowledge (New Brunswick: Transaction Publishers)

Benford, G. 2008, SETI for profit, *Nature*, 452, 1032

Benford, J. 2021, A Drake Equation for Alien Artifacts, *Astrobiology*, 21(6), 757–763

Bohensky, E.L., Reyers, B., van Jaarsveld, A.S. 2006, Future ecosystem services in a Southern African river basin: a scenario planning approach to uncertainty, *Conservation Biology*, 20(4), 1051–1061

Bradfield, R., Wright, G., Burt, G., Cairns, G., van der Heijden, K. 2005, The origins and evolution of scenario techniques in long range business planning, *Futures*, 37(8), 795–812

Charitidis, C.A. 2020, in: R. Iphofen (ed.), *Quest for Ethical and Socially Responsible Nanoscience and Nanotechnology*, Handbook of Research Ethics and Scientific Integrity (Cham: Springer Nature Switzerland), 983–997

Connors, M.M. 1976, The Role of the Social Scientist in the Search For Extraterrestrial Intelligence. Unpublished manuscript (Moffett Field: NASA, Ames Research Center)

Davies, P.C.W., Wagner, R.V. 2013, Searching for alien artifacts on the moon, *Acta Astronautica*, 89, 261–265

de Bono, E. 1985, Six Thinking Hats: An Essential Approach to Business Management (Boston: Little, Brown and Company)

Dick, S.J. 2018, in: S.J. Dick, *Discovery*, Astrobiology, discovery, and societal impact (Cambridge: Cambridge University Press), 37–64

Ehresmann, A., Tuomi, I., Miller, R., Béjean, M., Vanbremeersch, J.-P. 2018, in: R. Miller (ed.), *Towards a formal framework for describing collective intelligence knowledge creation processes that 'use-the-future'*, Transforming the future: Anticipation in the 21st century (Abingdon: Routledge), 66–91

Elsevier 2022, Research Futures 2.0: A new look at the drivers and scenarios that will define the decade (Elsevier)

Fletcher, A. and Benveniste, M. 2022, A new method for training creativity: narrative as an alternative to divergent thinking, *Annals of the New York Academy of Sciences*, 1512(1), 29–45

Gartrelle, G.M. 2015, Unknown caller: can we effectively manage the announcement of discovery of extraterrestrial life?, *International Journal of Astrobiology*, 14(4), 577–587

Gertz, J. 2016a, Post-Detection SETI Protocols & METI: The Time Has Come To Regulate Them Both, *Journal of the British Interplanetary Society*, 69, 263–270

Gertz, J. 2016b, E.T. Probes: Looking Here As Well As There, *Journal of the British Interplanetary Society*, 69, 88–91

Gidley, J.M. 2017, The Future: A Very Short Introduction (Oxford: Oxford University Press)

Harrison, A., Billingham, J., Dick, S.J., Finney, B., Michaud, M.A.G., Tarter, D.E., Tough, A., Vakoch, D.A. 2000, in: A. Tough (ed.), *The role of the social sciences in SETI*, When SETI succeeds: the impact of high-information contact (Bellevue: Foundation For the Future), 71–85

Harrison, A., Connell, K. 2001, Workshop on the Societal Implications of Astrobiology, held at NASA Ames Research Center, November 16-17, 1999 (Moffett Field: NASA, Ames Research Center)

Iden, J., Methlie, L.B., Christensen, G. E. 2017, The nature of strategic foresight research: A systematic literature review, *Technological Forecasting and Social Change*, 116, 87–97

Loeb, A. 2020, The Cosmic Dawn of Technology, *Scientific American*, [Online]. Available: `https://www.scientificamerican.com/article/the-cosmic-dawn-of-technology/`

Miller, R. 2018a, in: R. Miller (ed.), *Sensing and making-sense of Futures Literacy: towards a Futures Literacy Framework (FLF)*, Transforming the future: Anticipation in the 21st century (Abingdon: Routledge), 15–50

Miller, R. 2018b, in: R. Miller (ed.), *Futures Literacy Laboratories (FLL) in practice: An overview of key design and implementation issues*, Transforming the future: Anticipation in the 21st century (Abingdon: Routledge), 95–109

Poli, R. 2017, in: R. Poli (ed.), *Introducing Anticipation*, Handbook of Anticipation: Theoretical and Applied Aspects of the Use of Future in Decision Making (London: Springer), 1–14

Porter, M.E. 1985, The Competitive Advantage: Creating and Sustaining Superior Performance (New York: Free Press)

Postma, T.J., Liebl, F. 2005, How to improve scenario analysis as a strategic management tool?, *Technological Forecasting and Social Change*, 72(2), 161–173

Profitiliotis, G., Theologou, K. 2023, The monstrosity of the search for extraterrestrial life: Preparing for a future discovery, *Futures*, 147, 103117

Profitiliotis, G. 2024, in: S. Stelios and K. Theologou (eds.), *Moral Vistas to Xenic Beyonds: Fostering Moral Imagination to pre-empt Monsterization in future encounters with Extraterrestrial Life*, The Ethics Gap in the Engineering of the Future: Moral Challenges for the Technology of Tomorrow (Leeds, Emerald Publishing), 177–199

Ramirez, R., Wilkinson, A. 2016, Strategic reframing: The Oxford scenario planning approach (Oxford: Oxford University Press)

Roberson, T. 2023, Talking About Responsible Quantum: "Awareness Is the Absolute Minimum that... We Need to Do, *NanoEthics*, 17(2), 1–12

Smith, S., Ashby, M. 2020, in: S. Smith and M. Ashby, *Storytelling and prototyping: creating ways for others to engage with your future stories*, How to Future: Leading and sense-making in an age of hyperchange (London: Kogan Page Limited), 143–162

Socas-Navarro, H., Haqq-Misra, J., Wright, J.T., Kopparapu, R., Benford, J., Davis, R. 2021, Concepts for future missions to search for technosignatures, *Acta Astronautica*, 182, 446–453

Taylor, K. and Woods, S. 2020, Reflections on the practice of Responsible (Research and) Innovation in synthetic biology, *New Genetics and Society*, 39(2), 127–147

The Aerospace Corporation 2021, Strategic Foresight for the Space Enterprise (The Aerospace Corporation)

Urueña, S. 2021, Responsibility through Anticipation? The 'Future Talk' and the Quest for Plausibility in the Governance of Emerging Technologies, *NanoEthics*, 15(3), 271–302

van der Heijden, K. 2005, Scenarios: The art of strategic conversation, 2nd ed. (Chichester: John Wiley & Sons)

Voros, J. 2018, On a morphology of contact scenario space, *Technological Forecasting and Social Change*, 126, 126–137

Zwicky, F. 1957, Morphological Astronomy (Berlin: Springer-Verlag)

(Toward) Discovery of Life beyond Earth and its Impact
Proceedings IAU Symposium No. 387, 2026
H. Landt, M. Dominik & C. Oliver, eds.
doi:10.1017/S1743921325000183

Risk Communication and the Search for Life Beyond Earth: Challenges of Risk Amplification, Risk Information Seeking, and Risk Message Design for Research and Preparedness

Andreas Schwarz

Institute of Media and Communication Science
Technische Universität Ilmenau
Germany
email: andreas.schwarz@tu-ilmenau.de

Abstract. The growing international public attention to astrobiology and SETI, combined with the immense costs of space exploration and the potential outcome, the discovery of extraterrestrial life, will likely increase the pressure for public justification and the need to address societal concerns about the risks of searching for, finding, or being found by extraterrestrial life. Understanding these perceptions and concerns warrants a more systematic inclusion of risk communication research in studying pre- and post-detection scenarios. In this chapter, we review the state of the art in risk perception/communication research related to astrobiology and detection. Based on three major challenges (social risk amplification/attenuation, misguided risk information seeking, ineffective risk message design), we explore the contribution of risk communication theory (SARF, RISP, EPPM, IDEA) to future research and institutional preparedness for potential detections of extraterrestrial life.

Keywords. Risk communication, risk amplification, risk information seeking, astrobiology, SETI, post-detection

---

1. Introduction

Astrobiology has been defined by NASA as a multidisciplinary and interdisciplinary field that addresses the questions of how life begins and evolves, whether life exists elsewhere in the universe, and what the future of terrestrial and extraterrestrial life is (Des Marais et al. 2008). In this chapter, we consider the search for extraterrestrial intelligence (SETI) to be part of this field (Shostak 2004; Tarter 2004), while recognizing the controversies over disciplinary boundaries (Billings 2012). The search for extraterrestrial life (SETL) has been on the agenda of astrobiology, space exploration, but also the mainstream media for a long time (Garber 1999). Astronomical discoveries and space exploration missions with explicit astrobiological goals over the last three decades have given the field new momentum, including its public resonance (Mace & Schwalbe 2020). Among them was the launch of the Kepler Space Telescope which led to the discovery of thousands of exoplanets, some of which are Earth-sized and orbit stars in the habitable zone (Lissauer et al. 2014). Further examples include the controversial debate in academia and the media about the observation of the interstellar object 'Oumuamua' passing our solar system in 2017 that was interpreted by Harvard physicist Avi Loeb and others as a probe sent to Earth vicinity by an extraterrestrial civilization (Bialy & Loeb 2018). The discovery of phosphine in the atmosphere of Venus in 2020 (Greaves et al. 2021) sparked a debate about the potential existence of life on the planet. In 2021, ESA and NASA launched the James Webb Space Telescope to explore the potential for life in other solar systems, while the Perseverance rover landed on Mars to search for signs of ancient life.

These projects and observations have presumably increased the international public attention to astrobiology and SETI. In conjunction with the immense costs of space exploration, as well as the potential outcome, the discovery of extraterrestrial life, the pressure for public justification of these efforts as well as the need to address potential societal concerns about the risks of searching for, finding, or being found by extraterrestrial life, is likely to increase. Understanding these perceptions and concerns warrants a more systematic involvement of the social sciences in SETL (simple and intelligent life) in both pre-detection and post-detection scenarios. Explaining the science transparently and intelligibly throughout the process of SETL points to the importance of *science communication.* Analyzing, communicating, and controlling the societal risks of SETL before and after detection underscores the significance of *risk communication* as a field of study and practice. Calls for a more systematic inclusion of these disciplines in pre- and post-detection processes have been raised repeatedly in the SETL literature (Denning et al. 2019; Race et al. 2012).

In this chapter, we will follow this call and reflect on the role of risk communication in SETL and potential detection scenarios. The relevance of risk perception, risk communication, and risk management research, as well as some suggestions for incorporating them in the context of SETL or Messaging to Extraterrestrial Intelligence (METI), have been addressed previously (Korbitz 2014; Neal 2014). Particular emphasis was given to the role of planetary protection programs, especially Mars sample return missions (Lofstedt 2003; MacGregor et al. 1998; Race 2008; Race & MacGregor 2000; Race & Randolph 2002). There is a broad consensus among the authors that certain detection scenarios pose potentially serious or uncertain risks, and that space agencies, governments, and scientists should not only engage in regular science communication, but also prepare for detection scenarios in terms of risk communication with the media and the public in a two-way communication style, as opposed to top-down communication (experts to the lay public) (Lofstedt 2003; Race 2015).

However, to the best of our knowledge, neither a more systematic discussion and application of risk communication theory nor its use for data collection in (potential) detection scenarios is available. Discussions in that context mostly referred to risk *perception* as opposed to risk *communication* research. In addition, regular updates of findings on the perception and communication of risks of SETL are needed, as important aspects such as preferred and trusted information sources and institutional trust are constantly changing (Denning et al. 2019). Therefore, we will address the following questions in this article:

(1) What do we already know about (risk) perceptions and media narratives of SETL and detection scenarios?
(2) What are relevant insights and models from risk communication research for the SETL context and detection scenarios?
(3) How can risk communication models inform the preparation for detection scenarios at key institutions involved in space exploration, SETL, and detection response?

2. Perceptions and media representations of searching for life beyond Earth

Understanding and designing effective risk communication practices requires insights into risk perceptions among different stakeholders in society and the general public, but also an understanding of (social) media representations of risk. This applies to risk communication in general (Renn 2009), but also to the context of SETL, including both planetary protection and SETI (passive or active) (Jones 2013; Korbitz 2014; Lofstedt 2003). Therefore, we first review the available research based on survey data and media content analyses related to SETL, without claiming to be exhaustive.

Risk perception surveys have been conducted in the context of planetary protection and mars sample return missions. A study with experts and students as participants found that biological hazards associated with such missions were not perceived as a major risk relative to other terrestrial concerns (MacGregor et al. 1998). However, respondents indicated that Martian samples should be considered hazardous until proven otherwise and that NASA should take a rather cautious approach. They trusted in NASA to perform such missions, while some skepticism was found in terms of NASA's honesty and transparency when it comes to informing the public. Similar results were found in a survey of microbiologists (Race & MacGregor 2000). However, only half of the respondents considered the current methods of quarantining and biosafety to be adequate.

Other surveys have addressed SETL perceptions and public beliefs about the existence of ET or ETI more generally, without specific measures of risk perceptions (see the chapter by Schwartz (2020) for a condensed review of such surveys). A majority of respondents in these studies believes that ETL exists, while the support for space missions, SETL, and funding is somewhat mixed. A recent survey has addressed the perceptions of active SETI (METI), where respondents in the UK, for example, expressed some support for the idea of initiating contact with an ETI (56%) and only limited support of the notion that only scientists determine humanity's response (Hatfield & Trueblood 2021). The review of survey research by Harrison (2011, p. 666) led to the conclusion that "Any discovery of ETI is likely to produce a mix of emotions including fear, pandemonium, equanimity and delight but in North America and Europe neither [...] are likely to lead to widespread psychological disintegration and social collapse". In an attempt to develop psychometric measures of ETI perception,Vakoch and Lee (2000) found that more alienated, more religious, and less optimistic individuals were more likely to expect an ETI to be hostile, with some differences between U.S. and Chinese students. A recent Ipsos survey showed that 14% of the population in 33 countries considered an alien visitation of earth a global threat, with the highest proportions in India (30%) and Saudi Arabia (28%) and the lowest in most European countries (Ipsos 2021). One of the few studies that measured the reactions to an actual announcement of discovering extraterrestrial microbial life (not ETI), including risk perceptions, revealed that people feel more positively than negatively about such an announcement (Kwon et al. 2018). They emphasized more on reward than on risk, when confronted with the announcement. This also applied to people's assumptions about the overall reaction of humanity, although with a much lower positivity bias compared to the assumption about their individual reaction.

News media representations of space exploration or SETL were rarely empirically studied (Schäfer 2012) or limited to case-based observations and reflections of the news media's or social media's potential role in SETI and detection scenarios (Jones 2013; Oliver 1997; Oliver et al. 1999; Shostak 2004). Empirical research was mainly focused on space flight, space exploration, space technology, or national security (Byrnes 1994; Clark & Illman 2003; Kauffman 1994; Neal 2007). More recent studies have analyzed public perceptions or the coverage of Mars exploration (Dittmer 2007; Mace & Schwalbe 2020; Slobodian 2015). SETL as a subject of media reporting was of secondary interest, limited to aspects of Mars exploration, surfaced in the findings by accident, or was not addressed at all. A study by Neal (2007), for example, used a qualitative analysis of NASA's press materials and New York Times editorials to identify social constructions of human space flight. None of the resulting media frames was related to astrobiology or SETI. Another qualitative media analysis in the U.S. by Dittmer (2007) focused on the Mars Pathfinder mission (1996 to 1998) and identified the scientific advance and the search for life being an important part of the media narrative. Mace & Schwalbe (2020) also analyzed the coverage of Mars exploration in U.S. and UK news media between 2011 and 2016. Among the eight frames the authors identified, one ('life') referred to speculations about past or present life on Mars. The 'life' frame was the third most frequent and emerged mainly between 2012 and 2013 in the media. According to the authors, this frame was overly optimistic and sensationalized the science since life has not been found yet.

These studies of news media representations have in common that they rarely focus on science-related aspects of SETL, while SETI has been generally disregarded. SETL was usually limited to Mars exploration although a great deal of astrobiology is concerned with the study of exoplanets and/or the study of life on Earth. Studies of media content also lack the quantification of the occurrence and evaluation of relevant institutions in space exploration and SETL (e.g., SETI Institute, NASA, ESA, astrobiology centers). An exception is the study by Schwarz & Seidl (2023) who explicitly analyzed the news media representations of SETL and astrobiology in Germany between 2009 and 2022, including media references to risks of SETL. The three main media frames identified were beneficial space exploration (36%), unidentified anomalous phenomena/extraterrestrial intelligence (UAP/ETI) (36%), and SETI risk (28%). The beneficial space exploration frame mainly addressed the search for non-intelligent extraterrestrial life, often cited NASA and academic research institutions, and stressed the benefits of SETL. The SETI risk frame stressed the search for intelligent life, often cited the SETI Institute and academic research institutions. SETI was evaluated more often positively than negatively. However, articles using this frame contained an ambivalent assessment of societal risks and benefits since research institutions (e.g., social scientists), but also science celebrity Stephen Hawking, were often quoted with statements about the risks of SETI or METI. These reports often called for more regulation and risk governance. The UAP/ETI frame mainly addressed the sightings of UAPs (or UFOs) and the existence of ETI and cited a mix of U.S. government and academic actors, with the latter not being a frequent source. The topic was related to no benefit attributions and some societal risk. Altogether, the study showed that SETL reporting in Germany is scarce, but consistent, and mostly positive. In addition, there seems to be a clear differentiation between the search for simple versus intelligent life. While the search for simple life is mostly positively connotated and dominated by NASA and academic sources, SETI is evaluated in a more ambivalent way (positive and negative), stressing more on its societal risks, while encouraging more efforts in risk governance. The topic of UAPs was consistent as well and even growing in importance in more recent years, mostly related to the release of videos by the U.S. government and congressional UAP hearings.

In conclusion, we can state that risk-related research on the societal implications of SETL is scarce and was mainly conducted in Western countries. The few studies that have collected primary empirical data are scattered across publication outlets and longer time periods. Most of the publications focus on risk *perceptions* (if at all) as opposed to risk *communication* in terms of media use, messages/media content, message effects, processes, and networks. Therefore, in the next section we will explore selected approaches to risk communication and their potential contribution to understanding the societal implications of SETL and institutional preparedness.

3. Risk communication theory and implications for SETL

Authors in the SETL field have called for a more systematic application of risk and disaster communication as well as management frameworks to pre- and pos-detection concerns (Denning et al. 2019; Neal 2014). Although previous research has shown relatively low levels of concern among experts and the public about the potential threats posed by SETL, the situation could change rapidly as soon as certain anticipated scenarios become reality (Anton et al. 2024; Neal 2014). Even in cases of a false detection, risk communication can play a pivotal role in managing public perceptions to meet the public's information needs and counter misinformation. This is based on the notion that risk is a social construction and risk perception influences behavior independently from the objective likelihood, severity, or technical analysis of a risk. Perceived risk events can result in harm (e.g., backward contamination from Mars samples) and benefits (e.g., understanding origins of life) caused by risk creators (e.g., space agencies), borne by risk bearers (e.g., astronauts, astrobiologists, earth's population), and influenced by risk mitigators (e.g., emergency management agencies, government) (Heath & Palenchar 2016).

Given the immense (public) costs of SETL and its risks, risk communication and being prepared for risk scenarios play an essential role in the process. Risk communication was defined as a "process of exchanging information among interested parties about the nature, magnitude, significance, or control of a risk" (Covello 1992, p. 359). Others have emphasized the participatory nature of risk communication as "a social process by which people become informed about hazards, are influenced towards behavioral change and can participate in decision-making about risk issues in an informed manner" (Rohrmann 2008, p. 1). Thus, risk communication should create an understanding of risks among different stakeholder groups in society beyond experts, support the adaptation of behavior to minimize harm, strengthen trust towards responsible institutions, and enable dialogue to contribute to conflict resolution and the further democratic regulation of risks (Renn 2009).

Besides well-known best practices of effective risk communication (Seeger & Schwarz 2024), several paradigms and models have emerged in the field that potentially inform preparedness and communication management practices at SETL-related institutions. While a complete overview of these models is beyond the scope of this chapter, we selected three core problems that can be addressed with risk communication frameworks in pre- and/or post-detection scenarios:

1. Social risk amplification/attenuation: Presumed detection events or SETL projects may generate significant public attention and controversy, which may either socially amplify a low risk or non-risk in an unwarranted manner, or the event may result in a lack of attention to actual threats, leading to a social attenuation of risk. Such scenarios can be described, analyzed, and addressed by using the social amplification of risk framework, SARF (Kasperson et al. 1988; Kasperson et al. 2022).
2. Risk information seeking: Legitimate risks associated with presumed detection events or SETL projects are not sufficiently known to the public because people are unaware or unmotivated to actively seek out risk information, while others may turn to sources of misinformation. This issue can be addressed with models such as the risk information seeking and processing model (RISP) (Dong et al. 2024; Griffin et al. 1999).
3. Risk message design: Risk or crisis communication messages related to presumed detection events or SETL projects are not appropriately designed to facilitate comprehension, cognitive/affective learning, or behavioral adaptation (protective measures, information seeking) to the situation. Effective (instructional) risk communication can be informed by models such as the extended parallel process model (EPPM) (Roberto et al. 2009; Witte 1992) or the IDEA model (Sellnow et al. 2017; Sellnow & Sellnow 2019).

3.1. *Social risk amplification/attenuation*

SARF is not a predictive theory, but seeks to understand the interaction between risk and the social dynamics that influence individual and public risk perception, risk communication, and risk-related behavior (Kasperson et al. 1988, 2022). The framework explains how risks that are considered relatively low by experts become the focus of public concern and dynamic socio-political consequences (i.e., risk amplification). On the other hand, hazards that are deemed more serious by experts may receive less public attention and societal response (i.e., risk attenuation). Thus, the term 'amplification' implies its opposite, 'attenuation', since risk communication processes can amplify or attenuate public risk perceptions and behavior. Risk amplification/attenuation processes are influenced by 'social stations' such as government agencies, activist groups, opinion leaders, social media, or the news media through communication processes and information flows on different channels.

At the level of form and content of messages, the authors of the SARF proposed attributes of information that may influence social amplification/attenuation processes such as the volume of information, the degree to which information is disputed (e.g., by experts), the extent of dramatization (e.g., sensational media headlines), and the symbolic connotations of the information (Kasperson et al. 1988). However, the content dimension has not been well developed, which from a risk communication perspective is a notable weakness of the model since is it lacks a more detailed conceptualization of specific risk message characteristics and its effects on risk perceptions and risk-related behavior. More recent contributions have tried to fill this gap, for example by integrating aspects of framing theory (Rooke 2021; Rossmann et al. 2021) or fear appeal research in SARF (Liu & Li 2024; Schwarz & Unselt 2024; Zhao & Wu 2021). For the study of mediated societal risks of artificial intelligence, for example, the integration of SARF with the EPPM at the content-level was proposed to better understand risk amplification/attenuation processes on social media, resulting in the mediated risk amplification/attenuation of societal risk and risk governance (MASRisG) framework (see Figure 1) (Schwarz 2024). The details of EPPM will be further addressed in Section 3.3.

Depending on which information is circulated and how it is processed and transformed by organized and individual actors with their specific predispositions, norms, and interests, the SARF suggests several responses to the amplification/attenuation process such as risk-related changes of attitudes and behavior, political and social action, organizational responses, or social disorder/protest. As the risk messages spread to other societal domains and stakeholders ('ripple effects'), they can have impacts as for example regulatory actions, business effects, organizational change, changes of trust in institutions, or the further amplification/attenuation of physical risks because of feedback mechanisms that enlarge or lower the risk. These dynamics can lead to higher-order impacts that spread to further stakeholders, distant locations, or future generations.

The SARF was criticized for not offering any predictive framework to understand why specific risks or risk events lead to different degrees of social risk amplification/attenuation (Liu & Li 2024) and that it follows a somewhat simplistic transactional model of communication (Sellnow & Seeger 2021). However, since SARF emphasizes the dynamics of public communication and integrates processes at the individual (micro) and societal (macro) levels, it is a useful framework to study risk perception *and* communication dynamics following a presumed SETL/SETI detection event with considerable public resonance. This process may be influenced, but in no way determined, by expert confidence as quantified by the Rio/London scales (Almár 2011; Forgan et al. 2019), since experts and scientists are only a few of many social stations in the risk amplification/attenuation process.

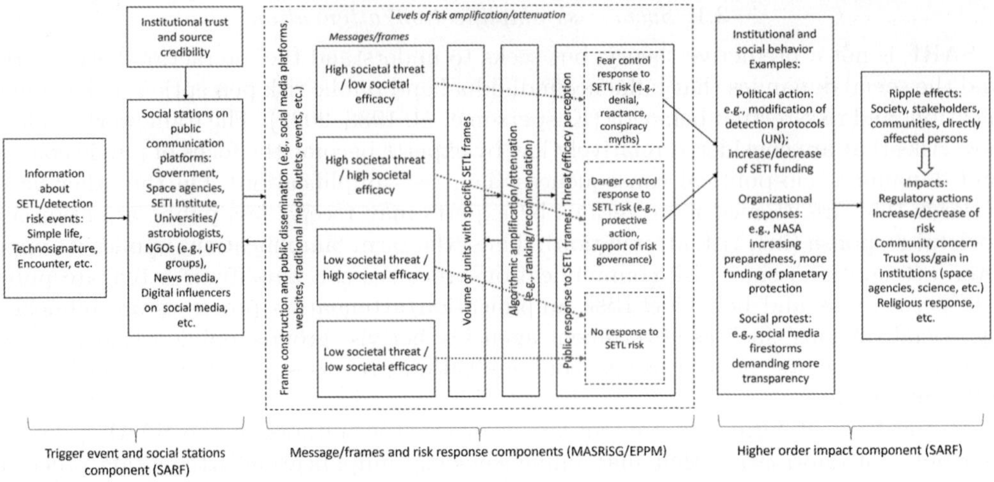

Figure 1. Mediated amplification of societal risk and risk governance (MASRisG) adapted to SETL and detection events based on Schwarz (2024) and Kasperson et al. (2022)

To date, humanity has not detected any solid evidence of extraterrestrial life. Therefore, institutional preparation for detection-related risk events needs to be based on available risk communication models and previous research in other risk contexts such as disasters, public health, or climate change. In addition, past cases of claimed detections such as ALH84001, EQ-Peg 121, Tabby's Star, or Oumuamua can be sources for understanding and anticipating risk amplification/attenuation processes (Denning et al. 2019). The SARF may serve as a framework to re-approach these case studies and design more systematic empirical research based on news media content analysis, social media data, in-depth focus group discussions, etc., to identify key social stations and their impact in the (risk) communication process and societal impacts. A previous news media analysis of SETL reporting in Germany (Schwarz & Seidl 2023), for instance, showed that astrobiology was often framed within the context of beneficial space exploration searching for simple extraterrestrial life, mostly stressing the benefits, while risks and planetary protection were almost irrelevant (i.e., risk attenuation). On the other hand, science celebrities such as Stephen Hawking had significant impact on the public discourse about active SETI/METI, which lead to an increased volume of news coverage and the emphasis on threats posed by a potentially hostile ETI becoming aware of the existence of Earth (i.e., risk amplification).

3.2. *Risk information seeking*

An important factor in SETL and detection contexts is how well people are informed about potential threats and how much they are motivated to seek and process accurate information (as opposed to misinformation) from credible sources to better understand these risks. The risk information seeking and processing model (RISP) aims to explain or predict why and how people actively seek or avoid risk information and whether they actively and systematically process that information as opposed to heuristic and less effortful information processing (Dong et al. 2024; Griffin et al. 1999). It is assumed that people who seek and process risk information more actively would adopt more stable beliefs, attitudes, and behaviors in response to a risk. Several factors have been identified in previous research that explain risk information seeking (vs. avoidance) and processing (systematic vs. heuristic), mainly in the context of risks related to public health, environmental hazards, and safety (Yang et al. 2014).

An important motivator of information seeking is the perceived level of confidence in having enough knowledge to deal with a risk (information sufficiency/insufficiency), with each individual having a certain information threshold for judgmental confidence. Higher information insufficiency would motivate people to seek more actively for information to cope with a given risk. Informational subjective norms refer to perceived social pressures from one's social environment (people perceived as personally relevant) to acquire sufficient risk information. Higher social expectations lead to more engagement with risk information according to RISP. People also hold different beliefs about the media or information channels they use to gather risk information. Such 'relevant channel beliefs' in terms of credibility, bias, sensationalism, or expectations about meeting information seeking goals by using a certain channel, can influence the risk information seeking process. However, RISP studies have produced mixed and inconsistent results with regards to channel beliefs (Dong et al. 2024). Perceived information gathering capacity was conceptualized as the perceived cost and ability of seeking and processing risk information. The more people perceive themselves to be able to access and process information about a given risk, the more likely to will actively seek and systematically process that information. Besides motivations and communication beliefs/behavior, perceived hazard characteristics (i.e., cognitive risk assessments) were included in the model. Specifically, subjective evaluations of probability and severity of a risk, institutional trust, personal control, and causal attributions were regarded as relevant dimensions. Most consistently, risks perceived as more likely and severe were related to more active risk information seeking and processing. Rather inconsistent or statistically less important influences were found for different individual characteristics such as personal risk experience, relevant values, demographic variables, etc. (Yang et al. 2014).

Besides immediate personal risks, the RISP model was found to be valid for impersonal risks such as climate change (Kahlor et al. 2006). Thus, it may be relevant in the SETL and detection context to study and practically facilitate risk information seeking among different stakeholder groups. The current practical implications of the RISP model for SETL are mainly relevant for pre-detection research to better understand different stakeholders, their level of knowledge to anticipate perceived information insufficiency, their prior risk perceptions about SETL and/or active SETI, their levels of trust towards institutions in the field (e.g., space agencies, astrobiologists, SETI research centers, governments, etc.), and relevant channel beliefs in terms of accessibility and credibility. Available research on these aspects, which is far from robust or representative, suggests that people lack knowledge about astrobiology or planetary protection (Lofstedt 2003; Schwartz 2020), have rather low risk perceptions (Kwon et al. 2018; Lofstedt 2003), and some distrust the (U.S.) government or NASA when comes to disclosing vital information about planetary protection risks or UAP detections (Kennedy & Lau 2021; MacGregor et al. 1998). However, much more research is needed using the RISP dimensions more systematically and beyond the U.S. and Europe to get a timely understanding of perceived information sufficiency, channel beliefs, risk perception, and institutional trust in SETL contexts.

In terms of preparedness for risk and detection scenarios the RISP offers several starting points, which can guide the development of risk communication programs and crisis plans, with the goal to facilitate active risk information seeking/processing in different stakeholder groups using credible and accurate information channels. Once specific risks of detection, space missions, or active SETI become more relevant, SETL institutions can use the RISP to inform their communication tactics. They can appeal to subjective norms ('many people in your community already follow us to get the latest information about the recent detection') or information insufficiency ('You think the probe we detected is too far away to be relevant for human contact? You should probably reconsider this, given the latest insights.'), for example.

3.3. *Risk message design and effects*

A great deal of research on risk and crisis communication has produced more or less evidence-based guidelines for best practices in form and content of communication (Seeger 2006; Seeger & Schwarz 2024), which may be relevant once SETL projects or detection events turn into crisis-like situations. Rather than reviewing this research, however, we will focus on selected models that have the potential to guide the design of risk messages before, during, and after risk-related detection events.

The extended parallel process model (EPPM) addresses the design of risk messages composed of information about threat and efficacy and the effects of these messages on perceived threat and efficacy. These perceptions were shown to influence individuals' behavioral dispositions to take protective action to cope with a risk (Cho & Witte 2005; Roberto et al. 2009; Witte 1992). Threat refers to message elements and perceptions of severity (magnitude of a threat) and susceptibility (likelihood of being affected). Efficacy includes message elements and perceptions of self-efficacy (the ability to perform recommended protective actions) and response-efficacy (the effectiveness of recommended actions to reduce a risk). The EPPM predicts (1) motivations for self-protective behavior to avoid danger (*danger control* process resulting from perceptions of high threat and high efficacy); (2) maladaptive behavior to *control fear* (e.g., denial, reactance, or turning to conspiracy beliefs; resulting from perceptions of high threat and low efficacy); or (3) *no response* to a risk (no threat perceived). The empirical relevance of threat and efficacy to predict dispositions of protective behavior was repeatedly demonstrated in different risk and crisis contexts (Bostrom et al. 2019; Kok et al. 2018; Peters et al. 2013). Later extensions of EPPM added societal dimensions with societal threat and societal/proxy efficacy (Bandura 2000;Evensen & Clarke 2012). Societal efficacy refers to messages and perceptions that a collective is able to reduce a threat and/or that trusted third parties (proxy agents) act on the collective's behalf to control a threat. Evidence suggests that societal efficacy is a relevant construct in media representations of risks (Evensen & Clarke 2012; Schwarz 2024; You et al. 2017). Media exposure was associated with societal-level risk perceptions and proxy efficacy can be more significant than self-efficacy in triggering danger control instead of fear control responses to a perceived threat (Li 2018).

Depending on the SETL-related risk or detection scenario, the EPPM can guide message development based on the intended outcome. When there is broad consensus, that a detection event does not pose any form of threat to people or specific groups of risk bearers, messages should contain compelling evidence and/or trusted sources explaining why the event does not pose a risk. Whenever threats reach a certain probability threshold (e.g., potentially infectious samples brought to earth, an ET probe approaching earth), risk communication should explain the severity and probability of that threat. In addition, it is crucial to transparently communicate what government authorities, space agencies, and the scientific community are doing to manage or control the risk (proxy efficacy) (e.g., strict biosafety measures, intense scanning and preparation for a probe encounter). In addition, people need to be informed about doable protective measures they can take to effectively reduce the risks for themselves and their loved ones (e.g., establishing open and up-to-date information channels, wearing masks, demonstrating cross-national and cross-disciplinary collaboration of competent/trusted institutions, etc.). Only if risk communication achieves both a sense of threat and efficacy among relevant publics (including the media), people are more likely to adopt recommended measures and/or trust in the institutional response to the threat on their behalf.

The IDEA (internalization, distribution, explanation, action) model uses insights from instructional communication research to develop evidence-based guidelines for effective risk and crisis communication (Sellnow et al. 2017; Sellnow & Sellnow 2019). By taking into account the high levels of uncertainty in many risk scenarios as well as varying levels of science literacy among publics, the model offers specific recommendations for designing risk messages that increase the likelihood for cognitive/affective learning about risks and compliance with recommended protective behavior. Internalization refers to message elements that motivate people to pay attention to a risk by expressing empathy for risk bearers and by highlighting personal relevance and proximity of a threat for individuals and their relevant social environment. Distribution of risk messages is effective when consistent and coordinated messages are disseminated through multiple credible channels used by relevant audiences. The explanation component refers to the use of trusted sources in the message content, complete information about a given risk, and intelligible translation of science or technical terms for audiences with varying levels of science literacy. The action component adds instructions for measures of protection and risk mitigation. These action instructions should be specific and perceived as doable. Research using the IDEA model has shown that risk messages are most effective when they include all four elements simultaneously, with even one element missing leading to a reduction of the intended outcome of motivation, comprehension, and compliance.

In the context of SETL and detection, the IDEA model offers specific recommendations for designing and distributing risk messages, especially for scenarios where individuals in certain stakeholder groups need to take protective action or support risk management institutions in their efforts. Such messages need to elaborate why a (potential) detection risk matters to people and their loved ones, express empathy to those who feel concerned about the risk, explain the science in a comprehensible non-technical language based on credible sources, give instructions for actions that are suitable and feasible to reduce the risk, and distribute these messages through communication channels that are accessible and trusted by the target groups.

4. Conclusions: Incorporating risk communication research into astrobiology and institutional preparedness

Despite significant advances in technology and space exploration, humanity has yet to find evidence of extraterrestrial life. Therefore, the risks of SETL, sample return missions, and the detection or encounter scenarios discussed in the literature are beyond reliable risk assessment. However, given the immense costs and increasing frequency of space missions and astronomical observation related to astrobiology, risk scenarios may become relevant sometime in the future. SETL-related institutions should prepare for situations in which either actual threats emerge or claimed discoveries lead to widespread public attention and (unjustified) risk perceptions among different audiences. Insights and models in the field of risk and crisis communication research, some of which were discussed in this chapter, can inform preparedness efforts by guiding pre-detection research as well as the preparation and updates of contingency plans or protocols at space agencies, SETL/SETI research centers, governments, or transnational bodies such as the UN. As soon as detection events occur, risk communicators at SETL-related institutions should assess potential threats and engage a broad range of societal stakeholders in a two-way risk communication process that is guided by evidence-based guidelines for effective risk and/or crisis communication. This needs to be done for scenarios that pose an actual threat, but even if, contrary to expert assessments (e.g., based on the Rio/London scales),

Table 1. Avenues for future empirical risk communication/perception research on SETL and detection events.

Theoretical framework	Research focus	Methods
Past claimed detections (EQ-Peg 121, Oumuamua, Tabby's Star, etc.)		
SARF: Amplification/ attenuation trends over time	News media coverage of detection events: • Frequency, cited sources, risk/benefit evaluations, evaluation of SETL institutions/ government, expert disagreement	Quantitative/qualitative content analysis
	Social media communications on detection events: • Most active accounts/ persons/ institutions and their networks (social stations) • Content of posts (risks, evaluations, sources, expert disagreement, controversy, etc.) • Engagement with different accounts and posts (views, comments, likes, shares, etc.)	Computational social media analysis (LLM, sentiment), network analysis
	Personal experience of experts/scientists, digital influencers,journalists, etc. with unfolding communication in the event and assessment of 'ripple' effects (regulation, loss of trust?)	Interviews/focus group discussions
EPPM/IDEA: Composition and impact of risk messages in terms of threat, efficacy,intelligibility, distribution	News media coverage of detection events: • Identification of threat/risk information in combination with societal efficacy information as media frames(EPPM) • Analysis of proximity, potential impact, intelligible translation of science (astrobiology/astrophysics), use of (credible) sources, etc. (IDEA)	Quantitative/qualitative content analysis
	Social media communications: • Content analysis of posts as in news media coverage (see above) • Identification of most active social media accounts and their overlap in risk messages (Distribution in the IDEA model)	Quantitative/qualitative content analysis, network analysis
Future claimed detections		
SARF: Amplification/ attenuation trends over time	News media coverage and social media communications on detection event → see above in past detection events	→ see above in past detection events
	Press releases/ press conference transcripts or videos of SETL-related institutions (government, space agencies, universities, government, space agencies, universities, SETI research centers, etc.): • Volume of released information units • Content: Expert disagreement, controversy, demands for institutional change/regulation, etc.	Quantitative/qualitative content analysis
RISP: Preconditions, process,and outcome of risk information seeking in different sociodemographic/cultural groups	Perceptions of informational subjective norms (peer pressures to be informed about detection), information insufficiency/knowledge about detection risks, channel beliefs (credible sources for detection information), perceived information gathering capacity (accessibility/cost of channels,perceived effort to process content, etc.), risk information seeking intentions, information processing	Quantitative survey, focus group discussions

Table 1. Continued

Theoretical framework	Research focus	Methods
EPPM/IDEA: Risk message composition and effect on cognitive-affective learning and behavioral dispositions in different sociodemographic/cultural groups	Perceptions of (societal) threat and (societal)efficacy related to the detection event, dispositions to take protective action,support of or trust in SETL-related institutions or government,maladaptive behavioral responses (denial, reactance,conspiracy beliefs)	Quantitative survey, focus group discussions
	News media coverage and social media communications on detection event → see above in past detection events	Quantitative/qualitative content analysis, network analysis
	Press releases/ press conference transcripts/ institutional websites or videos of SETL-related institutions (government,space agencies, universities, SETI research centers, etc.): • Identification of threat/risk information in combination with societal efficacy information as media frames (EPPM) • Analysis of proximity, potential impact, expressions of empathy, intelligible translation of science (astrobiology/ astrophysics), use of (credible) sources,action instructions, etc. (IDEA)	Quantitative/qualitative content analysis
Pre-detection research (general risk perceptions/communications – SETL, METI, planetary protection)		
SARF: Identification of most active social stations (institutional, media, laypersons) in terms of SETL information and their anticipation of risks and/or demands for regulation, preparation, etc.	News media coverage and social media communications on SETL, planetary protection → see above in past detection events, but focus on general SETL communications and anticipated risks	→ see above in past detection events
	Press releases/websites of SETL-related institutions (government, space agencies, universities, SETI research centers, etc.): • Volume of released information units with references to risk scenarios • Content: Anticipated risks, demands for institutional change/protocols/regulation, etc.	Quantitative/qualitative content analysis
RISP: Analysis of perceptions of different potential detection scenarios and ongoing SETL projects	Perceptions of ongoing projects in SETL, METI, planetary protection and attributed risks (psychometric risk perceptions), assessment of current knowledge in different sociodemographic/cultural groups about astrobiology/SETI etc., current channel beliefs (credible sources for SETL information) and perceived information gathering capacity (channels perceived as accessible and comprehensible in terms of SETL), social peer groups that are perceived to pressure individuals to be informed about SETL	Quantitative survey, focus group discussions
EPPM/IDEA: Impact of risk messages in different sociodemographic/cultural groups	Perceptions of risk messages and different theoretical dimensions (threat, efficacy, internalization,explanation, etc.) on cognitive-affective learning and behavioral dispositions	Social experiments (online, lab) with pre-designed stimulus materials on fictitious detection/risk scenarios

the media, specific communities (religious groups, UFO groups, etc.), or digital platforms socially amplify the risks of a detection event, risk communication will be critical to managing concerns and maintaining trust in SETL-related organizations or astrobiology/SETI in general. The effectiveness of risk communication and risk governance efforts will benefit from a more consistent involvement of risk and crisis communication experts as well as more funded and timely research in the field.

Although the outcomes of SETL and potential future threats are highly uncertain, future risk communication research should focus on public risk perceptions, mediated risk narratives (news media, social media, entertainment media), preferred and trusted sources of information, and the risk communication behavior of different social stations around the globe related to astrobiology and SETI. Using risk and crisis communication frameworks with some explanatory and predictive quality, the challenges we have discussed in this chapter (risk amplification/attenuation, risk information seeking/misinformation, risk message design and effects) can be effectively addressed if the appropriate research is conducted. Past (e.g., Tabby's Star, Oumuamua) and future claimed detections as well as present risk communication dynamics related to SETL projects (i.e., pre-detection research) offer a wide range of avenues for future empirical research. We have summarized several more detailed suggestions for topics, applicable frameworks, and research methods in Table 1. To date, the available data on these topics are scarce and fragmented. The research suffers from methodological limitations and is rarely grounded in risk communication theory. Lack of funding, reputational concerns of social scientists working on SETL topics ("giggle factor"), and lack of interdisciplinary collaboration are among the well-known reasons (Denning et al. 2019; Schwarz & Seidl 2023; Shostak 2004). However, the increasing inclusion of social scientists in SETL-related conferences, astrobiology/astronautics journals, and funding programs is a promising development, suggesting that the societal aspects of SETL are gaining relevance in the broader scientific community. Given the critical role of communication in pre- and post-detection scenarios, risk and crisis communication research should be one of the main pillars of social science contributions to the field.

References

Almár, I. (2011). SETI and astrobiology: The Rio Scale and the London Scale. *Acta Astronautica*, *69*(9), 899–904. https://doi.org/https://doi.org/10.1016/j.actaastro.2011.05.036

Anton, A., Elliott, J., & Schetsche, M. (2024). Meeting extraterrestrials: Scenarios of first contact from the perspective of exosociology. *Acta Astronautica*, *215*, 308–314. https://doi.org/https://doi.org/10.1016/j.actaastro.2023.12.003

Bandura, A. (2000). Exercise of Human Agency Through Collective Efficacy. *Current Directions in Psychological Science*, *9*(3), 75–78. https://doi.org/10.1111/1467-8721.00064

Bialy, S., & Loeb, A. (2018). Could Solar Radiation Pressure Explain 'Oumuamua's Peculiar Acceleration? *The Astrophysical Journal Letters*, *868*(1). https://doi.org/10.3847/2041-8213/aaeda8

Billings, L. (2012). Astrobiology in Culture: The Search for Extraterrestrial Life as "Science". *Astrobiology*, *12*(10), 966–975. https://doi.org/10.1089/ast.2011.0788

Bostrom, A., Hayes, A. L., & Crosman, K. M. (2019). Efficacy, Action, and Support for Reducing Climate Change Risks. *Risk Analysis*, *39*(4), 805–828. https://doi.org/10.1111/risa.13210

Byrnes, M. E. (1994). *Politics and Space: Image Making by NASA*. Greenwood Publishing Group.

Cho, H., & Witte, K. (2005). Managing fear in public health campaigns: a theory-based formative evaluation process. *Health promotion practice*, *6*(4), 482–490. https://doi.org/10.1177/1524839904263912

Clark, F., & Illman, D. L. (2003). Content Analysis of New York Times Coverage of Space Issues for the Year 2000. *Science Communication*, *25*(1), 14–38. https://doi.org/10.1177/1075547003255300

Covello, V. T. (1992). Risk Communication: An Emerging Area of Health Communication Research. *Annals of the International Communication Association*, *15*(1), 359–373. https://doi.org/10.1080/23808985.1992.11678816

Denning, K., Berea, A., Capova, K. A., Forgan, D., Korpela, E., Oman-Reagan, M. P., Tarter, J., & Tennen, L. I. (2019). SETI and Post-Detection: Towards a New Research Roadmap. 70th International Astronautical Congress (IAC), Washington D.C.

Des Marais, D. J., Nuth, J. A., Allamandola, L. J., Boss, A. P., Farmer, J. D., Hoehler, T. M., Jakosky, B. M., Meadows, V. S., Pohorille, A., Runnegar, B., & Spormann, A. M. (2008). The NASA Astrobiology Roadmap. *Astrobiology*, *8*(4), 715–730. https://doi.org/10.1089/ast.2008.0819

Dittmer, J. N.(2007). Colonialism and Place Creation in Mars Pathfinder Media Coverage.*Geographical Review*, *97*(1), 112–130. https://doi.org/10.1111/j.1931-0846.2007.tb00282.x

Dong, X.,Griffin, R. J., Yang, J., & Dunwoody, S. (2024). 23 Risk Information Seeking and Processing Model. In T. L. Sellnow & D. D. Sellnow (Eds.), *Communicating Risk and Safety* (pp. 413–452).De Gruyter Mouton. https://doi.org/10.1515/9783110752427-023

Evensen, D. T., & Clarke, C. E. (2012). Efficacy Information in Media Coverage of Infectious Disease Risks: An Ill Predicament? *Science Communication*, *34*(3), 392–418. https://doi.org/10.1177/1075547011421020

Forgan, D., Wright, J., Tarter, J., Korpela, E., Siemion, A., Almár, I., & Piotelat, E. (2019). Rio 2.0: revising the Rio scale for SETI detections. *International Journal of Astrobiology*, *18*(4), 336–344. https://doi.org/10.1017/S1473550418000162

Garber, S. J. (1999). Searching for Good Science - The Cancellation of NASA's SETI Program. *Journal of the British Interplanetary Society*, *52*(3), 3–12.

Greaves, J. S., Richards, A. M. S., Bains, W., Rimmer, P. B., Sagawa, H., Clements, D. L., Seager, S., Petkowski, J. J., Sousa-Silva, C., Ranjan, S., Drabek-Maunder, E., Fraser, H. J., Cartwright, A., Mueller-Wodarg, I., Zhan, Z., Friberg, P., Coulson, I., Lee, E. l., & Hoge, J. (2021). Phosphine gas in the cloud decks of Venus. *Nature Astronomy*, *5*(7), 655–664. https://doi.org/10.1038/s41550-020-1174-4

Griffin, R. J., Dunwoody, S., & Neuwirth, K. (1999). Proposed Model of the Relationship of Risk Information Seeking and Processing to the Development of Preventive Behaviors.*Environmental Research*, *80*(2), S230–S245. https://doi.org/10.1006/enrs.1998.3940

Harrison, A. A. (2011).Fear, pandemonium, equanimity and delight: human responses to extra-terrestrial life. *Philosophical Transactions of the Royal Society A: Mathematical, Physical and Engineering Sciences*, *369*(1936), 656–668. https://doi.org/10.1098/rsta.2010.0229

Hatfield, P., & Trueblood, L. (2021). SETI and democracy. *Acta Astronautica*, *180*, 596–603. https://doi.org/https://doi.org/10.1016/j.actaastro.2020.11.053

Heath, R. L., & Palenchar, M. J. (2016). Paradigms of Risk and Crisis Communication in the Twenty- first Century. In A. Schwarz, M. W. Seeger, & C. Auer (Eds.), *The Handbook of International Crisis Communication Research* (pp. 435–446). https://doi.org/10.1002/9781118516812.ch39

Ipsos.(2021).*Global predictions for 2022*. https://www.ipsos.com/en/global-predictions-2022

Jones, M.(2013). Mainstream Media and Social Media Reactions to the Discovery of Extraterrestrial Life. In D. A. Vakoch (Ed.), *Astrobiology, History, and Society: Life Beyond Earth and the Impact of Discovery* (pp. 313–328). Springer Berlin Heidelberg. https://doi.org/10.1007/978-3-642-35983-5_17

Kahlor, L., Dunwoody, S., Griffin, R. J., & Neuwirth, K. (2006). Seeking and Processing Information about Impersonal Risk. *Science Communication*, *28*(2), 163–194. https://doi.org/10.1177/1075547006293916

Kasperson, R. E., Renn, O., Slovic, P., Brown, H. S., Emel, J., Goble, R., Kasperson, J. X., & Ratick, S.(1988). The Social Amplification of Risk: A Conceptual Framework. *Risk Analysis*, *8*(2), 177–187. https://doi.org/10.1111/j.1539-6924.1988.tb01168.x

Kasperson, R. E., Webler, T., Ram, B., & Sutton, J. (2022). The social amplification of risk framework: New perspectives. *Risk Analysis*, *42*(7), 1367–1380. https://doi.org/10.1111/risa.13926

Kauffman, J. L. (1994). *Selling Outer Space: Kennedy, the Media, and Funding for Project Apollo, 1961–1963*. University of Alabama Press.

Kennedy, C., & Lau, A. (2021). Most Americans believe in intelligent life beyond Earth; few see UFOs as a major national security threat. https://pewrsr.ch/3w6Bgcx

Kok, G., Peters, G.-J. Y., Kessels, L. T. E., ten Hoor, G. A., & Ruiter, R. A. C. (2018). Ignoring theory and misinterpreting evidence: the false belief in fear appeals. *Health Psychology Review*, *12*(2), 111–125. https://doi.org/10.1080/17437199.2017.1415767

Korbitz, A. (2014). Toward understanding the active SETI debate: Insights from risk communication and perception. *Acta Astronautica*, *105*(2), 517–520. https://doi.org/https://doi.org/10.1016/j.actaastro.2014.07.005

Kwon, J. Y., Bercovici, H. L., Cunningham, K., & Varnum, M. E. W. (2018). How Will We React to the Discovery of Extraterrestrial Life? [Original Research]. *Frontiers in Psychology, 8*. https://doi.org/10.3389/fpsyg.2017.02308

Li, X. (2018). Media Exposure, Perceived Efficacy, and Protective Behaviors in a Public Health Emergency [RGC 21 - Publication in refereed journal]. *INTERNATIONAL JOURNAL OF COMMUNICATION*, *12*, 2641–2660. http://www.scopus.com/inward/record.url?scp=85100257840&partnerID=8YFLogxK https://www.scopus.com/record/pubmetrics.uri?eid=2-s2.0-85100257840&origin=recordpage

Lissauer, J. J., Dawson, R. I., & Tremaine, S. (2014). Advances in exoplanet science from Kepler.*Nature*, *513*(7518), 336–344. https://doi.org/10.1038/nature13781

Liu, Q., & Li, X. (2024). Identifying the impacts of different information access on emergency behaviors during COVID-19 pandemic. *Journal of Emergency Management and Disaster Communications*. https://doi.org/10.1142/s2689980924500118

Lofstedt, R. E. (2003). Public perceptions of the Mars sample return program. *Space Policy*, *19*(4), 283–292. https://doi.org/10.1016/j.spacepol.2003.08.002

Mace, M., & Schwalbe, C. B. (2020). From Robots to Humans: Newspaper Coverage of Mars in the United States and the United Kingdom 2011–2016. *Journalism Studies*, *21*(13), 1913–1932.https://doi.org/10.1080/1461670X.2020.1799238

MacGregor, D., Slovic, P., & Race, M. S. (1998). Lay and Expert Perceptions of Planetary Protection. *SSRN Electronic Journal*. https://doi.org/10.2139/ssrn.1562792

Neal, M. (2014). Preparing for extraterrestrial contact. *Risk Management*, *16*(2), 63–87. https://doi.org/10.1057/rm.2014.4

Neal, V. (2007). Framing the Meaning of Spaceflight in the Shuttle Era. In S. J. Dick & R. D. Launius (Eds.), *Societal Impact of Spaceflight* (pp. 67–88). National Aeronautics and Space Administration, Office of External Relations, History Division.

Oliver, C. (1997). *SETI and the Media* 48th International Astronautical Congress, Turin.

Oliver, C., Sim, H., & Shostak, S. (1999, 10/04/1999). *The Case of EQ Peg: Challenge and Response* International Astronautical Congress, Amsterdam.https://www.researchgate.net/publication/267763658_EQPegFinal

Peters, G.-J. Y., Ruiter, R. A. C., & Kok, G. (2013). Threatening communication: a critical re-analysis and a revised meta-analytic test of fear appeal theory. *Health Psychology Review*, *7*(Suppl 1), S8–S31. https://doi.org/10.1080/17437199.2012.703527

Race, M., Denning, K., Bertka, C. M., Dick, S. J., Harrison, A. A., Impey, C., & Mancinelli, R. (2012).Astrobiology and society: building an interdisciplinary research community. *Astrobiology*, *12*(10), 958–965. https://doi.org/10.1089/ast.2011.0723

Race, M. S. (2008). Communicating about the discovery of extraterrestrial life: Different searches, different issues. *Acta Astronautica*, *62*(1), 71–78.https://doi.org/10.1016/j.actaastro.2006.12.020

Race, M. S. (2015). Preparing for the discovery of extraterrestrial life: are we ready?: Considering potential risks, impacts, and plans. In S. J. Dick (Ed.), *The Impact of Discovering Life beyond Earth* (pp. 263–285). Cambridge University Press.https://doi.org/10.1017/CBO9781316272480.022

Race, M. S., & MacGregor, D. G. (2000). Integrating public perspectives in sample return planning.*Advances in Space Research*, *26*(12), 1901–1909. https://doi.org/10.1016/S0273-1177(00)00160-5

Race, M. S., & Randolph, R. O. (2002). The need for operating guidelines and a decision making framework applicable to the discovery of non-intelligent extraterrestrial life. *Advances in Space Research*, *30*(6), 1583–1591. https://doi.org/https://doi.org/10.1016/S0273-1177(02)00478-7

Renn, O. (2009). Risk communication: Insights and requirements for designing succesful communication programs on health and environmental hazards. In R. L. Heath & H. D. O'Hair (Eds.), *Handbook of risk and crisis communication* (pp. 80–98). Routledge.

Roberto, A. J., Goodall, C. E., & Witte, K. (2009). Raising the alarm and calming fears: Perceived threat and efficacy during risk and crisis. In R. L. Heath & H. D. O'Hair (Eds.), *Handbook of risk and crisis communication* (pp. 285–301). Routledge.

Rohrmann, B. (2008, 01/01). *Risk perception, risk attitude, risk communication, risk management: A conceptual appraisal* 15th International Emergency Management Society Annual Conference, Melbourne. https://tiems.info/dmdocuments/events/TIEMS_2008_Bernd_Rohrmann_Keynote.pdf

Rooke, M. (2021). Alternative media framing of COVID-19 risks. *Current Sociology*, *69*(4), 584–602. https://doi.org/10.1177/00113921211006115

Rossmann, K., Wegner, H., Stark, H., Grossmann, G., Jansen, A., & Frangoulidis, D. (2021). COVID-19 Pandemic & Bureaucracy: The Crisis Inside the Crisis [Article]. *Frontiers in Public Health*, *9*, 6, Article 665323. https://doi.org/10.3389/fpubh.2021.665323

Schäfer, M. S. (2012). Taking stock: A meta-analysis of studies on the media's coverage of science. *Public Understanding of Science*, *21*(6), 650–663. https://doi.org/10.1177/0963662510387559

Schwartz, J. S. J. (2020). 263 Myth-Free Space Advocacy Part IV: The Myth of Public Support for Astrobiology. In K. C. Smith & C. Mariscal (Eds.), *Social and Conceptual Issues in Astrobiology*. Oxford University Press. https://doi.org/10.1093/oso/9780190915650.003.0015

Schwarz, A. (2024). The mediated amplification of societal risk and risk governance of artificial intelligence: Technological risk frames on YouTube and their impact before and after ChatGPT. *Journal of Risk Research*, 1–23. https://doi.org/10.1080/13669877.2024.2437629

Schwarz, A., & Seidl, E. (2023). Stories of Astrobiology, SETI, and UAPs: Science and the Search for Extraterrestrial Life in German News Media From 2009 to 2022. *Science Communication*, *45*(6), 788–823. https://doi.org/10.1177/10755470231206797

Schwarz, A., & Unselt, J. J. (2024). Rage against the machine? Framing societal threat and efficacy in YouTube videos about artificial intelligence. *Risk Analysis*, *44*(10), 2377–2395. https://doi.org/10.1111/risa.14299

Seeger, M. W. (2006). Best practices in crisis communication: an expert panel process. *Journal of Applied Communication Research*, *34*(3), 232–244. https://doi.org/10.1080/00909880600769944

Seeger, M. W., & Schwarz, A. (2024). Best practices in crisis and disaster communication. In B. F. Liu & A. Mehta (Eds.), *Handbook of Risk, Crisis, and Disaster Communication* (pp. 65–82). Routledge. https://doi.org/10.4324/9781003363330-7

Sellnow, D. D., Lane, D. R., Sellnow, T. L., & Littlefield, R. S. (2017). The IDEA Model as a Best Practice for Effective Instructional Risk and Crisis Communication. *COMMUNICATION STUDIES*, *68*(5), 552–567. https://doi.org/10.1080/10510974.2017.1375535

Sellnow, D. D., & Sellnow, T. L. (2019). The IDEA model for effective instructional risk and crisis communication by emergency managers and other key spokespersons. *Journal of Emergency Management*, *17*(1), 67–78. https://doi.org/10.5055/jem.2019.0399

Sellnow, T. L., & Seeger, M. W. (2021). *Theorizing Crisis Communication* (2 ed.). Wiley Blackwell.

Shostak, S. (2004). SETI and the Media. *Symposium - International Astronomical Union, 213*, 535–541. https://doi.org/10.1017/S0074180900193854

Slobodian, R. E. (2015). Selling space colonization and immortality: A psychosocial, anthropological critique of the rush to colonize Mars. *Acta Astronautica, 113*, 89–104. https://doi.org/10.1016/j.actaastro.2015.03.027

Tarter, J. (2004). Life, the Universe, and SETI in a Nutshell. *Symposium - International Astronomical Union, 213*, 397–408. https://doi.org/10.1017/S007418090019360X

Vakoch, D. A., & Lee, Y. S. (2000). Reactions to receipt of a message from extraterrestrial intelligence: a cross-cultural empirical study. *Acta Astronautica, 46*(10), 737–744. https://doi.org/10.1016/S0094-5765(00)00041-2

Witte, K. (1992). Putting the fear back into fear appeals: The extended parallel process model.*Communication Monographs, 59*(4), 329–349. https://doi.org/10.1080/03637759209376276

Yang, Z. J., Aloe, A.M., & Feeley, T. H. (2014). Risk Information Seeking and Processing Model: A Meta-Analysis. *Journal of Communication, 64*(1), 20–41. https://doi.org/10.1111/jcom.12071

You, M., Joo, J., Park, E., Noh, G.-Y., & Ju, Y. (2017). Emerging Infectious Disease Content in Newspaper Editorials: Public Health Concern or Leadership Issue? *Science Communication, 39*(3), 313–337. https://doi.org/10.1177/1075547017705392

Zhao, S., & Wu, X. (2021). From Information Exposure to Protective Behaviors: Investigating the Underlying Mechanism in COVID-19 Outbreak Using Social Amplification Theory and Extended Parallel Process Model [Original Research]. *Frontiers in Psychology, 12*. https://doi.org/10.3389/fpsyg.2021.631116

(Toward) Discovery of Life beyond Earth and its Impact
Proceedings IAU Symposium No. 387, 2026
H. Landt, M. Dominik & C. Oliver, eds.
doi:10.1017/S1743921324001959

Extraterrestrial Intelligent Life: What The Public Believes

Chris Impey

Department of Astronomy, University of Arizona, Tucson, AZ 85721, United States
email: cimpey@as.arizona.edu

Abstract. Astrobiologists are moving purposefully towards the detection of extraterrestrial intelligence (ETI) through the radio and optical methods of SETI. Meanwhile, members of the public have formed their own opinions, only partially informed by science. In the United States, two thirds of all adults believe there is intelligent life beyond Earth, and a substantial minority believe we have already made contact. Opinions about the existence and nature of aliens are part of a complex landscape of beliefs that includes pseudoscience and superstition and extends to conspiracy theories. The widespread belief that unidentified flying objects (UFOs) represent visits by ETI is at the center of this landscape. Beliefs about ETI are also shaped by films, TV shows, and popular culture. This chapter summarizes what the public believes about ETI and presents possible explanations for these beliefs. Some international comparisons are presented. The author has surveyed the opinions and beliefs of tens of thousands of students on topics relating to life beyond Earth and the data gives insights into the modes of thinking of the college-age population. People are not "blank canvasses" when it comes to their beliefs and expectations about ETI, and this should be considered by the scientists who are preparing for eventual, actual contact.

Keywords. Extraterrestrial intelligence, pseudoscience, UFOs, science literacy, public opinion

---

1. Introduction

The detection of intelligent life beyond Earth would be one of the most profound discoveries in human history. After the first detection in 1995, detecting exoplanets has become routine, with nearly 5700 confirmed and another 7200 awaiting confirmation (National Aeronautics and Science Administration 2024). The definition of habitability is still under debate, but hundreds of exoplanets have the appropriate size and temperature to host life (Mendez 2022), projecting billions of habitable worlds in the Milky Way galaxy. Biology has not been detected yet on any exoplanet, but the methods to measure the biosignatures of microbial life are advancing (Schwieterman 2018). With billions of years for biology to advance to the level of an intelligent species with technology on a fraction of the habitable planets, the search for extraterrestrial intelligence using radio and optical techniques is well motivated (Anton & Schetsche 2023). But after over half a century of listening, no artificial signal has been detected. Meanwhile, in the absence of evidence, members of the public have formed their own opinions about ETI. This paper will summarize this survey data and give a context for the origin of the associated belief systems.

2. Surveys of Beliefs

The largest global survey of beliefs about life in the universe was carried out by Glocalities, a marketing consulting company based in Amsterdam (Glocalities 2024).

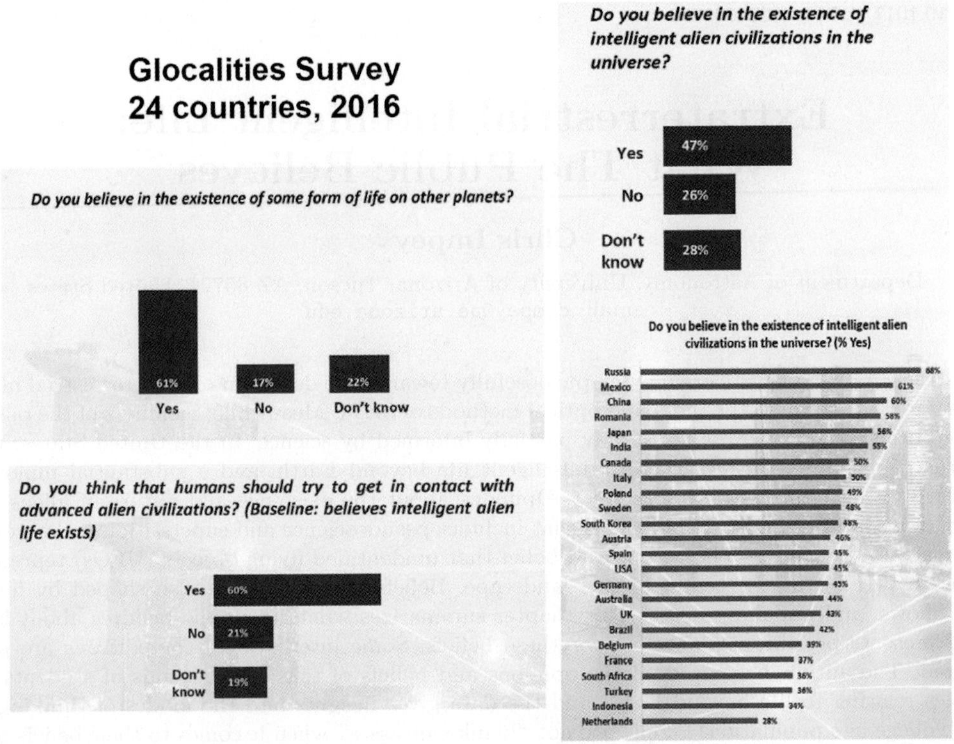

Figure 1. Survey by Glocalities in 2016 of 26,000 adults from 24 countries on their beliefs about life in the universe, including a breakdown by country on belief in ETI (Glocalities 2024)

In 2016, they polled over 26,000 adults from 24 countries with a series of questions about life in the universe, the existence of intelligent alien civilizations, the advisability of trying to make contact, and the panspermia hypothesis. Just over 60% believe in some form of life on other planets, and 47% think that intelligent alien civilizations exist. Belief in ETI is highest in Russia and lowest in the Netherlands, with no obvious geographical or other trend underlying these very large variations. Among those who believe in ETI, 60% support the idea of SETI. (Figure 1). Believers are characterized by being open-minded and tolerant, having an anti-authoritarian mindset, being engaged in politics, and having faith in technological progress.

The belief in ETI is not abstract. A 2022 survey of over 24,000 adults in 36 countries asked if it's likely that aliens would visit Earth in 2023. Globally, 18% said yes, with by far the highest percentage in India (43%), the United States near the average (17%), and Japan at the low end (8%). Despite a continuing flurries of UFO reports, no alien visit occurred, however the public seems primed for it to happen (Ipsos 2023). Again, there is no obvious trend behind the order of beliefs (Figure 2). In the United States, most people are confident that ETI exists. In a 2020 poll of 1000 adults, 66% said there is life on other planets, while a slightly smaller 57% said there is intelligent life and civilizations beyond Earth. Nearly half (45%) said UFOs exist and have visited the Earth and a quarter (26%) thought alien and human societies would be similar (Ipsos 2020). Another layer of the American belief system is revealed by a Politico survey of 2000 registered voters in 2022. Among the 62% of respondents who believed in the existence of extraterrestrial life, 82% said that their government has concealed the existence of UFOs (Politico 2022). This conspiracy theory is not unique to the United States. Among British adults surveyed in 2021, 49% said their government was concealing UFO information (YouGov 2021).

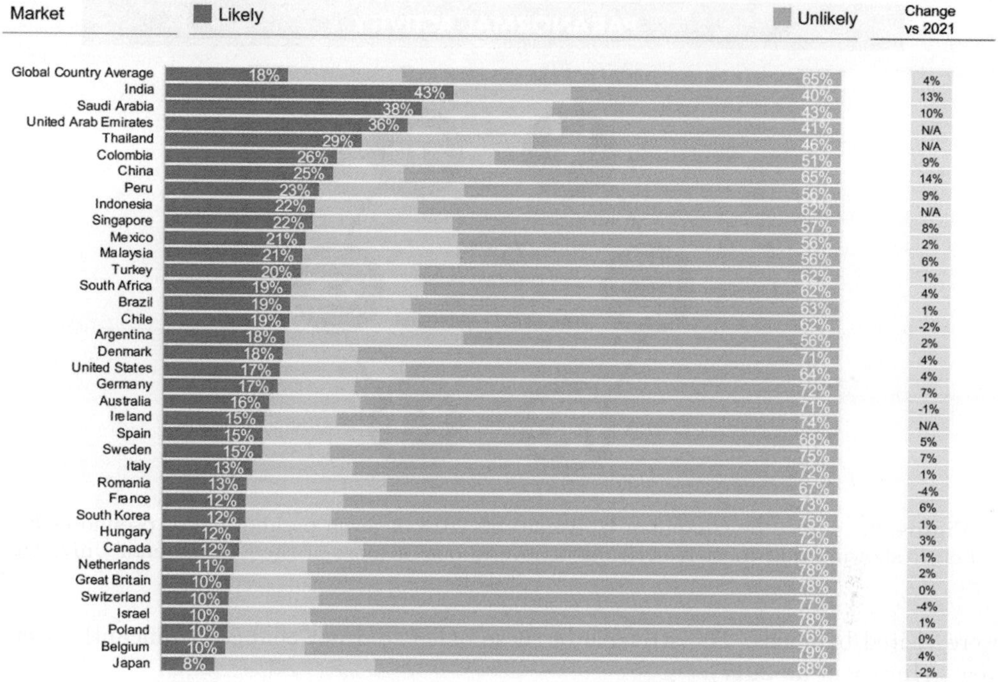

Figure 2. Survey by Ipsos in 2022 of 24,000 adults from 36 countries asking if they thought aliens would visit in 2023. Most countries saw an increase relative to 2021 (Ipsos 2020).

The backdrop for ETI belief is a modern world awash in misinformation, on subjects ranging from politics to science (Scheufele & Krause 2019). A particularly pernicious form of misinformation is pseudoscience: beliefs, statements, or practices that claim to be factual, but which are incompatible with the scientific method. Pseudoscience literally means "false science." Differentiating pseudoscience from science matters because many legal, medical, and public policy decisions depend on a good understanding of the natural world. Astrology or UFOs might be considered a harmless belief system, but in health, misinformation and anti-scientific mindsets can be fatal (Hotez 2021). Drawing the boundary between science and pseudoscience is not always straightforward and there is no litmus test for distinguishing them (Pigliucci 2010). Falsification as a criterion is problematic. String theory is a legitimate scientific framework, but it is not currently falsifiable. Meanwhile, the hypothesis that intelligent extraterrestrials exist is scientifically plausible, but falsifying it would require inspection of billions of habitable worlds beyond the Solar System (Cockell 2014). Philosophers have explored these demarcation and definition issues in detail (Grodin 2021).

At the University of Arizona, we have surveyed undergraduate non-science majors for thirty years, with a sample size of 13,000. Pseudoscience beliefs run high. About 40% think that the positions of the planets affect everyday life, and the same percentage think some people are psychic. A quarter believe in faith healing and a similar fraction trust in lucky numbers. Strikingly, none of these beliefs correlate with science literacy (Impey et al. 2011). Beliefs and attitudes were measured with 24 Likert-scale items that were subject to a factor analysis (Impey et al. 2012). One factor included pseudoscience items on astrology, psychic powers, and lucky numbers. Another item included the belief that UFOs are alien spacecraft visiting Earth and the belief that ancient civilizations

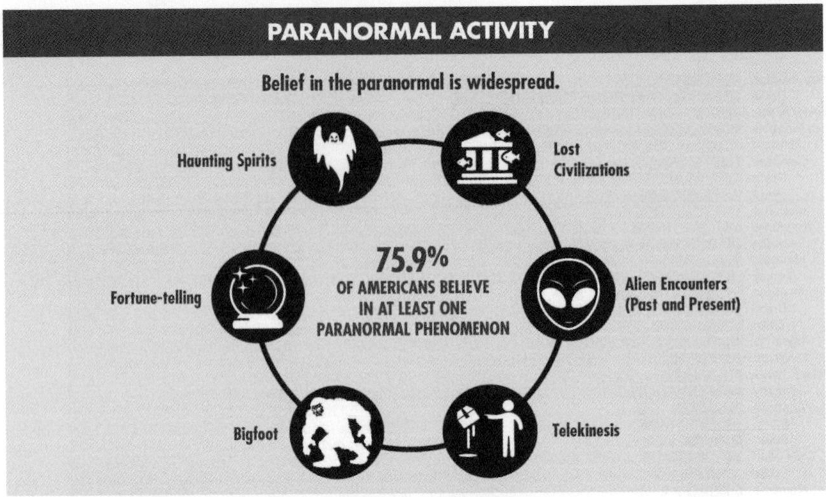

Figure 3. Survey by Chapman University in 2018 of seven paranormal or pseudoscience beliefs showing that most Americans believe in one or more of them (Chapman University 2018).

were visited by aliens. The strength of these beliefs did not change significantly over 27 years (Impey et al. 2017).

A benchmark on pseudoscience beliefs in the United States comes from biennial reports called "Science and Engineering Indicators" (National Science Board 2020). In surveys spanning forty years, the public has been polled about ten topics: extrasensory perception, haunted houses, ghosts and spirits, telepathy, clairvoyance, astrology, channeling, communicating with the dead, witches, and reincarnation. Of these items, only a quarter of Americans believed in none, 32% believed in four, and 22% believed in five or more. These beliefs rose through the 1990s before stabilizing (Newport & Strausberg 2001). Survey responses show widespread misunderstanding of how science works and how scientists judge evidence (Losh et al. 2003). Similar results were found in a survey by Chapman University of paranormal or pseudoscience beliefs (Chapman University 2018). In 2018, 41% of respondents said aliens have visited in our ancient past and 35% said aliens have come to Earth in modern times. Among seven of these paranormal beliefs, only a quarter believed in none, a third believed in four or more, and the belief levels in all seven were rising (Figure 3). The paranormal has become a firmly rooted strain of American culture (Bader et al. 2017).

3. UFOs and Aliens in Books

Google Books is a project to scan large numbers of books, convert the data into text using optical character recognition, and store it in a digital database. Analysis of this corpus allows cultural trends to be tracked quantitatively (Michel 2017). The database holds over 40 million books going back five hundred years, including nearly two trillion words and a third of the titles ever published. Google also provides a tool called Ngram Viewer which counts the occurrence of strings of up to five words, including wildcards and searches for parts of speech (Google Books 2024). Only matches found in 40 or more books are returned and the data are normalized by the number of books published in each year.

It was soon recognized that Google Books could provide cultural insights by tracking trends in word usage across decades and centuries (Bohannon 2010). The tool has led to research in diverse fields like economics, medicine, political science, and literature.

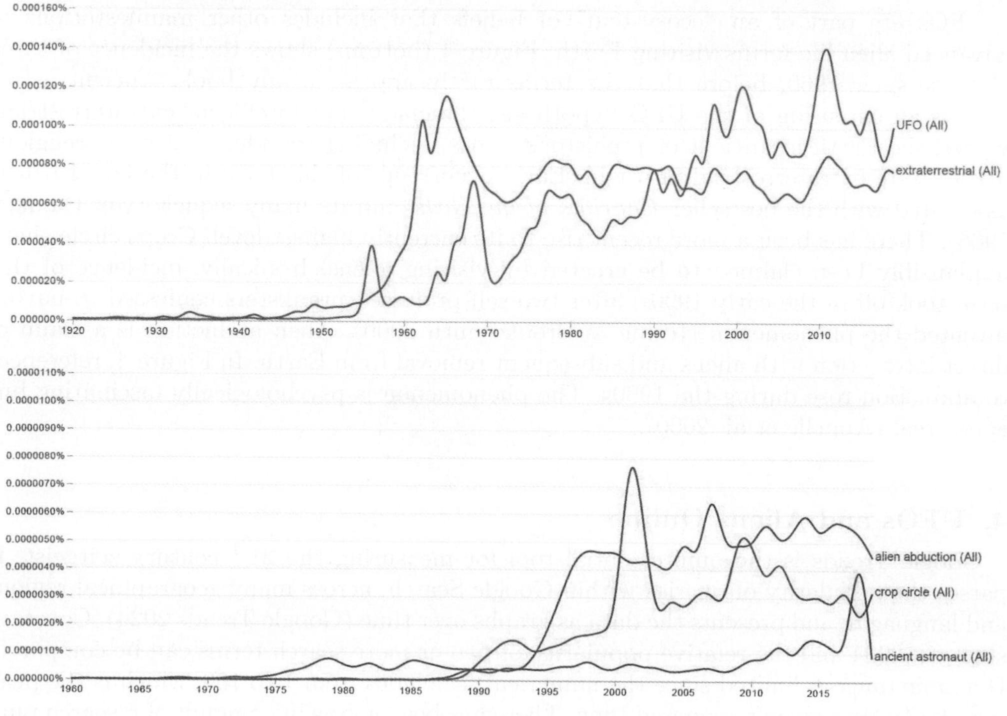

Figure 4. (Top) Incidence of the acronym UFO and the word extraterrestrial in Google Books since 1920. Both terms have negligible occurrence before 1940 but show a rise after the 1947 Roswell incident. (Bottom) Incidence of the phrases alien abduction, crop circle, and ancient astronaut in Google Books since 1960. The first two of these phrases show increasing use after 1990 with the growth of the Internet (Google).

However, there are caveats and limitations. The early data are sparse. From 1500 to 1800, there are few published titles. By 1800, the corpus grows to 100 million words per year, then it rises rapidly, by 1900 to two billion, and by 2000 to 11 billion. Also, Google Books is essentially a library. A single, prolific author can have disproportionate weight, and any book appears once, regardless of whether it has been read once or a million times. With the growth of science in the 20th century, and the fact that many of the scanned books are from academic libraries, scientific terms have become an increasing part of the database (Pechenick et al. 2015). These effects and biases can be understood and mitigated (Younes & Reips 2019).

UFOs have migrated from being a fringe belief to the mainstream of American popular culture (Pasulka 2019). The many Earth-like exoplanets being discovered make it plausible that there is intelligent life beyond the Solar System, and the subject gets some scholarly attention (Andresen & Chon Torres 2022), However, scientists find reports of extraterrestrials visiting the Earth to be unconvincing (Fraknoi 2010). Figure 4 (top) shows the incidence of the term UFO (the acronym for Unidentified Flying Object) since 1920, along with extraterrestrial, referring to life beyond Earth more generally. References in books to UFO started to appear in the early 1950s, a few years after the Roswell incident (Saler et al. 2010). Use rose steadily until 2000 and flattened after that. Use of extraterrestrial rose strongly until 1964 and has been relatively flat since. This rise in references to UFOs tracks publicly reported sightings (National UFO Reporting Center 2022). Sightings began at the time of the Roswell incident, climbed to dozens of reports per month in the 1970s and 1980s, and have reached several hundred per month in the past few years. In this paper UFOs are used as a proxy for belief in ETI.

UFOs are part of an "ecosystem" of beliefs that includes other manifestations of advanced alien life forms visiting Earth. Figure 4 (bottom) shows the incidence of three of these since 1960; before that the terms rarely appear in any book. Ancient astronaut is an extension of the UFO hypothesis, the claim that intelligent extraterrestrials visited the Earth in ancient or prehistoric times, including the idea that some religious deities were extraterrestrial in origin. Early occurrence of this term in the late 1970s is associated with the bestseller *Chariots of the Gods*, and its many sequels (von Daniken 1968). There has been a more recent rise to its currently highest level. Crops circles have implausibly been claimed to be created by visiting aliens. Ironically, incidence of this term took off in the early 1990s, after two self-professed pranksters confessed to having invented the phenomenon (Irving & Brookesmith 2009). Alien abduction is a claim of direct interaction with aliens and subsequent removal from Earth. In Figure 4, references to abduction rose during the 1990s. The phenomenon is psychologically fascinating but is not real (Appelle et al. 2000).

4. UFOs and Aliens Online

Google Trends is the quintessential tool for measuring the 21st century zeitgeist. It parses the popularity of queries within Google Search, across many geographical regions and languages, and presents the data as graphs over time (Google Trends 2024). Coverage starts in 2004 and the relative popularity of two or more search terms can be compared. Dynamic range is limited since the numerical scale goes from 1 to 100, which is the peak popularity for a given region and time. There has been a growing amount of research published using Google Trends. The data can illuminate human behavior (Mavragani et al. 2018). It has also been useful for tracking interest in the environment (**?**), and for forecasting economic indicators (Choi & Varian 2012) and financial markets (Preis et al. 2013). Google Trends often outperforms survey-based indicators (Vosen & Schmidt 2011).

Search engine data must be used with caution. Inconsistencies can easily arise when considering intervals of less than a year (Behnen et al. 2020). Unlike Google Books, where the scientific content has been increasing, the fraction of all Google searches for scientific terms has been decreasing as the Internet is used increasingly to track leisure and politics and to explore the popular culture (Ficetola 2013). For example, between 2004 and 2022 searches for coffee and weather increased by factors of 12 and 8 relative to searches for science. Similarly, one of the National Science Foundation's biennial reports (National Science Board 2016) found searches for science, engineering, and technology declining from 2004 to 2014. Little has been published previously on Google Trends tracking interest in science and pseudoscience (Baram-Tsabari & Segev 2009).

The search frequency of UFO relative to astronomy is shown in Figure 5 (top). Unlike the case for books, web searches for extraterrestrial are very rare, 20 to 30 times less frequent than searches for UFO. Astronomy searches have declined, rivaling UFO searches around 2004 and dropping to 4-5 times less recently. Searches for astronomy dip every summer, an effect seen in other science searches, indicating student activity. UFO searches show sharp peaks, correlating with media coverage of a single sighting, and a recent surge in searches started in April 2020 as the U.S. Government released three Navy videos of UFOs (U.S. Department 2020). States with the most UFO searches per capita are New Mexico, Arizona, and Nevada, all in the desert Southwest with mostly dark skies. Relative search frequencies of alien abduction, crop circle, and ancient astronaut are also shown in Figure 5 (bottom). All are much less common than UFO searches and their peaks do not line up with UFO peaks or with obvious news stories. The searches for ancient astronaut are particularly sparse but searches for alien abduction and crop circle have increased by a factor of four relative to astronomy since 2004.

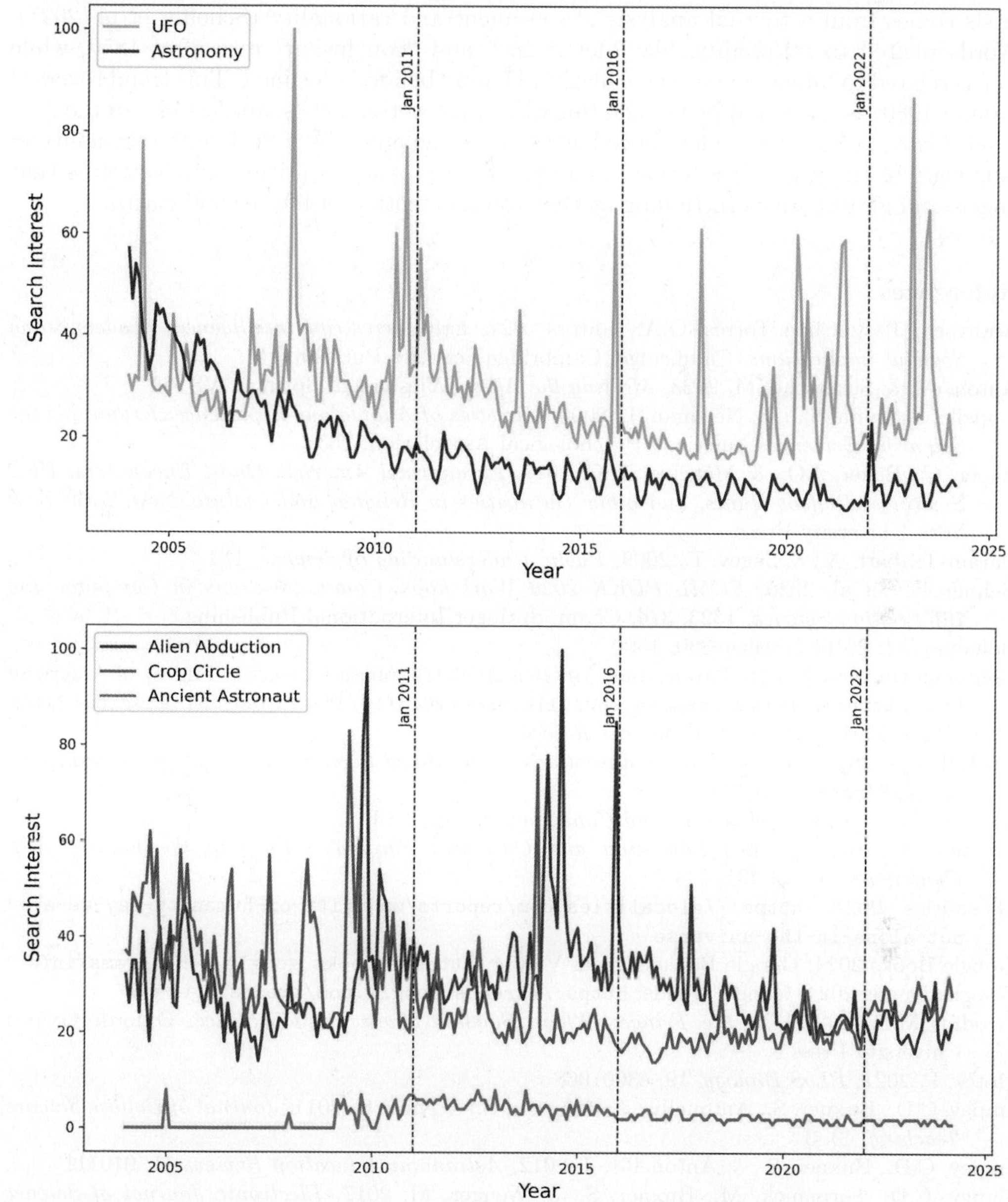

Figure 5. (Top) The relative occurrence of the search terms UFO (yellow) and astronomy (blue) in Google Trends, showing a stronger decline for astronomy. (Bottom) The relative occurrence of alien abduction (purple), crop circle (green) and ancient astronaut (red) in Google Trends. Sharp peaks are often associated with news reports online (Google)

5. Discussion

Pervasive public belief in the existence of life beyond Earth is in accord with the large number of habitable exoplanets, although a life detection experiment has not yet occurred. However, the similar level of belief in the existence of intelligent aliens and their visits to the Earth is decoupled from any scientific evidence. The explanation requires a look at the broad cultural predisposition toward superstition, pseudoscience beliefs, and conspiracy theories, especially in the United States. One sign of this science literacy crisis comes from a textual analysis of sentiment and rationality (Scheffer et al. 2021). Words related to rationality, like "determine" and "conclusion" rose after 1850, while words related to human experience, like "feel" and "believe" declined. This trend reversed around 1980, as measured by Google Book Ngrams, web searches, and articles in the *New York Times.* A rise in pseudoscience has been accompanied by a decline in rationality in language. In the way the Internet is commonly used, the dispiriting implication is that angels and ghosts are as intriguing as the actual detection of life beyond Earth.

References

Andresen, J., & Chon Torres, O.A., editors 2022, *Extraterrestrial Intelligence: Academic and Societal Implications.* Cambridge: Cambridge Scholars Publishing

Anton, A., & Schetsche, M. 2023, *Meeting the Alien*, Wiesbaden: Springer VS, 65

Appelle, S., Lynn, S.J., & Newman, L. 2000, *Varieties of Anomalous Experience: Examining the Scientific Evidence.* American Psychological Association, 253

Bader, C., Baker, J.O., & Mencken, F.C. 2017, *Paranormal America: Ghost Encounters, UFO Sightings, Bigfoot Hunts, and other Curiosities in Religion and Culture.* New York: New York University Press

Baram-Tsabari, A., & Segev, E. 2009, *Public Understanding of Science*, 1, 1

Behnen, P., et al. 2020, *ECML PDKK 2020 Workshops. Communications in Computer and Information Science,* 1323, 374. Cham: Springer International Publishing.

Bohannon, J. 2010, *Science*, 330, 1600

Chapman University 2018, Paranormal America 2018: Chapman University Survey of American Fears, `https://blogs.chapman.edu/wilkinson/2018/10/16/paranormal-america-2018/`

Choi, H., & Varian, H. 2012, *Economic Record*, 88, 2

Cockell, C.S. 2014, *Philosophical Transactions of the Royal Society A*, 372, `https://doi.org/10.1098/rsta.2013.0082`

Ficetola, G.F. 2013, *Biodiversity and Conservation*, 22, 2983

Fraknoi, A. 2010, *Science Education and Outreach: Forging a Path to the Future, ASP Conference Series*, 431, 514

Glocalities 2024, `https://glocalities.com/reports/majority-of-humanity-say-we-are-not-alone-in-the-universe`

Google Books 2024, Google Books Ngram Viewer, `https://books.google.com/ngrams/info`

Google Trends 2024, Googe Trends, `https://trends.google.com/trends/?geo=US`

Grodin, M.D. 2021, *On the Fringe: Where Science Meets Pseudoscience.* Oxford: Oxford University Press

Hotez, P. 2021, *PLoS Biology*, 19, e3001068

Impey, C.D., Buxner, S., Antonellis, J., Johnson, E., & King, C. 2011, *Journal of College Science Teaching*, 40, 3

Impey, C.D., Buxner, S., & Antonellis, J. 2012, *Astronomy Education Review*, 11, 010111

Impey, C.D., Formanek, M., Buxner, S., & Wenger, M. 2017, *Electronic Journal of Science Education,* 21, `http://ejse.southwestern.edu`

Ipsos 2020, Majority of Americans Believe There is Intelligent Life and Civilizations on Other Planets, `https://www.ipsos.com/en-us/news-polls/majority-believe-intelligent-life-exists-on-other-planets`

Ipsos 2023, Global Advisor Predictions, `https://www.ipsos.com/en/ipsos-global-predictions-2023`

Irving, R.,& Brookesmith,P.2009,*Smithsonian Magazine*, `https://www.smithsonianmag.com/arts-culture/crop-circles-the-art-of-the-hoax-2524283/`

Losh, S.C., Tavani, C.M., Njoroge, R., Wilke, R., & Mcauley, M. 2003, *Skeptical Inquirer*, 27, 30

Mavragani, A., Ochoa, G., & Tsagarakis, K.P. 2018, *Journal of Medical Internet Research*, 20, e270

Mendez, A. 2022, *New Frontiers in Astrobiology*, edited by R. Thombre and P. Vaishampayan, Amsterdam: Elsevier, 179

Michel, J.-P., Shen, Y.K., Aiden, A.P., et al. 2011, *Science*, 331, 176

National Aeronautics and Science Administration 2024, *NASA Exoplanet Archive*, `https://exoplanetarchive.ipac.caltech.edu/`

National Science Board 2016, *Science and Engineering Indicators 2016*. NSB-2016-1. Alexandria, VA: National Science Foundation.

National Science Board 2020, *Science and Engineering Indicators 2020*. NSB-2020-1. Alexandria, VA: National Science Foundation.

National UFO Reporting Center 2022, *UFO Report Index by Month*, `https://www.nuforc.org/webreports/ndxevent.html`

Newport, F., & Strausberg, M. 2001, Gallup Poll News Service, `http://poll.gallup.com/content/default.aspx?ci=4483&pg=1`

Pasulka, D.W. 2019, *American Cosmic: UFOs, Religion, Technology*. Oxford: Oxford University Press

Pechenick, A.E., Danforth, C.M., & Dodds, P.S. 2015, *PLoS ONE*, 10, e0137041

Pigliucci, M. 2010, *Nonsense on Stilts: How to Tell Science from Bunk*. Chicago: University of Chicago Press

Politico 2022, Most Alien Believers Think Government Has Kept UFO Existence from the Public, `https://pro.morningconsult.com/instant-intel/ufos-extraterrestrial-life-government-secrets-survey`

Preis, T., Moat, H.S., & Stanley, H.E. 2013, *Scientific Reports*, 3, 1684

Saler, B., Ziegler, C.A., & Moore, C.B. 2010, *UFO Crash at Roswell: The Genesis of a Modern Myth*. Washington, DC: Smithsonian Institution

Scheffer, M., van de Leemput, I., Weinans, I., & Bollen, J. 2021, *Proceedings of the National Academy of Sciences*, 118, e2107848118

Scheufele, D.A., & Krause, N.M. 2019, *Science*, 116, 7662

Schwieterman, E.W. et al. 2018, *Astrobiology*, 18, 663

U.S. Department of Defense 2020, Statement by the Department of Defense on the Release of Historical Navy Videos, `https://www.defense.gov/News/Releases/Release/Article/2165713/statement-by-the-department-of-defense-on-the-release-of-historical-navy-videos/`

von Daniken, E. 1968, *Chariot of the Gods*. New York: Penguin Books

Vosen, S., & Schmidt, T. 2011, *Journal of Forecasting*, 30, 565

YouGov 2021, Half of Britons Think Aliens Exist – And 7% Claim to Have Seen a UFO, `https://yougov.co.uk/politics/articles/36610-half-britons-think-aliens-exist-and-7-claim-have-s`

Younes, N., & Reips, U.-D. 2019, *PLoS ONE*, 14, e0213554

(Toward) Discovery of Life beyond Earth and its Impact
Proceedings IAU Symposium No. 387, 2026
H. Landt, M. Dominik & C. Oliver, eds.
doi:10.1017/S1743921324003211

The storytelling brain and connecting public audiences with SETI

Carol A. Oliver

Australian Centre for Astrobiology
School of Biological, Earth, and Environmental Sciences
University of New South Wales, Sydney, Australia
email: carol.oliver@unsw.edu.au

Abstract. The Search for Extraterrestrial Intelligence (SETI) is, at its core, a grand endeavour in communication. While we hope to detect signals from intelligent civilisations beyond our solar system, searching for cosmic company has profound implications for our species' curiosity, technological capabilities, and innate need to connect socially. It gives pause for thought that while we focus the search and post-detection, we undervalue what evolutionary psychologists assert, neuroscientists have demonstrated, science communicators know works in connecting science with society, and journalists employ: We are hardwired for storytelling by evolution, remaining our most effective form of communication. What if the most important message is not from the stars but what the search tells about wanting to know what is in the last chapter of one of the most profound stories ever told? Should the experiment succeed, the stories of survival of intelligence from a distant civilisation may be of most interest.

Keywords. SETI, astrobiology, science communication, storytelling

1. Introduction

For tens of thousands of years, without cave drawings or writing, storytelling around the campfire was the only way to pass on information in a memorable way—the turning of the seasons, when and how to hunt, what dangers to avoid, and the shared beliefs and values of a group. Humans have an innate ability to tell and understand stories, which evolved as a means of social cohesion and survival. We need to feel connected to others and learn from what others do and what happens to them.

According to evolutionary psychologists, we are hardwired for storytelling by evolution, selecting storytelling as an evolutionary adaption (Storr 2019). We can track this through stories handed down through time in Australian Aboriginal songlines that carry laws, stories, and directions across more than 80,000 years (Nunn & Reid 2016). The oldest story may be 100,000 years old. Astronomers wondered why only six bright stars could be seen in the Pleiades when stories across cultures over time talk about the seven sisters. Constellations move imperceptibly, but there were seven bright stars, not the six bright members we see today 100,000 years ago (Norris 2020).

As soon as we took our first steps in conveying information through writing, the first story – a poem - appeared. The Epic of Gilgamesh is a rip-roaring 4,000-year-old tale of King Gilgamesh's adventures. He was supposedly one-third man and two-thirds a god who slew monsters in the quest for immortality (History Channel 2018). The story is "...a myth-making biography of a legendary king, a love story, a comedy, a tragedy, a cracking adventure" or because of Biblical-like stories reminiscent of the Ark and the

Garden of Eden origin stories (Pryke 2017). The story has several moral themes, including what we are driven to do (figuratively) in the quest for love and long lives.

Storytelling is the most compelling science communication tool the SETI community can employ to connect with public audiences and prepare for the possibility of success. This paper explores how and why SETI should use storytelling to build trust and credibility through the three basic rules of good science communication: Know your audience, tell a good story, and know why you are telling that story (Cormick 2019).

The social brain

American neuroscientist Paul Zak pointed out that we can watch a man, and his young son walk in a park. That is all they do, and we quickly lose interest. However, then we learn that the worried look on the man's face is because his son, as happy as he is now, does not have long to live. He does not know how to play with his son, knowing the truth. Even reading these words, a neurochemical is flowing, prompting empathy (Zak 2013). As we are drawn into the story, dopamine begins to flow – the neurochemical that leaves us wanting more. It is like eating chocolate and experiencing the feeling of wanting more. If the storyteller can elicit empathy for the character, then oxytocin - the love chemical – hits the brain. Zak's work also demonstrated what science communicators are aware of – we can empathise with people.

We cannot do it with inanimate things like a telescope or the data it spits out because they do not have feelings. We can empathise with a person – the emotions of Jill Tarter in the real world or Jodie Foster in the imaginary one, when (in the latter) a promising signal appears. In the context of astrobiology and SETI, our imaginings of what it means to be human connect us to the search for others among the stars. A significant body of research links the emotion and empathy generated in storytelling to learning and memory.

Why is storytelling so compelling? Martinez-Conde & Macknik (2017) argue that we process information through emotion:

> *"...our mammalian brains continue to assess incoming data in an emotional context. Information is meaningful insofar as it evokes emotion".*

The example they offer is a story that begins with:

> *"The king died, and then the queen died",*

The statement does not emotionally engage in the same way as the following:

> *"The king died, and then the queen died of grief."*

Damaisio (1994) noted that the neural circuits that drive everyday thinking are embedded in our evolution and, as such,

> *"...emotion is integral to the process of reasoning."*

Our mind is a story processor, according to Haidt (2012). We are always in story mode, with our brains creating a world of our own in which to live. "Story is what the mind does" (Storr, 2019). Hollywood spells out the power of storytelling (Ettinger 2020). J.J. Abrams, co-creator of the television series "Lost", says of the series that it is the opening of mystery boxes:

> *"... the mystery is the catalyst for imagination"* (TED talk, 2008).

Mystery and challenge drive de Paulis's SETI story 'A Sign in Space' in which a message was bounced oN Mars to be received and decrypted on Earth. Time has passed, and the message remains un-decrypted (De Paulis et al. 2023). It is a compelling story in our quest to understand whether we share cosmic company.

(Dahlstrom 2014) asserts that there is a categorical difference between paradigmatic and narrative processing. Narratives can be read twice as well and twice as fast. Stories are more straightforward to comprehend, engaging, influential, and persuasive. Therefore, they remain a critical tool for the toolkit of science communicators. Scientists are at the forefront of researching our natural world. They have vivid experiences of the processes of science that most ordinary people will never experience, as do SETI scientists.

Advances in neuroscience are also peeling back the layers of why our brains react differently to stories than information presented in any other way. There are neurobiological reasons why our brains remember the content of the former but struggle with the latter. Science communication, science education, cognitive psychology and neuroscience contribute to the widely accepted concept that storytelling effectively communicates ideas and influences how we think and learn.

Zak's work underscores that the most memorable stories are about people – what happens to them, how they face adversity, and how they succeed. The media focuses on seeking 'the story' in the research and personalising it because, as a species, we are inherently interested in what people have done and how they did the research. The increasing focus on personalisation has changed science communication practices to focus on presenting research as a personalised story to make science stories more attractive to the media (Elliott 2012). Dahlstrom (2014) agrees, pointing out that most public audiences do not have a university education, so they get most of their science information from the mass media, which aims to resonate with them by using storytelling to cut through the noise. Stories are easier to understand than scientific communication, which aims to provide data and facts when attempting to engage with public audiences.

Storytelling enables a sense of community, belonging, and shared beliefs and values. Individuals have beliefs, values, attitudes and behaviours, and changing the latter requires changing beliefs or values. For example, in cancer testing communication, storytelling enables patients to overcome barriers to screening, such as a simple at-home test for bowel cancer (Green 2006). Transporting the reader, listener or viewer into narrative worlds can change beliefs and motivate actions, as a large body of research across many fields suggests.

It was a tack that the astronomer Dr Carl Sagan knew how to use, attracting hundreds of millions worldwide with his TV series Cosmos. He was just as adept with the media. As a journalist in the 1980s, Sagan took me across the cosmos in a one-on-one hour interview, opening the door to imagining a giant spaceship built in orbit by arch-enemies Russia and the US that would take people to Mars. While the international space station will not go beyond Earth's orbit, it is a multinational effort. Russians and Americans, among other nationalities, work happily together in orbit despite wars and political division on Earth, writing an alternative story of cooperation in the dangerous environment of space, but seldom, if ever, told (Oliver 1988).

The neuroscience of storytelling

The emotion generated in storytelling is related to learning and memory (Oliver 2023). The hippocampus and amygdala work together to accept and remember stories (Phelps 2004). While the amygdala and hippocampus are independent memory systems, they act in concert when emotion meets memory, particularly in storytelling, and there is "abundant evidence that memories for emotional events have persistence and vividness the other memories seem to lack" (ibid, p14). A joke is an example of how emotion plays a role in memory. It generates attention (people pay attention to the joke), emotional response (they laugh) and rehearsal (if the joke is retold, it must be exact to prompt the same reaction) (Negrete 2021).

Researcher Uri Hasson et al. (2012); Hasson (2021) says that storytelling enables us to transfer ideas, thoughts or emotions into the brains of others, and he has tested this in his laboratory at Princeton University. He uses an MRI machine to track subjects' brainwaves as they watch films. He calls his work *neurocinematics.* Neural coupling occurs between a speaker and listener - in a,

> *"...shared neural substrate that exhibits a temporally aligned response pattern across communicators."*

Earlier studies have seen similar responses in storytelling, such as when watching a movie or listening to a story. Stephens et al. (2010) reported that neural mechanisms play a role in communicating with audiences.

Storytelling engages us more than on an emotional level. We can mimic the actions of a character in our minds. For example, the mention of the crunch of icy snow beneath the character's feet is mirrored in the brain in the same areas that would be activated as if walking on frozen snow. Rizzolatti and Craighero called this neurophysiological reaction the mirror neuron mechanism:

> *"...it appears to play a fundamental role in both action understanding and imitation"* (2004, p169).

It raises the question of how SETI stories could activate these mirror neurons.

Imagining aliens

Imagine we are one tribe of many among the stars. They have self-awareness, can change their environment, are curious, can love and enjoy, and can cry and feel the pain of loss. However different their morphology will be, intelligence will likely have the common features that drive our social nature.

Jill Tarter was asked by journalists covering Project Phoenix at the Parkes Radio Telescope about how she imagined aliens, and she responded that she did not care if they were little green men or big blue women, reflecting that scientists naturally resist speculating on what they do not know (Oliver 1994). However, it is what we humans want to know, and science fiction authors and scriptwriters tell endless stories about aliens. It is not speculation, though, to imagine biological beings or whatever morphology or colour to be influenced by the physical components of various potential biospheres.

There are things we can know about the communication civilisations we seek. For example, we use radio and optical telescopes to search for communicating civilisations – and, as Frank Drake pointed out on several occasions, they would be living on the land, curious about the heavens, and can build radio and optical telescopes. We are searching for civilisations like our own, with social cohesion and survival instincts. We might also know other physical constraints – the gravity required to build radio telescopes might be quite wide from a Mars-like one-third of Earth's gravity to at least one full Earth gravity. How far beyond that? Mars-like beings would be thin and not require a bone structure to withstand one full Earth gravity. We may be able to put constraints on the potential biospheres of communicating civilisations that have radio telescopes, making fertile ground for the imagination of public audiences. It may be intriguing for audiences to consider what aliens may look like from a different evolution but a converging path to building radio telescopes and building the challenges of searching for such civilisations into that story.

Astrobiology and science fiction

Science fiction is pure storytelling, but it goes hand in hand with the interface between science and science fiction. For example. Emma Puranen's PhD study of the influence of the discovery of unusual and unimagined extrasolar worlds impacts the portrayal of other worlds in science fiction (2024). This influence flows in both directions in terms of impact on technology. Television and movie series like Star Trek may have influenced designs to some degree of early flip-lid mobile phones (remember the flip-lid communicators in the earliest Star Trek series?) and iPads (the often-quoted smart pad used by Captain Picard in The Next Generation in 2012). Science fiction, in part, imagines alien worlds and beings.

We see it in real stories, too, from Carl Sagan's Cosmos series to humans landing on the surface of the Moon and the Mars rock ALH84001. These set the world on fire, engaging hundreds of millions and even billions.

How to write a SETI story

The Greeks knew the power of storytelling, and so did philosophers like Socrates, Plato, and Aristotle. Walk in Aristotle's leather sandals in 335 BC on the hot paving stones of Athens to a place in the shade where he sits with his students, expounding on the influential nature of orations. This is the home of Greek tragedy and comedy, and orators aimed to create change by prompting their listeners' emotions. Aristotle said orations have three key ingredients - ethos (the credibility to speak on a subject), logos (the facts, the data), and pathos (eliciting an emotional response). These elements are key to telling a good story: The media seek credible people to speak about SETI – so building that pool of credible people is crucial. They must be eloquent and willing to prompt an emotional response by making the communication personal– that there is a character, a challenge, and a resolution, and enables the reader, listener, or viewer to identify with and grasp the data in the story.

German author and playwright Gustav Freytag first identified 150 years ago that all real or imagined stories have a dramatic story arc. The story arc has an introduction (exposition), a rising action, a climax, a falling action and a resolution or catastrophe (denouement). There is a plot (what, when, where, why, to whom and how it is happening), a conflict, the transformation through conflict, and a resolution. The story stirs our emotions - happiness, sadness, love, hate, wonder, and rage, among other feelings. Memories, experiences, characters, events and a sense of place create a connection so the listener *feels* something. Stories can be real, imagined, or mixed – think of what is known about an Earthlike exoplanet, applying the data to the probable physical constraints of its imagined inhabitants – what they look like, the world they inhabit, and the stories they may be able to tell.

The ABT (And, But, Therefore) formula is a shorthand for the dramatic story arc. Scientist turned movie maker Randy Olsen identified the ABT (And, But, Therefore) model and describes it in his book *"Houston, We Have a Narrative"* (2015). Olsen's book is unique because it focuses on the storytelling elements concerning science communication.

The ABT book followed *"Don't Be Such a Scientist"* (2010); Olsen identified many scientists' problems in connecting with public audiences and then offered a possible solution with the ABT approach. As a scientist, he was used to presenting information as *And, And, And* or *Despite, However, Yet,* which makes information boring, just like watching people walking in Zak's imaginary park, doing nothing but walking. Olsen connected the dots by talking with the well-known cartoon series *South Park producers. Before airing, producers* check all *"South Park"* episodes for the ABT elements. ABT is a story formula scaffold with the words serving as a prompt to ensure the elements are present and are then taken away - much like the scaffolding used to build a new house.

Seminal science papers have been written using the ABT formula, such as Watson & Crick's 1953 Nature paper proposing the double helix structure of DNA. Researchers have put forward a three-stranded model of the structure of DNA. However, there are several reasons why that model is unsatisfactory. *Therefore,* we propose a radically different model of two helical chains coiled around the same axis (Watson & Crick 1953). It has fundamentally changed our understanding of biology.

What makes a SETI story media-worthy?

The media know what the audiences they serve want to read, hear, or view. They are in business to make money, so paying attention to what the audience wants matters. It also matters when connecting with audiences via the web and social media. What are the elements that make a 'good story'?

- It is new
- It is out of the ordinary.
- It is a discovery
- It is controversial
- It is intriguing
- It is novel in some way
- It is of high public interest
- A mystery in some way
- It has proximity to the audience
- It is relevant to the audience

Here is a (real) SETI story: Frank Drake visited the Parkes Radio Telescope in Australia as part of a SETI conference. Of course, Frank was the "Father of SETI" visiting Australia. I asked Frank to say something relevant to Australia, which he graciously did. He first spoke with the Sydney Morning Herald (which is syndicated in all Australian cities), pointing out that if ETI were ever discovered, it would most likely be from Australia because Australia – unlike the Northern Hemisphere nations – can see the galaxy's centre where there is more 'real estate'. It was an audience-relevant 'hook' personalised by the Father of SETI saying it. The audience could identify with the story. The interview requests continued for a week across Australia in print, radio and television (pre-internet in 1998). There could have been a few Australians who had not heard Frank's views (Oliver 1998).

Building trust and credibility

Leading practice science communication in building public trust in SETI and therefore confer the credibility to be the best source of information in the event of post-detection.

The intimately linked triad of understanding uncertainty in science, public trust in science, and the credibility of scientists as sources of information is the basis for risk communication strategies. In his book The Science of Communication Science, Cormick (2019) concludes that communicating uncertainty in science is essential.

Deputy Director of the Centre for the Public Awareness of Science at the Australian National University in Canberra, Rod Lamberts 2017 article in *The Conversation* that at least seven ways of communicating science-related risks could fail – and some apply to SETI, including problems among public audiences in understanding the term 'statistically significant', and being confused by numbers, including percentages. According to Lamberts, public audiences are also influenced by how risk is communicated.

Public perceptions of scientific uncertainty are broad, covering trust, relevance, too little or too much information, and the degree of wanting to know (Friedman et al. 1999). Terms used in SETI may present the same language barrier between other science and the public. However, accurately communicating uncertainty can increase trust (Fischoff

& Davis 2014). Trust is a mental shortcut that public audiences use to assess the complexities of science, scientists, and scientific research institutions, but knowing that the message has been received is challenging.

As the playwright and Nobel Laureate George Bernard Shaw noted:

> *"The single biggest problem in communication is the illusion that it has taken place."*

In 1992, British science communication scholar Brian Wynne wrote a seminal paper on issues in the social relationship between the public and scientists. Although the article is over three decades old, it still challenges the persistent idea that public audiences absorb scientific information without other inconvenient factors like beliefs, values, and expert lay knowledge. The case study Wynne covers is long-gone history, but the elements of the interaction between the Cumbria hill sheep farmers in the UK and scientists could be, at least partly, echoed today.

Wynne's point is that surveys of public attitudes towards science and public scientific literacy do not reflect the actual situation in the social negotiation between science and the public when the actors meet. The livelihoods of the farmers are at stake.

> *"Whether the beliefs are measured quantitatively or qualitatively, and whether they are about electrons, genes or trustworthiness of social actors such as scientists, they are considered unambiguously real"* (Wynne 1992, p. 282).

Public trust in science is measured by something other than what the public knows. Trust and credibility are generated in the relationships between people, audiences and scientists, and audiences and science-related organisations. These relationships make sense if you think about your friends. You are likely to trust those with whom you have a closer personal relationship rather than a more distant friend and part of a larger group. You know those closest to you better than those more distant. As Wynne notes in his paper, there is a:

> *"...complex and multi-dimensional social basis of trust and credibility as a central factor in the reception of 'understanding'..."*

Public responses to scientific knowledge surveys should be seen through the lens of

> *"...social relationships, networks and identities.*

Unlike any other scientific endeavour, communicating uncertainty gets lost in our common natural desire to believe we are not special. Smith 2011 American Scientist article that the public wants to believe aliens exist, and scientists and the press fuel it by adopting "these sensationalist attitudes". While that is just one view, most of the public across 24 nations believes we are not alone in the universe (Geocities 2017; Ipsos 2020). Public figures like Neil deGrasse Tyson and Brian Cox say the sheer vastness of the universe, the number of stars and the likely even more significant number of planets means we will likely have cosmic company. As Sagan rephrased Laplace:

> *"Extraordinary claims require extraordinary evidence."*

While the universe could be teeming with life, we do not know that it is. Numbers alone do not make life. We can engage public audiences with that reality by telling stories about the relatively narrow band of possibilities when the search is restricted to our anthropocentric search for communicating civilisations elsewhere. It serves as a risk communication strategy in preparing for silence or any discovery SETI may make.

Testing the effectiveness of storytelling

Sharing SETI's stories with public audiences undoubtedly impacts public audiences, but we don't know because it is rarely measured.

In 2001, Rick Borchelt led a study for the NASA Marshall Spaceflight Centre on its education and outreach activities. The report was damming. It found that despite half a century of education and public outreach, there needed to be more evidence of its impact. Scientific endeavours – of which SETI is one – are evidence-based. Borchelt pointed out that the same standard of proof should be required for science communication in connecting with public audiences. Evidence is much more than counting the number of attendees or the number of schools or students engaged in a project. These do not measure impact, only that an event was attended. How that impacted attendees was never measured. The R2 Report concluded that since science is a data-driven enterprise, it is surprising that science does not demand evidence of the effectiveness of its science communication (Borchelt 2001). Sless & Shrensky 2001, and subsequently widely quoted in the literature because it remains true today:

> *"The evidence for the effectiveness of (science) communication is about as strong as the evidence of linking rainmaking ceremonies to the occurrence of rain."*

How can the effectiveness of SETI communication be measured? A simple approach is to conduct short entry and exit surveys with Likert answers (strongly agree, agree, neutral, disagree, strongly disagree) to be analysed quantitatively and at least one open answer box, which can be analysed using standard qualitative software like Nvivo. For media interactions, a measure might be the number of other media requests that result. In social media, it is easy to like or dislike, but it takes time to make comments and impressions of those comments, making it a better measure. One of my past PhD students, Isabelle Kingsley, found that qualitative analysis of survey data enables a more significant understanding of the effectiveness of outreach than quantitative analysis alone Kingsley et al. 2019.

A communication plan has an overall goal and a series of goals contributing to that overarching direction, underpinned by the target audience or audiences (knowing your audience is part of Cormick's earlier mentioned three rules of good science communication). Each goal has one or more SMART objectives (Specific, Measurable, Achievable, Relevant, and Timebound). Each objective has one or more strategies to achieve the objective, and each strategy has a tactic (what will occur to meet the strategy that supports the objective and the goal). For example, a goal might be to build awareness of the SETI group, an objective to achieve that goal might be an event (the success of which depends on SMART), such as the launch of a website intended for public audiences), a strategy might be to build awareness of the new site through social media. The tactic is building an audience on the most appropriate social media platform.

What examples of science communication planning are there? One is a group of researchers demonstrating the benefit of planning at a research station above the Arctic Circle. Their overall goal was to make public audiences aware of the impact of climate change above the Arctic Circle. They built a substantial audience in just three years, telling stories about the processes of the work in their journey to discover new information about the Arctic region (Pavlov et al. 2018), which they could demonstrate was more effective than simply creating press releases. The audience identified with the researcher and their journey.

Even well-established SETI websites should consider their goals and whether they meet them, or the organisation is unsure if the website is still meeting its goals. Another example is Earthscope, which used social media to build an audience for its website (Bohon et al. 2013).

Conclusion

Storytelling is the most effective form of science communication, and coupled with goal and objective setting, any SETI group can establish how and why it will engage with public audiences.

The most immediate impact of SETI is in the present: The reality is that without any evidence (yet) of life beyond Earth, we could be alone. In a world on fire with wars and extreme political and cultural division, we are still a social species unable to bear loneliness individually, and perhaps in the face of the currently silent cosmos, as a species alone in the vastness of space. SETI is about us and our place in the universe. As University of Cambridge evolutionary psychologist Garcia-Pelegrin et al. (2021) says:

> *"Storytelling is a primary driver in our need to understand our place in the universe and where we are travelling next"* (p2).

Storytelling is the access to public audiences, whether SETI succeeds or not. As science fiction author Arthur C. Clarke noted:

> *"Two possibilities exist: either we are alone in the universe, or we are not. Both are equally terrifying".*

Communication is especially relevant in preparing for a post-detection scenario, but most need help understanding the foundational concept that communication is about people, not events or discoveries. People manage events, and they make discoveries. The most effective form of communication is storytelling and building an evidence-based communications plan that tracks that effectiveness. These are two fundamental tools in the toolkit of science communicators that the SETI community should consider in its interactions with public audiences.

References

Abrams, J.J. (2008). *The Mystery Box.* TED talk, `https://www.ted.com/talks/j_j_abrams_the_mystery_box?subtitle=en`

Bohon, W., Robinson, S., Arrowsmith, R. & Semken, S. (2013). Building an effective social media strategy for science programs. *EOS,* American Geophysical Union, 94(27) pp 237–224.

Borchelt, R. (2001). Communicating the future: Report of the research roadmap panel for the public communication of science and technology in the twenty-first century. *Science Communication* 23(2) 194–211.

Cormick, C. (2019). *The Science of Communicating Science,* CSIRO Publishing.

Dahlstrom, M.F. (2014). Using narratives and storytelling to communicate science with non-expert audiences. *PNAS,* 111 (suppl. 4). September 16. `www.pnas.org/cgi/doi/10.1073/pnas.1320645111`.

Damaiso, A.R. (1994). Descartes' error and the future of human life. *Scientific American* (271), 144.

De Paulis, D., Ferri, P., Shostak, S., Estevez, D., Montebugnoli, S., Bianchi, G., Melis, A. & Scuka, D. (2023). A sign in space: An interdisciplinary exploration of the potential reception of an extraterrestrial signal. *Acta Astronautica,* 212, 436–440.

Elliott, R. (2012). *The Medialization of Regenerative Medicine: Frames and Metaphors in UK News Stories.* In: Rödder, S., Franzen, M., Weingart, P. (eds) The Sciences' Media Connection – Public Communication and its Repercussions. Sociology of the Sciences Yearbook, vol 28. Springer, Dordrecht. `https://doi.org/10.1007/978-94-007-2085-5_5`.

Ettinger, J. (2020). What Hollywood can teach researchers about scientific storytelling. *Nature,* https://www.nature.com/articles/d41586-020-01731-9

Fischoff, B. & Davis, A. (2014). Communicating scientific uncertainty. *PNAS,* September 16, 111 (supplement 4) 13664–13671. Available online https://www.ncbi.nlm.nih.gov/pmc/articles/PMC4183175/

Friedman, S. M., Dunwoody, S., & Rogers, C. L. (Eds.). (1999). *Communicating uncertainty: Media coverage of new and controversial science.* Lawrence Erlbaum Associates Publishers.12

Garcia-Pelegrin, E., Wilkins, C. & Clayton, N.S. (2021). The ape that lived to tell the tale: The evolution of the art of storytelling and its relationship to mental time travel and the Theory of Mind. Frontiers in Psychology, 12, article 755783. 10.3389/fpsyg.2021.755783

Geocities. (2017). https://glocalities.com/reports/majority-of-humanity-say-we-arenot-alone-in-the-universe

Green, M. C. (2006). Narratives and cancer communication. *Journal of Communication,* (56)s1, August, pp s163–183. https://doi.org/10.1111/j.1460-2466.2006.00288.x

Haidt, J. (2012). *The Righteous Mind.* Allen Lane, p281.

Hasson, U., Ghazanfar, A.A., Galantucci, B., Garrod, S. & Keysers, C. (2012). Brain-to-brain coupling: A mechanism for creating and sharing a social world, *Trends in Cognitive Science,* 16(2), 114-121.

Hasson, U. (2019). Storytelling and memories: How the act of storytelling shapes our minds. *iBiology,* https://www.youtube.com/watch?v=CTsStZqxPwY (accessed 21 January 2021).

History Channel. (2018). *What is the oldest piece of literature?* (22 August), https://www.history.com/news/what-is-the-oldest-known-piece-of-literature (accessed 18 January 2021).

Ipsos. (2020). Majority of Americans believe there is intelligent life and civilsations on other planets. https://www.ipsos.com/en-us/news-polls/majority-believe-intelligentlife-exists-on-other-planets

Kingsley, I., Oliver, C.A. & Slavich, E. (2019). Hidden in the figures: What students are telling us about the effectiveness of astrobiology outreach. *Astrobiology,* 19(9): 1103-1116.

Lamberts, R. (2017). Why don't people get it? Seven ways the communicating risk can fail. The Conversation, January, https://theconversation.com/why-dont-people-get-itseven-ways-that-communicating-risk-can-fail-68912

Negrete, A. (2021). Remembering rhythm and rhyme: Memorability of narratives for science communication. *Geoscience Communication,* 4, 1–9.

Norris, R. (2020, December 22). The world's oldest story? Astronomers say global myths about the 'seven sisters' stars may reach back 100,000 years. Phys.Org. Accessed January 19, 2021, from https://phys.org/news/2020-12-world-oldest-storyastronomers-global.html.

Nunn, P.D. & Reid, N.J. (2016). Aboriginal memories of the inundation of the Australian coast date back more than 7,000 years. *Australian Geographer* 47(1), 11–47.

Martinez-Conde, S. & Macknik, S. L. (2017, August). Finding the plot in science storytelling in hopes of enhancing science communication. *PNAS,* 114(31), 8127–8129. https://www.pnas.org/content/114/31/8127

Oliver, C.A. (1988). Personal communication – one-on-one interview with Dr Carl Sagan.

Oliver, C.A. (1994). Personal communication with Jill Tarter and Parkes Radio Telescope was during the visit to Project Phoenix.

Oliver, C.A. (1998). Personal communication, Frank Drake, University of Western Sydney and Parkes Radio Telescope.

Oliver, C.A. (2023). The Social Brain and the Neuroscience of Storytelling. In: Rowland, S., Kuchel, L. (eds) *Teaching Science Students to Communicate: A Practical Guide.* Springer, Cham. https://doi.org/10.1007/978-3-030-91628-2_4

Olsen, R. (2015). Houston, we have a narrative. University of Chicago Press. ISBN-10:022627084X.

Pavlov, A.K., Meyer, A., Rosel, A., Cohen, L., King, J., Itkin, J., Negrel, S., Gerland, S., Hudson, S., Dodd, P.A., de Steur, S., Mathisen, S., Cobbing, N. & Granskog, M.A. (2018). Does your lab use social media? *Bulletin of the American Meteorological Society,* 99(6), 1135–1146.

Phelps, E. A. (2004). Human emotion and memory: Interactions of the amygdala and hippocampal complex. *Current Opinion in Neurobiology*, 14(2), 198–202.

Phillips, D. (2017). The magical science of storytelling. TEDx Stockholm. Accessed May 2, 2021, from `https://www.youtube.com/watch?v.Nj-hdQMa3uA`

Pryke, L. (2017). Guide to the Classics: The Epic of Gilgamesh. The Conversation. Accessed January 19, 2021, from `https://theconversation.com/guide-to-the-classicsthe-epic-ofgilgamesh-73444`

Puranen, E. (2024). Exoplanets in Science Fiction. PhD thesis, University of St Andrews. `https://research-repository.st-andrews.ac.uk/bitstream/handle/10023/30349/Thesis-Emma-Puranen-complete-version.pdf?sequence=8`

Rizolatti, G., & Craighero, L. (2004). The mirror-neuron system. *Annual Review of Neuroscience*, 27, 169–192.

Sless, D. & Shrensky, R. (2001). Conversations in and landscape of science and magic: Thinking about science communication. Chapter 6, *Science Communication in Theory and Practise.* Eds Stocklmayer, S., Gore, M.M. and Bryant, C.R., Kluwer Academic Publishers.

Smith, H.A. (2011). Alone in the Universe: Despite the growing catalog of extrasolar planets, data so far do not alter the estimates that we are effectively on our own. American Scientist, 99(4) pp 320–327.

Stephens, G.C., Silbert, L.J. & Hasson, U. (2010). Speaker-listener neural coupling underlies successful communication, PNAS, `https://www.pnas.org/content/pnas/107/32/14425.full.pdf`.

Storr, W. (2019). *The Science of Storytelling.* 4th Estate.

Watson, J.D. & Crick, F.H.C. (1953). Molecular structure of nucleic acids: A structure for Deoxyribose Nucleic Acid. *Nature*, 171, 737–738.

Wynne, B. (1992). Misunderstood misunderstanding: social identities and public uptake of science. Public Understanding of Science, 1(3), 281–304. `https://doi.org/10.1088/0963-6625/1/3/004`

Zak, P. (2013). The future of storytelling. Accessed January 20, 2021, from `https://www.youtube.com/watch?v.DHeqQAKHh3M&feature.emb_logo`

(Toward) Discovery of Life beyond Earth and its Impact
Proceedings IAU Symposium No. 387, 2026
H. Landt, M. Dominik & C. Oliver, eds.
doi:10.1017/S1743921324001820

Teaching by imagining *Life Beyond Us*: Science fiction for outreach and education

Julie Nekola Nováková[1,2,3]

[1]Institute of Physics, Czech Academy of Sciences; novakovajn@fzu.cz

[2]Department of Philosophy and History of Science, Faculty of Science, Charles University

[3]European Astrobiology Institute
email: novakoju@natur.cuni.cz

Abstract. Astrobiology is generally considered to be a promising gateway into science, and storytelling techniques have been getting increasingly more traction lately in education and outreach because of their potential benefits for learning. At the European Astrobiology Institute, we combined these two pathways by creating two anthologies of astrobiology-inspired science fiction stories accompanied by science fact essays, *Strangest of All* and *Life Beyond Us*. Furthermore, we created educational materials that build on the anthology project and acquaint students with the issue of planetary protection. Here we introduce the project and planned future directions in story-driven astrobiology outreach and education.

Keywords. astrobiology, narrative-based learning, science education, science fiction, science outreach

---

1. Introduction

Astrobiology has frequently been considered to be a good pathway to science literacy for a variety of audiences (Carrapico et al. 2002, Oliver & Fergusson 2007, Taber 2010, Kwok & Kolb Kwok 2018, Kingsley et al. 2019, Impey 2021), given its interdisciplinary nature and a lot of pre-existing public interest in the question of whether there is life beyond Earth.

The use of narratives, storytelling, in outreach and education has also gained a lot of traction lately but in fact, it has a long tradition: Herbert George Wells, the famous science fiction writer behind *The War of The Worlds, The Invisible Man* and other novels (and also a prolific science writer and a biology teacher), advocated for popular science books to adopt more storytelling techniques in his Nature column in 1894. More recently, storytelling has been shown to be a promising approach in education for varying target audiences, especially thanks to helping build more connection to the subject and integrate knowledge (e.g. Engel et al. 2018, Walan & Enochsson Walan 2019, DeMirci & Okur 2021, Kiselyov & Shunn Kiselyov 2022). However, it should also be noted that the term "storytelling" encompasses vastly different techniques that can be passive as well as active, and include narrative nonfiction, fiction, video- or board games, movies, etc., and each has its specifics in terms of involvement, short- and long-term efficacy and suitable audiences. For a more detailed overview of the forms and benefits of storytelling in outreach and education, we refer the reader to Oliver (this volume).

Here we describe a project that merges storytelling and astrobiology through the use of science fiction (SF) stories, science fact essays and related educational materials. Astrobiology often features heavily in SF, but there has been little systematic research into the relationship between SF and science in general, much less specifically astrobiology, and the ways SF can sway public perceptions of the field or even directions of scientific research and technological innovation, even though tentative claims and attempts at practical uses of the SF-science relationship, e.g. in 'design fiction' but also educational contexts, exist (Chirlesan & Chirlesan 2014, Hasse 2015, Graham et al. 2015,Raven 2017, Moezzi et al. 2017, Michaud & Appio Michaud 2022, Lockard & Goggin 2022, Eichmeier et al. 2023).

At the European Astrobiology Institute with its project team 'Science Fiction as A Tool for Astrobiology Outreach and Education', we combined fiction and nonfiction narratives in publishing two anthologies of astrobiology-themed science fiction stories accompanied by science fact essays: e-book *Strangest of All* (ed. Nováková 2020), essays and classroom tips by the editor, and *Life Beyond Us* (ed. Nováková, Law and Forest, Nováková 2023) with 27 stories and the same number of essays, each written by a different expert.

Both books are intended for science outreach for general young-adult and adult audiences, but they can be used in higher-level education as well through using the stories and essays as starting points for deeper discussions, literature dives or projects. *Strangest of All* was created as a freely available e-book during one of the first Covid waves, but this would have been untenable for a print anthology so that while *Life Beyond Us* is available for purchase, we decided to create extra educational materials available for free download. The first set, focused on planetary protection, had been released on October 5, 2023, for World Teachers' Day, and we are working on more thematic sets to be released in late 2024.

2. Sample educational materials: Planetary protection

The insofar released educational materials (Nováková 2023, available at `europeanastrobiology.eu/life-beyond-us/`) have been inspired by the opening story and essay of Life Beyond Us, Eric Choi's "Hemlock on Mars" and Giovanni Poggiali's "Planetary Protection: Best Practices for the Safety of Humankind (and all those aliens out there)".

In the story, extremely resistant bacteria were found in a clean room where a life-seeking mission en-route to Mars had been assembled. The spacecraft is likely clean - but can the mission leaders afford to take the risk that it is contaminated? The spacecraft is supposed to revisit Labeled Release, one of the life-searching experiments flown to Mars on the Vikings in the 1970s that yielded controversial results. Microbial contamination could compromise the results of the mission, but more importantly might threaten indigenous Martian life, if there is any. With so many unknowns, how should experts decide whether to go on with the mission, even if the contamination risk is found to be low? Is there a truly "right" decision to make? The story depicts the reality of space mission operations, sterilization procedures and decision making down to tiny details. Giovanni's essay then introduces the issue of invasive (= alien) species on Earth, such as rabbits in Australia, and summarizes the question of planetary protection (contamination avoidance) since the advent of the Space Age until the present-day.

The high school-level section of the materials contains two prompts for a classroom discussion: whether the pupils consider the eventual decision of the mission manager right or wrong, and how do they view a crewed Mars mission from the point of view of planetary protection. In addition, it contains three activity suggestions:

1. finding out some invasive species on Earth, especially in the pupils' own region, and discussing their impacts;

2. trying to think of a system to monitor invasive species in the area, looking up existing initiatives e.g. in the form of citizen science, and potentially getting the class involved in them;

3. imagining potential extraterrestrial life forms and the best ways to search for them. The exercises include tips and suggestions for the educators, such as a list of some of the European citizen science projects.

The university-level discussion prompts include both high school-level prompts where the discussion can go into greater detail and existing scientific studies and space missions, plus an extra prompt on whether the students consider the life-seeking mission in the story a good one, and how they would go about searching for extant life on Mars. It also contains three activities: 1. reading the paper "Clean room microbiome complexity impacts planetary protection bioburden" (Hendricksen et al. 2021) and answering several related questions; 2. dividing students into small groups and assigning each group a life-detection method to list its pros and cons and environments where it may be most useful; 3. dividing students into two groups to let each one summarize arguments for or against more stringent planetary protection. These prompts and activities, too, include directions for the educators.

The materials have been designed to promote critical thinking, connecting information across fields and considering science and technology in synergy with ethics. However, so far, the materials have been introduced at a university astrobiology seminar, but still await formal evaluation.

3. Future directions

The work conducted on *Strangest of All* and *Life Beyond Us*, and the educational materials introduced above, is only the beginning. As noted above, we are planning to release further materials in fall 2024. We are also going to formally evaluate them in both high school-level and university-level environment to assess their impact.

In addition, we are preparing several conference workshop sessions on storytelling for space sciences outreach and education, the first of which is going to take place at the International Astronautical Congress (IAC) in Milan on October 1, 2024. Subject to funding, we would like to organize specialized cross-disciplinary meetings for scientists, artists, journalists and educators to facilitate closer collaboration and interdisciplinary projects (such as science art, science-inspired fiction, movies, podcasts or games, outreach storytelling projects, etc.). Finally, we are planning to conduct and help facilitate research on the history of science fiction-astrobiology relationship and its implications for education.

Based on the handful of existing studies and much anecdotal evidence, science fiction appears to be a good gateway to science, or at least curiosity and interest in science, and a resource for science education and outreach. Our own experience with using science fiction for outreach supports this claim. However, more careful evaluation (both quantitative and qualitative, since some effects may only be discerned in the latter approach e.g. Kingsley et al. 2019) and further studies are needed to assess the impacts of science fiction in various forms of media on education and outreach for different audiences and purposes. We hope that the work of the European Astrobiology Institute's project team 'Science Fiction as A Tool for Astrobiology Outreach and Education' will facilitate more such initiatives.

References

Carrapico, F. J., Loureno, A., Fernandes, L., & Rodrigues, T. (2002). Journey to the origins: the astrobiology paradigm in education. In Instruments, Methods, and Missions for Astrobiology IV (Vol. 4495, pp. 295–300). SPIE.

Chirlesan, G., & Chirlesan, D. (2014). Incorporating Science Fiction literature in teaching sciences. In INTED2014 Proceedings (pp. 2299–2308). IATED.

Demirci, T., & Okur, S. (2021). The Effect of Teaching Science through Storytelling on Students' Academic Achievement, Story Writing Skills and Opinions about Practice. Education Quarterly Reviews, 4(2), 562–578.

Eichmeier, A. A., Bao, L., Xenos, M. A., Brossard, D., & Scheufele, D. A. (2023). Fictional scenarios, real concerns: science fiction and perceptions of human genome editing. Journal of Science Communication, 22(1), A08.

Engel, A., Lucido, K., & Cook, K. (2018). Rethinking narrative: Leveraging storytelling for science learning. Childhood Education, 94(6), 4–12.

Graham, G., Greenhill, A., Dymski, G., Coles, E., & Hennelly, P. (2015). The Science Fiction Prototyping Framework: Building Behavioural, Social and Economic Impact, and Community Resilience. Future City & Community Resilience Network.

Hasse, C. (2015). The material co-construction of hard science fiction and physics. Cultural Studies of Science Education, 10, 921–940.

Hendrickson, R., Urbaniak, C., Minich, J. J., Aronson, H. S., Martino, C., Stepanauskas, R., ... & Venkateswaran, K. (2021). Clean room microbiome complexity impacts planetary protection bioburden. Microbiome, 9, 1–17.

Impey, C. (2021). Astrobiology education: Inspiring diverse audiences with the search for life in the Universe. Astrobiology: Science, Ethics, and Public Policy, 135–156.

Kingsley, I., Oliver, C., & Slavich, E. (2019). Hidden in the figures: What students are telling us about the effectiveness of astrobiology outreach. Astrobiology, 19(9), 1103–1116.

Kiselyov, K., & Schunn, C. D. (2022). Storytelling as a tool to enhance conceptual knowledge in cell biology. Journal of Microbiology & Biology Education, 23(2), e00308–21.

Kwok, S., & Kolb, V. M. (2018). Astrobiology as a medium of science education. Handbook of Astrobiology. Boca Raton, FL: CRC Press Taylor & Francis Group, 45–48.

Lockard, J., & Goggin, P. (2023). Teaching Mars Literature. Science & Education, 32(3), 821–844.

Michaud, T., & Appio, F. P. (2022). Envisioning innovation opportunities through science fiction. Journal of Product Innovation Management, 39(2), 121-131.

Moezzi, M., Janda, K. B., & Rotmann, S. (2017). Using stories, narratives, and storytelling in energy and climate change research. Energy Research & Social Science, 31, 1–10.

Nováková, J. (ed.) (2020) Strangest of All. European Astrobiology Institute. Available at `https://europeanastrobiology.eu/strangest-of-all/`

Nováková, J., Law, L., & Forest, S. (eds.) (2023). Life Beyond Us. European Astrobiology Institute & Laksa Media. ISBN 1988140471.

Nováková, J. (2023) Life Beyond Us: Resources for Educators. Available at https://europeanastrobiology.eu/life-beyond-us/

Oliver, C. A., & Fergusson, J. (2007). Astrobiology: A pathway to adult science literacy? Acta Astronautica, 61(7-8), 716–723.

Raven, P. G. (2017). Telling tomorrows: Science fiction as an energy futures research tool. Energy research & social science, 31, 164–169.

Taber, K. S. (2010). Learning about astrobiology: a challenge for the public understanding of science. In S. Hegeds & J. Csonka (Eds.), Astrobiology: Physical Origin, Biological Evolution and Spatial Distribution (pp. 1–25). New York: Nova.

Walan, S., & Enochsson, A. B. (2019). The potential of using a combination of storytelling and drama, when teaching young children science. European Early Childhood Education Research Journal, 27(6), 821–836.

Wells, H. (1894) Popularising Science. Nature 50, 300-301.

(Toward) Discovery of Life beyond Earth and its Impact
Proceedings IAU Symposium No. 387, 2026
H. Landt, M. Dominik & C. Oliver, eds.
doi:10.1017/S1743921324002680

Discussion: The future is our future

Hermine Landt[1], Eleni Chatzichristou[2], James R. A. Davenport[3], Michael G. Edmunds[4] and David H. Grinspoon[5]

[1]Department of Physics, Durham University, UK
email: hermine.landt@durham.ac.uk

[2]European Research Council, Belgium

[3]DiRAC Institute, Department of Astronomy, University of Washington, USA

[4]School of Physics and Astronomy, Cardiff University, UK

[5]NASA Headquarters, Washington DC, USA

Abstract. The last day of the Kavli-IAU symposium featured a fourth discussion, which was about the future. The discussion focused on the changing landscape of scientific research and the allocation of funding and their implications for the ideal scientist of tomorrow. It was chaired by H. Landt, who was joined by the four panellists E. Chatzichristou, J. R. A. Davenport, M. G. Edmunds and D. H. Grinspoon.

Keywords. research funding, science communication, multi-disciplinarity, data science, astrobiology

1. Introduction

Maybe not all of you but hopefully some of you are aware that things are changing in the way we do science, and also in the way funding, not just in the sciences, is distributed and allocated in order to encourage certain activities. The way we do science is changing because we have many expensive facilities producing large data sets that cannot be satisfactorily exploited by just a few people. The direct consequence is that large consortia need to be established, first in order to get the funding for these facilities and then to exploit them. By necessity, these consortia are both international and multi-disciplinary. Since they also work like an 'organism', a most positive outcome of this development is that it leads to a continued 'democratisation of science'. A similar development can also be observed in the work of international collaborations that create global networks of smaller facilities and in the movement towards Open Science. Open Science, which aims at providing researchers with an unprecedented amount of publicly available data, has the potential of making the scientific process more transparent, inclusive and democratic. The UNESCO Recommendation on Open Science provide an internationally agreed definition as well as a set of shared values and guiding principles for it.

At the same time the landscape of academia is changing. Because governmental funding agencies are encouraging more and more interactions with the private sector, a further emphasis is placed on 'multi-disciplinarity'. The motivation is that academia should increasingly participate in the economic cycle in order to lead to economic growth for the benefit of society, thus justifying the investment of tax-payers money made in them and their often large projects. Therefore, the CV of the ideal scientist is changing. Experts are now sought after who have crossed disciplines during their careers and have gained experience in different sectors such as, e.g., industry, academia and policy-making. It is believed that encouraging such cross-pollination of expertise between different kinds of working environment will be beneficial to all in the long-term.

The way funding is allocated is also changing. An increased emphasis is put on public engagement and science communication as a measure of the impact of scientific research. However, at the same time, we see more and more philanthropic funding being distributed by more and more individuals. We have an increased number of wealthy individuals in society who would like to make their money available to science. The scientific research funded in this way is usually high-risk/high-reward and is always non-democratic because its allocation is determined solely by the funder.

What do these developments mean for the future of science and for the future of the scientist? Dr Eleni Chatzichristou (Research Policy Analyst at the European Research Council (ERC)), Prof James Davenport (Associate Director of the DiRAC Institute at the University of Washington), Prof Michael Edmunds (Emeritus Professor of Astrophysics at Cardiff University and President of the Royal Astronomical Society (RAS)) and Dr David Grinspoon (Senior Scientist for the Astrobiology Strategy at NASA) were invited to debate this question.

2. The panel discussion

Hermine Landt (HL): I'd like to start with the non-funders. So, David, you and Eleni are the funders on this panel, and Mike and Jim are the scientists and learned society representatives. So we start with you, Jim. You work at the DiRAC Institute, which is an inter-disciplinary data-intensive astronomy research institute. At the same time, you are heavily involved in the LSST, the Legacy Survey for Space and Time, which will produce 500 petabytes of data. For those who cannot grasp this large number, it is equivalent to 500 million gigabytes. And the SKA will then make us enter into the exabyte era. We will only be able to extract useful information with AI methods. I would like to pose a provocative question: Does the future belong to the machine? But please talk us also through the workings of the DiRAC.

James Davenport (JD): Thank you. So, DiRAC stands for Data-intensive Research in Astrophysics and Cosmology. We are a hybrid institute in terms of our funding strategy. We do have private philanthropic funding sources, and then, of course, like all others, we also apply for public funding through NASA and the National Science Foundation (NSF). As such I actually do believe that I'm very keen to talk to the funders and talk to you all about the methods of funding and how they influence the work we do. One of the advantages we've seen with a private funding sometimes leading the way is that it allows for a lot of possibilities, whereas the NSF or the massive funding sources can be critical to doing big projects. They are relatively slow. And even though I think we've seen a lot of advances in taking high-risk/high-reward projects forward or looking for exotic objects, they can be very difficult to pursue out of fears of time and where they might lead. Whereas, having the control of an endowment or private funding resource can give you a level of flexibility that is really key.

Your question was about AI, and "Is the machine in control of our future?". I think, it is definitely true that such as in our inter-disciplinary institute we see a huge value in our students learning to become better programmers, partnering with data scientists and computer programmers and such. We see a lot of return on behind the questions they ask, the data they can interrogate and the amount of work we're able to do. We're still a relatively small institute, with only about 40 people in total, and the majority of that is students, of course. What I haven't seen any evidence of is that machines ask very good questions. And the thing that I'm most drawn to in this area is that Machine Learning (ML) is very boring, and AI is extremely stupid. It's extremely good at answering the question you ask it, or interpolating the data that you give it, or clustering things into clumps. But it doesn't ask good questions. This may change, I don't know, I am not a computer scientist. I love using ChatGPT to sketch our ideas, to mock up code and to even write my topics. It's brilliant, but it doesn't ask good questions. And so I'm extremely heartened by the idea that we can shoot through petabytes and terabytes of data per night in real-time data streams from the Rubin observatory in the search for a million interesting objects per night that we don't have to sit through. There's no way that the capacity of all of us in this room or all of the astronomers on this planet can get through that data. So it's an exciting frontier. But, what will not change, and what we have to speak for us humans, is that we can ask good and better new questions of these data with this new hammer we have in our toolbox.

HL: Thank you very much. I would like to stay on this subject of 'multi-disciplinarity' and go next to Mike. Science was historically developed by polymaths. The current drive to multi-disciplinarity seems to bring researchers, as a group, back to the kind of individuals they once were. What do you think about this concept, from the point of the learned society and from the science history that you know? Are we moving backwards to the future - in a positive way?

Michael Edmunds (ME): I don't know. I'll just make an observation. It's obviously in the history, certainly of theoretical physics, that you do get particular individuals who seem to be particularly good at pushing boundaries forward. That's established. And I assume within science that will always be true. There'll be certain individuals, of a particular intelligence or whatever it is, maybe a bit of ego and in certain cases also luck, who make very fundamental discoveries. But the majority of research will be, I might say, at a lower level, perhaps also less inspired, but it's still very necessary and useful to reinforce those views, theories, and so on.

The drive that you're talking about, towards inter-disciplinary research, I think, is a different question and I'm not sure where that will go. You could argue that one way to get an inter-disciplinary scientist is to give them a physics degree, then give them a chemistry degree, then give them a biology degree, then give them a linguistics degree, probably a philosophy degree, by which time they're ready to retire. Obviously, it is a good thing to have some transition between different roles, as you say, interact through pollination, but I'm not sure that that's the best way to do inter-disciplinary research. I had experience with very large inter-disciplinary projects. And that worked, not because everybody became an expert in the different fields, i.e. everybody tried to become an expert in A, B, C and D, but because we had experts in A, experts in B, experts in C, experts in D, and for once we were prepared to talk to each other. And it seems to me that for inter-disciplinary studies the best thing is, if you can get people who know a bit about the science but are not experts and they'll get the expert advice from the expert, and others will rally around and do the communication. That's the way I think that you do inter-disciplinary science and I suspect that will continue. Because if you look at the big projects, just read the list of names, not all of those scientists can be inter-disciplinary. There are probably all sorts of people in there, experts and non-experts. But it's rather a matter of having different talents, with the most important talent actually being able to communicate to each other.

HL: That's that's very good point. So let's now go over to the funder representatives here on the panel. Eleni, 'Life Beyond Earth', as I personally see it, is a high-risk/high-reward scientific endeavor; high-risk in the sense that we do not know if there is life beyond Earth and high-reward in the sense that, if we do find it, then it will be a most profound discovery. The European Research Council (ERC) say that they are funding high-risk/high-reward science and thus differentiate themselves from the rest of the funding program within the Horizon Europe programme. I would like to know your take on this. Would 'Life Beyond Earth' fall directly into the ERC remit? So, should we rush and write Advanced Grants and Synergy Grants on this subject? We need your advice on this.

Eleni Chatzichristou (EC): Okay, thank you. First of all, when we talked about 'high-risk/high-gain' and you gave the interpretation of what this could be, that's exactly the problem that the ERC has eventually addressed. That is, they changed this formulation, because, indeed, there was a mis-interpretation. So, as of the Work Programme 2024, this does not exist anymore in the formulation of the 'high-risk/high-gain' and it has became actually something that is more easy to agree upon, which is 'excellence'. Then, it's 'excellence' as far as research projects go, and 'excellence' as far as the PIs go. So, instead of the words 'high-risk/high-gain', now we talk about 'groundbreaking research', about 'foundation research' and about 'physical research', which, again, needs to be explained in a sense. And then, as far as the PIs are concerned, the 'excellence' is about intellectual capacity, about creativity, and about commitment, as you already know, probably. The CV of the PIs is not really only judged upon, say, by publication record or research age. But it's actually a more holistic way of viewing the series of the applicants. We have this actually, in two levels, a generic level at the stage stage one and then a more qualitative, let's say, approach of this. In this respect as well as in the respect of inter-disciplinarity that you were talking about before, I think that the assumption is that you really know a field, actually a combination of fields, obviously within the remits where the PIs are funded.

Space Sciences by ERC is funded within two panels in the domain of Physical Sciences & Engineering (PE), namely, P9 (Universe Sciences) and P10 (Earth System Sciences). I did a very quick search on terms such as, for example, 'Life beyond Earth' and astrobiology. And then, when you combine these keywords, we end up with a pretty high number of proposals funded in this area since the beginning of the ERC's existence. I can't remember now, but I think we had more than 80 projects in the PE domain alone, and then we had another 10 or 12 in the Life Sciences domain, and then we even had some proposals with these keywords in the Social Sciences & Humanities domain.

HL: I would like to stay on the ERC. The ERC is a relatively new institution. It was incepted in 2007, so I would like to call it 'new'. But its budget has steadily increased and has been now doubled. Your website states that it currently stands at 2.3 billion Euro per year. But, as I just said, we are moving from the individual to the consortia in many of the sciences, and definitely in astronomy. Can you have a take on this or even give an explanation? What is the reason that the budget has increased significantly for individual research? Because the ERC, as you can expect, is not funding consortia, it is funding individuals who have a large team. Can you comment on where the movement is from here?

EC: So, basically there are two questions in one. Regarding the budget, you are right. So, we started with 7.5 billion, then we went up to 13 billion and now it's at 16 billion over the entire Horizon Europe period (2021 - 2027). But this always represented something like 16% or 17% of the overall funding programme. And, indeed, this amounts to more or less 2 billion a year. The point is that from the very beginning of the ERC until now, we have funded over 30,000 PIs with teams that included another 100,000 researchers and professionals and covered over 900 host institutions. If you know a little bit about the evaluation process, you know that we have two steps, except for the Synergy Grants, where you have three separate steps. The proposals that make it through the first step, i.e. they are scored A, let's say, then move on to the second stage. And there you actually proceed with a more detailed evaluation of the whole proposal, according to the criteria I was mentioning before. And then you end up funding some projects. The point is that, even with this quite huge number, I think, of standards for funding proposals, we leave outside many, many proposals scored with A, which means that they are excellent. This is simply because we do not have enough funds. This is the reason why we're requesting the doubling, not only of the ERC funding, but of that for the entire programme to exceed 200 billion. So, the ERC is actually keen to fund more research, more excellent projects, and that is the reason why we're looking for higher funding. Also, another problem is that the amount of grants remains more or less the same over the years.

The second part of your question was about individual research funding versus consortia. So, I do think that the ERC totally recognises the fact that nowadays more and more science is inter-disciplinary. That's why the funds are hardly ever disciplinary, i.e. they don't have hard disciplinary boundaries. But you also have the Synergy Grants. So, the Synergy Grants exactly address the issue of having more than one PI. Actually, you need to have 2 - 4 PIs, and 4 PIs is the maximum that can be funded. This, in a sense, addresses the issue of consortia. And, also within the so-called Individual Grants, when a project is intensively across panels, then the PI has the possibility to ask for different expertise to be funded from the different panels.

HL: Thank you. So, let's go over to David now. Effective communication of research results is encouraged, in particular if you want to write some funding proposals. Now, especially for astrobiology, communication is a vital component in its research endeavor. I would like to claim that we cannot have one without the other. We know that in early March NASA conducted under your auspices a workshop called "Communicating Discoveries in the Search for Life in the Universe", a three-day workshop, for particularly this reason, because more skills need to be developed within the community. I would like to invite you to give us a briefing on that. Thank you very much.

David Grinspoonn (DG): Sure. Thank you, first of all, for inviting me to participate. And it's a lovely meeting, and I really regret not being able to be there in person. Various events in Washington this week prevented me. I have been sort of ducking in and ducking out and participating when I can, and following the Slack chats. And it's obviously a vibrant and really interesting meeting. And I appreciate that you're doing that. And I appreciate you're including me like this.

So, we've had this longstanding goal at NASA, which very much dovetails to the goals that you've been describing and working on at your meeting, so there's a lot of overlap, for trying to pro-actively think about how we should communicate about discoveries in the area of extraterrestrial life and possible signs of extraterrestrial life, and how we characterize partial or incremental discoveries, and categorize them both within the scientific community and then for the public, and avoid some of the pitfalls that one can see coming. You know, to some extent one can't prevent over-sensationalizing or the public getting burnt out on thinking they've heard the same announcement over and over again. So, there was this past activity a couple of years ago that you've probably heard of, the "Biosignatures Standards of Evidence Workshop", a virtual worskhop that involved a lot of members of the astrobiology community. One of the upshots of it was to come up with a list of questions to ask and categorize a discovery, e.g. 'Have you detected an authentic signal?', 'Have you adequately identified the signal?' and so forth. I'm not going to go through all of them, since I think you're probably familiar with this. I think it was a productive exercise. There was some pushback and critique, which is good of the way this was done. In particular, we had a National Academy of Sciences review of this report and that report was mostly positive, but had some critique as well, which was much appreciated. One of the critiques was pointing out the necessity to really separate out how we think about communicating with the public and how we think about communicating within the scientific community, and another questioned the legitimacy or seriousness of the standards of evidence, or what is sometimes called the CoLD scale, that came out of that, i.e. this idea of a numerical scale to categorize how definitive or non-definitive is a potential discovery. Putting that on a scale, there was some pushback on that. And the idea that came out of this National Academy of Sciences report was that perhaps that scale is more appropriate for how we talk within science to each other about discoveries, but wasn't really adequate for thinking about how we talked to the press. And there was a recommendation in this report that we involve more journalists and more communicators in our deliberations about this, and really try to understand that sort of ecosystem of scientists working with communicators, so that we can anticipate and develop best practices and develop more understanding of how these communities work together. So, on the basis of that recommendation, we planned this recent workshop "Communicating Discoveries in the Search for Life in the Universe", which actually happened last month.

What was different and innovative, I think, about this workshop, which was a multi-day virtual workshop, was the way that we incorporated the participation of a lot of journalists and other professional people in the communication space with the participation from scientists. And I'm happy to note that there's some good overlap between the participants of our meeting and the participants of your Kavli-IAU symposium there, both in terms of some of the journalists and some of the scientists, since it's important that we treat this as a coherent effort across oceans and decades in order to do this well. We had roughly a hundred participants, with about half of them professional scientists in astrobiology and related fields, and the other half journalists and other people in the communication space. We set it up so that we had a sort of modified Chatham House rules so that you could not report on the sessions. If you were a journalist, you were participating in those sessions not as a journalist, you were participating as somebody who was there to learn and listen and teach mutually. And that led to a really extraordinary set of conversations. And I've gotten a lot of great feedback about this, because it was kind of unusual for the journalists to be there not as journalists, sort of by rule. By the way, they were allowed to interview people outside of the sessions and make connections and so forth, you could just not report on what was said in the actual sessions. And there's already been some reporting, e.g. Sarah Scoles just had a piece that came out a couple of days ago in Scientific American about this. So, it wasn't like there's got to be media silence about this meeting. It was just simply the sessions themselves, or you're not on the record. And that meant that people were really able to let their hair down. And I saw that working. And I think a lot of other people experienced that working a kind of magic where, even though a lot of these journalists had spoken to a lot of the scientists before, it had always been in this sort of more carefully guarded way of 'You're interviewing me. I'm being interviewed by you. I'm gonna watch what I say. We're gonna sort of follow these rules.' And in this setting that did not apply. There was a lot of discussion about the mutual needs and expectations of the two groups and I just saw, and other people reported, the value of that kind of getting those groups together with that kind of understanding that that was what was going on.

We split it up and had a session of a day of past case histories, reviewing famous relevant examples like the ALH 84001 Martian meteorite discovery and reporting and fallout, and the arsenic life episode, and we did one on relevant examples from the climate debates and communication and miscommunication, and how all that played out. Then we had a day of imagining future scenarios where we played out a few different kinds of discoveries; one had to do with bringing samples back from Mars, and that was fun, one had to do with an exoplanet discovery, and so forth. And we even did some role playing where the scientists got to play journalists and the journalists got to play scientists, and, you know, had some fun, and I just saw a lot of light bulbs going on and off and good connections being made. Then we had a day of starting to write a report, and the report writing is still under way, and so I can't give you a neat set of bullet points of what are the results. And I want to be careful to not even try, because, as the NASA ex-officio person on the organising committee, I am determined that this should be a community-based report. We facilitated these conversations, but I deliberately tried to kind of shut up as much as possible during the actual discussions, being from NASA, because this is not a NASA report, it's a community report. So, the writing is under way and we expect it to come out this summer. We're pretty excited about that, and we'll publish it either in a journal and/or widely on-line. Obviously, some of the themes emerging are advice for agencies like our own, best practices for journalists and best practices for scientists. One of my particular concerns is how we avoid some of the mistakes that NASA has made in the past with sort of over-hyping results in what seems to be a repetitive way. So, there's the old joke about 'NASA discovered water on Mars, again'. And we don't want to have that kind of pitfall where every possible biosignature found somewhere is reported in such an over-hyped way that, by the time we find something that's really more significant, the public sees the story and thinks 'Oh, we've already heard that ten times.' Of course, we recognize that we don't have control over stories once they're out there in the mediaverse. But there are aspects of the process that we, meaning 'we' as NASA, 'we' as a scientific community, and 'we' as a broader community of scientists and journalists, do have control over. And so it seems like a really worthwhile effort. I'd love to be more specific about the results, but I'm deliberately letting the process play out. So that's probably all, in terms of specifics, that I can report about that workshop now. But I can tell you that the feedback we've received is that it was a really good experience. And just the experience of doing it and getting that community convened and the connections made, and the efforts already happening to continue those conversations in various ways seems like its own reward. And then, hopefully, the report about the actual results that comes out will add to that.

HL: I have a question, if I may. You just said that, although you want to change some of the things that NASA did wrong in the past, you still cannot speak now about this. So it seems that some of the NASA procedures are still there and unchanged. Now, we had our Kavli-IAU symposium and we do not have that kind of environment and we can freely speak. And what has come out of this symposium is that in the communication with journalists they do not want to see another scientific paper. They want a very short answer like 'Yes, we found it', 'No, we didn't find it.', 'Yes, it has three eyes.', 'No, it has four eyes.', so very, very concise. So, I think it will be very interesting to read your report. Can you tell us if a similar result was reached during your workshop, since this is something that I, as a scientist, didn't know about science journalism, and I don't mean 'science communication' but 'science journalism'. I wonder if similar results have come out. Can you just say 'yes' or 'no'?

DG: Let me make sure that I understand. So, you're saying that you're hearing from journalists that they don't want you to give them a paper and say 'Read this, it has our results.' They want you to tell them in a few sentences 'Here's what we found', and simplify with 'Here's the headline'. Is it that?

HL: Correct.

DG: Yes, I think we did hear some of that. Honestly, there was probably a self-selection of the particular kinds of journalists who chose to spend their time at this meeting. Maybe this is the failing of what we did. It would be interesting to do a meta-analysis of what we left out. But, I think, there's a selection effect that the people who showed up are the people who are concerned with these questions. So, probably they have a somewhat more sophisticated understanding and the desire to dive deeply and tell the stories and have some past experience with the subject. So, I wouldn't say that I heard a ton of that at our meeting. And so it's interesting. I'd love to talk more about the differences and how we selected, who showed up and the way they participated, and how that perhaps led to somewhat different messages being received. Because that would be a valuable thing to learn about as well.

HL: Yes, I think summer is also a good point in time for us to talk about it, because our proceedings will be published soon thereafter. So there's another space race, but this time it's about who publishes first. I will go now to the next question, but I'll come back to you later and ask you more questions. Jim, back to you. We are entering the era where we can now make discoveries with outlier science. We always allowed for serendipity, but I think now serendipity will take more and more centre stage. But we still don't allow ourselves to name 'serendipity' or 'outlier science' as the driver for funding, because we believe no one will want to fund this. However, as we said, we have all these large data sets, we will make a lot of the discoveries with the help of AI, probably even automatically and not knowingly and the scientist will just show up and say 'Oh, what's that? That's interesting. I'll publish it.' Do you think that the funding landscape will need to adjust to the way we will make significant discoveries with the help of the machine? And, in this respect, do you think that we need to redefine what we call the 'freedom of research' and the boundary to 'crazy research', if the human factor gets more and more pushed out of the discovery process?

JD: I quite like what Eleni had said about redefining from 'high-risk/high-reward' to something more about 'excellence'. I like this idea to use the probability of feasibility, one of the categories which out of the many categories is especially unique about the 'Search for Life beyond Earth'. I really dislike the term 'high-risk/high-reward', because the search for SETI, which I'm involved in with my colleagues at Breakthrough Listen, is a search for outlier science and as such is extremely low-risk. But, although the likelihood of finding the one particular signal is perhaps low, the likelihood of finding something interesting is very, very high. And, in fact, that's what we have learned every time we step into a new wavelength regime or time coverage or area of the sky coverage, anytime you deploy a new telescope, we are guaranteed to make easy discoveries that the Universe has not yet gotten. And so it's very low-risk, and it's a great endeavor for students that way.

That being said, it can be very difficult to be specific since you don't know what you will find and only that you will find it. Therefore, it can be very difficult to get funding. And this is again right back to the comment that non-profit funding can be leveraged here to take forward high-risk projects, or maybe more ambitious or more nebulous projects. But we have seen, I think, progress, at least from the grant writers perspective, in terms of NASA being more open to the good measurement of this, if you will, open form of collaborative funding structures. This, I think, is really good. What remains the challenge for us is to convince our colleagues. Well, here's my hot take for the panel; I think there's no evidence that scientists are any good at picking out which grant proposals are good. We've done a whole lot of effort on this, you know, split it in several streams from 'garbage' to 'spectacular'. But other than that, it's bit of a dice role. I think that if we embrace that sort of futility, then we can become a little more generous with our funding, or maybe have a little more 'social spare funding' or some other model. There needs to be a little less focus on 'excellence' and more focus on 'equity'.

HL: Very good. Thank you very much, Jim. I would like to go now to Mike and ask another question. Now, we do expect that the first Life beyond Earth will be detected in the UK and that's why we're doing the Kavli-IAU symposium in the UK. And we have heard earlier from Chris Impey that it's very likely that aliens do speak English. So now, let's assume this has happened, and, of course, the UK government will come to the Royal Astronomical Society, bypassing the Astronomer Royal, and ask you for advice as a learned society rather than as an individual, although they will ask both sources, if they're wise. You've learned from the Kavli-IAU symposium quite a bit, I hope. Would you like to give us a summary of what you would tell the UK government or warn about?

ME: Well, no. I'm afraid I'll have to shoot you if I told you! Seriously though, I don't think I am in a position to be able to tell you what we would say. What I'm interested in at the moment is setting up a mechanism for deciding how we would work out such a strategy. And it would be impossible to be definite, since we don't know what the threat is going to be. What we can do, though, and I think we need to do, is to think about - for when the government does ask for advice - how are we going to give that?, what is going to be our mechanism for providing such advice? It may be that we'll be able to write some of it down before anybody asks. It may be just a matter of saying "Oh well, we're going to contact A, B, C and D, and that will be the small panel, or whatever, who will work in the same circumstances that would apply if we were to see the antenna waving outside our window." I think one would hesitate (after this symposium) to say what alien life going to be like. Should we have the nukes ready to blast it? We just don't know. I'm sceptical that we have any consensus on the nature of alien life, and I suspect it would be just as difficult to reach a consensus among ourselves for action if actually confronted with an outside threat. But simply getting a consensus on what would constitute a confirmation that extraterrestrial life exists would be a fundamental advance.

So, I'm backing off from saying I could immediately give advice. What I think what we need to decide what is the best strategy is for working out what that advice would be. And, hopefully, that's not something we will need in a hurry. I get the feeling, this is just a gut feeling, that it is not that likely that the alien will knock on our door next week and provide an existential threat. It looks to me as if the biomarkers are not going to be nearly as reliable as we thought they were going to be. Any confirmation of a discovery of Life beyond Earth is going to take a fair bit of time. That means we have time to work on the reception of it, or the perception of it, which we can do. And I do think we need to set up something so that we know how to give advice.

But I will ask you one question back, if I may, which to me is an elephant in the room to some extent. We heard a statistic earlier that about 60% of the public believe there are aliens. Well, what do you want them to believe? Do you want there to be aliens, or do you not want there to be aliens? I'm asking because you're wanting money to go and detect them. So, what is it, in your overall public education or whatever you call it (and I might call it coercion) that you want to achieve? Now, what I should perhaps have said is that 60% of people believe that aliens have visited or are still here, because then you could work on that in a in a different way. But I don't think we've really established what it is that you want this overall education program to do. My suspicion is that we would like people to keep open minds, watch the evidence, make judgment on the evidence, and not to believe that we have found it yet. But I'm open to questions on that.

HL: Well, because of this reason the public probably thinks "Oh, yes, but I already knew that." if you indeed had a discovery of a Life beyond Earth and were telling them about it. But I see that you have not mentioned in your response the Astrobiological Society of Britain, who are, I wouldn't say a rival, under your auspices at the moment. Would you go and talk to them?

ME: Just to be clear, nobody has asked us anything. If the Government did suddenly want to ask somebody, the first people they are going to ask are the Royal Society. It would certainly ask the Astronomer Royal, then it might cascade down to us, I don't know. But again, it's those relationships that are not set in stone yet. I'll make some inquiries after this meeting as to what, for example, the Department of Science and Technology had by way of subcommittees, developed committees, that it would call in such an event. We will offer advice. We will say that we would like to help with this, if we can. But obviously, as you say, that would hopefully involve several societies. As you mentioned, the astrobiology one would be an obvious one. So far as I know, it isn't set up at the moment.

HL: So, contrary to the U.S., we do not have an Astrobiology Strategy in the UK, which brings me back to you, David. Now, you do have an Astrobiology Strategy and you are the senior scientist of it. But probably funding within NASA is also an issue, as we see cuts, more recently at JPL, for example. Can you imagine that in the future NASA will outsource astrobiology, maybe to some philanthropic funder who would like to take it on and make a name for themselves?

DG: That's a really interesting question. I certainly can't imagine our entire astrobiology programme being outsourced. There are obviously roles for private companies that are working with NASA, and that's very high profile now. But, of course, that's always been the case throughout the entire history of the space programme that we've worked with contractors and companies and so forth. And there are, you know, some obviously very high-profile launch companies that we've become sort of co-dependent upon, for better or worse. And we see that some of these private endeavors, like the Venus mission by Rocket Lab, fill an interesting niche. And, ideally, companies like that are doing things that NASA can't quite do, but is supportive of. And so, for instance, with the Rocket Lab mission, we, NASA, are helping them out in various ways with the deep space network, with their entry, descent and landing technology and so forth. So, that's really a private astrobiology mission that we are supportive of and trying to help, and very much hope that they can succeed. I can see more happening in that category. I can't see our very large flagship or new frontiers class missions becoming outsourced. I think those are very key to NASA and our process for selecting and funding large missions. I can't see us just turning over some big contract to some company, but certainly, again, the involvement of contractors and all that has always been there, and it will continue and will shift. So, I guess I do see an increasing role for private companies at the fringes of astrobiology and some of the missions. And, of course, that raises some concerns. If companies are actually doing missions, then how do we ensure that they keep to our standards of planetary protection, for example? So, I see potential there and I see some concerns that we have to try to stay on top of.

HL: Thank you very much. I think all panellists have been very honest in this discussion and I would like to thank them all with a big applause before we go on to the audience.

3. Audience engagement

Instead of taking questions for the panellists, the audience was invited to say what they wished for the future.

Martin Dominik (University of St Andrews, UK): I think it might be worth having a fundamental rethink of how we assess applications for responsive-mode funding. It is extremely striking that some funders who recently switched part of their decisions to a lottery system found that working better. There's very little evidence to show that certain procedures that we use actually meet the intended goals. If it cannot be demonstrated that they do better than a lottery, then they don't have any justification. I would like to question whether disciplinary panels are good idea. To what extent do we even need experts to assess those things? What is it that experts actually should form an opinion on? We should not forget that the only person who tends to fully understand all details of the research proposal is the applicant. I think there are some very interesting jobs for social scientists. I think we can do much better than we do right now.

Julie Nekola Nováková (Charles University, Czech Republic): I wish for more citizen science, greater transparency and accessibility, more engagement and greater trust in science and greater funding for this research. And I'm prepared to try to work on all of that.

Erik Persson (Lund University, Sweden): I actually wish for time to be able to do research.

Kathryn Denning (York University, Canada): I wish for greater integration of social sciences and humanities and science communication research, and all these things, because we can accomplish so much if we actually work together on these issues that are very pressing for the intersection of astrobiology and technosignatures and society. What matters is that we have the expertise, we just need to be marshalling it. What I really like to see is sustained, focused attention to safety for scientists and science, journalists and science communicators, and everybody who's in the public eye because of the interesting work that they do. So that means sustained institutional attention to keep everybody in the community safe.

Jacob Haqq-Misra (Blue Marble Space Institute of Science, USA): I'm an optimist in the room usually, and I felt all the concerns that everyone has raised. But we're studying astrobiology. This is amazing. We had a whole great week. So I want to do more of what we're doing. We could do some things better, but let's find life, let's keep SETI. I think this is great. I'm having so much fun.

Leen Decin (KU Leuven, Belgium): I wished that more people would have attended this kind of multi-disciplinary conference. Since I'm listening, I have been motivated to go beyond my own horizons.

Conor Kennedy (Dublin City University, Ireland): Just a comment. If it was the case that 62% of the public are scared in their beds at night for fear of the monsters that live below, then it is the responsibility of all of us to turn the lights on.

(Toward) Discovery of Life beyond Earth and its Impact
Proceedings IAU Symposium No. 387, 2026
H. Landt, M. Dominik & C. Oliver, eds.
doi:10.1017/S1743921324001510

Public responses towards the "Life on Venus?" claim

Sze-leung Cheung

Stardusts
email: cheungszeleung@gmail.com

Abstract. In this study, we examine the social media response statistics to a recent astrobiology headline about the discovery of phosphorus on Venus, which led to a media outbreak of 'Life on Venus?' claims. We estimate its impact scale by comparing it with other events.

Keywords. Science Communication, life on Venus

1. Introduction

The goal of this study is to evaluate the public interest in astrobiology, particularly when a possible discovery of extraterrestrial life is announced. On September 14, 2020, a study about the discovery of phosphorus on Venus was published (Greaves *et al.* (2021)), receiving some coverage through press releases and subsequently picked up by different media outlets. Some media associated the discovery with the 'Life on Venus' claim. We are using this as an example to evaluate public interest.

2. Social media analysis

2.1. *The data*

The table in Figure 1 records the engagement numbers (i.e., likes, comments, and shares) across multiple social media platforms covering the news. We compare the differences between public-oriented mass media and science audience-focused scientific media. The engagement rate in the table is the number of engagements compared to the total number of subscribers or followers of that channel.

The data was collected 24 hours (+/- 10 minutes) after the press release embargo time. The original purpose of the data collection was to understand the performance of different organizations' social media. Now, we are using it to understand the public interest in the issue itself.

The engagement rate is largely affected by the performance of individual channels and cannot be used to justify the public interest level. Other factors, such as the news publication time, time-zone issues, and readers viewing the same news multiple times, also affect the result.

For benchmarking, we calculated that

(1) The average engagement number on the New York Times Facebook page in 2023 was 1,213 per post.
(2) The engagement number for a space-related news post about the *SpaceX Starship IFT2 test* was 2,229 (as of April 8, 2024) on the New York Times Facebook page.

Total Engagment (Engagment rate)

		Facebook		Twitter		Instagram	
Astronomical media	NARIT (National Thai observatory)	72,199	14.60%	2413	28.27%	65	6.67%
	Planetary Society	475	0.06%	1317	0.56%	6439	1.41%
	ESO	1727	0.63%	927	0.81%	2131	2.96%
	Astronomy Magazine	4316	0.32%	85	0.04%	-	-
	Nature Astronomy	-	-	4338	13.39%	-	-
	Royal Astronomical Society	1550	9.46%	1636	3.39%	517	8.86%
	SETI Institute	259	0.05%	811	0.11%	1991	1.78%
	Space.com	193	0.01%	499	0.06%	2207	1.61%
	Astronomy Now	75	0.01%	307	0.26%	-	-
	East Asian Observatory	44	1.44%	181	5.81%	66	2.12%
	BBC Sky at Night Magazine	62	0.22%	111	0.26%	87	3.37%
	Sky & Telescope	-	-	52	0.04%	-	-
	Spaceth.co (Thai)	13,759	5.46%	-	-	-	-
Science media	New Scientist	224	0.01%	17729	0.45%	8068	2.51%
	MIT Technology Review	94	0.01%	184	0.02%	9277	0.84%
	Scientific American	257	0.01%	419	0.00%	3544	1.83%
	MIT	417	0.03%	806	0.07%	1600	0.53%
	IFL Science	1280	0.10%	306	0.14%	-	-
	Futurism	487	0.03%	109	0.05%	-	-
Mass media	BBC News	3265	0.01%	7956	0.02%	345000	2.33%
	NewYorkTimes	4610	0.03%	99700	0.20%	42857	0.42%
	UNILAD	2024	0.00%	47	0.02%	58049	1.35%
	The Independent	6400	0.06%	2887	0.08%	3473	0.61%
	Metro	1211	0.05%	11300	3.38%	-	-
	CNN	921	0.00%	5016	0.01%	-	-
	NBC News	2681	0.03%	1644	0.02%	-	-
	Sky News	1546	0.02%	371	0.01%	-	-
	Reuters	851	0.02%	555	0.00%	-	-
	The Atlantic	376	0.01%	447	0.02%	-	-
	thairath (Thai)	63,415	0.49%	-	-	-	-
	The Standard (Thai)	11,156	0.57%	-	-	-	-
	The MATTER (Thai)	6,870	0.62%	-	-	-	-

Figure 1. Engagement number and engagement rate on different channels after 24 hours

2.2. *Findings*

(1) Based on the engagement rate between the public and the science-focused audience, there is no significant difference in their reactions to the news, except in the case of Thailand, which we conclude is due to the high performance of the organization's social media in general.

(2) Based on the New York Times Facebook page, we rechecked the number of engagements for the 'Life on Venus' post, which was 13,872 as of April 8, 2024. This is 10 times more than the average engagement number it typically receives.

(3) The 'Life on Venus' post attracted 6 times more engagement than the *SpaceX Starship IFT2 test*, which belongs to a similar category of interest and target audience.

3. Public Interest Over Time: A Google Trends Analysis

Using Google Trends, we can explore how people's interests have evolved. Figure 2 and 3 recorded the behavior of search interest in different topics: "extraterrestrial life," "astronomy," "Unidentified flying objects" (UFOs), and "Venus." relative to the highest point on the chart for the given region and time. Here's what we observed:

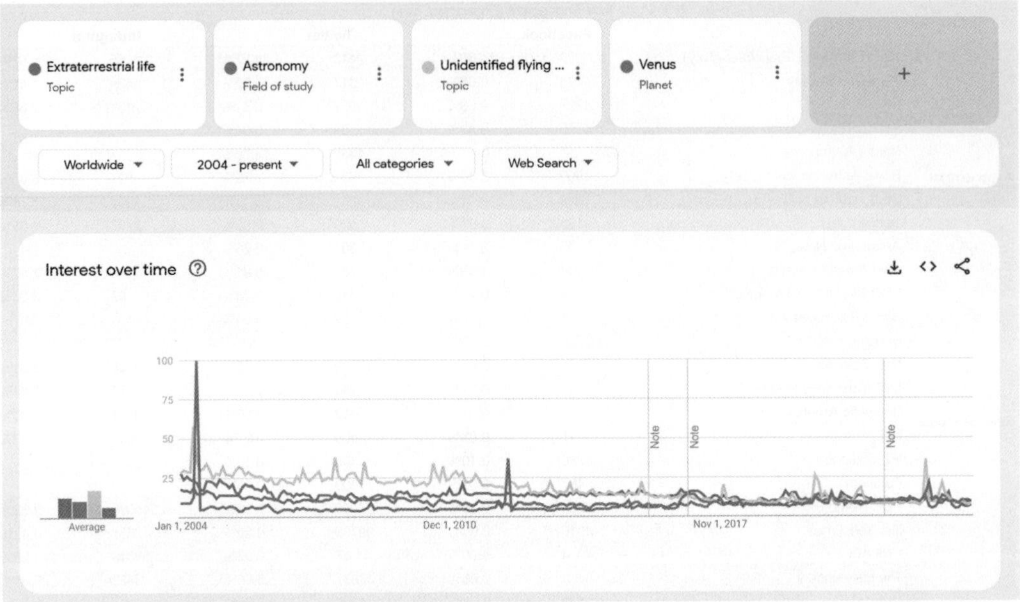

Figure 2. Search interest using different keywords during 2004 to 2024

(1) Extraterrestrial Life vs. Astronomy: The public interest in "extraterrestrial life" closely mirrors that of "astronomy," sometimes even surpassing it.
(2) UFOs vs. Extraterrestrial Life: Before 2013, the interest in UFOs was significantly higher than that in "extraterrestrial life." However, after 2013, the two interests became more comparable, indicating a shift in focus.
(3) Phosphorus Discovery on Venus: On September 14, 2020, the detection of phosphorus on Venus was announced. The interest in "extraterrestrial life" and "Venus" only experienced a slight rise during this press release. Perhaps the discovery of phosphorus didn't capture the public's imagination as much as one might expect.
(4) Transit of Venus vs. Phosphorus Discovery - During the Transit of Venus in 2004 and 2012, interest in Venus peaked significantly. This astronomical event generated greater interest than the phosphorus discovery on September 14, 2020.

4. Discussions

While there was notable public interest in the potential existence of extraterrestrial life following the announcement about phosphorus on Venus, it is crucial to clarify that this finding does not constitute a definitive discovery. Consequently, it did not create significant ripples among the public.

Beyond the scientific implications, we must also consider the ethical aspects of communicating astrobiology-related news. The challenge lies in ensuring that the public can differentiate between groundbreaking discoveries and ongoing research. Unfortunately, some media outlets exacerbate this issue by using sensational language. For instance, declaring unequivocally that life exists on Venus can be misleading.

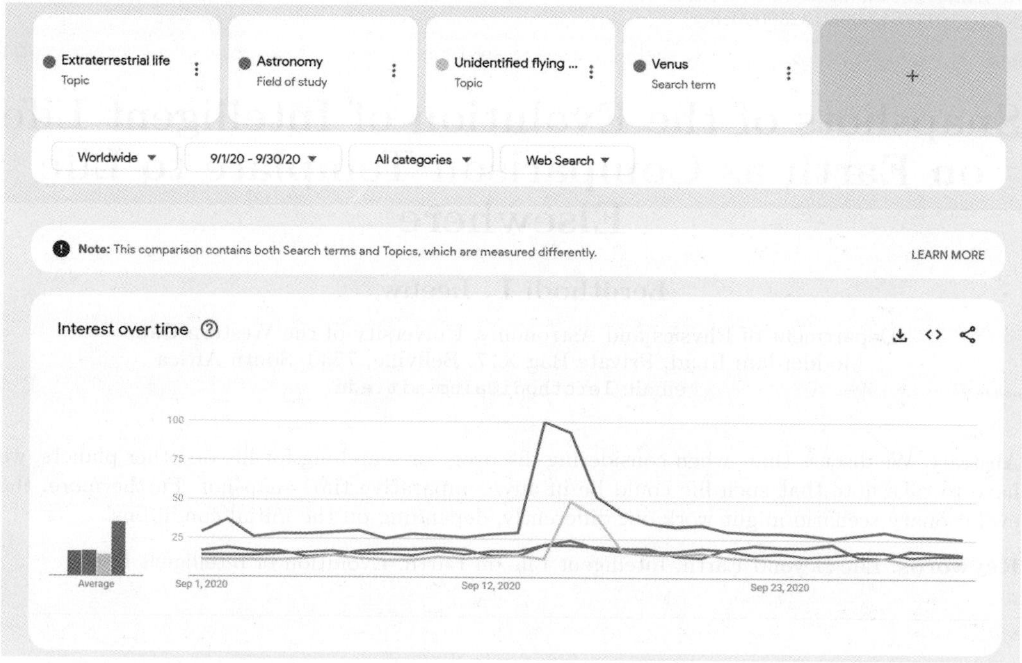

Figure 3. Zoom-in view of search interest using different keywords during 2020

Addressing communication ethics in the context of astrobiology is crucial for fostering accurate understanding and preventing unwarranted hype. As we move forward, finding a balanced approach-one that informs without sensationalism-will be essential.

Reference

Greaves, J. *et al.* 2021 Nature Astronomy, 5, 655

(Toward) Discovery of Life beyond Earth and its Impact
Proceedings IAU Symposium No. 387, 2026
H. Landt, M. Dominik & C. Oliver, eds.
doi:10.1017/S1743921325000018

Snapshots of the Evolution of Intelligent Life on Earth as Comparison Template to Life Elsewhere

Lerothodi L. Leeuw

Department of Physics and Astronomy, University of the Western Cape
Modderdam Road, Private Bag X17, Bellville, 7530, South Africa
email: `lerothodi@alum.mit.edu`

Abstract. We discuss that, when considering discovery or searching for life on other planets, we have to take note that such life could be in any comparative time snapshot. Furthermore, the evolutionary scenario might work out differently, depending on the initial conditions.

Keywords. Life Beyond Earth, Intelligent Life on Earth, Evolution of Intelligent Life

1. The Framework with Selected Snapshots of Intelligent Life on Earth

We use snapshots of the evolution of intelligent life on Earth, as shown in Fig. 1, as comparison template to life elsewhere. The snapshots are in episodes on logarithmic scale from modern to the depths of time just under 2 million years ago; and, for simplicity and focus, they use selected key moments related to astrophysical and physical science activities shown in representative images from literature. It is noted the evolutionary scenario discussed in Fig. 1 is just one particular convergent example, which might not work out the same way in case of substantial differences in initial conditions and/or sequences of catastrophic or otherwise events. Thus this scenario is presented well aware that life elsewhere could have taken infinitely many other different evolutionary paths different from here on Earth.

As shown by Leeuw (2024), the framework shown in Fig. 1 allows a global view of science and human development over many years in a single picture, and that power is exploited in this presentation. The modern times include a depiction of this current meeting about the possible discovery of life elsewhere, while the earliest time go back just under 2 million years ago to the early homo sapiens. It's worth noting that the period covered here, even though it is 2 million years long, is a very small time fraction of the age of our planet Earth, that is some 4.54 billion years old (Dalrymple (2001)). It goes without saying that, even though 2 million year covered here is only a fraction of the age of our planet Earth, incredibly a lot happened in that time frame, so that going one factor from one episode to another produced huge shifts in development. One example in the figure is of Cape Town 200 years ago largely made of huts with no street lights and dark night skies, and 20 years ago with major light pollution from both population growth and the development of infrastructure, that would most probably be recognizable in views of Earth at night from Space by another civilization, perhaps assessing life and its development on our planet.

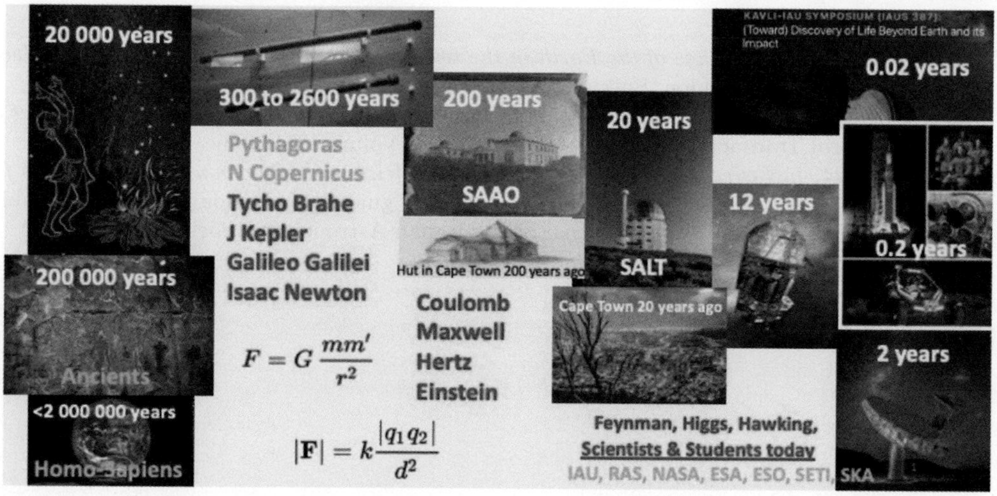

Figure 1. Timeline depicting snapshots of the evolution of intelligent life on Earth, as comparison template to life elsewhere. The snaps are in episodes on logarithmic scale from modern times to just under 2 000 000 years ago and use selected key moments related to astrophysical and physical science activities and representative images from literature.

2. The Snapshot Comparative Outlook to Life Beyond Earth

Planets beyond Earth and possible life on them are most probably not going to be synchronized with the age of Earth, or of evolutionary or developmental periods of Earth. When considering discovery or searching for life in other planets, we therefore have to remember such life could be in any comparative time snap here on Earth, such as shown in Fig. 1; and, that can be anywhere between now and the 2000 years ago of the Pythagorean contemporaries and the technologies that subsequently developed, or to 200 000 years ago when humans first tamed fire and manipulated it to its many subsequent exploitation in our development (Gowlett (2016)), or indeed to just under 2 million years ago at the appearance of homo sapiens and their ensuing evolution to modern humans. This timeline and discussion doesn't include the possible evolution of life on Earth prior to the 2 million years ago and its different scenarios nearly 4.54 billions years ago.

3. The Takeaway Messages

The main takeaway messages from this short presentation are the following. Firstly, when considering discovery or searching for life in other planets, we have to take note that such life could be in any comparative time snapshot as discussed above, not only 2 million years as shown in Fig. 1 but anytime back to Earth's formation nearly 4.54 billions years ago. Secondly, the evolutionary scenario discussed in Fig. 1 is just one particular convergent example of infinitely many, which might not work out the same way in case of substantial differences in initial conditions and/or sequences of catastrophic or otherwise events. These takeaway messages are also important for when talking about this topic to the public or developing related policy and strategies to it, as is the subject of this meeting.

Acknowledgement

Fig. 1 uses a compilation of images from literature as acknowledged in Leeuw (2024) and other sources cited therein.

References

Dalrymple, G.B. 2001, *The age of the Earth in the twentieth century: a problem (mostly) solved*, Geological Society, London, Special Publications, Vol. 190, Issue 1, pp. 205–221

Gowlett, J.A.J. 2016, *The discovery of fire by humans: a long and convoluted process*, Philosophical Transactions of the Royal Society B, Volume 371, Issue 1696

Leeuw, L.L. 2024, *Inclusive astronomy and science in Africa and other non-western regions for the global benefit*, EAS Congress: The hitchhiker's guide to astronomy education, public outreach and communication, Journal of the Italian Astronomical Society, Vol 95, n. 2

(Toward) Discovery of Life beyond Earth and its Impact
Proceedings IAU Symposium No. 387, 2026
H. Landt, M. Dominik & C. Oliver, eds.
doi:10.1017/S1743921324001819

Summary of KAVLI-IAU Symposium (IAUS 387): (Toward) Discovery of Life Beyond Earth and its Impact

Andjelka B. Kovačević[1] **and Nigel J. Mason**[2]

[1]University of Belgrade-Faculty of Mathematics, Department of Astronomy, Belgrade, Serbia,
email: andjelka.kovacevic@matf.bg.ac.rs

[2]School of Physics and Astronomy, University of Kent, Canterbury, CT2 7NH, United Kingdom,
email: N.J.Mason@kent.ac.uk

Abstract. The KAVLI-IAU Symposium (IAUS 387), held on April 15-19, 2024 at University of Durham, UK, brought together specialists from a range of complementary fields to discuss the probability of extraterrestrial life, methods of detection and the ramifications of its detection for humanity, The Symposium underscored the need for strong cooperation between scientists, ethicists, theologians, and journalists. Critical outcomes included the exploration of detection protocols to elevate the credibility of the Search for Extraterrestrial Intelligence (SETI), and the need to increase outreach activities on the possibility of life detection (or non-detection). The Symposium's Discussion panels focussed on the importance of defining life, extending the concept of habitability, and exploring the Great Filter. The Symposium concluded with a strong advocacy for integrating these perspectives into global sustainability policies, highlighting the benefits of astrobiology for understanding both the cosmos and Earth.

Keywords. astrobiology, extraterrestrial life, exoplanets, biosignatures,technosignatures, Great Filter

1. Introduction

From ancient stargazers to modern scientists, the question of life beyond Earth has captivated human imagination (see historical overview in Mix 2019). The KAVLI-IAU Symposium (IAUS 387):(Toward) Discovery of Life Beyond Earth and its Impact † built upon this enduring curiosity, brought together experts from diverse fields to discuss the current state of humanity's search for life beyond Earth, the technological challenges that such studies pose and the social implications of what life we might find beyond our planet (see also Praet 2023).

The Symposium theme was rooted in the legacy of scientific inquiry that began with astronomers and philosophers who pondered the existence of life beyond Earth from old Greek culture. As space exploration expanded, so did our questions (see also Mix 2018). *Where is everybody?* physicist Enrico Fermi famously asked, pointing to the tension between the anticipated high probability of extraterrestrial life and the apparent lack of evidence for it (Webb 2002). Unlike in any other research field, despite the undetection of a signal, the Fermi paradox does not receive a straightforward answer; instead, it has spurred extensive research (Brin 1983; Gray 2015; Hart 1975; Tipler 1980; Haqq-Misra *et al.* 2020). The Drake equation provides a formulaic approach suggesting that

† Official website https://kavli-iau-2024.durham.ac.uk/wordpress/

plausible values for its variables yield that our past light cone contains intentional (e.g. Oberhaus 2019) or unintentional (e.g., Dyson 1960) indices of extraterrestrial civilizations, a stark contrast to the zero evidence observed, thus intensifying the enigma of the Fermi Paradox ‡.

Since the first confirmed exoplanet in 1995 (Mayor & Queloz 1995), over 5,632 exoplanets have been identified †, suggesting planets are common across the galaxy (Seager & Bains 2015). This realization has reinvigorated the search for life, with the Symposium examining the tension between null detections and advancing methods to detect life signs (Lubin 2016; NASA Technosignatures Workshop Participants 2019). In addition to scientific discussions, the Symposium addressed the broader context of defining life and the implications of astrobiology in our quest for extraterrestrial intelligence (Mix 2015). The absence of signals may serve as a "null signal," guiding humanity towards sustainable futures (Haqq-Misra & Baum 2009; Chaisson 2016). The enduring silence from the cosmos might suggest that civilizations capable of long-term survival likely harness stellar energy while minimizing detectable outputs, reflecting a cosmic selection principle favoring sustainability (Baum 2010; Naganuma 2009). As the Symposium highlighted, sustainability isn't merely political (Kates 2010), but an evolutionary necessity for advanced civilizations (Frank & Sullivan 2014; Sullivan & Baross 2007).

Thus, as we look deeper into the implications of the Fermi Paradox and our ongoing search for extraterrestrial life/intelligence, it becomes increasingly clear that these endeavors are not just about discovering others in the universe but also about refining our strategies for sustainable existence. This Symposium explored how these insights from astrobiology could inform and transform our approaches to energy use, technological development, and overall planetary stewardship, guided by the lessons implied by the null signals in our search for life beyond Earth.

2. Overview of invited talks

The Symposium featured eleven invited talks‡. The transcendental aspects of astrobiology, which are woven into all Symposium topics, were introduced by Lucas Mix from the University of Durham in his opening lecture. He compared astrobiology and cathedrals as cultural and transcendental endeavors of humanity (see Figure 1). This side-by-side imagenery also reflects on conscious apperception (see Brook & Wuerth 2023), as humanity becomes both the subject and the object of astrobiology studies, as well as both the designer and the design of research.

Frances Westall presentation emphasized the complexity of detecting extraterrestrial life, urging realistic scientific expectations (see Figure 2). She highlighted the importance of geological contexts, captured in her slogan, *No Minerals, No life; No rocks + hot water, No life* (Westall *et al.* 2015; Agić 2021). Evidence of early plate tectonics from Jack Hills zircons (Gamaleldien *et al.* 2024; Jiang *et al.* 2024) suggests a key role for tectonic activity in life's evolution, though more global samples are needed. Westall also stressed that active, communicative civilizations (ACCs) likely require continents and oceans (Stern & Gerya 2024), with a revised Drake equation proposing a very low probability of intelligent life due to rare planetary conditions. She concluded by advocating for a rigorous, evidence-based approach to detecting biosignatures on Mars, icy moons, or

‡ Based on Sainsbury (2009), a paradox is an unacceptable conclusion derived from seemingly acceptable reasoning and premises. Since the acceptable cannot logically lead to the unacceptable, we must conclude either the conclusion is actually acceptable, or there is a hidden flaw in the premises or reasoning (see Sainsbury 2009).

† `https://exoplanetarchive.ipac.caltech.edu/`

‡ Book of Abstracts `https://kavli-iau-2024.durham.ac.uk/wordpress/?page_id=272`

Figure 1. Astrobiology is like a cathedral; It is a massive, interdisciplinary endeavour, requiring thousands of people and likely to last centuries. It synthesizes knowledge; it inspires people; and it shapes the intellectual skyline. Location: Durham Cathedral, Durham UK. Image attribution: Lucas Mix.

exoplanets, highlighting the need for patience and advanced technology in the search for life (Werth 1998; Bohlmann & Bürger 2018; Döbler *et al.* 2023).

Stephen Webb's lecture, 'The Great Silence reconsidered: Reexamining the Fermi Paradox in the age of discovery," (Webb 2002), explored three major scenarios to explain the Fermi Paradox: intelligent life is rare, civilizations don't last long enough to be detected, or they remain undetected due to technological limitations. The rarity of intelligent life could stem from either the scarcity of favorable physical conditions or the difficulty in evolving from simple to complex life. Despite many exoplanets in the habitable zone (HZ), none of those studied to date have shown the presence of free oxygen. The second scenario proposed that technological civilizations may be fleeting and therefore hard to detect. The third scenario tackled the limits of our technology, with Arthur C. Clarke's famous quote, "Any sufficiently advanced technology is indistinguishable from magic," suggesting that extraterrestrial technologies may be beyond our current understanding. Webb also referenced Steven J. Dick's "Intelligence Principle" (Dick 2003), suggesting civilizations rapidly evolve into post-biological forms (PBCs) reliant on artificial intelligence. This could explain why advanced civilizations remain undetectable. These discussions raised questions about focusing solely on biological life in SETI. Finally, these considerations linked the Fermi Paradox to the importance of energy management, where civilizations that efficiently harness solar energy may survive longer, highlighting the implications for our own sustainability.

Jacques Arnould addressed the Symposium with a provocative question about the impact of discovering extraterrestrial life on human theology and ethics in the talk "The challenging implications for theology and ethics" (see e.g., Arnould 2021, and references therein). Arnould pointed that discovering extraterrestrial life could serve as a critical juncture for theological and ethical thought. He concluded that such a discovery would

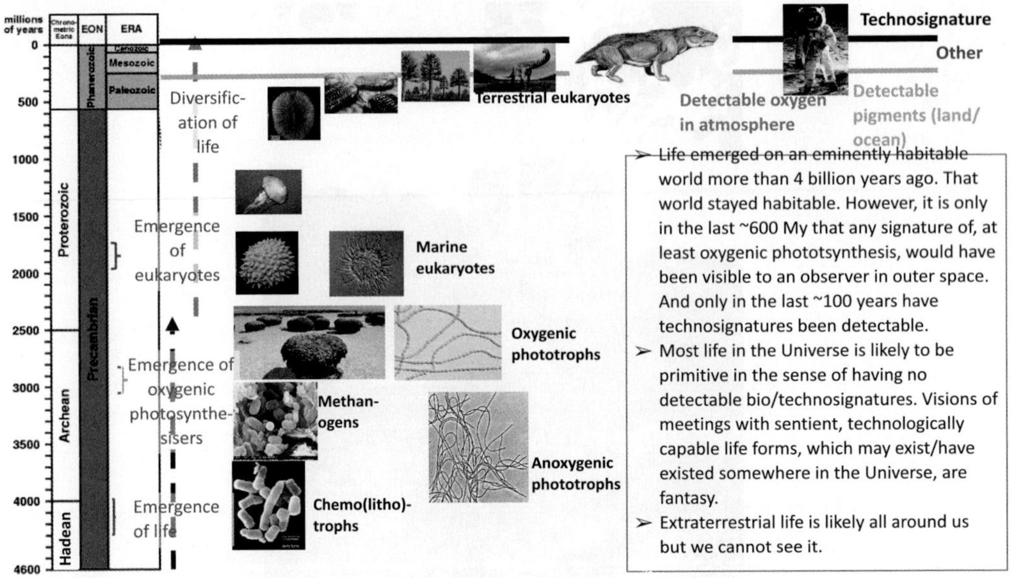

Figure 2. The emergence and diversification of life on Earth. Timeline highlights key milestones such as the emergence of life, oxygenic photosynthesizers, and eukaryotes. Detectable oxygen in the atmosphere appeared approximately 600 million years ago, while detectable technosignatures have only emerged in the last 100 years. This suggests that while primitive life may be common in the Universe, advanced, technologically capable life forms are exceedingly rare and likely undetectable with current methods. Image attribution: Frances Westall.

not only challenge existing religious and ethical frameworks but potentially enrich them by integrating new cosmic perspectives. This could lead to a profound reevaluation of humanity's place in the universe, urging religious communities and ethical thinkers to contemplate broader existential questions and embrace a more universal perspective.

Arnould transcendental perspective seamlessly led into the artistic and imaginative realms explored by Stephen Baxter. Stephen Baxter captivated the audience during the "Anticipating the Alien: Creating Visionary Science Fiction" session with his engaging tales illustrating how science fiction can lay the groundwork for scientific exploration and help humanity prepare for the unforeseen. In his works, to mention the Xeelee series and the novel "The Time Ships" (see Baxter 1995), Baxter created or co-created at least six to seven fictional species. He emphasized the unique role of science fiction in bridging the gap between our current scientific knowledge and the vast possibilities of the future, including the potential discovery of extraterrestrial life. He concluded that science fiction serves as a vital preparatory tool, helping society to conceptualize and emotionally prepare for profound changes. Baxter emphasized that through narratives, we can explore potential impacts and ethical dilemmas in a controlled and thoughtful manner, making science fiction an essential element in the cultural preparation for new scientific realities.

The ancient Greek word 'techne' literally translates to 'art' or 'skill'. As technology continues to evolve at an ever-accelerating rate, the integration of art, culture, and technology becomes increasingly profound. This set the stage for the SETI visionary talk by Steve Croft, who transformed these imaginative relation into tangible scientific pursuits. Steve Croft's presentation on the advancements and future of SETI technology was a highlight of the conference, addressing its core theme of exploring life beyond Earth. As a project scientist for the Breakthrough Listen (BL) initiative, Croft provided a resume on the use of advanced deep learning techniques in a comprehensive approach to SETI, known as "semi-unsupervised learning." This innovative methodology identified eight signals originating from five different stars located between 30 and 90 light-years away from Earth. These signals were promising candidates for genuine technosignatures. Croft explained that these signals exhibit distinct characteristics: they were only detectable when observing the star and disappear when looking away, distinguishing them from constant local interference. Additionally, the frequency of these signals shifted over time in a manner suggesting they are extraterrestrial in origin (details in Ma *et al.* 2023). However, Croft also urged caution. Given the vastness of the datasets, which contain millions of signals, the unique traits of these signals could potentially occur by chance. While these eight signals closely matched the expected profile of technosignatures, further evidence is needed to conclusively link them to extraterrestrial intelligence. This necessitated repeated observations and dedicated follow-up investigations to validate these findings. He offered a concrete look at the current and upcoming deep learning tools that could make the detection of extraterrestrial signals possible in electromagnetic domain, emphasizing the scientific backbone of the search for extraterrestrial intelligence. Croft's insights were crucial for understanding the practical aspects of how we might eventually confirm the presence of extraterrestrial beings, using various facilities, including upcoming optical synoptic surveys such as Vera C. Rubin Legacy of Survey of Space and Time.

In the talk titled "The Problems of Preparation: On ambiguity, proliferating possibilities, Eureka's shadows, and unspectacular pragmatism," Kathryn Denning addressed the need for careful, well-considered preparation in anticipation of discovering extraterrestrial life. She discussed the potential scientific breakthroughs and societal impacts, highlighting both the complexities and significant ramifications of such a discovery (see Race *et al.* 2012). Denning suggested that before we can understand an alien civilization, it might be useful to understand our own. As an anthropologist, Denning studied the ways scientists, engineers, and the public think about space exploration and the search for alien life. From the fantasies of "Star Trek" to the scientific endeavors of SETI, our modern world constantly presents futures not previously imagined, where we engage with bizarre alien races. Denning noted that when people discuss these futures, they often reference the past, yet they frequently demonstrate a poor grasp of historical context. She concluded that global collaboration and comprehensive planning are essential to responsibly manage the discovery of extraterrestrial life. Denning stressed the importance of establishing protocols and frameworks capable of handling the diverse outcomes of such discoveries, ensuring that they benefit society and enhance our understanding of the universe.

As a science reporter, Nadia Drake covered a wide array of topics, from the discovery of exoplanets to the research in Earth's extreme environments as an analogues for conditions on icy moons in our solar system. She also chronicled the ongoing challenges faced by the SETI in gaining scientific legitimacy. Her research offered an insight into the media's decision-making process regarding the reporting of science stories–what constitutes a compelling story, and how it is constructed (see also Drake 2022). In her talk titled "The Media: Friend or Foe? Exploring the Cosmic Narrative from a Science Journalist's Perspective," Nadia Drake explored the media's role in both educating the public and shaping perceptions about potential discoveries of extraterrestrial life. Drake argued that responsible journalism is essential, particularly in areas susceptible to sensationalism such as the search for extraterrestrial life. Emphasizing the necessity for journalists to uphold scientific accuracy and integrity, Drake highlighted the critical role of the media in preventing misinformation and enabling a well-informed public to engage thoughtfully with these significant topics. Lastly, she discussed the challenges of communicating discoveries in the search for life beyond Earth, looking both at present difficulties and potential future scenarios where evidence may suggest that we are not alone.

The SETI Post-Detection Hub, established in November 2022 at St. Andrews University and affiliated with Durham University's Law School, represents a major multidisciplinary initiative (Elliott *et al.* 2023). Dr John Elliott introduced its goals, emphasizing transparency and the need to openly share any contact with extraterrestrial intelligence. He highlighted the potential for contact through alien artifacts or microbial life, noting that the latter, while less dramatic, would still be profound evidence of life's ubiquity. Elliott speculated that AI-equipped space probes would be most likely to discover extraterrestrial life due to lower risks compared to manned missions. He also warned of the possibility that alien civilizations may self-destruct at early stages of technological advancement, providing a cautionary tale for humanity. Elliott outlined the necessity of a structured response to mitigate the societal and technological impacts of such a discovery, proposing protocols to bridge the gap between scientists and policymakers. In conclusion, he echoed the need for a global framework to manage information dissemination and international relations, as advocated by Nadia Drake, ensuring public response is managed responsibly.

Elliott's focus on preparation aligned with Jacob Haqq-Misra's exploration of civilization sustainability in "Escaping the Great Filter" (Hanson 1998; Haqq-Misra *et al.* 2020). Haqq-Misra proposed that technosignature discoveries could reveal whether the Great Filter is behind or ahead of us. Detecting technosignatures would suggest that civilizations have surpassed key evolutionary hurdles, meaning the Filter is in the past. However, the absence of technosignatures might indicate future obstacles, like self-destruction. Haqq-Misra stressed the importance of sustainable development and energy use (Meurer *et al.* 2024), echoing Arwen Nicholson's themes in Gaia theory. He also noted that humanity's growing energy consumption and uncertain population stabilization could present a test of the Great Filter, particularly as we approach a Kardashev type-I civilization. The search for technosignatures, conducted through ground- and space-based observatories, is essential not only for testing the Great Filter hypothesis but also for helping humanity navigate potential future challenges and achieve sustainable growth. Tied to this, recent James Webb Space Telescope (JWST) observations of K2-18b, a habitable-zone sub-Neptune, revealed significant levels of CH_4 and CO_2 in its atmosphere (Madhusudhan et al. 2023). It has been interpreted as an evidence of a habitable 'Hycean' world with a thin H_2-dominated atmosphere overlying a liquid ocean, potentially supported by a methane-producing biosphere (Madhusudhan et al. 2023). In contrast, Wogan et al. (2024) proposed that K2-18b is a gas-rich mini-Neptune, where the observed methane (4%) and CO_2 (0.1%) are produced thermochemically, aligning with JWST data without the need for a biosphere. These differing interpretations highlight the challenges in definitively classifying exoplanetary atmospheres and assessing their habitability, illustrating the complexities of the fourth paradigm of science (Figure 3). AI-driven models trained on extensive datasets of planetary atmospheres, could be crucial in resolving these challenges. By analyzing complex data interactions, AI has the potential to refine atmospheric models, enhance accuracy, and possibly detect biogenic activities overlooked by conventional methods. This AI approach could significantly advance our search for habitable worlds and technosignatures, aligning with the goals of the Fifth Paradigm of scientific research (Hey *et al.* 2009; Szalay 2019).

Arwen Nicholson's talk, building on "Selection by Survival" (SBS) (Lenton *et al.* 2018; Arthur & Nicholson 2023) and "Sequential Selection" (SS) (Betts & Lenton 2008), discussed how life stabilizes Earth's environment and the Gaian relationship between life and planet. Nicholson explored the likelihood of Gaia emerging and the conditions that prevent "anti-Gaian" dynamics, where life destabilizes habitability. Using a dice game analogy, she illustrated how planetary biospheres' stability could be modeled, linking these concepts to the Gaian bottleneck theory, which suggests that early life is essential

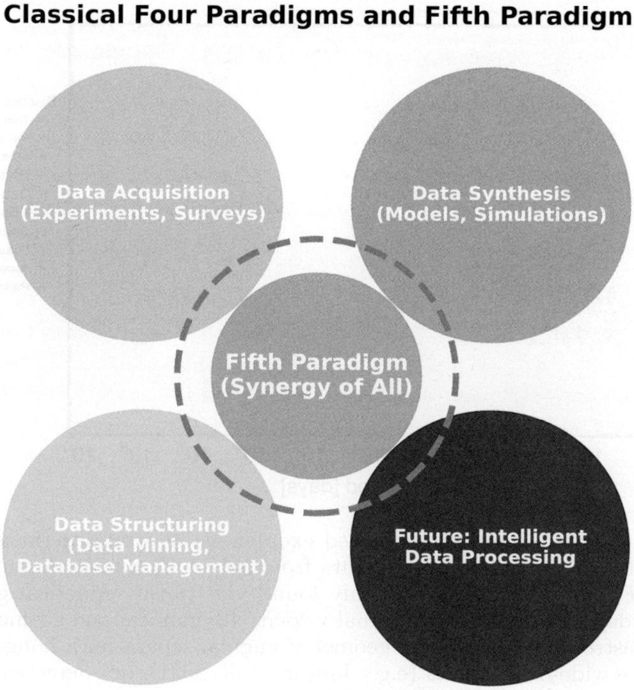

Figure 3. Classical four science paradigms from the perspective of interaction with information, moving from Data Acquisition (lower left) to Data Synthesis, Data Structuring, and culminating in Future Intelligent Data Processing (lower right). The Fifth Paradigm, at the center, represents the synergy of these paradigms, integrating advanced capabilities such as predictive analytics, seamless data integration, ethical data governance, and technological innovations like quantum computing, spatial computing, and sensor-enabled technologies.

for maintaining long-term habitability (Arthur & Nicholson 2023). These principles are crucial for understanding planetary stability and habitability in astrobiology.

Giovanna Tinetti delivered an overview presentation titled "Exoplanet Atmospheres in the Era of JWST and ARIEL," offering a synoptic view of both current and future astronomical facilities for observing exoplanets. Tinetti expressed her excitement about the transformative potential of the James Webb Space Telescope (JWST, see the conecept in Beichman & Greene 2018) and the Atmospheric Remote-sensing Infrared Exoplanet Large-survey (ARIEL) mission (e.g., Pace 2019). Tinetti highlighted that we are indeed living in a golden age for the study of exoplanets, with the statistical expectation that there are as many planets as stars in our galaxy–a galaxy that is one among billions. Currently, $\sim 5,632$ planets have been identified†. Statistically, there appeared to be a higher prevalence of smaller planets than larger ones, such as Jupiter or Saturn, or even intermediate-sized planets like Neptune and Uranus. Observations from missions like TESS and Kepler, as well as ground-based facilities, confirmed this bias, particularly noting that most observed planets have orbital periods $\lesssim \mathcal{O}(10^2)$ days (Bergsten et al. 2022), see Figure 4. The orbital period informs our understanding of the evolution of the planetary systems, whereas current observational technologies are strongly sensitive to planetary orbital distance from the host star and therefore to the orbital period (Kopparapu *et al.* 2013, 2017).

The distribution of relatively small planets is bimodal (see e.g. Owen & Wu 2013; Fulton & Petigura 2018), with one peak around Earth-like sizes and another for those

† `https://exoplanetarchive.ipac.caltech.edu`

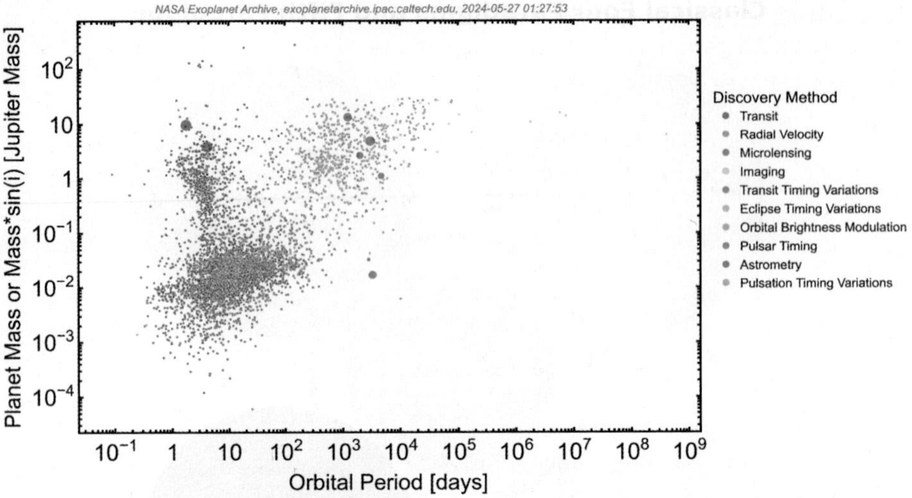

Figure 4. Stacked distributions of confirmed exoplanets by orbital periods and masses, with point sizes scaled by distance from Earth. Data from NASA Exoplanet Archive (May 27, 2024). Exoplanets with $P \lesssim \mathcal{O}(10^2)$ days are mainly found via transit (spherical geometry). Periods up to $P \leq \mathcal{O}(10^5)$ days are detected via radial velocity (dynamical and geometric), microlensing (relativistic), and astrometry (Euclidean geometry, angular separation). Long-period exoplanets, $\sim \mathcal{O}(10^6)$ days, like wide-orbit giants (e.g., Janson *et al.* 2021), are detected by direct imaging (Euclidean geometry, angular separation, brightness contrast).

between Earth and Neptune planets of a size which we know very little about. Tinetti also discussed observations from the Atacama Large Millimeter/submillimeter Array (ALMA) facility, which examines protoplanetary disks primarily composed of hydrogen gas, with about 1% of the material being dust. These observations revealed a diversity of disk structures, such as rings and spirals, suggesting that the variety we observed in exoplanets may stem from their formative processes in these diverse disks. Looking forward, Tinetti outlined the core questions that the next decade of observational research aims to address: understanding why our solar system is structured as it is, the chemical characterization of exoplanets, and the origins of planetary diversity. She explained that by studying the chemistry of such planets, we could infer which chemical and physical processes occurred during their history. For instance, if a planet were the size of an apple, the atmosphere of small planets would be thinner than the apple's crust, highlighting the technological challenge of extracting information from such minuscule chemical environments (Tinetti *et al.* 2018). As the Principal Investigator of ARIEL, Tinetti provided insights into the mission's objectives. ARIEL, a space telescope part of the European Space Agency's Cosmic Vision program, is set to launch in 2029 aboard an Ariane 6 rocket along with the Comet Interceptor (see Tinetti *et al.* 2018; Lueftinger *et al.* 2022). This telescope, though smaller than JWST, will have more observation time available for planetary characterization, studying the chemical composition and thermal structures of ~ 1000 known exoplanets using the transit method. In her conclusion, Tinetti emphasized that these missions might soon provide key evidence of life beyond Earth by potentially identifying biosignatures in the atmospheres of distant worlds. This prospect could significantly expand our knowledge of the universe and redefine the search for life in the cosmos. She underscored the distinctiveness of our solar system and noted that no exoplanetary systems identical to ours have been discovered, which may be more a reflection of our current observational limits than a testament to our system's uniqueness. Tinetti also highlighted the need for rethinking our approach to constructing large observational facilities, advocating for the sustainability and effectiveness of smaller, more targeted instruments. This strategic shift could enhance our ability to detect Earth-sized planets in stable, year-long orbits and larger gas giants in extended orbits, thereby enriching our understanding of the diverse planetary systems that populate the universe. Additionally, as a co-founder and co-director of Blue Skies Space Ltd., she is committed to creating new opportunities for science through space satellites, ultimately advancing our exploration of the cosmos and reexamining our understanding of life within it.

3. Overview of contributed talks

The contributed talks covered a wide range of topics from the tantalising search for signs of life on Mars to the philosophical considerations of encountering extraterrestrial intelligence, and extension of the term of HZ. In organizing this collection of presentations, we identified four fundamental pillars, each representing a unique aspect of astrobiological investigation. Moreover, these pillars could be seen as key frameworks through which modern astrobiologist examine the enigmas surrounding extraterrestrial life.

Pillar One: Probing Extraterrestrial Habitability

Talks under this pillar collectively explored the conditions necessary for life beyond Earth, encompassing both theoretical frameworks and experimental simulations. Armando Azua-Bustos discussed the geological analogs on Earth, offering insights into Mars' past habitability (Azua-Bustos *et al.* 2023). Eva Mateo-Martí extended this discussion through laboratory simulations, examining the potential habitability of planetary surfaces (see also Mateo-Marti *et al.* 2019). Raghav Narasimha Ponnaganti proposed using high-powered lasers to move water-rich planets or icy objects into the HZ of planetary system (see also Narasimha *et al.* 2023; Lubin 2016; NASA Technosignatures Workshop Participants 2019), requiring immense power in the zetta-Watt (10^{21} W) or quetta-Watt (10^{30} W) range. This would be significantly less energy-intensive than dismantling a Jupiter-sized planet for constructing a Dyson Sphere (see Suazo *et al.* 2024, for Dyson sphere candidates). They suggested searching for Strange Exoplanetary Architectures (SEAs) in the hunt for biosignatures and technosignatures, hinting at complex systems and advanced civilizations. In the future, SETI efforts may encompass seeking out modified or arranged systems that imply the existence of multiple habitable planets, termed "Service Worlds," along with advanced communication and infrastructure linking them. Amri Wandel further expanded this discourse by exploring the extension of HZs through various planetary conditions, enhancing our understanding of where life might thrive in the universe. His work demonstrated how atmospheres and liquid water could persist on planets orbiting close to M-dwarf stars, expanding the traditional HZ boundaries (see Wandel 2023, and references therein). Subglacial liquid water could further extend this zone outward. Recent findings, such as the probable detection of atmospheric water vapor on GJ 486 b, alongside previous detections on similar planets, provided empirical evidence for the potential habitability of such worlds. The study suggested that water on these planets may exist in subglacial layers. By applying this model to various exoplanets, water detection could offer insights into their atmospheric properties and potential habitability.

Pillar Two: Detecting Extraterrestrial Intelligence

This contigent of talks showed the methods and challenges associated with detecting signs of intelligent life beyond our planet. Niklas Döbler explored the exopsychology, focusing on how humans think about extraterrestrials and how extraterrestrials might think, feel, and behave. By applying the concept of "affordances," which are action possibilities shaped by abilities and customs, the talk examined the relationship between information processing, the body, the environment, and material engagement in space exploration and SETI. The speaker argued against using "intelligence" in SETI, suggesting that successful contact requires understanding species-specific backgrounds (see details in Döbler & Carbon 2023). Chris Impey investigated public perceptions and beliefs regarding extraterrestrial intelligence, highlighting the societal implications of SETI endeavors (Impey 2023). Luke Sellers proposed a novel approach using gravitational wave (GW) signals to search for technologically advanced civilizations, offering a complementary method to traditional SETI efforts (see Sellers *et al.* 2022). As the SETI operates under the principle that everything should be done with the minimum expense of total energy, then to detect a signal through GW that indicates technological intelligence, a massive amount of energy is required; for instance, moving a Jupiter-mass object at high velocity is necessary to produce a detectable signal. In contrast, electromagnetic radio waves are much more energy-efficient; they can be generated with just a fraction of the energy needed for higher energy waves like gamma waves. Gravitational waves, as a method for SETI, could more likely detect leakage signals where some massive structure is being manipulated inadvertently. This methodology aligns with the concerns that civilizations venturing into space might prefer not to broadcast their existence widely, to avoid attracting attention from potentially hostile forces (see also Cocconi & Morrison 1959; Drake 1961; Loeb & Zaldarriaga 2007; Tellis & Marcy 2015). The challenge with leakage signals is their indistinguishable nature from noise until their information content is decoded. As the sensitivity of detectors improves, we can anticipate the ability to observe a broader range of gravity-distorting events in the universe, some of which may extend beyond the limits of our current scientific understanding. Kevin Knuth focused on using simulations and the nested sampling algorithm to predict the characteristics of interstellar civilizations capable of encountering Earth, based on our understanding of planetary populations and complex organic molecules in space. Saeed Jaffari drawing upon our understanding of interspecies communication on Earth, highlighted the challenges of decoding alien languages and emphasizes the need for a 'Rosetta Stone' equivalent to bridge the transmission gap.

Pillar Three: Exploring Astrobiological Frontiers

Talks in this pillar pushed the boundaries of astrobiological exploration, engaging with interdisciplinary perspectives and speculative scenarios. David Clements provided updates on the detection of phosphine on Venus, underscoring the importance of interdisciplinary collaboration and critical evaluation in astrobiology (see details in Clements 2022). Although the recent reports challanged detection of a trace of phosphine in the clouds of Venus (e.g., Lincowski et al. 2021; Trompet et al. 2021), the interest for its detectability in exoplanet atmospheres was increased, with the subsequent development of a new space observatory projects (see details in Angerhausen et al. 2023). Anthony Milligan looked into the philosophical and methodological considerations surrounding potential encounters with extraterrestrial artifacts, challenging prevailing notions of evidence and discovery. Julie Nekola Nováková presented innovative educational strategies using science fiction narratives to enhance public engagement with astrobiology, bridging the gap between scientific inquiry and popular imagination (Nováková 2021). Shirin Shater Zadeh Yazdi emphasized the need for public awareness and engagement through effective science communication and media strategies, drawing insights from past outreach projects and current decision-making processes.

Pillar Four: Understanding the Nature of Life

This pillar focused on conceptual frameworks and methodological approaches for understanding the nature of life in the universe. Sohan Jheeta discussed the formation of organic molecules in space, shedding light on the chemical processes that may have contributed to the origin of life on Earth. Pallavi Kajrekar explored the stability of biomolecules in extreme astrophysical environments, offering insights into the resilience of life's building blocks beyond Earth. Philipp Spillmann provided a philosophical analysis of the concept of "weird" life, challenging conventional distinctions and highlighting the complexities of defining life beyond Earth. Peter Vickers addressed the challenges of assessing uncertainty in the search for extraterrestrial life, advocating for rigorous standards of evidence and uncertainty assessment akin to climate science methodologies (see also Vickers 2022). The intersectional dynamics between social and public engagement, as well as interdisciplinary public engagement, were elaborated by Dr. David Whitehouse † (see e.g., Whitehouse 2024). He analyzed in details the ongoing tension between scientists and journalists. Exoplanet discoveries, in particular, are susceptible to misinterpretation and exaggeration (see commentary in Seager 2012). The media and the public often favor sensational stories, even when the scientific reality is far more nuanced. This chaotic dynamic can easily lead scientists into the trap of overstating findings. Whitehouse principles offered a framework that can be directly applied to the communication strategies in astrobiology:

Simplicity and clarity: Focus on the core message and 'omit unnecessary details,' using simple language to explain complex discoveries.

Appropriate use of language: 'English is better than Latin'–use clear, everyday language to ensure understanding, avoiding jargon.

Strategic use of clichés: 'Avoid clichés like the plague,' unless they effectively simplify complex ideas for the audience.

Avoid overhyping: 'The road to hell is paved with adverts'–present discoveries factually to maintain public trust and avoid unrealistic expectations.

4. Brief Overview of discussion panels, workshops and public lecture

The Symposium hosted four robust panel discussions, each featured as a separate publication led by their respective chairs. Nigel J. Mason of the University of Kent, UK, chaired the first panel on *Our definition of 'life' and its impact on the evidence* (see Mason et al. 2025). Panelists included Sohan Jheeta of NoRCEL UK, Juan Pérez-Mercader of Harvard University USA, Dirk Schulze-Makuch of TU Berlin Germany, and Terence Kee of the University of Leeds UK. Various definitions and criteria used to characterize life were discused and reflect the ongoing debate and complexity in the astrobiology and biology (see Figure 5). The Panel discussed how the emergence of life as we know it could be informed by estimates of the probability of randomly synthesizing RNA polymers with the correct structure and sufficient length to self-replicate (Gilbert 1986; Orgel 2004; Robertson & Joyce 2012). Recent studies (Totani 2020) using polymer physics and inflationary cosmology (Guth 1981, 2007; Liddle & Lyth 2000) suggested that extraterrestrial RNA worlds (Joyce & Orgel 1993), and thus the emergence of life in an inflationary universe, could be statistically rare. This contrasts with the demonstrated, though not definitively explained, polymerization of long RNA chains in water (Costanzo et al. 2009). It has been hypothetized (Johnson 2022) that water nanoclusters can catalyze prebiotic RNA synthesis, thereby increasing the likelihood of RNA worlds and the emergence of life wherever water is present, but in a non-inflationary universe. Models requiring a multiverse (Guth 2007) vastly larger than the observable universe are untestable beyond the parameter space of the visible universe (Mix 2018). Therefore, the panel focused on pragmatic and testable models and ideas. The main conclusion of the panel was that there is very high uncertainty in defining life (see details in Mason et al. 2025; Bartlett & Wong 2020; Schulze-Makuch & Irwin 2004).

The second panel, *The Great Filter* (see Wilman 2024), chaired by Richard Wilman of Durham University UK, emphasized the need for more data in understanding this

† http://davidwhitehouse.com/

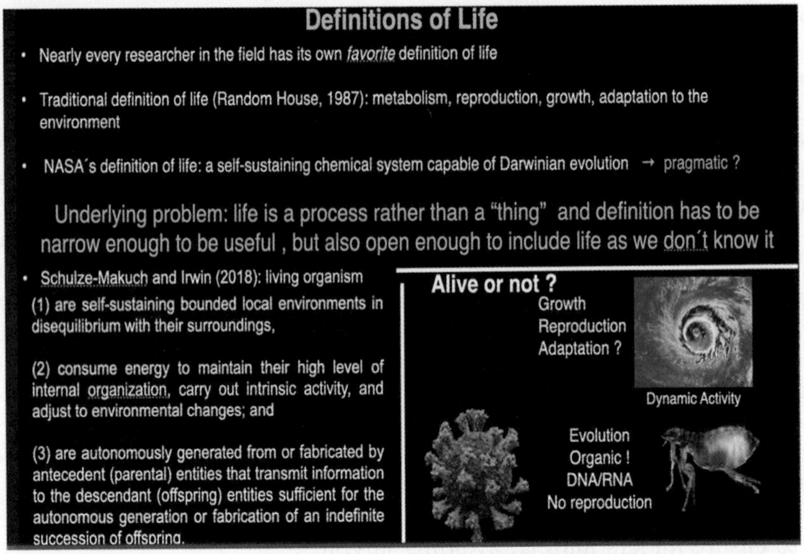

Figure 5. Slide from discussion panel on *Our definition of 'life' and its impact on the evidence* (see Mason et al. 2025), presented by Dirk Shulze-Makuch. Image attribution: Dirk Schulze-Makuch.

cosmic hypothesis, with insights from Arwen Nicholson of the University of Exeter UK, and Jacob Haqq Misra of the Blue Marble Space Institute of Science USA. The timing of events in the history of life on Earth suggests that there were perhaps several hard steps that life had to go through to reach the present level, and that most planets are unlikely to achieve this level (Hanson *et al.* 2021). To observe advanced life like ours, so-called loud aliens are better suited for empirical study than quiet ones (Hanson *et al.* 2021). This distinction ties into the Great Filter concept. However, the discussion panel focused instead on the broader implications of the Great Filter, including the likelihood of technological civilizations surviving long enough to communicate, the role of environmental and societal factors in their development, and the potential filters that might prevent civilizations from reaching an advanced technological stage.

The third panel, *Life as a Cosmic Phenomenon* (see Dominik et al. 2025), was jointly chaired by Martin Dominik of the University of St. Andrews UK, Istvan Praet of Durham University UK, and Mazlan Othman of the Academy of Sciences Malaysia, with contributions from Ricky Lee of the University of Notre Dame Australia, Olli-Pekka Vainio of the University of Helsinki Finland, Shoaib Malik, an independent researcher from UAE, Erik Persson of Lund University Sweden, Perig Pitrou of CNRS France, and Daniela de Paulis of SETI Institute USA. The panel discussed our present moment in cosmic history (see Figure 6) as a crucial phase in the context of the Great Filter, suggesting Earth's early emergence of life in the universe's timeline of habitable epochs (Hanson *et al.* 2021). This contrasts with the universe's cooling trend (see Figure 6), where Earth's premature habitability forecasts more favorable conditions ahead. However, the cooling universe's impact on habitability, influenced by astrochemistry and environmental stability, remains uncertain. The dark energy-driven exponential expansion, $a(t) \propto \exp(H_0\Omega_{\Lambda,0}t)$, with H_0 at 73km/s/Mpc from late measurements and 67.7km/s/Mpc from early techniques (Freedman & Madore 2023, see), predicts that galaxies outside the Local Group will be unreachable in $\sim \mathcal{O}(10^2)$ Gyr (see e.g., Hooper 2018; Freese & Kinney 2003), complicating future cosmic interaction, and even our existence. This necessitates a multidisciplinary

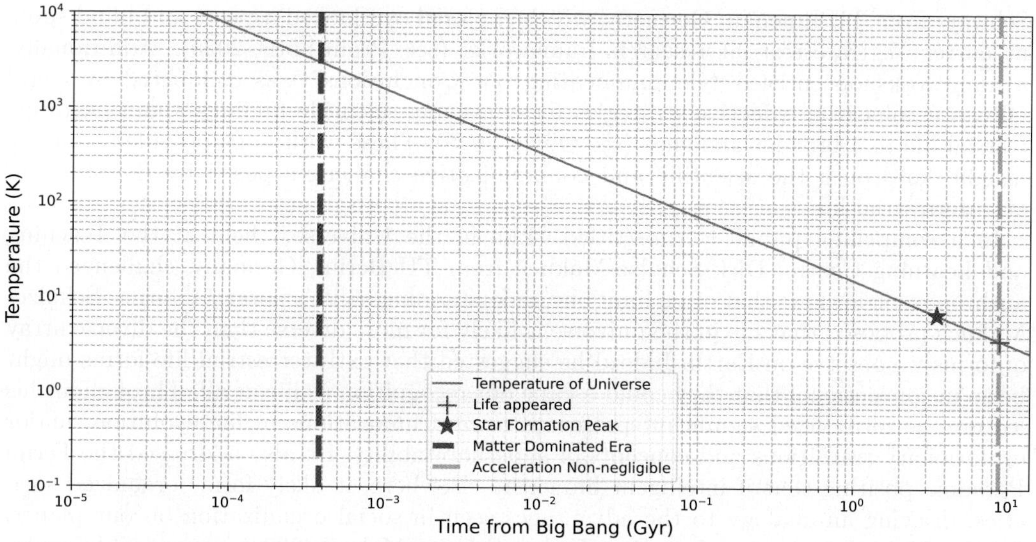

Figure 6. Temperature evolution of the universe and key cosmological events. Life on Earth (green cross) emerged well after the peak of star formation (blue star), which occurred 7 to 9 billion years ago and declined by a factor of ~ 10. Life's appearance coincides with the universe's accelerated expansion (yellow dash-dotted line), beginning around 5 billion years ago when dark energy began dominating (Frieman *et al.* 2008). Recent findings from the Dark Energy Spectroscopic Instrument (DESI) suggest that dark energy's influence may be slowly diminishing (DESI Collaboration et al. 2024).

approach integrating cosmology, environmental science, and astrobiology to explore life's future distribution in the universe (Haqq-Misra *et al.* 2018; Wandel 2023).

The fourth panel discussion, titled *The Future is Our Future* (see Landt et al. 2025), chaired by Hermine Landt of Durham University, UK, and Federica Bianco of the University of Delaware, USA, featured a multi-disciplinary group of experts including David Grinspoon from NASA, Mike Edmunds from the Royal Astronomical Society, UK, Eleni Chatzichristou from the European Research Council, Belgium, and James Davenport from the University of Washington, USA. Through the lens of the Great Filter, this panel critically examined the technological, societal, and environmental constraints that currently limit our civilization's expansionary capabilities (Hanson 1998; Hanson *et al.* 2021)†. The panelists debated whether humanity will remain constrained by these factors or evolve to overcome them, aligning their discussion with the broader perspective of the necessity for multidisciplinary collaboration in addressing the complex existential questions arising from our place in the universe and the future of interstellar exploration and settlement. The conversation referred back to earlier discussions on multidisciplinary collaboration, highlighting how diverse scientific inputs–from political science to deep-space astronomy–are crucial in addressing the existential questions posed by our place in the universe and the future of interstellar exploration and settlement.

The conference also featured several engaging workshops, each designed to simulate scenarios involving the detection of extraterrestrial life and the engagement of various stakeholders. The *Futures I - Resilience and Flexibility* workshop was led by George Profitiliotis from TU Delft, the Netherlands, focusing on the preparedness and adaptability of our responses to potential extraterrestrial discoveries. He also facilitated *Futures II - What are we missing?* alongside Carol Oliver from UNSW, Australia, and Andreas

† A scenario where civilizations expand from their origin planet at a significant fraction of the speed of light, making visible and lasting changes.

Schwarz from TU Ilmenau, Germany, which explored overlooked aspects and blind spots in current extraterrestrial research frameworks (see Profitiliotis 2025). Additionally, a mini-workshop entitled "Communicating our Conclusions" was conducted by Carol Oliver, emphasizing effective strategies for conveying scientific findings about extraterrestrial life to a broader audience (see Oliver 2025). These workshops provided valuable insights into both the scientific and communicative challenges posed by the study of extraterrestrial life.

In a compelling public lecture titled "Life in the Universe," held at the Teaching and Learning Centre, Dirk Schulze-Makuch from TU Berlin, Germany, challenged the anthropocentric views that often limit our understanding of life's potential forms. Schulze-Makuch proposed that life in the universe is likely far more diverse than the diverse array of forms we observe on Earth. Indeed he suggested that extraterrestrial life forms might be so unconventional that they could escape our recognition as life at all. Throughout his lecture, he introduced significant pathways toward innovations in life evolution, major evolutionary transitions and concepts of superhabitability. He also addressed the Fermi Paradox, positing that if intelligent life exists elsewhere, it likely forms organized societies, drawing an analogy to the advantages seen in social organization on our planet. Schulze-Makuch suggested (see Crawford & Schulze-Makuch 2024) two possible explanations for why we might not see any: either there truly are no aliens (see Hart 1975; Tipler 1980), or there are reasons we haven't detected them yet, one of which could be the "Zoo Hypothesis" (Brin 1983) where advanced extraterrestrials deliberately conceal themselves from us, observing life on Earth like animals in a zoo without our knowledge. He concluded that by the continued exploration of the Solar system and Universe, distinction between these two alternatives maybe reached within the next half-century.

5. Forward-Looking Perspectives

We present a comprehensive overview of the invited and contributed talks from the *KAVLI-IAU Symposium (IAUS 387): (Toward) Discovery of Life Beyond Earth and its Impact*, organizing our analysis into two subsections that address the symposium's key themes: defining and detecting life, societal impacts, communication strategies, and media engagement in astrobiology and posters in Virtual Gallery. First, we apply the concept of a dynamical network to explore these discussions. In the second subsection, we extend this analysis by framing it within the context of the Fermi Paradox and the Great Filter, offering insights into the barriers to the emergence and detectability of intelligent civilizations. Paying attention to the caution that 'it is always dangerous to prophesy, particularly, as the Danish proverb says, about the future' (see Hill 1956), we also present some key future perspectives that have emerged from the Symposium.

5.1. *Network analysis of Symposium Discussions*

Table 1 highlights the synergy between astrobiology research and philosophical introspection from both invited and contributed talks. Scientific advancements, such as detection technologies and exoplanet diversity, drive philosophical reflection on human existence, ethics, and the implications of extraterrestrial life. Meanwhile, philosophical inquiries into humanity's place in the universe and the ethical dimensions of contact with other civilizations guide scientific exploration. This interdisciplinary approach at the Symposium underscores the importance of integrating science with philosophy to address the profound questions surrounding the search for life beyond Earth.

Figure 7 illustrates the research themes of invited speakers, showing key interdisciplinary connections. Each node represents a speaker or keyword, and edges indicate shared interests. Clusters around topics like SETI, energy sustainability, and exoplanet

Table 1. Symposium synergy between science and philosophy: combined highlights of invited and contributed talks.

Pillar	Scientific Highlights	Philosophical Highlights
One: Probing extraterrestrial habitability	Advances in SETI-e.g., JWST, and ARIEL: Improving detection of extraterrestrial signals. Geological analogs and lab simulations: Probing habitability through Earth and Mars comparisons.	Defining life and its boundaries: Challenges in understanding habitability. Impact on worldviews: How potential discoveries challenge human perspectives.
Two: Detecting extraterrestrial intelligence	SETI and GW signals: New methods for detecting intelligent life. Public perceptions of extraterrestrial life: Societal implications.	Philosophical implications of contact: Reflection on human uniqueness. Science fiction as preparation: Its role in shaping expectations.
Three: Exploring astrobiological frontiers	Organic molecules and biomolecules: Insights into life's origins and resilience. Science fiction in education: Connecting science with public imagination.	Limits of life: Exploring survival in extreme conditions. Ethical frameworks: Need for global cooperation on extraterrestrial discoveries.
Four:Understanding nature of life	Standards of evidence for alien artifacts: Redefining discovery criteria. Weird life and uncertainty: Philosophical and scientific challenges in detecting extraterrestrial life.	Defining life and its diversity: How new life forms challenge our understanding. Existential reflections: Impact of discoveries on humanity's future.

exploration reflect core discussions and potential for collaboration. Significant interactions, such as between "energy consumption," "sustainability," and "technosignatures," highlight innovative dynamics linking civilization's technological impact to the search for intelligent life. The connections within the network suggest the following potential collaboration clusters:

• SETI and Technologies: Keywords like "SETI," "technosignatures," and "big data" highlight the integration of computational methods with astrobiological research, suggesting a robust area for collaboration.

• Planetary Science and Life Detection: Keywords like "biosignatures," "early Earth," and "Mars exploration" indicate shared methodologies for detecting extraterrestrial life, pointing to practical areas of research synergy.

Additionally, some keywords serve as bridges between different research fields, showing how interdisciplinary efforts can be developed:

• "Ethics" and "Communication": These connect technical space exploration with socio-political considerations, emphasizing the need for ethical frameworks in expanding space research.

• "Energy Consumption" and "Sustainability": These link environmental concerns with planetary sustainability and the "Great Filter" hypothesis, tying them to broader concepts like the Gaia hypothesis and exoplanet exploration.

These connections reveal untapped opportunities for innovation, particularly in understanding alien ecosystems by applying Earth-based concepts like Gaia. Given these patterns, the following strategic recommendations emerge:

• Foster cross-disciplinary meetings to strengthen interactions between clusters and bridging nodes.

• Develop targeted grant programs that promote high-impact research at the intersection of key clusters and nodes.

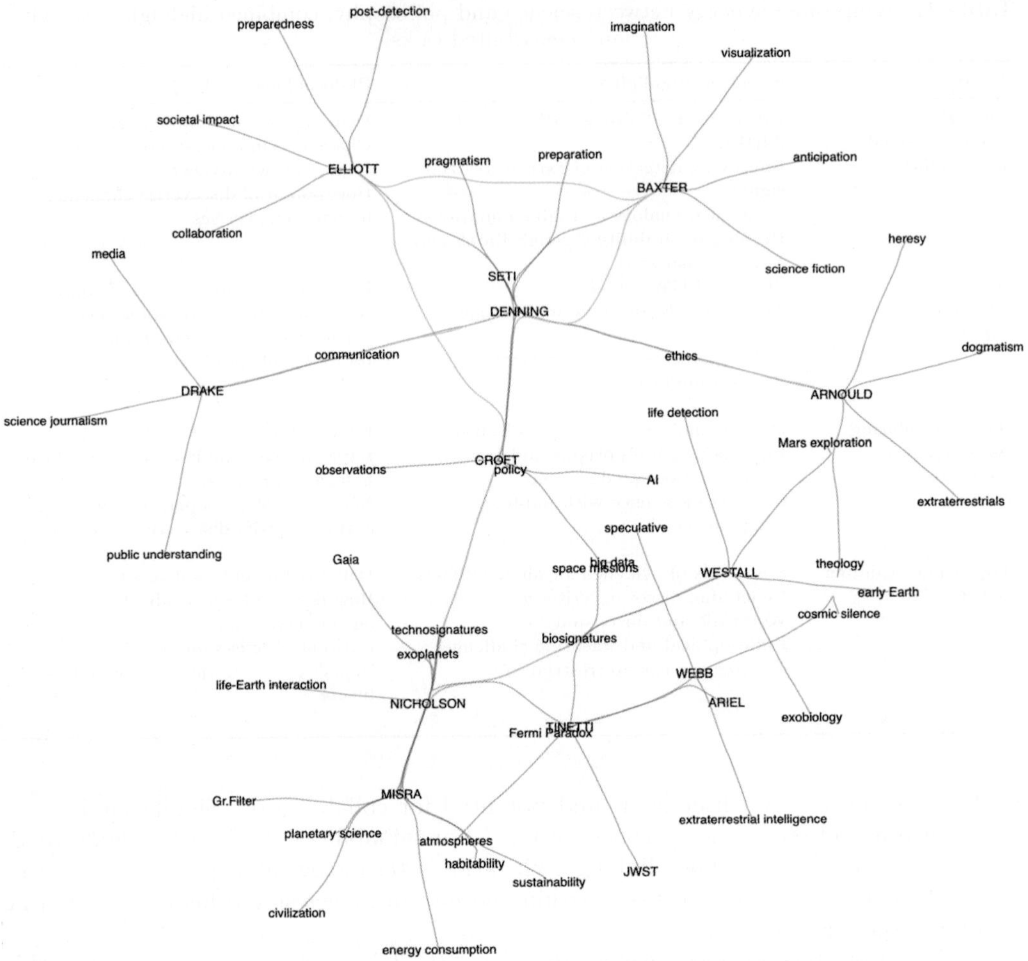

Figure 7. Thematic network of KAVLI Symposium invited talks and collaborative potential, based on the Kamada-Kawai layout algorithm (Kamada & Kawai 1989). The thickness of lines indicates the level of correlation between speakers' topics.

• Enhance public engagement and policy discussions to address the societal implications of new discoveries.

Similarly, the visualization of key terms from the contributed talks shows a complex network with varying connectivity (Figure 8). Topics like "Mars," "life," and "biosignatures" are highly connected, indicating their central importance, while terms like "extraterrestrial life" and "communication" also stand out as core subjects. Specialized topics with fewer connections suggest areas that could benefit from more collaboration. The network reveals possible thematic clusters:

• Astrobiology and life detection: Focuses on detecting life, with keywords like "biosignatures" and "planetary simulation."

• Technological advances: Topics like "nested sampling" and "spectroscopic fingerprints" highlight analytical methodologies.

• Social and public engagement: Emphasizes the role of public interaction in bridging science and society.

Moreover, potential collaboration clusters can be inferred from closely connected nodes, indicating complementary expertise and shared research interests:

Figure 8. Thematic network of KAVLI Symposium contributed talks and collaborative potential. The thickness of lines indicates the level of correlation between speakers. Due to the complexity of the network, some subnodes are overlapping and have long branches.

• Life detection and technological advances: Combining expertise to improve extraterrestrial life detection.

• Interdisciplinary public engagement: Enhancing societal impact by integrating research with public outreach.

Based on the dynamics observed in the contributed talks, several strategies can enhance the potential of the astrobiology network:

• Enhance inter-cluster collaboration: Encourage initiatives that connect researchers across different thematic areas, fostering cross-disciplinary projects that address shared challenges.

• Support emerging research and young researchers: Provide funding and platforms for less-represented topics, promoting exploration and innovation while ensuring a diverse and inclusive research community.

• Expand public engagement: Develop outreach programs using media and interactive experiences to engage the public, emphasizing the excitement and importance of space research.

Analysis of keyword intersections among speaker pairs revealed strong overlap in topics like exoplanet research and biosignature detection, where shared methodologies and research priorities suggest maturity. However, this could lead to research saturation, limiting innovation. Conversely, lower keyword intersections, such as between biosignature experts and SETI researchers, highlight the untapped potential for interdisciplinary research. This balance between diversity and consensus is critical for advancing the field. Emerging trends show how new technologies and interdisciplinary approaches, like AI integration and astroecology, are expanding traditional research paths. Arik Kershenbaum's talk emphasized that complex ecological interactions are essential to life, prompting the development of astroecology to address how ecosystems constrain life (Kershenbaum 2021; Meurer *et al.* 2024).

Based on overview of invited and contributed talks several recommendations can be formulated to guide future research in the field of extraterrestrial life investigation:

• Promote interdisciplinary collaborations: Facilitate environments that encourage dialogue and partnerships across different scientific disciplines. This could be achieved through interdisciplinary research grants, collaborative think tanks, and joint conference sessions designed to merge diverse scientific approaches.

• Invest in emerging technologies: Prioritize funding and support for incorporating advanced technologies such as AI, and big data analytics into astrobiological research. These tools offer the potential to revolutionize our detection capabilities and analytical precision.

• Ethical and societal engagement: Develop comprehensive frameworks to address the ethical, societal, and philosophical implications of discovering extraterrestrial life. This involves not just the scientific community but also ethicists, policymakers, and the general public, ensuring a well-rounded approach to the societal impact of such discoveries.

• Enhance educational outreach: Expand public engagement and education initiatives to foster a broader understanding of astrobiology and SETI. This could include curriculum integration in schools, public lecture series, and media collaborations that make the science accessible and engaging to a wider audience.

• Strengthen global cooperation: Encourage international collaboration by fostering global research projects and partnerships. Shared missions and data exchange protocols can help unify the global scientific community towards common goals in the search for extraterrestrial life.

5.2. *Analysis of Symposium Discussions within the Framework of Great Filter*

Thank you for confirming! Since the versions are already submitted, I'll clean up the documents by removing the highlights and send you the final submitted versions shortly.

The Fermi Paradox and the Great Filter were central theoretical frameworks explored throughout the Symposium. The Fermi Paradox can be viewed as a polylemma †. Similarly, the Drake Equation (Drake 1961) can be interpreted as an elaboration of the Great Filter equation (see also Aldous 2012), which is why we analyzed the invited and contributed talks using the Great Filter framework. The Great Filter equation can be explicitly rewritten by substituting terms from the Drake Equation as follows:

$$(R^* \times f_p \times n_e) \times (f_l \times f_i) \times (f_c \times L) - 1 \leq 0 \qquad (1)$$

where R^* is an average rate of star formation in our galaxy, f_p is a fraction of those stars that have planets, n_e is and average number of planets that could potentially support life, per star that has planets, f_l is a fraction of planets that could support life, f_i is a fraction of planets with life that actually go on to develop intelligent life, f_c is a fraction

† See this URL `https://philarchive.org/rec/REFPVC`

Table 2. Focus of invited and contributed talks with respect to the Great filter parameters.

Lecturer	Focus (Increase/Decrease)	Parameter Affected
Arnould, Tinetti, Clements, Jafari	Increase	p & q
Baxter, Croft, Denning, Drake, Elliott, Misra, Döbler, Impey, Knuth, Milligan, Sellers, Yazdi Milligan, Sellers, Yazdi	Increase	q
Westall, Nicholson, Azua-Bustos, Mateo-Martí, Jheeta, Kajrekar, Ponnaganti, Nekola Nováková, Spillmann, Vickers, Wandel	Increase	p
Stephen Webb	Neutral/Contextual	p & q

of civilizations that develop technologies that are detectable, and L is a timebaseline such civilizations release detectable signals into space.

In this equation, the terms in the first set of parentheses represent the number of Earth-like planets in the galaxy. The terms in the second set constitute p, the probability that intelligent life will arise on such a planet at some point. The terms in the last set correspond to q, the probability that such a species would survive and be detectable to other galactic species for an appreciable length of time. By exploring both p and q, the Symposium contributed to a broader understanding of humanity's position relative to the Great Filter and offered insights into how we might interpret our place in the cosmos (see Table 2).

Furthermore, in the left panel of Figure 9, we visualize based on talks, an evident tension between the nonlinear deterministic parameters, such as R^*, f_p, n_e and the stochastic parameters like f_l, f_i, f_c and L. The divide between these two regimes, thoroughly discussed in Symposium, represents an uncharted territory in the field of astrobiology.

An early example of the concept of 'bridging' uncharted territories was found in the equation for microlensing event rates devised by Dominik et al. (their Eq. 4, 2008). By accounting for the nonstationary observer, this equation showed a clear connection between observer and exoplanets in a three-dimensional space defined by two galactic spatial coordinates and relative velocities (see Molaverdikhani & Tabeshian 2009, and their Figure 5). This approach showcases a non-anthropocentric mapping of the relative velocities of point sources, structured to increase the likelihood of detecting microlensing events.

To illustrate further a potential realization of interactions across the uncharted territory between deterministic and stochastic parameters in the Great Filter, we conceptualize a simple Gaussian Process (GP) model (right panel in Figure 9), with combined Multi-Layer Perceptron (MLP) and Matérn kernels. The first kernel captures nonlinearity whereas second accounts for stochasticity of considered parameters. The yellow region, typically near the diagonal, where parameters strongly correlate with themselves, gradually transitions to blue, indicating lesser correlation between parameters that are further apart in sequence or nature. A reduction in covariance values near the center of the diagonal, could represent a transitional phase in the astrobiological process where deterministic factors give way to more stochastic factors. This area might signify a critical threshold in the Great Filter where predictability based on earlier stages starts to decline, and uncertainty in outcomes begins to increase. Also areas of low covariance might identify stages where traditional predictive models fail, thus requiring more innovative approaches or new data as pointed out by Jacob Haqq-Misra and Arwen Nicholson during the panel discussion on Great Filter.

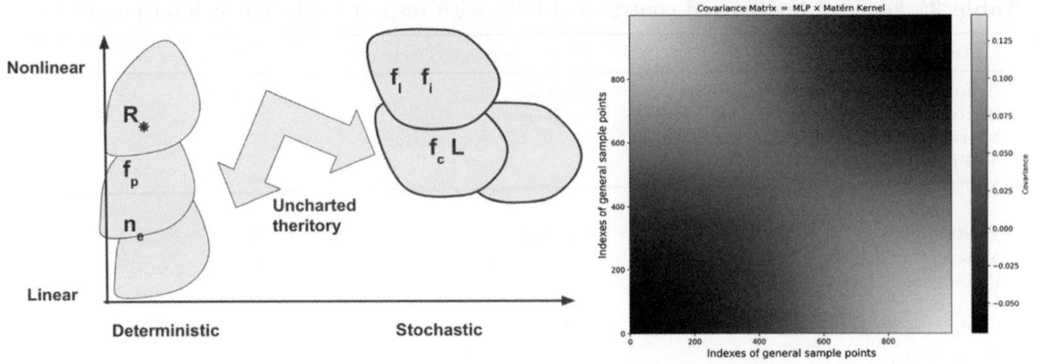

Figure 9. Coarse view of astrobiology's uncharted territory. *Left*: divide between nonlinear deterministic, and stochastic parameters within Great filter theory. *Right*: a toy model of a covariance within the uncharted territory based on Gaussian process. The covariance matrix, created using a combined MLP (simulating nonlinear process) and Matérn (simulating stochastic processes) kernel, diminishes in the center and off-diagonal areas. This reduction in covariance suggests where predictability based on earlier stages declines, uncertainty increases, and traditional predictive models become less effective, signaling the need for more innovative models and/ or new data.

The main implication of the discussions from the invited and contributed talks, with respect to the Great Filter relevant parameters (see Table 2), revolved around the potential barriers that may prevent life from arising (parameter p), sustaining itself, and becoming detectable across the cosmos (parameter q). Many of lectures suggested increased trajectory of p, which might suggest that intelligent life is more common than previously thought, which in turn could imply that the Great Filter lies ahead of us, rather than behind. Here is a synthesis of these implications:

• Pre-life conditions: Enhancing our understanding of how life-supporting conditions are established and maintained could help identify planets that might bypass early filters in the progression from chemical precursors to complex life.

• Post-life emergence: Identifying and supporting the factors that allow life to flourish and become technologically advanced (capable of signaling existence) addresses later stages in the Great Filter.

• Technological and observational capability: Improving detection technologies and methodologies addresses the very end of the Great Filter, enhancing our ability to observe and identify civilizations that have successfully passed all prior filters.

The Symposium mostly debated anthropomorphic SETI communication channels (Wilson 1984) based on implicit assumption that predominant species of communication is electromagnetic. However, Sellers *et al.* (2022) presented a new possibility of searching for technosignatures through gravitational wave signals. The history of 60 years of null detection in SETI suggested, based on Bayesian reasoning, that the expected rate of technoemissions across the entire galaxy is $\lesssim 1-5$ per century (Grimaldi 2023). If the Earth is located in *a coomunication void zone*, it might not receive any technosignals for 60-1800 years, with a probability of 0.5 (Grimaldi 2023). This is highlighting the importance of expanding SETI sampling of universal communication set to include all species of communication, such as quantum communication and physical artifacts (see details in Wilson 1984). Figure 10 (left panel) represents the potential communication signal universe, highlighting areas that remain uncharted. In the past 60 years SETI experiments have focused narrowband microwave (GHz) signals (Hippke 2021), while accounting for Doppler drift $\lesssim \pm 20\,\mathrm{Hz\,s^{-1}}$ (see e.g., Price et al. 2020; Sheikh et al. 2020). A smaller number of programs have targeted the optical domain, with the focus on pulsed lasers, with expected pulses of $\sim 100\,\mathrm{photons}\,m^{-2}10^{-9}\mathrm{s}$ (see e.g., Tellis &

Marcy 2015). Advanced civilizations might use quantum communication (see details in Wilson 1984; Berera & Calderón-Figueroa 2022). Some studies postulate that the black holes are the most efficient capacitors of quantum information (Dvali & Osmanov 2023). Due to this, the alien quantum computers may employ black holes and will radiate in ordinary particles such as neutrinos and photons within the range of potential sensitivity of our detectors. This opens a new avenue for SETI, e.g., including the civilizations entirely composed of hidden particles species interacting with our world exclusively through gravity (Dvali & Osmanov 2023).

The intersection of electromagnetic and quantum communication channels might be indicative of a extraterrestrial transitional technology phase or the use of hybrid systems. For example Hippke (2021) suggested that collected electromagnetic signals might in fact include a weaker, underlying signal based on the quantum communications. Some astrophysical phenomena, such as black hole mergers (de Mink & King 2017; Dong-Páez et al. 2023), neutron star mergers (Abbott et al. 2017) may produce coincident electromagnetic radiation and gravitational waves. Detecting coincident electromagnetic and gravitational wave signals of massive structures or events engineered by advanced civilizations is the next uncharted territory (see right panel in Figure 10). The communication that combine quantum and gravitational wave signals could represent highly advanced technological civilizations, capable of manipulation of spacetime. The coincident electromagnetic, quantum and gravitational wave communication signals would correspond to extraordinarily advanced civilization capable of employing all known physical communication methods. This is an ultimate uncharted territory, a realm that we do not know if human technology could ever reach.

In relation to the above, and in the context of quantum computing becoming a reality (Arute et al. 2019), it is clear that exploring the integration of quantum computing with machine learning algorithms should be a primary focus in the next decade for astrobiology. Machine learning is already a common strategy for data processing, but its computational complexity increases exponentially with the dimensionality of datasets. So far, quantum machine learning has been applied to the binary classification of pulsar data, a hybrid quantum support vector machine for classifying galactic morphology, and variational quantum computing for dark matter cosmological simulations (see Brunet et al. 2024, and references therein). Assuming that the growth of quantum computers will be similar to classical computers as described by Moore's law, achieving an increase from 100 to 10^6 qubits would take approximately 30 years. Quantum computers are expected to simulate quantum mechanical systems with exponential speedup over classical computers, such as in galaxy formation and black hole accretion. Quantum algorithms can be used to search databases, perform clustering and classification tasks, detect anomalies, and optimize the analysis of astrobiological data. It is also expected that exoplanets could be observed directly by employing quantum hypothesis testing procedure to analyse telescope images (Huang & Lupo 2021).

The Symposium discussions analyzed within the framework of the Great Filter have led to the following specific recommendations, addressing key challenges and opportunities for future research in astrobiology:

• Improve understanding of planetary conditions that support life: Focus on refining models and observational methods to better identify planets with conditions that could potentially bypass early-stage filters (e.g., chemical precursors to life), such as through the study of exoplanet atmospheres and surface environments using advanced detection techniques like space born observatories.

• Support research on the factors that influence the survival of intelligent civilizations: Enhance studies of factors that contribute to the development and longevity of technological civilizations (e.g., environmental sustainability, energy consumption) that could

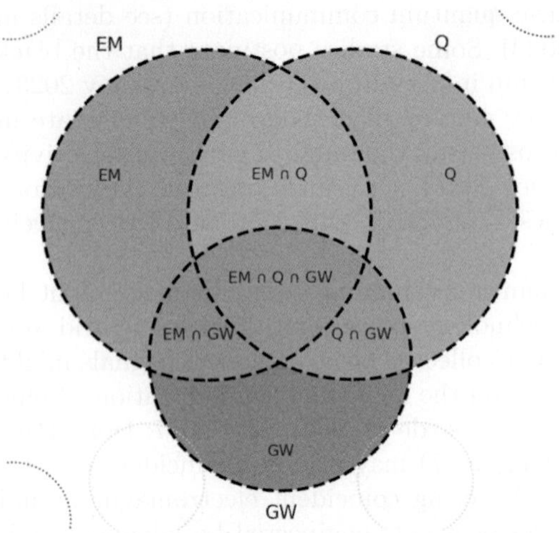

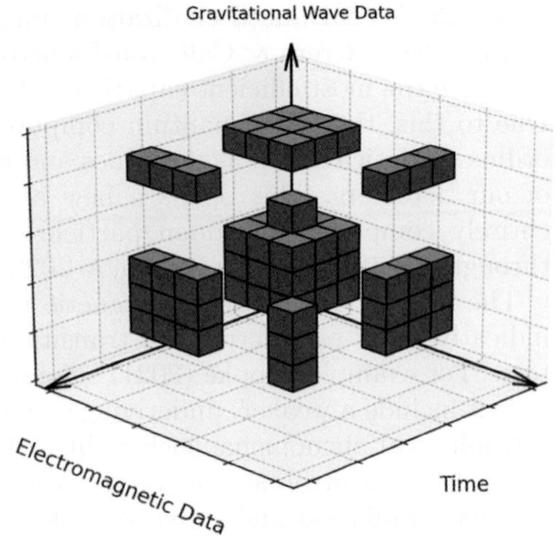

Figure 10. Slice of the communication universe. *Left:* EM, Q, and GW represent possible electromagnetic, quantum, and gravitational wave communication species. A SETI attempt targets subsets generated by intersecting random communication events based on prior knowledge and assumptions (see Wilson 1984). Two coincident species suggest advanced civilizations, while the usage of all three may be beyond their reach. *Right:* Example of a data tensor within the EM and GW intersection over time, potentially containing images, time series, spectra, etc.

pass through the later stages of the Great Filter, focusing on the planetary and social conditions that allow for sustained technological advancement.

• Develop and implement advanced detection technologies for technosignatures: Prioritize the development of next-generation detection technologies that go beyond traditional electromagnetic signals, including gravitational wave detection, quantum communication channels, and multi-wavelength observational methods. This will allow us to search for civilizations that may use non-traditional communication methods or hidden signals.

• Integrate quantum computing with machine learning for large-scale astrobiological data analysis: Given the exponential growth in data from observational platforms, the integration of quantum computing with machine learning techniques should be pursued to enhance the efficiency and accuracy of processing astrobiological data, including the detection of faint signals or anomalies that may indicate the presence of technosignatures.

Acknowledgements. A.B.K. expresses gratitude to the Organizers of the *KAVLI-IAU Symposium (IAUS 387): (Toward) Discovery of Life Beyond Earth and Its Impact*, the University of Durham, and the University of Kent for their support. A.B.K. acknowledges funding provided by the University of Belgrade - Faculty of Mathematics (contract 451-03-66/2024-03/200104) through grants by the Ministry of Science, Technological Development, and Innovation of the Republic of Serbia. N.J.M. recognises the support of the Europlanet 2024 RI which has received funding from the European Union's Horizon 2020 research and innovation programme under grant agreement No 871149.

References

Abbott B. P., Abbott R., Abbott T. D., Acernese F., Ackley K., et al., 2017, ApJL, 848, L12
Agić, H. 2021, Origin and Early Evolution of the Eukaryotes: Perspectives from the Fossil Record. In: A. Neubeck, S. McMahon (eds.), *Prebiotic Chemistry and the Origin of Life*, Advances in Astrobiology and Biogeophysics, Springer, p. 255
Aldous, D. J. 2012, *Mathematical Scientist*, 37, 55
Angerhausen, D., Ottiger, M., Dannert, F., *et al.* 2023, *Astrobiology*, 23(2), 183–194

Arnould, J. 2021, Who Goes There? When Astrobiology Challenges Humans. In: O. A. Chon Torres, T. Peters, J. Seckbach, R. Gordon (eds.), *Astrobiology*, John Wiley & Sons, Ltd, p. 79

Arthur, R., & Nicholson, A. 2023, *Astrobiology*, 23, 1238

Arute F., Arya K., Babbush R., Bacon D., Bardin J. C., Barends R., et al. 2019, *Nature*, 574, 505

Azua-Bustos, A., Fairén, A. G., González-Silva, C., *et al.* 2023, *Nature Communications*, 14, 808

Bartlett, S., Wong, M. L. 2020, Defining Lyfe in the Universe: From Three Privileged Functions to Four Pillars. *Life*, 10(4), 42

Baum, S. D. 2010, *Sustainability*, 2(2), 591

Baxter, S. 1995, The Time Ships, Harper Collins Publishers.

Beichman, C. A., Greene, T. P. 2018, Observing Exoplanets with the James Webb Space Telescope. In: H. J. Deeg, J. A. Belmonte (eds.), *Handbook of Exoplanets*, Springer International Publishing, Cham, p. 1283

Berera, A., Calderón-Figueroa, J. 2022, *Phys. Rev. D*, 105,123033

Bergsten, G. J., Pascucci, I., Mulders, G. D., Fernandes, R. B. 2022, *AJ*, 164(5), 190

Betts, R. A., & Lenton, T. M. 2008, Second chances for lucky Gaia: A hypothesis of sequential selection, Met Office.

Bohlmann U. M., Bürger M. J. F. 2018, *Ac. Aau*, 143, 163.

Brin, G. D. 1983, The Great Silence–the controversy concerning extraterrestrial intelligent life, *QJRAS*, 24, 283-309

Brook, A., Wuerth, J. 2023, Kant's View of the Mind and Consciousness of Self. In: E. N. Zalta, U. Nodelman (eds.), *The Stanford Encyclopedia of Philosophy*, Spring 2023, Metaphysics Research Lab, Stanford University, `https://plato.stanford.edu/archives/spr2023/entries/kant-mind/`

Brunet, T., Tolley, E., Corda, S., Ilic, R. 2024, *Astronomy and Computing*, 47, 100796

Chaisson, E. J. 2016, Internalizing null extraterrestrial "signals": An astrobiological app for a technological society. In: S. J. Dick (ed.), *The Impact of Discovering Life beyond Earth*, p. 324

Clements, D. L. 2022, *Contemporary Physics*, 63, 180

Cocconi, G., Morrison, P. 1959, *Nature*, 184, 844

Costanzo, G., Pino, S., Cicinello, F. and DiMauro, E. 2009, *Journal of Biological Chemistry*, 284, 33206

Crawford, I.A., Schulze-Makuch, D. 2024, *Nat. Astron.*, 8, 44

Dick, S. J. 2003, *Internat. Jour. Astrobio.*, 2, 65

DESI Collaboration, Adame A. G., Aguilar J., Ahlen S., Alam S., Alexander D. M., et al. 2024, arXiv, 2404.03002

DeVincenzi, D. L. 1984, NASA's Exobiology Program, *Origins of Life and Evolution of the Biosphere*, 14, 793

Dominik, M., et al. 2025, Life as a cosmic phenomenon facing human culture, Proceedings IAU Symposium No. 387, H. Landt, M. Dominik, C. Oliver (eds.) doi:10.1017/S1743921325000274

Dominik, M., Jorgensen, U. G., Horne, K., *et al.* 2008, arXiv e-prints, arXiv:0808.0004

Dong-Páez C. A., Volonteri M., Beckmann R. S., Dubois Y., et al., 2023, A&A, 676, A2

Drake, F. D. 1961, *Physics Today*, 14, 40

Drake, N. 2022, Space. In: D. Blum, A. Smart, T. Zeller Jr. (eds.), *A Tactical Guide to Science Journalism: Lessons From the Front Lines*, Oxford University Press, p. 202

Döbler, N. A., Carbon, C.-C. 2023, *Acta Astronautica*, 210, 350

Döbler, N.A., El Amri, Y., Carbon, C.C. 2023, *Discov. Psychol.*, 3, 21

Dvali, G., Osmanov, Z.N. 2023, *Intern. Jour. of Astrob.*, 22(6), 617

Dyson, F. J. 1960, *Science*, 131, 1667

Elliott, J., Genevieve, K., Dominik, M., Denning, K., *et al.* 2023, The SETI post-detection hub: preparing for discovery, 74th International Astronautical Congress

Frank, A., Sullivan, W. 2014, *Anthropocene*, 5, 32

Freedman, W., Madore, B. 2023, *Jour. of Cosmo. and Astropart. Phys.*, 11, 1–50
Freese, K., Kinney, W. 2003, *Phys. Lett.B*, 558, 1
Frieman, J. A., Turner, M. S., Huterer, D. 2008,*ARAA*, 46, 385
Fulton, B. J., Petigura, E. A. 2018, *ApJ*, 156, 264
Gamaleldien, H., Wu, L.-G., Olierook, H. K. H., Kirkland, C. L., et al. 2024, *Nat. Geosci.*, 17, 560
Gilbert, W. 1986, *Nature*, 319, 618,
Gray, R. H. 2015, *Astrobiology*, 15, 195
Grimaldi, C. 2023, *Acta Astronautica*, 212, 95
Guth, A. H. 1981, *Phys. Rev. D*, 23, 347
Guth, A.H. 2007, *Jour. of Phys. A*, 30, 6811
Hanson, R., Martin, D., McCarter, C., Paulson, J. 2021, *ApJ*, 922, 182
Hanson, R. 1998, Great Filter–Are We almost Past It, `https://mason.gmu.edu/~rhanson/greatfilter.html`
Haqq-Misra, J., Baum, S. 2009, *Journal of the British Interplanetary Society*, 62, 47
Haqq-Misra J., Kopparapu R. K., Wolf E. T. 2018, *Inetrn. Journal Astrob.*, 17, 77
Haqq-Misra, J., Kopparapu, R. K., Schwieterman, E. 2020, *Astrobiology*, 20, 572
Hart, M. H. 1975, *QJRAS*, 16, 128
Hey, T., Tansley, S., Tolle, K. 2009, *The Fourth Paradigm: Data-Intensive Scientific Discovery*, Redmond: Microsoft Research
Hill, B. 1956, Journal of the Royal Statistical Society. Series A (General), "Proceedings of the Meeting", [Speaker: Bradford Hill],119, 146, Blackwell Publishing for the Royal Statistical Society.
Hippke, M. 2021, ApJ, 162, 1
Hooper, D. 2018, *Phys. of Dark Univ.*, 22, 74
Huang, Z., Lupo, C. 2021, *Phys. Rev. Lett.*, 127, 130502
Impey, C. 2023, *Worlds Without End: Exoplanets, Habitability, and the Future of Humanity*, The MIT Press
Janson, M., Gratton, R., Rodet, L., *et al.* 2021, *Nature*, 600(7888), 231
Jiang, J., Zou, X., Mitchell, R.N., Zhang, Y., et al. 2024, *PNAS*, 121, e2405160121
Johnson, K. 2021, *Inter. Jour. of Astrob.*, 20(2), 111
Joyce, G. F., Orgel, L. E. 1993, Prospects for understanding the origin of the RNA world. In: Gesteland, R. F., Atkins, J. F., editors. The RNA World, Cold Spring Harbor Laboratory Press:Cold Spring Harbor, NY. 1-22.
Kamada, T., Kawai, S. 1989, Inf. Process. Lett., 31, 7
Kates, R. W., ed. 2010, *Readings in Sustainability Science and Technology*, CID Working Paper No. 213, Center for International Development, Harvard University
Kershenbaum, A. 2021, *The Zoologist's Guide to the Galaxy*, Penguin
Kopparapu, R.K., Ramirez, R., Kasting, J.F., *et al.* 2013, *ApJ*, 765, 131
Kopparapu, R. K., Wolf, E. T., Arney, G., *et al.* 2017, *ApJ*, 845, 5
Landt, H., et al. 2025, Discussion: The future is our future, Proceedings IAU Symposium No. 387, H. Landt, M. Dominik, C. Oliver (eds.) doi:10.1017/S1743921324002680
Lenton, T. M., Daines, S. J., Dyke, J. G., Nicholson, A. E., et al. 2018, Trends in Ecology & Evolution, 33(8), 633
Liddle, A. R., Lyth, D. H. Cosmological Inflation and Large-Scale Structure, ambridge University Press, 2000
Lincowski, A. P., Meadows, V. S., Crisp, D., *et al.* 2021, *ApJ*(Letters), 908, L44
Loeb, A., Zaldarriaga, M. 2007, *Journal of Cosmology and Astroparticle Physics*, 2007, 20
Lubin, P. 2016, *REACH*, 1, 20-45
Lueftinger, T., Tinetti, G., Eccleston, P., Salvignol, J.-C. 2022, *BAAS*, 54, 502
Ma, P. X., Ng, C., Rizk, L., Croft, S., *et al.* 2023, *Nature Astronomy*, 7, 492
Madhusudhan, N., Sarkar, S., Constantinou, S., *et al.* 2023, *ApJ*(Letters), 956(1), L13
Mason, N.J., et al. 2025, The definition of life and its role in astrobiology and the search for evidence of life in the universe, Proceedings IAU Symposium No. 387, H. Landt, M. Dominik, C. Oliver (eds.), doi:10.1017/S1743921324001248

Mateo-Marti, E., Prieto-Ballesteros, O., Muñoz Caro, G., *et al.* 2019, *Life*, 9, 72

Mayor, M., Queloz, D. 1995, *Nature*, 378, 355

Meurer, J. C., Haqq-Misra, J., Mendonça, M. de Souza, Jr. 2024, *Internat. Journ. Astrobio.*, 23, e3

de Mink S. E., King A., 2017, ApJL, 839, L7

Mix, L. J. 2015, *Astrobiology*, 15, 15

Mix, L. J. 2018, *Inetrnat. Journ. Astrobio.*, 17, 189

Mix, J. L. 2019, *Life Concepts from Aristotle to Darwin*, Palgrave Macmillan Cham

Molaverdikhani, K., & Tabeshian, M. 2009, arXiv e-prints, arXiv:0911.4424

Naganuma, T. 2009, *Sustainability*, 1, 827

NASA Technosignatures Workshop Participants. 2019, NASA and the Search for Technosignatures: A Report from the NASA Technosignatures Workshop, arXiv:1812.08681

Narasimha, R., Safonova, M., Sivaram, C. 2023, arXiv e-prints, arXiv:2309.06562

NASEM 2022, Decadal Survey cover: Origins, Worlds and Life: A Decadal Strategy for Planetary Science and Astrobiology 2023-2032, Washington, DC: The National Academies Press

Nováková, J. 2021, Worlds Without End: Exoplanets, Habitability, and the Future of Humanity, European astrobiology institute

Oberhaus, D. 2019, Extraterrestrial Languages, The MIT Press

Oliver, C. A. 2025, The storytelling brain and connecting public audiences with SETI, Proceedings IAU Symposium No. 387, H. Landt, M. Dominik, C. Oliver (eds.) doi:10.1017/S1743921324003211

Orgel, L. E. 2004, *Crit. Rev. Biochem. Mol. Biol.*, 39, 99

Owen, J. E., Wu, Y. 2013, *ApJ*, 775, 105

Pace, E. 2019, ARIEL: a space telescope to investigate the atmospheres of extrasolar planets. In: *Frontier Research in Astrophysics - III. 28 May - 2 June 2018. Mondello (Palermo)*, p. 76

Praet, I. 2023, Anthropology and Contemporary Space Exploration, with a Note on Hopi Ladders. In: J. F. Salazar, A. Gorman (eds.), *The Routledge Handbook of Social Studies of Outer Space*, p. 54

Price D. C., Enriquez J. E., Brzycki B., Croft S., et al., 2020, AJ, 159, 86

Profitiliotis, G., 2025, Rehearsing Post-Detection Futures: Building Futures Literacy vis-á-vis Post-Detection Considerations for a Responsible Search for Extraterrestrial Life, Proceedings IAU Symposium No. 387, H. Landt, M. Dominik, C. Oliver (eds.) doi:10.1017/S1743921324001595

Race, M., Denning, K., Bertka, C. M., *et al.* 2012, *Astrobiology*, 12, 958

Robertson, M. P., Joyce, G. F. 2012, *Cold Spring Harb. Perspectives Biol.*, 4

Sainsbury, R. M. 2009, Paradoxes, Cambridge: Cambridge University Press

Schulze-Makuch, D., Irwin, L. N. 2004, Definition of Life. In: D. Schulze-Makuch, L. N. Irwin (eds.), *Life in the Universe: Expectations and Constraints*, Springer Berlin Heidelberg, Berlin, Heidelberg, 8

Seager, S. 2012, Planetary science: The search for Earth's twin, *Nature*, 490, 479

Seager, S., Bains, W. 2015, *Sci. Adv.*, 1, e1500047

Sellers, L., Bobrick, A., Martire, G., Andrews, M., & Paulini, M. 2022, arXiv e-prints, arXiv:2212.02065

Sheikh S. Z., Siemion A., Enriquez J. E., Price D. C., et al., 2020, AJ, 160, 29

Stern, R.J., Gerya, T.V. 2024, *Sci. Rep.*, 14, 8552

Suazo, M., Zackrisson, E., Mahto, P. K., *et al.* 2024, *MNRAS*, 531, 695

Sullivan, W. T. III, Baross, J. (Eds.). 2007, *Planets and Life: The Emerging Science of Astrobiology*, Cambridge: Cambridge University Press

Szalay, A. 2019, The Era of Surveys and the Fifth Paradigm of Science, American Astronomical Society Meeting Abstracts #233, 233, 400.01

Tellis, N. K., Marcy, G. W. 2015, *PASP*, 127, 540

Tinetti, G., Drossart, P., Eccleston, P., *et al.* 2018, *Experimental Astronomy*, 46, 135

Tipler, F. J. 1980, *QJRAS*, 21, 267

Totani, T. 2020, *Sci. Rep.*, 10, 1671
Trompet, L., Robert, S., Mahieux, A., *et al.* 2021, *A&A*, 645, L4
Vickers, P. 2022, *Identifying Future-Proof Science*, Oxford University Press
Wandel, A. 2023, *AJ*, 166, 222
Wandel, A., Gale, J. 2020, *Inter. Journal of Astrobio.*, 19, 126
Webb, S. 2002, If the universe is teeming with aliens...where is everybody?, Springer International Publishing
Westall, F., Foucher, F., Bost, N., *et al.* 2015, *Astrobiology*, 15, 998
Werth, L.F. 1998, *Jour. of Appl. Phil.*, 15, 83
Whitehouse, D. 2024, *The Alien Perspective: A New View of Humanity and the Cosmos*, 2nd ed., Icon Books
Wilman, R. 2024, The Great Filter. In: H. Landt-Wilman, M. Dominik, C. Oliver (eds.), *KAVLI-IAU SYMPOSIUM (IAUS 387) Proceedings*
Wilson, T. L. 1984, *Q. Jl. RAS*, 25, 435,
Wogan, N. F., Batalha, N. E., Zahnle, K. J., *et al.* 2024, *ApJ*(Letters), 963(1), L7

Author index